THE Business WRITER

John Van Rys
Redeemer University College

Verne Meyer
Dordt College

Pat Sebranek
The Write Source and UpWrite Press

Houghton Mifflin
Boston   New York

Editor-in-Chief: Suzanne Phelps Weir
Development Manager: Sarah Helyar Chester
Sponsor: Michael Gillespie
Assistant Editor: Anne Leung
Senior Project Editrix: Aileen Mason
Editorial Assistant: Susan Miscio
Composition Buyer: Chuck Dutton
Senior Art & Design Coordinator: Jill Haber
Manufacturing Manager: Karen Banks
Senior Marketing Manager: Cindy Graff Cohen
Marketing Associate: Wendy Thayer

Cover image: © Stockbyte/Getty Images

CREDITS: p. 88 *Figure 5.2* Reprinted by permission from Microsoft® Corporation. **p. 137** *Figure 7.2* Reprinted by permission from Sirsi Corporation. **p. 139** *Figure 7.3* Reprinted by permission of OCLC Online Computer Library Center, Inc. FirstSearch and WorldCat are registered trademarks of OCLC Online Computer Library Center, Inc. **p. 142** *Figure 7.4* © 2004 EBSCO Publishing. All rights reserved. **p. 143** *Figure 7.5* © 2004 EBSCO Publishing. All rights reserved. **p. 147** *Figure 7.6* © 2005 LII.ORG. Reprinted with permission. **p. 148** *Figures 7.7 and 7.8* © 2005 LII.ORG. Reprinted with permission. **p. 150** *Figure 7.9* © 2005 Ask Jeeves, Inc. **p. 197** Reprinted by permission of HowStuffWorks, Inc. **p. 583** *Figure 36.3* Reprinted by permission from Key Tronic Corporation. **p. 583** *Figure 36.4* Reprinted by permission from Karmaloop LLC. **p. 584** *Figure 36.5* Courtesy oktrucking.org for reprint. **p. 585** *Figure 36.6* Reprinted by permission of Oregon Employment Department, Workforce & Economic Research Division.

Copyright © 2006 by Houghton Mifflin Company. All rights reserved.

No part of this work may be reproduced or transmitted in any form or by any means, electronic or mechanical, including photocopying and recording, or by any information storage or retrieval system without the prior written permission of Houghton Mifflin Company unless such copying is expressly permitted by federal copyright law. Address inquiries to College Permissions, 222 Berkeley Street, Boston, MA 02116-3764.

Printed in the U.S.A.

Library of Congress Control Number: 2005924908

ISBN: 0-618-37087-0

1 2 3 4 5 6 7 8 9-QWT-09 08 07 06 05

Preface

> "In the workplace, you don't write for a grade—you write for a living."
>
> —Jim Franken, past president
> of The Harbor Group

Jim is right. As president of an electrical engineering firm with both domestic and international contracts, he knows that people in the workplace must communicate well in order to get jobs, keep jobs, and advance in their careers. That's why we wrote this no-nonsense handbook. We want to help busy college students gain the writing, reading, speaking, and listening skills that they need to thrive in the twenty-first century workplace.

FEATURES OF *THE BUSINESS WRITER*

Quick-Reference Format

The Business Writer is written in an easy-access, handbook style with features that help your students get in the book quickly, find what they need, and get out:

- Each chapter's introduction briefly presents the topic and lists what follows.
- Most material is presented in self-contained, one-or two-page spreads.
- Bulleted or numbered lists, brief explanations, summary boxes, and graphic organizers deliver instruction concisely and precisely.
- Succinct, trait-based guidelines help students analyze the writing situation, spell out their writing goals, and work through the writing process (the traits are shown on page v).
- Authentic workplace models illustrate real writing situations and show the finished document.

Preface

- Introductions for each model describe the writer's analysis of the purpose, audience, and format.
- Side notes point out key elements, such as the model's three-part structure (opening, middle, closing), stylistic devices, and writing strategies.
- End-of-chapter checklists help students effectively benchmark the quality of their writing, using the seven-traits criteria.
- The seven-traits vocabulary facilitates clear conversations about writing, suggests specific strategies for revising writing, and provides clear criteria for grading writing.

These quick-reference features make *The Business Writer* a book that students will find efficient and valuable—both in college today and in the workplace tomorrow.

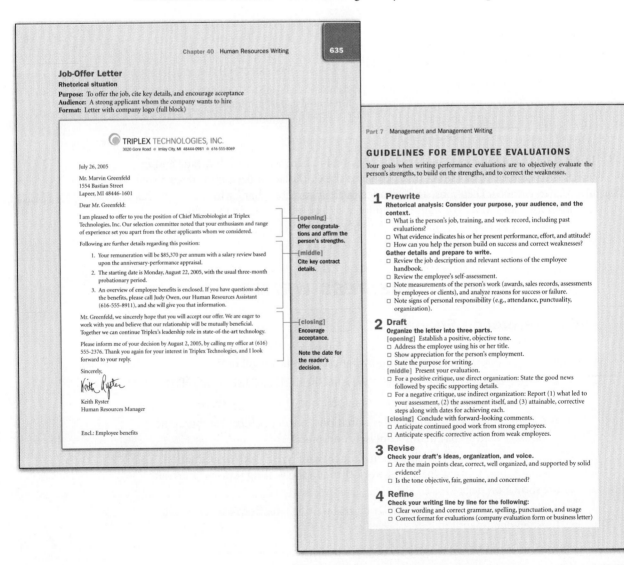

Seven Traits of Effective Writing

Throughout the book, we discuss the seven traits (or qualities) found in all good writing—including workplace writing: **strong ideas, logical organization, conversational voice, clear words, smooth sentences, correct copy,** and **reader-friendly design.** In both writing instructions and models, we demonstrate these traits and explain how to develop them.

For example, Part 2, "Benchmarking Writing with the Seven Traits," describes each trait, shows how to develop it, and explains how to test for it. In addition, each guidelines page explains how to develop the traits in a specific form of writing, and each writing checklist helps students measure the quality of their writing in relation to the traits.

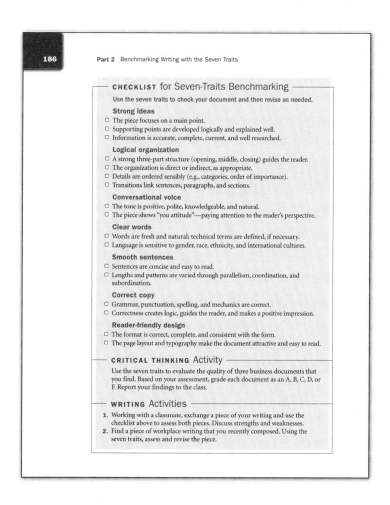

Preface

Reader-Friendly Design

Because document design (trait 7) is such an important quality in today's visually oriented workplace, *The Business Writer* addresses the topic in seven ways.

- With its strong design elements (e.g., colors, boxes, bulleted lists) and attractive page layouts, the book itself illustrates a reader-friendly design—one that supports the message and makes reading easy.
- Chapter 5, "Using Graphics in Business Documents," explains what graphics are and shows how to use them.
- Chapter 16, "Trait 7: Reader-Friendly Design," presents basic design principles, along with practical strategies for crafting attractive, readable documents.
- Every model in the book appears in a full-page format that illustrates an effective design for that form of writing.
- Each guidelines page and checklist addresses page design.
- Chapter 36, "Writing for the Web," presents principles, guidelines, and models for developing effective Web sites and Web pages.
- The book's Web site includes additional guidelines, tips, and models for creating well-designed business documents.

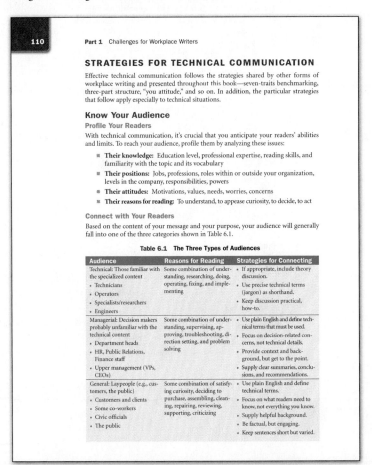

Complete Coverage for Using Technology for Research and Writing

The Business Writer helps students with one of the biggest challenges in workplace writing: using technology to research, write, and design effective documents.

- Chapter 2, "Writing and Technology," includes instructions on using writing software and electronic resources, including databases and the Internet.
- Chapter 5, "Using Graphics in Business Documents," explains how to design and use graphics, including tables, graphs, charts, maps, and photographs.
- Chapter 6, "Communicating Technical Information," explains how to relay specialized or complex information (often related to an applied science or trade) in ways that readers can understand and use the message.
- Chapter 7, "Conducting Research for Business Writing," addresses topics such as completing a Web search, evaluating online sources, and documenting sources.
- Chapter 23, "Writing E-Mail Messages and Sending Faxes," shows how to use these important media in accordance with professional business standards.

> **WEB Links**
> - Visit our Web site http://college.hmco.com/english for information about specific cultural practices, links to intercultural resources, and advice about tailoring Web sites for intercultural readers.
> - To learn more about another country's cultures and business practices, find quality Web sites from that country. For example, information on Canada can be accessed by starting with the federal government's Web site at www.canada.gc.ca.

- Chapter 36, "Writing for the Web," explains how to address the challenges of communicating in a Web environment.
- Chapters 17 through 36, 39, and 40 show how to communicate technical information in a variety of writing forms including letters, proposals, reports, and bids.

In addition, throughout the book are helpful Web links that direct students to interactive exercises, helpful bibliographies, additional writing models, and information on planning and pursuing their careers.

Coaching Tone

We authors are all seasoned professionals with a combined 50 years in business, 25 years in college teaching, and two dozen co-authored books on writing. We know from experience—and from consulting extensively with other businesspeople—that writing is a process, that learning to write takes time, and that every writer needs a colleague to respond to his or her writing.

This book is our effort to be your students' writing colleague. In these pages, we describe the kinds of writing challenges that we face in business, and that students will also face. To use their time well, we introduce each topic as briefly as we can, lay out the key issues, and then follow up with practical writing strategies that will help them produce their best writing in the least amount of time.

Because some students might have special challenges such as little writing experience, weak grammar skills, or difficulty with the English language, we offer suggestions throughout the book that will help them address those challenges. Our bottom line is this: **We want your students to succeed**—and this book is our effort to help them do that.

Exceptional ESL (English as a Second Language) Coverage

We address ESL issues in three ways:

- *The Business Writer's* design includes elements such as succinct presentations, clear headings, and bulleted lists that help ESL learners read and understand the book's contents.
- Chapter 4, "Writing for Diversity," explains to all students how to write for individuals from different cultures or with limited English-language skills.
- Chapter 50, "Addressing ESL Issues," offers ESL readers extra help with sentence structure, word choice, grammar, and mechanics.

ORGANIZATION AND USE

The Business Writer is designed for students in business writing courses at four-year colleges, at two-year colleges, and in online programs. The book is organized into nine parts, which generally move from broad issues to more specific topics. This organization will work well with many different approaches to teaching writing.

Parts 1 and 2

Parts 1 and 2 introduce the craft of workplace writing. For example, Part 1, "Challenges for Workplace Writers," examines eight potential problems faced by writers, including using the writing process, writing with technology, writing in teams, writing for diversity, using graphics, communicating technical information, conducting research, and addressing ethical issues. Each chapter describes one of the challenges and then carefully explains how to address it. Part 2, "Benchmarking Writing with the Seven Traits," presents seven traits (or qualities) of effective workplace writing: strong ideas, logical organization, conversational voice, clear words, smooth sentences, correct copy, and reader-friendly design. Each chapter explains and illustrates a trait, offers strategies for developing the quality, and then provides a checklist for assessing writing for that quality.

Parts 3 Through 7

Parts 3 through 7, destined to be the most heavily used sections of the book, explain how to compose 110 different forms of writing. For each form, students will find a brief introduction, clear guidelines that walk the writer through the writing process, an annotated model, and a writing checklist. For example, Part 3, "The Application Process and Application Writing," includes instructions for writing all application-related documents, including résumés, inquiries, requests, application essays, and thank-you's. Part 4, "Correspondence: Memos, E-Mails, and Letters," includes instructions for writing a full range of good-news, bad-news, neutral, and persuasive messages. Part 5, "Reports and Proposals," begins with an overview of report and proposal writing, and then offers specific instructions for writing different types of reports and proposals. Similarly, Part 6, "Special Forms of Writing," explains how to compose public-relations forms, instructions, and Web writing. Finally, Part 7, "Management and Management Writing," presents the principles of effective management and then explains how to develop management-writing forms, including human-resources writing. Together, Parts 3 through 7 offer students a wealth of workplace-writing information that is so clear they can complete the writing assignments with little or no additional instruction.

Parts 8 and 9

Parts 8 and 9 address additional topics. Part 8, "Speaking, Listening, and Giving Presentations," explains how to perform a variety of speaking and listening tasks. Finally, Part 9, "Proofreader's Guide," serves as a quick reference for all forms of writing. This part of the book includes thorough instructions related to grammar, sentence structure, punctuation, capitalization, and usage. The last chapter of that part gives more detailed information on issues that challenge ESL learners.

SUPPLEMENTS FOR STUDENTS

The Business Writer **Web site** for students at college.hmco.com/english provides support directly related to the book. The site contains an almanac, additional writing assignments and exercises, real-world Web links, a career-planning tutorial, and additional model documents, including sample résumés.

The Houghton Mifflin English Student Homepage at college.hmco.com/english/students provides additional support for coursework. For example, the **Internet Research Guide** contains Learning Modules on the purpose of research; e-mail, listservs, newsgroups, and chat rooms; surfing and browsing; evaluating information on the Web; building an argument with Web research; and plagiarism and documentation.

The **e-Library of Exercises** is full of self-quizzes that give students the opportunity to increase their grammar and writing skills in thirty areas. As they sharpen their skills and strengthen their knowledge with more than 700 exercises, students can work at their own paces wherever they want—home, computer lab, or classroom.

Additional resource passkeys may be packaged with *The Business Writer,* including **SMARTHINKING™,** a password-protected online tutoring service that connects students to experienced writing tutors during primetime homework hours for one-on-one writing assistance. Students can also submit papers for feedback and questions about writing twenty-four hours a day, seven days a week, and receive a writing tutor's response within twenty-four hours. In addition, students can consult the extensive collection of additional educational services available at the SMARTHINKING site.

Finally, **WriteSpace,** a powerful new interactive writing program, is a dynamic technology tool for writers at all levels. The online program, which resides in Houghton Mifflin's Eduspace (www.eduspace.com) site, includes diagnostic tests; self-paced exercises for grammar, mechanics, and punctuation; a wide variety of writing modules (tutorials) and writing assignments; and instant access to SMARTHINKING for real-time online tutoring. Incorporated into the writing environment, the *Digital Keys 4.0 Online* handbook links students to grammar instruction in a single click. WriteSpace also enables collaborative writing with Word and communications tools inside the program.

SUPPLEMENTS FOR INSTRUCTORS

The *Instructor's Resource Manual* contains many helpful elements including the following:

- Sample course syllabus
- Suggestions for using multiple-intelligence theory to enhance students' writing
- An analysis of workplace writing in relation to academic writing and writing across the curriculum
- A summary of each chapter
- Tips for introducing each chapter
- Additional activities for each chapter

***The Business Writer* Web site** for instructors at college.hmco.com/english includes a downloadable version of the print *Instructor's Resource Manual,* a PowerPoint presentation of the features of the book, and more.

The **Instructor's Resource Center,** located at the Houghton Mifflin Web site college.hmco.com/english/instructors, provides numerous and varied sources of assistance.

WriteSpace, Houghton Mifflin's new online classroom-management platform and student-writing program, offers diagnostic tests linked to exercises that instructors prescribe for students based on their individual needs. WriteSpace also includes a variety of helpful classroom-management and communication tools.

ACKNOWLEDGMENTS

The authors express their gratitude to the following people who have contributed their valuable time, energy, and ideas in the development of *The Business Writer*:

Nancy G. Barron, Northern Arizona University

E. Wallace Coyle, Boston College – Carroll School of Management

David Dedo, Samford University

Donald F. Doell, Pima Community College, East Campus

Sheila Getzen, Baruch College, CUNY

Sydney Gingrow, Pellissippi State Technical Community College

Nancy D. Kersell, Northern Kentucky University

William Kleing, University of Missouri – St. Louis

Robert McEachern, Southern Connecticut State University

Benjamin Meyer, Montana State University

Kevin R. Swafford, Bradley University

David Sudol, Arizona State University

John Van Rys
Verne Meyer
Pat Sebranek

Contents

Preface iii

PART 1 Challenges for Workplace Writers 1

INTRODUCTION
The Business of Writing 3

The Practice of Workplace Communication 4
The Transition from Academic to Workplace Writing 9
Workplace Writing: First Principles 10
The Business Writer's Code of Ethics 11
   Checklist for Communicating 12
   Critical Thinking Activities 12
   Writing Activities 12

CHAPTER 1
Using the Writing Process 13

To Speak, Write, or Do Both 14
The Process of Writing: An Overview 16
Prewriting 18
Drafting 24
Revising 28
Refining 32
One Writer at Work 37
Beating Writer's Block 42
   Checklist for the Writing Process 43
   Critical Thinking Activities 44
   Writing Activities 44

CHAPTER 2
Writing and Technology 45

Strategies for Learning New Software 46
Word-Processing Software 47
Special Applications 49
Digital Resources: Databases and the Web 51
   Checklist for Writing and Technology 52
   Critical Thinking Activities 52
   Writing Activities 52

CHAPTER 3
Teamwork on Writing Projects 53

Using Teamwork to Strengthen Documents 54
Using Peer Review for an Early Draft 56
Using Peer Editing for a Later Draft 60
Working on a Group-Writing Project 63
Testing Documents with Readers 68
   Checklist for Peer Review and Editing 70
   Critical Thinking Activity 70
   Writing Activity 70

CHAPTER 4

Writing for Diversity — 71

Strategies for Intercultural Communication 72
Writing to an Intercultural Audience 74
Showing Respect for Diversity 78
Effective Attention to Diversity: A Model 83
- Checklist for Intercultural Communication 84
- Critical Thinking Activities 84
- Writing Activities 84

CHAPTER 5

Using Graphics in Business Documents — 85

Guidelines for Designing Graphics 86
Parts of Graphics 87
Using the Computer to Develop Graphics 88
Integrating Graphics into Text 90
Choosing the Right Graphics 92
Tables 93
Graphs 94
Charts 99
Visuals 102
- Checklist for Using Graphics in Business Documents 106
- Critical Thinking Activities 106
- Writing Activities 106

CHAPTER 6

Communicating Technical Information — 107

Getting Technical: An Overview 108
Ineffective Versus Effective Technical Communication 109
Strategies for Technical Communication 110
Features of an Effective Technical Style 112
- Checklist for Technical Communication 116
- Critical Thinking Activities 116
- Writing Activities 116

CHAPTER 7

Conducting Research for Business Writing — 117

Research Overview: A Flowchart 118
Planning Your Research 119
Managing Your Project: Note-Taking Strategies 123
Doing Primary Research 130
Doing Library Research 137
Doing Internet Research 145
Organizing Your Findings 151
Using and Integrating Sources 152
Avoiding Plagiarism 156
Following APA Documentation Rules 158
APA References List 161
- Checklist for Research 167
- Checklist for a Research Report 167
- Critical Thinking Activities 168
- Writing Activities 168

CHAPTER 8

Business Writing Ethics — 169

Strategies for Ethical Writing 170
Information Ethics 171
Persuasion Ethics 174
- Checklist for Business Writing Ethics 178
- Critical Thinking Activities 178
- Writing Activities 178

PART 2

Benchmarking Writing with the Seven Traits — 179

CHAPTER 9

The Seven Traits at Work — 181

Traits of Ineffective Writing 182
Assessing an Ineffective Document 183
Traits of Effective Writing 184
Assessing an Effective Document 185
- Checklist for Seven-Traits Benchmarking 186

Critical Thinking Activity   186
Writing Activities   186

CHAPTER 10
Trait 1: Strong Ideas   187

Stating Ideas Clearly   188
Supporting Ideas Effectively   190
Thinking Creatively   192
Thinking Logically   194
Using Thinking Patterns (From Describing to Evaluating)   195
- Checklist for Ideas   214
- Critical Thinking Activities   214
- Writing Activities   214

CHAPTER 11
Trait 2: Logical Organization   215

Strategies for Getting Organized   216
Foolproof Organization Strategies   218
Structuring Documents Through Paragraphing   225
- Checklist for Organization   232
- Critical Thinking Activities   232
- Writing Activities   232

CHAPTER 12
Trait 3: Conversational Voice   233

Weak Voice   234
Strong Voice   235
Making Your Writing Natural   236
Making Your Writing Positive   238
Developing "You Attitude"   240
- Checklist for Voice   242
- Critical Thinking Activities   242
- Writing Activities   242

CHAPTER 13
Trait 4: Clear Words   243

Cutting Unnecessary Words   244
Selecting Exact and Fresh Words   248

Avoiding Negative Words   255
- Checklist for Word Choice   258
- Critical Thinking Activities   258
- Writing Activities   258

CHAPTER 14
Trait 5: Smooth Sentences   259

Smooth Sentences: Questions and Answers   260
Rough Problems and Smooth Solutions   261
Combining Choppy Sentences   262
Energizing Tired Sentences   266
Dividing Rambling Sentences   270
Sentence Smoothness in Action   271
- Checklist for Sentences   272
- Critical Thinking Activities   272
- Writing Activities   272

CHAPTER 15
Trait 6: Correct Copy   273

Basic Terms: A Primer for Correctness   274
Correcting Unclear Wording   275
Correcting Faulty Sentences   278
Correcting Punctuation Marks   284
Correcting Mechanical Difficulties   286
- Checklist for Errors   288
- Critical Thinking Activities   288
- Writing Activities   288

CHAPTER 16
Trait 7: Reader-Friendly Design   289

Weak Versus Strong Design   290
Understanding Basic Design Principles   292
Planning Your Document's Design   293
Developing a Document Format   294
Laying Out Pages   297
Making Typographical Choices   302
- Checklist for Document Design   304
- Critical Thinking Activities   304
- Writing Activity   304

PART 3: The Application Process and Application Writing 305

CHAPTER 17
Understanding the Job-Search Process 307

Overview of the Job-Search Process 308
Assessing the Job Market 309
Guidelines for Career Plans 312
Conducting a Job Search 315
Researching Organizations 316
Using Web Resources 317
   Checklist for Career Planning 318
   Critical Thinking Activities 318
   Writing Activities 318

CHAPTER 18
Developing Your Résumé 319

Guidelines for Résumés 320
   Checklist for Résumés 326
   Critical Thinking Activity 326
   Writing Activities 326

CHAPTER 19
Writing Application Correspondence 327

Guidelines for Application Letters 328
Guidelines for Recommendation-Request Letters 330
Guidelines for Application Essays 332
Guidelines for Job-Acceptance Letters 334
Guidelines for Job-Rejection Letters 336
Guidelines for Thank-You and Update Messages 338
   Checklist for Application Correspondence 340
   Critical Thinking Activity 340
   Writing Activities 340

CHAPTER 20
Participating in Interviews 341

Interviewing for a Job or Program 342
Inappropriate or Illegal Questions 344
Common Interview Questions 346
Guidelines for Interview Follow-Up Letters 348
Interviewing a Job Applicant 350
   Checklist for Interview Follow-Up Letters 352
   Critical Thinking Activities 352
   Writing Activities 352

PART 4: Correspondence: Memos, E-Mails, and Letters 353

CHAPTER 21
Correspondence Basics 355

Writing Successful Correspondence 356
E-Mail, Memo, or Letter: Which Should It Be? 357
Three Types of Messages 358
Correspondence Catalog 361
   Checklist for Correspondence Basics 362
   Critical Thinking Activities 362
   Writing Activities 362

CHAPTER 22
Writing Memos 363

Guidelines for Memos 364
Basic Memo 365
Expanded Memo 366
   Checklist for Memos 368
   Critical Thinking Activity 368
   Writing Activity 368

CHAPTER 23
Writing E-Mail Messages and Sending Faxes 369

Guidelines for E-Mail Messages 370
E-Mail Model and Format Tips 371

Choosing and Using E-mail   372
E-Mail Etiquette and Shorthand   373
Faxing Documents   374
- Checklist for E-Mails and Faxes   376
- Critical Thinking Activity   376
- Writing Activities   376

CHAPTER 24
Writing Letters   377

Guidelines for Letters   378
Professional Appearance of Letters   379
Basic Letter   380
Expanded Letter   382
Letter Formats   384
Letters and Envelopes   388
Forms of Address   391
- Checklist for Letters   396
- Critical Thinking Activity   396
- Writing Activity   396

CHAPTER 25
Writing Good-News and Neutral Messages   397

The Art of Being Direct   398
Guidelines for Informative Messages   400
Guidelines for Routine Inquiries and Requests   406
Guidelines for Positive Responses   410
Guidelines for Placing Orders   414
Guidelines for Accepting Claims   416
Guidelines for Goodwill Messages   418
- Checklist for Good-News and Neutral Messages   424
- Critical Thinking Activities   424
- Writing Activities   424

CHAPTER 26
Writing Bad-News Messages   425

The Art of Being Tactful   426
Guidelines for Denying Requests   428

Guidelines for Rejecting Suggestions, Proposals, or Bids   436
Guidelines for Explaining Problems   440
Guidelines for Resigning   444
Guidelines for Making Claims or Complaints   446
- Checklist for Bad-News Messages   450
- Critical Thinking Activities   450
- Writing Activities   450

CHAPTER 27
Writing Persuasive Messages   451

The Art of Persuasion   452
Guidelines for Special Requests and Promotional Messages   454
Guidelines for Sales Messages   460
Guidelines for Collection Letters   466
Guidelines for Requesting Raises or Promotions   470
- Checklist for Persuasive Messages   472
- Critical Thinking Activities   472
- Writing Activities   472

CHAPTER 28
Writing Form Messages   473

Guidelines for Form Messages   474
Standard Form Message   475
Menu Form Message   476
Guide Form Message   477
- Checklist for Form Messages   478
- Critical Thinking Activities   478
- Writing Activities   478

PART 5
Reports and Proposals   479

CHAPTER 29
Report and Proposal Basics   481

Writing Successful Reports and Proposals   482

Types of Reports and Proposals 483
  Checklist for Report and Proposal Basics 484
  Critical Thinking Activities 484
  Writing Activities 484

CHAPTER 30
Writing Short Reports 485

Guidelines for Incident Reports 486
Guidelines for Investigative Reports 490
Guidelines for Periodic Reports 496
Guidelines for Progress Reports 500
Guidelines for Trip or Call Reports 504
  Checklist for Short Reports 508
  Critical Thinking Activities 508
  Writing Activities 508

CHAPTER 31
Writing Major Reports 509

Guidelines for Major Reports 510
  Checklist for Major Reports 520
  Critical Thinking Activities 520
  Writing Activity 520

CHAPTER 32
Writing Proposals 521

Guidelines for Proposals 522
Operational Improvement Proposals 524
Sales or Client Proposals 530
Grant and Research Proposals 533
  Checklist for Proposals 538
  Critical Thinking Activities 538
  Writing Activity 538

CHAPTER 33
Designing Report Forms 539

Guidelines for Designing Report Forms 540
  Checklist for Designing Report Forms 544
  Critical Thinking Activity 544
  Writing Activities 544

PART 6
Special Forms of Writing 545

CHAPTER 34
Public-Relations Writing 547

Guidelines for News Releases 548
Guidelines for Flyers and Brochures 552
Guidelines for Newsletters 560
  Checklist for Public-Relations Writing 564
  Critical Thinking Activities 564
  Writing Activities 564

CHAPTER 35
Writing Instructions 565

Types of Instructions 566
Tips for Writing Instructions 567
Guidelines for Instructions 568
  Checklist for Instructions 574
  Critical Thinking Activities 574
  Writing Activity 574

CHAPTER 36
Writing for the Web 575

Web Page Elements and Functions 576
Strategies for Developing a Web Site 578
Sample Web Sites and Pages 582
  Checklist for Developing Web Pages and Sites 586
  Critical Thinking Activities 586
  Writing Activities 586

PART 7
Management and Management Writing 587

CHAPTER 37
Managing Your Time and Manners 589

Managing Your Time 590
Evaluating Your Time-Management Skills 592

Practicing Workplace Etiquette   594
Polishing Your Etiquette   596
Eating and Drinking   597
   Checklist for Appropriate Dress   598
   Critical Thinking Activity   598
   Writing Activity   598

CHAPTER 38
Managing Effectively   599
Managing Writing Tasks   600
Delegating Work   601
Solving Problems   602
Sustaining a Supportive Work Climate   603
Developing Successful Employees   604
Dealing with Discrimination   607
   Checklist for Managing Effectively   608
   Critical Thinking Activities   608
   Writing Activities   608

CHAPTER 39
Management Writing   609
Guidelines for Mission Statements   610
Guidelines for Position Statements   612
Guidelines for Policy Statements   614
Guidelines for Procedures   618
Guidelines for Company Profiles (or Fact Sheet)   620
   Checklist for Management Writing   624
   Critical Thinking Activities   624
   Writing Activities   624

CHAPTER 40
Human Resources Writing   625
Guidelines for Job Descriptions   626
Guidelines for Job Advertisements   630
Guidelines for Employer's Follow-Up Letters   632
Guidelines for Employee Evaluations   636
Guidelines for Employee Recommendations   640

   Checklist for Human Resources Writing   644
   Critical Thinking Activities   644
   Writing Activities   644

PART 8
Speaking, Listening, and Giving Presentations   645

CHAPTER 41
Communication Basics   647
Speaking Effectively   648
Listening Effectively   649
Giving and Taking Instructions   650
Giving and Taking Criticism   651
Understanding Conflicts   652
Resolving Conflicts   653
   Checklist for Communication Basics   654
   Critical Thinking Activity   654
   Writing Activity   654

CHAPTER 42
Communicating in a Group   655
Beginning a Group   656
Working in a Group   657
Making Decisions   659
Listening in a Group   660
Responding in a Group   661
Roles in a Group   662
Disagreeing in a Group   663
   Checklist for Communicating in a Group   664
   Critical Thinking Activity   664
   Writing Activities   664

CHAPTER 43
Communicating in Meetings   665
Formal Versus Informal Meetings   666
Formal Meetings   667
Order of Business for a Meeting   668
Making Motions   669
Officers and Their Responsibilities   670
Guidelines for Minutes   672

Contents

Checklist for Meeting Minutes 676
Critical Thinking Activity 676
Writing Activity 676

CHAPTER 44
Writing and Giving Presentations 677

Giving Presentations 678
Planning Your Presentation 679
Organizing Your Presentation 680
Writing Your Presentation 683
Writing with Style and Motivational Appeals 684
Using Visual Support 690
Developing Computer Presentations 691
Practicing Your Delivery 692
Overcoming Stage Fright 693
Checklist for Writing Presentations 694
Critical Thinking Activity 694
Writing Activities 694

PART 9
Proofreader's Guide 695

CHAPTER 45
Understanding Grammar 697

Noun 697
Pronoun 700
Verb 704
Adjective 710
Adverb 711
Preposition 712
Conjunction 713
Interjection 713

CHAPTER 46
Constructing Sentences 715

Using Subjects and Predicates 715
Using Phrases 718

Using Clauses 720
Using Sentence Variety 721

CHAPTER 47
Using Punctuation 723

Period 723
Question Mark 724
Exclamation Point 725
Parentheses 725
Comma 726
Apostrophe 731
Colon 733
Semicolon 734
Ellipsis 735
Quotation Marks 736
Hyphen 738
Dash 740
Brackets 741
Diagonal 741
Italics (Underlining) 742

CHAPTER 48
Checking Mechanics 743

Capitalization 743
Plurals 747
Numbers 749
Abbreviations 751
Acronyms and Initialisms 753
Spelling 754

CHAPTER 49
Using the Right Word 755

CHAPTER 50
Addressing ESL Issues 767

The Parts of Speech 767
Understanding Sentence Basics 779
Sentence Problems 780
Numbers, Word Parts, and Idioms 782

Index 787

PART 1

Challenges for Workplace Writers

	Introduction: The Business of Writing
1	Using the Writing Process
2	Writing and Technology
3	Teamwork on Writing Projects
4	Writing for Diversity
5	Using Graphics in Business Documents
6	Communicating Technical Information
7	Conducting Research for Business Writing
8	Business Writing Ethics

In this chapter
The Practice of Workplace Communication **4**

The Transition from Academic to Workplace Writing **9**

Workplace Writing: First Principles **10**

The Business Writer's Code of Ethics **11**

Checklist for Communicating **12**

Critical Thinking Activities **12**

Writing Activities **12**

INTRODUCTION

The Business of Writing

Each day, business people write, read, speak, and listen to do their work—sending e-mail messages, reporting on projects, making presentations, and developing proposals. The success of their work often depends on the quality and speed of the communication. In fact, in a *Job Outlook 2003* poll, employers rated oral and written communication skills as the most important ones that job candidates can possess. Unfortunately, employers also noted that these skills are the ones that candidates most often lack. A potential hire might be a great engineer, be a marketing guru, or possess a strong legal mind, but if that candidate can't communicate, he or she is handicapped.

It's in this context that you are seeking to improve your business-writing skills. As a budding business writer, you may be making the transition from college to full-time employment, or perhaps you have been in the workplace and are seeking to strengthen a valuable skill. Whatever your situation, you need to prove yourself through good writing. Strong writing will help you advance the organization's mission, promote efficiency, ensure quality, build a sense of team, and strengthen customer relations. By contrast, poor communication may stall work, destroy important ties, or even lead to injury or death.

In your business-writing class and in the workplace, *The Business Writer* will strengthen your skills. It introduces you to the transactional context of workplace writing, offers writing principles adaptable to any workplace practice, provides instruction on all the common forms of workplace writing, and supplies quick reference help on grammar. In short, *The Business Writer* will help you write better and faster.

THE PRACTICE OF WORKPLACE COMMUNICATION

Business writers don't work in a vacuum. Rather, workplace writing happens within the broader context of workplace communication—communication aimed at promoting the organization's mission and advancing its operations. What, then, is workplace communication? What makes it work, and what makes it fail? What is its role in the company? How can you participate positively in this dynamic activity? And where specifically does your writing fit? That's the focus of the next few pages.

Essentially, workplace communication refers to the exchanges of information and ideas that help people get work done. The Latin root of *communication* is *communis,* meaning "common." In other words, people in the workplace communicate (or share things) so that they have ideas and information "in common." Workplace communication is always a transaction between flesh-and-blood people—a dynamic exchange or dialogue involving whole people, with all their personal and cultural characteristics. A successful transaction helps produce clear understanding, informed decisions, and productive actions.

Communication: The Little Picture

In brief, the communication process involves a sender and a receiver, a message and a medium. Working within a given context, the sender develops and encodes the message in a given language, verbal or otherwise, and sends the message through one or more media (e.g., sound waves, paper, e-mail). The receiver, in turn, accesses the medium and decodes the message. Some form of feedback (e.g., head nodding, a reply e-mail, a follow-up phone call) is often built into the process. In fact, the process can be so dynamic that sender and receiver simultaneously send, receive, and respond in a highly dialogic interaction. Figure I.1 pictures how the communication process works.

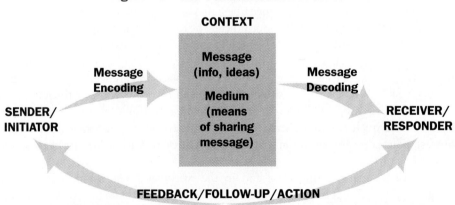

Figure I.1   The Communication Process

Note the following insights concerning workplace communication:

The process is multimedia. At its core, communication is thought transmitted through—and stimulated by—a range of media.

- **Writing and reading.** The communication exchange can be rooted in written texts using one or more languages. Such texts can be paper or electronic, from product labeling to hyperlinked Web pages.
- **Speaking and listening.** Communication can be oral—relying on voices and ears, whether through live conversations, phone calls, or recordings.
- **Imaging and viewing.** Communication can involve two categories of visual images and cues. First, an exchange can include graphics (e.g., tables, charts, photos, drawings), icons, color-coding, and so on—images used to share ideas and information. Second, an exchange can include physical cues and actions—demonstrations, physical setting and space (inside versus outside, room size and layout, furniture types and arrangements, proximity of sender and receiver), clothing (type, formality, colors), and body language (facial expressions, eye contact, gestures, posture, hair styles).

Communication can be "rich" or "lean." When communication is truly multimedia, involving both verbal and nonverbal elements as well as a high degree of interaction, it is considered "rich." When nonverbal elements and interaction are minimized, as with a formal written document, such communication is considered "lean." A specific message can fall anywhere between these two extremes.

Communication can be intentional or unintentional. Because communication is frequently multimedia, it can involve senders and receivers who are conveying both intended and unintended messages. Sometimes, the intended and unintended messages are consistent; at other times, they aren't. While words may convey one message, tone, word connotations, document appearance, and body language may convey the opposite message.

The process always takes place in a concrete context. Workplace communication is not an abstract activity. Rather, it happens in a particular place and time between real people for a particular reason with real consequences. In addition, each communication act is linked to other acts of communication—whether we think of those links as a chain, a web, or a wheel.

The process may be hampered by "noise." Communication is easily disrupted by various forms of interference—what's generally called "noise." Such interference can be literal noise that makes holding a conversation difficult, or it can be poor reproduction of a document, inappropriate word choices, the speaker's tone, or the receiver's biases.

The process is a dance. Communication has been called a loop, implying that the exchange is a tidy, circular process. Perhaps a dance is a better analogy, because it involves partnership, leading and following, direction, dynamic movement, response to the musical context, continuous exchange, and skill—all crucial to all forms of communication.

Communication: The Big Picture

To understand workplace communication, you need the big picture of how businesses and other organizations order themselves, their employees, and their work. Figure I.2 shows the basic structure of an organization. Understanding how communication works in such structures requires understanding several key factors.

Figure I.2 Organizational Structure

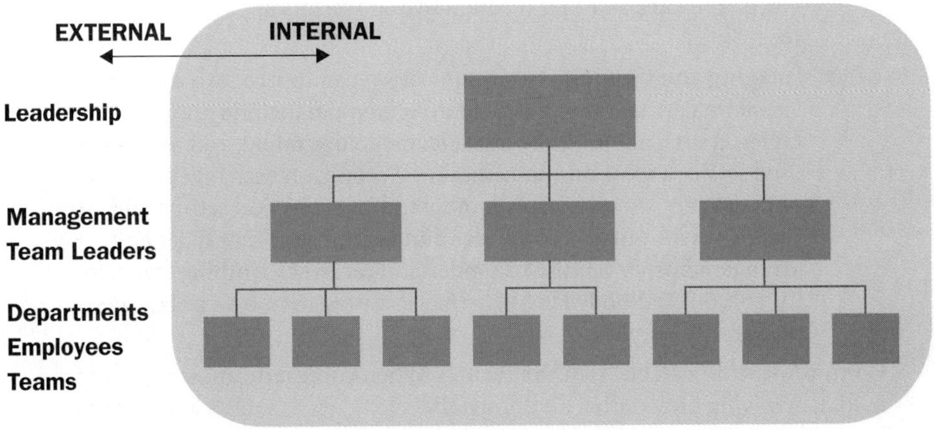

EXTERNAL ←→ INTERNAL

Leadership

Management
Team Leaders

Departments
Employees
Teams

Chain of Command

The lines in Figure I.2 generally indicate authority and accountability in the organization. Communication typically flows along these lines.

Levels

Companies with few levels are considered flat, with the result that the communication chain can be shorter, more informal, and more dynamic. Companies with many levels are considered tall and tend to be more hierarchical, formal, and rigid in their communication—with a chain that is primarily vertical.

Company Culture

Like people, organizations have personalities. This company culture is determined partly by the levels described above and partly by the values, beliefs, rituals, procedures, and work environment that members share. Factors affecting the culture may include the types of products or services the company offers, the company's size, its leadership style and philosophy, its location, its history, and so on. While each company's culture is unique, cultures can range from traditional to contemporary:

- The traditional company is strongly hierarchical, with an emphasis on authority, top-down decision making, limited communication, and narrowly defined responsibilities.
- The contemporary company tends to be more creative and dynamic, offering workers greater freedom and responsibility in terms of carrying out the company's mission, making decisions, solving problems, and communicating.

Cultural Diversity

While company culture refers to what is shared, organizations are also characterized by internal and external diversity—gender, ethnic, racial, regional, and so on. From its dealings with its own diverse workforce to its participation in a global community, today's organization must address such diversity in its communication practices.

Communication Media

To develop an efficient communication system, organizations use a wide range of media, from face-to-face meetings to e-mail to the Web.

Direction and Flow

Within any organization, communication moves in four directions:

- *Downward* communication moves from superiors to subordinates. Its function is to inspire, direct, establish policies, inform, request, or give orders.
- *Upward* communication moves from subordinates to superiors. Its function is to provide required information on work or to offer suggestions.
- *Lateral* communication moves sideways between colleagues on the same level—fellow managers, team members, and so on. It aims to share information that builds harmony, advances projects, and promotes efficiency (no duplication).
- *Diagonal* communication moves both between different levels and between different areas or departments. Its general aim is to promote efficient information sharing, reliance on expertise, and cross-divisional cooperation.

Channels

Any organization generally has two parallel, interrelated communication systems:

- The *formal system* includes official channels for communicating with colleagues, clients, and others, such a memos, reports, customer-service bulletins, the company Web site, intranet postings, and so on.
- The *informal system* refers to unofficial channels, also known as the company grapevine—water-cooler conversations, e-mail gossip.

Inside Versus Outside Communication

Where the message originated and where it goes are crucial factors in the communication process. Generally, the exchange can be considered internal or external:

- *Internal messages* involve an exchange between colleagues within the company—whether on the factory floor, across the hall, or between offices around the globe. Whether a conversation, an e-mail, or a newsletter, an internal message focuses on pushing work forward. Because colleagues are "insiders," they rely on inside knowledge and use an inside voice.
- *External messages* flow across the organization's boundaries—from outside to inside, or the reverse. Such communication might be with clients, suppliers, potential employees, the public, or government officials. Whether in ads, phone calls, letters, or Web pages, you portray and promote the company.

Part 1 Challenges for Workplace Writers

Communication: Challenges and Strategies

Workplace communication clearly involves complex challenges. To face these challenges successfully, study and use the strategies below.

Work within the company context. Understand your company's mission, values, strategies, benchmarks, policies and procedures, and preferred communication practices. Identify how human diversity affects the company. Map out your various audiences in concentric circles around your own position, both within and outside the organization, near and far. Understanding this context, you can commit your communication to advancing your company's work.

Strengthen key communication skills. Through study, training, and practice, work on critical thinking, creative thinking, empathy, listening, cooperation, negotiation, persuasion, and plain-English writing.

Acknowledge your own perspective. Understand where you are coming from: your gender, education, values, priorities, religious stance, ethnic identity, citizenship, and so on. What are the strengths and limits of your perceptions and perspectives? How do these relate to and differ from those of your audience?

Anticipate and prevent breakdowns. In your communication, look for and remove barriers that prevent successful exchanges:

- Language barriers: Problems with verbal clarity, such as the use of complex jargon, words with negative connotations, and rambling sentences.
- Media barriers: Problems with the choice of medium and the successful transmission of a message (e.g., sending a print document to someone who is blind, posting an important notice on a hard-to-find Web site, using e-mail for a highly sensitive message).
- Cultural barriers: Problems that arise because of cultural gaps, such as the generation gap, the gender gap, and so on.

Work for open communication. Contribute to a productive environment:

- Build credibility and trust by sending consistent, thoughtful, controlled messages. Avoid sending mixed messages, showing favoritism, and creating gaps between words and actions. Do not instill fear through your communication or create confusion about your values. Be aware of and beware of turf wars, egos, and office politics: Work to deflate these.
- Create efficiency and clarity by minimizing long lines of communication and preventing grapevine gossip: Both lead to message distortion.
- Make lots of space for feedback and follow-up in your communication. Create good habits with respect to response time, interrupting others, and dealing with interruptions.
- Use communication technologies (e-mail, fax machines, cell phones, the Web) to promote efficient and productive communication.
- Create an information-rich environment. To work effectively and participate fully in an organization, people need information about decisions, successes, failures, trends, and developments. Unless information is sensitive or data overload is a problem, seek ways to keep people informed.

THE TRANSITION FROM ACADEMIC TO WORKPLACE WRITING

Writing is a critical component of workplace communication. But how do you make the transition from academic to workplace writing, from writing for a grade to writing for a living? Think about your experiences in college and at work. Imagine a typical essay and a common business letter. How are academic and workplace writing similar and different? Study Table I.1, noting that (1) academic-writing skills are foundational for workplace writing, and (2) workplace-writing tasks expand upon and refocus college-writing tasks.

Table I.1   College Versus Workplace Writing

	College	Workplace
Context	• Coursework: activities related to learning content and demonstrating learning • Classroom: a group of peers supervised by an expert	• Business or nonprofit work: selling products and providing services • Office, home office, factory, field, Internet: sites where colleagues combine efforts to serve clients
Goals	• To learn important ideas, concepts, principles, and practices • To share learning with peers and instructor • To be evaluated or graded	• To get and keep a job • To inform and/or persuade • To be evaluated, but also to evaluate, sell, order, report, propose, instruct, promote, and build goodwill
Readers	• Primarily instructors: experts in the subject matter • Sometimes peers: classmates looking at the same or related topics	• Diverse audience: experts but also laypeople in relation to the topic • Audience may be multiple and/or complex: colleagues and clients; managers, technicians, and production workers; English and ESL readers
Topics and content	• Academic subjects: focus on ideas, research, evidence, in-depth discussion • Topics primarily related to your major or specialization	• Work-related subjects: focus on practical applications of ideas, on decisions and actions • Topics related to your expertise, but also topics related to all aspects of business
Format and presentation	• Traditional essays or reports • Follow academic format (MLA, APA, CBE, or some other style) • Submitted on paper; occasionally e-mailed or posted	• Highly varied formats: reports, correspondence, Web pages, manuals • Distributed and exchanged through e-mail, interoffice mail, postal and courier services, intranet and Internet postings
Ethical challenges	• Honest research, including research with natural and human subjects • Proper use of information sources, including avoiding plagiarism	• Honesty, accuracy, and sensitivity • Adhering to policies, regulations, and laws • Promoting safety and guarding life • Proper use of information, including avoiding plagiarism, copyright infringements, and privacy violations
Skills required	• Thinking skills: exposition, analysis, synthesis, evaluation, argumentation • Essay, paragraph structure • Academic English: clear word choice, mature sentences, correct grammar • Academic format, quality research, and documentation	• Thinking skills: exposition, analysis, synthesis, evaluation, argumentation; practical common sense • Document, paragraph, list structure • Plain, conversational English: clear word choice, smooth sentences, correct grammar • Reader-friendly design and format fitting for type of document

WORKPLACE WRITING: FIRST PRINCIPLES

Keeping in mind the complex communication context, where do you start on improving your workplace writing? Below is a set of first principles, which are more fully elaborated elsewhere in this handbook and threaded throughout its pages.

Set positive goals. To complete each document successfully, do the following:

- Consider timing—reasonable deadlines, the situation's urgency, and efficient practices.
- Know exactly what you want to accomplish—the measurable outcome.
- Build goodwill by creating and maintaining strong work relationships.
- Align your writing goals with the company's mission and values.

Write directly to readers. That may sound obvious, but it's tough. Think about each reader's knowledge, motives, position, and power. How will he or she react to your message? How can you build common ground? In particular, consider the diversity of your business audience—co-workers, clients, and communities at home and abroad. Then write respectfully to that audience. (See page 18.)

Master the writing process. Don't expect to achieve perfection in one shot. To get it right, do the necessary work—prewriting, drafting, revising, and refining. Develop effective and efficient writing habits that adapt the process to the document. (See pages 16–36.)

Follow good models. If possible, avoid starting each document from scratch. Keep samples of good writing, check the models in this book, and develop document templates. Then copy, paste, and modify models to fit the given situation.

Benchmark with the seven traits. Practice quality control by measuring each document against these traits:

- **Strong Ideas:** Quality content that is accurate, precise, and complete.
- **Logical Organization:** An internal structure ordering the document into sensible patterns.
- **Conversational Voice:** A friendly but professional tone that fits the document and reader.
- **Clear Words:** Precise and fresh language; plain English, not "business English."
- **Smooth Sentences:** Easy to read, concise, and varied in pattern and length.
- **Correct Copy:** No glaring grammar, punctuation, usage, mechanics, and spelling errors.
- **Reader-Friendly Design:** Attractive, accessible format, layout, and typography.

Package it professionally. Sloppy documents are about as appealing as poppy seeds in your teeth. Use appropriate stationery, folds, inks, fonts, logos, envelopes, and so on.

Be a team writer. Work together so that the project, company, and reader all win. Test important documents before sending or distributing them. Ask for and give honest feedback that improves each piece of writing. (See page 53 for more.)

THE BUSINESS WRITER'S CODE OF ETHICS

As a business writer conducting exchanges with real people in a real company, you will face many ethical challenges. To guide your effort to address these challenges, adopt a code of ethics like the one shown below. The code offers guidance with respect to language choices, reasoning, and behavior, helping you do the right thing. Study the code, embrace it, and practice it throughout your career.

To write ethically, I shall do the following:
1. Be sensitive to human diversity in my word choices, assumptions, and tone.
2. Use tact to guard the reader's ego by avoiding anger, sarcasm, bluntness, and bad humor.
3. Avoid gobbledygook—dense, difficult, obscure writing.
4. Construct objective, sound, and balanced arguments that legitimately blend rational and emotional appeals.
5. Avoid manipulation and groupthink, especially in persuasive writing (e.g., sales letters).
6. Work efficiently to develop quality documents, practicing good stewardship of company resources while meeting the information needs of colleagues, clients, and others.
7. Take account of my document's likely effects and possible side effects.
8. Strive to be honest by avoiding distortion, deception, half-truths, lies, exaggeration, and errors—never using language to evade responsibility or to mislead.
9. Guard sensitive information and respect privacy where appropriate.
10. Meet obligations to my readers, company, colleagues, and other communities (e.g., the government, the public) by treating all people with respect and avoiding gossip and slander.
11. Negotiate competing interests by striving for positive, fair outcomes.
12. Value long-term goodwill, trust, and credibility over short-term gain.
13. Observe all legal requirements related to issues such as discrimination, labor and safety standards, environmental conditions, harassment, and copyrights.
14. Give credit where credit is due—for information, ideas, and assistance.

WRITE Connection

Something may be legal, but not ethical. In your writing, seek to do not only what is legally correct but also what is ethically right. For a fuller treatment of ethics, see Chapter 8, "Business Writing Ethics." Attention to ethical issues has also been integrated throughout this book.

CHECKLIST for Communicating

Periodically use this checklist to monitor progress on improving your communication skills.

- ☐ I try to make my communication open, clear, and productive.
- ☐ I respect my company's culture, its chain of command, and my own position.
- ☐ I make sensible choices of communication media (speaking, writing, images) and communication technologies (paper, e-mail, telephone).
- ☐ I anticipate and seek to prevent "noise" and other communication barriers.
- ☐ I am aware of the strengths and limits of my perceptions and perspectives.
- ☐ I respect others' perceptions and perspectives.
- ☐ I seek to build credibility and trust.
- ☐ I use fitting strategies for upward, downward, lateral, and diagonal communication.
- ☐ I use formal and informal channels effectively.
- ☐ I promote my organization in fitting ways with those outside it.
- ☐ When I put my message in writing, I do so for good reasons—both to advance work and to strengthen working relationships.
- ☐ My writing shows fitting seven-traits characteristics.
- ☐ My writing (and all my communication) adheres to a strong code of ethics.

CRITICAL THINKING Activities

1. Describe communication at your college: What are the components of the formal and informal systems? What are examples of downward, upward, lateral, and diagonal communication, as well as internal and external communication? What media are available and how are they used? How would you characterize the college's structure and culture?
2. List organizations that you have belonged to (businesses, clubs, religious institutions, nonprofit organizations). Did or does communication work well in these organizations? Why or why not? Be specific and concrete.
3. Collect three samples of workplace writing and analyze their chief characteristics. Are these messages successful or unsuccessful "transactions"?

WRITING Activities

1. In writing, introduce yourself to your classmates and instructor, explaining who you are and why you are in this class.
2. Develop a one- to two-page report about a professional in a career related to your major. In the report, develop a profile of the individual that will help fellow students understand what such a career entails.

In this chapter

To Speak, Write, or Do Both 14
The Process of Writing: An Overview 16
Prewriting 18
Drafting 24
Revising 28
Refining 32
One Writer at Work 37
Beating Writer's Block 42
Checklist for the Writing Process 43
Critical Thinking Activities 44
Writing Activities 44

CHAPTER

1

Using the Writing Process

When it comes to writing, business people face competing pressures. First, they must get the job done soon (the pressure of deadlines). Second, they must get the job done right (the pressure of results).

Moreover, business writers constantly face choices, variables that change from one situation to the next. Is writing a good choice for the situation? Perhaps a better medium would be some form of speech, such as a phone call, or perhaps a combination of oral, written, and graphical communication, such as a PowerPoint presentation. Even if writing *is* the better medium, which form is best—an e-mail message, a letter, or something else? All this thinking precedes applying fingers to keyboard or pen to paper.

When their fingers do move, business writers must take sensible steps that will lead to a quality document. That's where the writing process comes in. Process decisions include strategies for prewriting (generating ideas), drafting (developing ideas), revising (strengthening content), and refining (fixing style and grammar problems). Smart business writers use this process, with flexibility, because they know that it leads to positive results. This chapter offers a set of transferable strategies that you can adapt to any workplace writing situation.

TO SPEAK, WRITE, OR DO BOTH

Whenever you have a message to share, you also have decisions to make: Should you speak, write, or do both? Which medium should you use—a face-to-face talk, a phone call, or an e-mail message? Poor choices can waste time, create confusion, cause unnecessary follow-up, and make trouble. Three steps will help you make sensible decisions.

Analyze the Situation

Understanding the context is the first step toward a sensible choice:

Audience. Who is your receiver? One person or several? People you know well or little? Someone near or far? A group that is uniform or diverse? Someone with strong or weak language skills?

Purpose. What do you want to accomplish? What goals or outcomes are you hoping to achieve? What specific understanding, decision, or action should result?

Pressures. What is the time frame for accomplishing these goals? How complex, urgent, or sensitive is the message? What ethical challenges do you face?

Response. What kind of reaction do you expect? What must readers be able to do with the message? Save it? Copy, share, critique, or revise it?

Protocols. What rules or routines (company, cultural, or other) should you follow? What record of the message do you or others need?

Resources. What communication technologies do you and your readers have available? What are your restrictions?

Weigh Advantages and Disadvantages

For your specific situation, list the strengths and weaknesses of speaking versus writing.

Speaking advantages
- Can be immediate, quick
- Can reveal emotion
- Can be spontaneous
- Can be dramatic
- Can be interactive (rapid feedback)
- Can be tactful and personal

Speaking disadvantages
- May be too hasty or inefficient
- May reveal unwanted emotions
- May seem unprepared, chaotic
- May seem threatening
- May be unfocused, disorganized
- May lack tact or polish

Writing advantages
- Offers stable message
- Provides permanent record
- Can be clear, concise, detailed
- Can be revised for focus, organization
- Can be distributed widely
- Can include graphics
- Can look official, weighty
- Can be dynamic, colorful, persuasive

Writing disadvantages
- May be time-consuming to create
- May be incriminating, "haunting"
- May be vague, confusing
- May ramble if poorly revised
- May fall into the "wrong hands"
- May cost much to produce
- May seem impersonal
- May be easy to ignore, discard

Select a Medium

Once you've analyzed the situation and decided whether to speak or write, choose one or more channels for sending your message. In many cases, a combination of speaking and writing will get the best results. For example, an informative e-mail about a policy change might be followed by a group discussion about procedures for implementing the policy. A summary of points agreed to in a phone call or meeting might be e-mailed or faxed to participants. The key is deciding how best to sequence and combine oral and written communication to get the best results. Here are the main media to consider:

Have a private conversation when you need immediate feedback, the topic is sensitive, or you need to go "off the record."
Benefits: immediate response to words, body language, tone of voice, attitude

Have a group discussion when you need to brainstorm, reach consensus, or build team spirit.
Benefits: shared decisions and idea development

Make a phone call when you need to exchange ideas and information quickly.
Benefits: makes physical presence unnecessary, allows rapid feedback, and includes tone (but not body language)

Leave a voice-mail message when your message is casual, a return call is adequate, you want to check your message before sending it, or you want to contact several people in person.
Benefits: convenient for both speaker and receiver

Make an oral presentation when you want to reach a large group, use audio-visual aids, and get group feedback.
Benefits: opportunity to shape the message in response to an audience

Send an e-mail when you want to contact one or many people with a simple or routine message that they can conveniently read, respond to, delete, or save.
Benefits: rapid, flexible, powerful

Write a memo or letter when you need to officially communicate important or complex information, policies, requests, or decisions.
Benefits: creates a written record, promotes clear understanding, can reach many readers conveniently

Write a report, proposal, procedure, or other business document when you need to create a record for a project, persuade readers to act, or share information with specific groups of people.
Benefits: special forms share detailed information in formats that meet special needs

F.Y.I.

A *paper trail* or an *e-trail* is written documentation of statements, actions, or decisions. Create a trail if someone will hold you accountable for events and outcomes—especially if legal action is possible. See Chapter 8, "Business Writing Ethics."

THE PROCESS OF WRITING: AN OVERVIEW

It's easy to feel overwhelmed by a writing project. However, using the writing process will relieve some of that pressure by breaking the task into a series of manageable steps: prewriting, drafting, revising, and refining.

The Big Picture

In Figure 1.1, the inside terms and gray arrows indicate the steps usually needed to develop a document that achieves your purpose. They show that writers often move back and forth between steps. The outside arrows stress writing direction: Documents get written in the context of writers and readers working toward results.

Figure 1.1   Developing a Document

The Writer
Whether you are initiating the communication or responding to previous communication, you may have one of several reasons to write:
- To address a task (e.g., instructing readers how to safely use a pressure washer)
- To solve a problem (e.g., alerting your superiors to unsafe vehicle-undercoating methods)
- To pursue an opportunity (e.g., following up a convention contact)
- To promote an idea (e.g., sharing an insight about how your organization can better serve minorities)

The Results
You and your reader come together through the written message to accomplish something and complete (or continue) the communication. Potential results include:
- Understanding (e.g., production employees comprehend a new no-beard policy)
- Feedback (e.g., an e-mail reply confirms an appointment)
- Decision (e.g., management chooses to replace internal-combustion forklifts with electric ones)
- Action (e.g., the customer safely uses the pressure washer)
- Follow-up (e.g., committee members complete tasks assigned in the meeting minutes)

The Reader
Consider these factors:
- Sometimes your readers *must* read your document—as part of their job, for example.
- At other times, you must *convince* your readers to read your message—to open the envelope, to visit the Web site, to read past the first sentence.
- Much of the time, readers will be *skimming* for what they want or need.
- Typically, readers are wondering how the message relates to their work, their company, their life.
- Readers read to perform a task, solve a problem, pursue an opportunity, or understand a valuable idea.

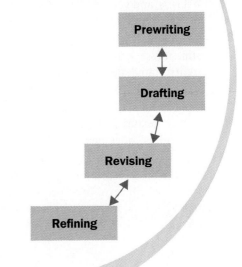

Process Guidelines

To develop successful workplace documents, you must make the writing process work practically for you. Follow these guidelines.

Focus on different traits at different stages. During prewriting, drafting, and revising, focus on global traits—ideas, organization, voice, and overall design. While refining, pay attention to local traits—word choice, sentence smoothness, correctness, and design details. Note the process–traits connections in the following list.

Prewriting   **Goal:** Find your focus and prepare to draft.

Clarify your purpose, think about your reader, choose a format, plan your design, and list the information you need to share. Then do necessary research, review models and templates, and develop an outline.

Drafting   **Goal:** Record and develop your ideas.

Flesh out your outline in sentences and paragraphs that speak in a fitting tone.

Revising   **Goal:** Fix any content problems in the first draft.

Test the quality and clarity of the ideas, organization, voice, and design. Add, cut, clarify, condense, and redesign as needed.

Refining   **Goal:** Fine-tune the piece so that it's ready to send.

Edit and proofread by checking details of word choice, sentence smoothness, grammar, punctuation, spelling, format, and design.

Adapt the process to the piece. For short, informal e-mails, compress the process. For big projects with a lot at stake—bound reports, sales proposals, news releases, Web pages, and so on—use the process more thoroughly.

Use time-saving techniques. Write efficiently by relying on models like those presented in this book. Save the messages you write and use them as templates for other messages. Reuse material through copy and paste.

Invest in prewriting. By doing the heavy lifting up front in planning, brainstorming, outlining, and designing, you'll be more likely to get the document right the first time—and avoid costly rewriting and embarrassing errors.

Fit the process to your personality. Try different strategies, and use those that work best for you. For example, listing may help you generate ideas, whereas clustering may not.

Collaborate with co-workers. From a quick read-through to a full-scale document review, colleagues' feedback at different stages while you are developing a document can help ensure positive results.

WRITE Connection

For tips and instructions on working through the writing process as part of a team project, see Chapter 3, "Teamwork on Writing Projects."

PREWRITING

Using prewriting strategies jump-starts your writing. Techniques from brainstorming to freewriting get you ready to draft by helping you collect information and focus your thinking. By investing time in prewriting, you save time and avoid trouble later in the process; in fact, a few minutes of prewriting may save hours of writing and rewriting.

Tips and Techniques

At the start of any writing project, you need to analyze the rhetorical situation—to consider your purpose, your readers, and your context.

Clarify your purpose. Consider your overall purpose and your specific goal.

- What result do you want to see from your writing?
- Is the desired goal or outcome realistic and measurable, not vague and general?
- Do you need to inform or persuade to achieve results?

  Informing tasks: state, clarify, outline, list, record, report, analyze, compare, describe, define, explain

  Persuading tasks: request, sell, recommend, convince, apologize, evaluate, complain, turn down, promote

Profile your reader. Consider who your reader is in relation to your goal.

- Is your reader within or outside your organization?
- How well do you know your reader?
- How much does your reader know about the topic?
- What are your reader's values, needs, and priorities? Is your message likely to be received positively or negatively?
- Will your reader be making a decision or doing something?
- What authority and responsibility does your reader have?
- Will there be a number of readers? If so, are they diverse in age, occupation, gender, education, culture, and language skills? How can you accommodate this diversity?
- Aside from your main or primary readers, will you have secondary readers—people who might eventually see the document if it is copied, shared, or forwarded? How might this potential audience affect your document?

Know the context. Consider requirements related to format, timing, and follow-up.

- What type of document are you writing? What are its design requirements? How will you create, send, and store this document?
- When must the document be finished and sent?
- What information do you have available? What additional information do you need to gather?
- What future contact or action might be needed by you, the reader, or others?

Gathering Information

You can identify key ideas, gather reliable material, and organize information by using a variety of prewriting techniques like those described below and on the next pages. At this stage in the process, you should not be overly concerned with style, grammar, or mechanics.

Using a Planner

Consider using a planner similar to the one in Figure 1.2 to clarify your writing task. This kind of prewriting planner can help you better understand the writing situation by focusing your attention on the five W's and H: who, what, when, where, why, and how. Planners can be especially helpful for longer projects such as reports or proposals.

Figure 1.2 Prewriting Planner

Prewriting Planner

1. Why am I writing? _____
2. Who is my reader? _____
3. What do I want the reader to think or do? _____
4. What are her or his needs, biases, questions? _____
5. How can I get the desired results? _____
6. What information do I need? _____
7. Where can I find this information? _____
8. What form of writing should I use? _____
9. When and where do I need to send or submit this? _____

WRITE Connection

If collecting information for your documents involves research, go to Chapter 17, "Conducting Research for Business Writing," for help.

Using Diagrams

Diagrams help you generate and record information graphically. Select the diagram that seems most closely related to your topic and task (see Figures 1.3–1.10). For example, if you are writing an incident report (see pages 486–489), you might use the five W's chart to organize the important facts about what happened (Figure 1.3).

Figure 1.3 Five W's Chart

Use this chart to collect the *Who? What? When? Where?* and *Why?* details for your document.

Examples: sales bid, incident report, news release

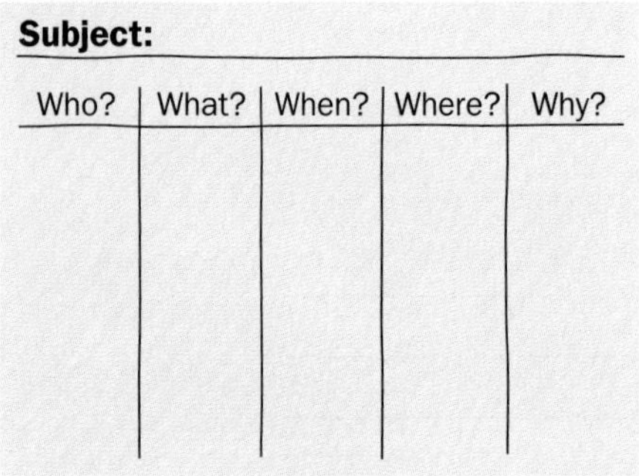

Subject:				
Who?	What?	When?	Where?	Why?

Figure 1.4 Compare and Contrast Chart

Use the overlapping circles, called a Venn diagram, to examine similarities and differences between two or more items.

Examples: investigative report, policy change, justification proposal

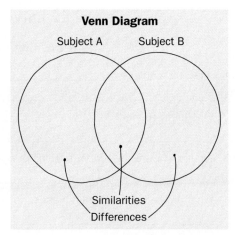

Figure 1.5 Flowchart

Use a flowchart to map out events, processes, or time lines.

Examples: procedure, instructions, itinerary, research or grant proposal

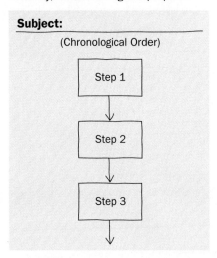

Figure 1.6 Problem/Solution Diagram

Use a problem/solution diagram to define a problem, break it into parts, specify its causes, clarify its effects, and develop solutions.

Examples: troubleshooting proposal, sales letter, complaint message

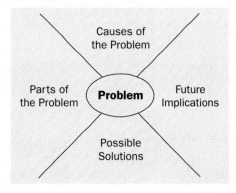

Chapter 1 Using the Writing Process

Figure 1.7 Definition Diagram
Use a definition diagram to map out different ways of clarifying and explaining a term.
Examples: policy, technical documentation

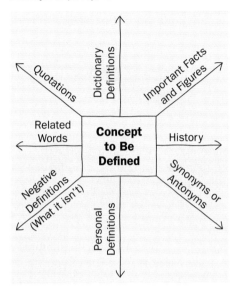

Figure 1.8 Classification Diagram
Use a classification diagram to break down a general topic into subtopics and details.
Examples: major report, company profile

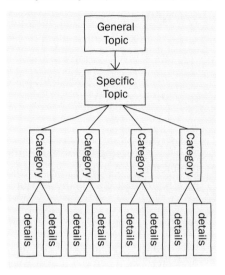

Figure 1.9 Cause/Effect Diagram
Use a cause/effect diagram to clarify logical relationships between events and results.
Examples: incident report, justification proposal, marketing and sales report

Subject: _____

Causes (Because of . . .)	Effects (. . . these conditions resulted)
•	•
•	•
•	•
•	•

Figure 1.10 Evaluation Table
Use an evaluation table to judge topics, plans, recommendations, and solutions.
Examples: employee evaluation, bid rejection, proposal acceptance

Subject: _____

Points to Evaluate	Supporting Details
1.	
2.	
3.	
4.	
5.	

Freewriting

Writing whatever comes to mind on a particular topic is called freewriting. Freewriting is really brainstorming on paper (or on the computer). It helps you record, develop, and understand your thinking.

1. Write nonstop about your subject or project, following rather than directing your thoughts.
2. Resist the temptation to stop and judge or edit what you write.
3. If you get stuck, switch directions and follow a different line of thinking.
4. When you finish, reread your material and highlight useful passages.

F.Y.I.

Try turning down the brightness on your monitor while you freewrite. If you are unable to see what you are typing, you are more likely to go with the flow of your ideas and less likely to stop, judge, rethink, and edit.

Clustering

Clustering creates a web of connections between a general topic and specific subtopics. A cluster, like that shown in Figure 1.11, allows you to see the structure of your topic and plan your writing.

1. Write a key word or phrase in the middle of your page.
2. Record or "cluster" related words and phrases around this key word or phrase.
3. Circle each idea and link it to related ideas in your cluster.
4. Continue recording related ideas until you have covered the topic.

Figure 1.11 Sample Cluster

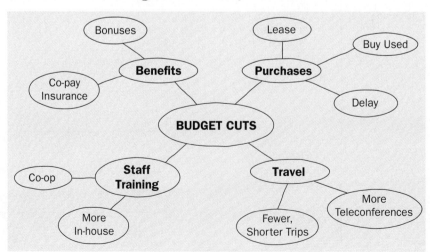

Getting Organized

Before you draft, take time to organize your thoughts—to find your focus and construct an outline.

Find Your Focus

Each piece of writing needs a focus—a main point or controlling idea that directs the message. To find your focus, review your prewriting material and use this formula:

Main point = *specific topic* + <u>statement or claim about topic</u>

> Join the Chamber of Commerce <u>because it gives you a voice in the community's development and access to promotional materials.</u>

> *A $100,000* line of credit with Cottonwood Hills Greenhouse <u>would allow us to place larger orders more frequently.</u>

> *Only an integrated program of cockroach control* <u>can effectively manage the infestation in the Cherryhill Apartment Complex.</u>

Make an Outline

A good outline provides a working plan for developing your main point. Follow these steps:

1. Circle or highlight key ideas in your notes that develop your main point.
2. Connect supporting details to these key ideas. Try listing, circling, drawing lines between items, using file cards, or numbering.
3. Arrange your key ideas and supporting details sensibly by considering your reader's likely response, questions, and concerns, as well as the order that best advances your main point.

Sample outline

I. An integrated program of cockroach control will manage the infestation in the Cherryhill Apartment Complex.
   A. Education
      1. Tenants will learn how an infestation starts.
      2. Tenants will be provided with a checklist for roach-proofing their units.
   B. Cooperation
   C. Treatment
      1. Building manager will contract with an exterminator for monthly visits.
      2. Trash area in basement of building will be upgraded and cleaned weekly.

WRITE Connection

For more tips on getting organized, see Chapter 11, "Trait 2: Logical Organization."

DRAFTING

When you draft, you convert prewriting raw material (such as your notes, clusters, and freewriting) into coherent sentences and paragraphs. Drafting involves your best first effort to develop your ideas effectively and state them clearly.

F.Y.I.
Drafting on a computer allows you to concentrate on ideas, copy-and-paste prewriting notes, and share your work easily with co-workers and colleagues. As you draft, periodically save your work to avoid accidental losses of material.

Tips and Techniques

Create productive conditions. Before you begin drafting, gather your materials and set up your writing space.

- Make your writing area efficient and comfortable.
- Set aside a block of uninterrupted time, if possible. Close your door and screen phone calls, if necessary.
- Place your writing tools and prewriting material within easy reach.

Use your prewriting material. Review your notes to find the information and approach you need.

- Draw on your planning, clustering, freewriting, or outlining.
- Review other documents related to your message.
- Highlight statements, facts, examples, and graphics to include in your draft.

Draft material in an order that makes sense. No rule says that you must start with your document's introduction and work toward the closing. Start where you feel comfortable—with the middle or the ending, if that works best.

Focus on ideas, organization, and voice. Concentrate on the big picture by keeping the content-related traits in mind.

- Expand and connect ideas, always focusing on your purpose—presenting information, clarifying what it means, and explaining why it's important.
- Use your outline as a map, but remain open to new ideas.
- Establish and maintain a natural, polite, and professional voice or tone.

Develop a logical flow. Determine the best way to present your information. Common approaches include moving from overview to close-up, from background to discussion, from problem to solution, from question to answer, from main idea to explanation, or from familiar topic to new information.

Think paragraphs. Group related thoughts into traditional paragraphs.

- Present each main idea in a single sentence (a topic sentence), and then support that idea in a single, well-developed paragraph.
- Make openings and closings brief paragraphs separate from the body of your message. Openings introduce your topic, while closings restate your main point and offer conclusions or recommendations for action.

Drafting the Opening

Your opening should quickly set a clear direction for your writing. An effective opening should do some or all of the following:

- Indicate the document type (e.g., report, proposal, instructions).
- State your topic, your reason for writing, and your main point (if appropriate).
- Get the reader's attention and interest.
- Establish a tone of voice.
- Provide background and context.
- Preview the content.
- Provide a summary (if appropriate).

Sample Openings

Simple Opening (e.g., letter) — *context*

main point — Thank you for requesting a credit account at Cottonwood Hills Greenhouse and Florist Supply. We are pleased to extend you $100,000 in credit based on Dale's Garden Center's strong financial condition. Congratulations! — *positive tone*

Complex Opening (e.g., report, proposal) — *writing form and topic*

background — Subject: Report on Investigation of Cockroach Infestation at 5690 Cherryhill

During the month of July 2003, 26 tenants from the 400-unit building at 5690 Cherryhill informed the building superintendent that they had found cockroaches in their units. On August 8, the Management–Tenant Committee authorized our group to investigate these questions:

1. How extensive is the cockroach infestation?
2. How can the cockroach population best be controlled?

objective tone — We investigated this problem from August 9 to September 11, 2003. This report contains (1) a summary, (2) an overview of our research methods and findings, (3) conclusions, and (4) recommendations. — *preview of content*

F.Y.I.

Here are two tips on developing strong introductions for business documents.

- Many documents (memos, e-mails, reports) start with a subject line or title. Your reader should be able to "get the point" at a glance, so subject lines and titles should be direct and precise, not vague or confusing. (See page 221 for more on creating high-information subject lines, titles, and headings.)
- For major documents, it often makes sense to draft the introduction last—after you have a keen sense of the middle and the closing—so that you can effectively put the document in context and summarize its main points.

Drafting the Middle

In the middle part of your message, you need to develop strong, clear paragraphs that advance the main idea presented in your introduction. Follow your outline and tackle one paragraph at a time. If necessary, go back to your prewriting material or rework your outline.

An effective middle should do the following:

- Link back to the opening through a clear transition
- Include well-written paragraphs that have strong opening sentences
- Provide readers with necessary information
- Present information in easy-to-access formats

To achieve these goals, follow these guidelines:

- Develop each main point in a separate paragraph.
- Keep each paragraph fairly short, three to eight lines in length.
- Use a topic sentence to signal each paragraph's direction.
- Expand and connect ideas, always keeping your reader in mind.
- Incorporate headings, lists, and graphics when appropriate.
- Maintain the tone established in the opening.
- Prepare readers for the closing by building toward it.

Sample Middle Paragraphs (e.g., letter)

clear topic sentence — As you requested, I have enclosed a list of Home Builders' affiliates in Missouri. Each affiliate schedules and handles its own work groups. Call or write directly to any of them for more information about local offerings.

prepare reader for closing — Because you are in college, I have also enclosed some information on Home Builders' Campus Chapters. The pamphlet explains how to join or start a Campus Chapter and discusses service learning for academic credit. Feel free to contact

helpful tone throughout — Ben Abramson, the Campus Outreach Coordinator for your area, at the address on the enclosed material.

WRITE Connection

As you draft the middle of your message, consider these patterns of development as choices for ordering your discussion: alphabetical order, cause to effect, effect to cause, chronological order (sequence), classification into categories, comparison and/or contrast, order of importance (most to least or least to most), partitioning according to spatial order, problem–solution, and question–answer. Examples and instructions for such patterns can be found on pages 195–213. Similarly, for examples of more complex middle development, check the model reports and proposals in Chapters 30–32.

Drafting the Closing

Close your writing logically and naturally by recalling your goal and main point. Your closing should help your reader understand and act on the message. In other words, ask what your reader needs to take away from the message.

An effective closing should do some or all of the following:

- Restate your main point and offer conclusions
- Provide recommendations, propose the next step, or offer help
- Focus on appreciation, cooperation, and future contact
- Include all details needed for follow-up and action
- End with energy by using clear, strong verbs

F.Y.I.
In your closing, avoid stating the obvious; rambling; focusing on the negative; being apologetic, defensive, or wishy-washy; or concluding abruptly (leaving your reader hanging).

Sample Closings

Simple Closing (e.g., memo, e-mail, letter)

With short-term solutions and long-term cooperation, together we will keep Premium Meats operating and prospering. — *focus on cooperation*

If you have any questions or suggestions, please speak to your immediate supervisor or a member of the Quality Task Force. (Task Force information is supplied in the attachment.) — *recommendation and follow-up*

Complex Closing (e.g., report, proposal)

Our proposal is to replace our internal-combustion forklifts with electric lifts. — *restatement of main point*
This change will involve a higher initial cost but will reap two important benefits: (1) CO levels will drop below OSHA's recommended 25 ppm, ensuring employee safety and product quality; and (2) in the long run, the electric lifts will prove less costly to own and operate. — *focus on outcomes*

I recommend that Rankin management approve this plan, to be phased in — *recommendation*
over the next four years as outlined above. Improved safety and product quality, combined with total operating costs over the long run, outweigh the plan's initial costs. If you wish to discuss this proposal, please call me at extension 1449. — *action, future contact*

F.Y.I.
If appropriate, present possible outcomes or long-term implications of what has been discussed. In other words, what's the real bottom line? What's the "so what?" of your message?

REVISING

Your focus while drafting was simply recording and developing your ideas. When revising, you step back and look at the big picture. That is, you check the overall content of the draft. Where can the ideas, information, and explanations be strengthened? How? Keep reworking your draft until it says exactly what you want it to say.

F.Y.I.
Computers make it easy to revise through delete, copy, and paste. But effective revising still requires careful thought. Take the time to do it right.

Tips and Techniques

Set aside the draft. Let the draft cool a bit by setting it aside for a few minutes (if it's a simple e-mail) or a few days (if it's a major document like a report or proposal). Time will give you more objectivity about the draft.

Print a hard copy. Seeing the entire message on paper helps you assess the overall effectiveness of your writing, test your logic, and map out changes. Consider double-spacing the draft to make studying it and marking it up easier.

Review your purpose and audience. Have you done what you set out to do with this reader? Use these questions to conduct your test:

- Does my message stay focused from start to finish?
- Have I answered my reader's potential questions: *Why did I receive this? What's the point? Why is it important?*

Read the draft aloud. Listen to your writing from your reader's perspective. Doing so allows you to identify problems in content and tone.

Go through your draft with pencil in hand. Examine and change the ideas, organization, and voice whenever you recognize weaknesses.

- Test for complete information, coherent order, logical flow, and polite voice.
- Write comments in the margin, cross out sentences, move sections, and add material as needed. Transfer these changes to your electronic draft, but keep the first draft in case you need material from it later.

Try the cut–clarify–condense system. Read your draft with an eye toward making it as clear and complete as possible.

- [Bracket] material that could be cut.
- <u>Underline</u> material that needs to be clarified.
- Put (parentheses) around material that needs to be condensed.

Get input from others. Share your draft with a colleague or a potential reader. Fresh eyes can look at the messages from a more objective, reader-centered perspective.

WRITE Connection

For specific instruction on peer revision, see pages 56–59.

Revising for Ideas

If your first draft lacks focus, substance, or clarity, you need to revise your content. The following guidelines and models should help you do so.

Focus

Check the focus of the whole document and of each paragraph. Try cutting or moving material that rambles or is unrelated to the point being developed.

> **Original Passage**
>
> Thursdays are generally light days for truck undercoating. However, during a routine inspection, I found an undercoater struggling to breathe under a truck. It was March 21 around 10:45 a.m. because it was the same date that we received our long-overdue order of oil filters.
>
> **Revised Version**
>
> During a routine inspection on Thursday, March 21, at 10:45 a.m., I found undercoater Bob Irving struggling to breathe under the truck he was working on.

Content

Strengthen the content in your draft where necessary by expanding explanations, offering examples, and adding concrete details.

> **Original Passage**
>
> Maintainer Corporation provides suits to protect undercoaters from this dangerous sealant. However, this isn't good enough.
>
> **Revised Version**
>
> Maintainer Corporation provides oxygen suits to protect undercoaters' skin and oxygen supply from sealants that (a) produce noxious fumes, (b) cause choking if swallowed, and (c) injure skin upon contact. However, the incident shows that our safeguards are inadequate. First, because the oxygen suits are ten years old, the meters and air tubes frequently malfunction. . . .

Clarity

Check your writing for clarity and ease of reading. Where necessary, add background and context, define terms, and rewrite confusing passages.

> **Original Passage**
>
> We should do something about the problem. The facts point toward an investment in new equipment and in undercoaters working together.
>
> **Revised Version**
>
> To protect undercoaters from sealant hazards, I recommend the following actions:
>
> 1. Purchase oxygen suits and equipment that meet the 2005 OSHA standards. In particular, air-hose locks should have emergency-release latches.
> 2. Raise trucks on a lift so that undercoaters may stand.

Revising for Organization

To make sure the reader is able to follow and understand your writing, test your draft's organization and flow from start to finish. (See pages 28–29.)

The Opening

Does the opening clearly announce your purpose, provide background information, and then point the way forward?

Original Passage

As you know, insurance is important. How are you doing with coverage?

Revised Version

Periodically, insurance companies review the cost of offering their policies and then make changes where needed. When changes are made, it's my responsibility as an agent to inform my clients and help them make necessary adjustments.

The Middle

Does the middle cover your key points in a smooth, logical way? Does each point follow logically from the previous one? Are transitions used between sentences and paragraphs? Does the information flow at a steady pace?

Original Passage

Hawkeye Casualty has chosen not to renew you. To keep their premiums low, some insurance companies will not cover high-risk drivers. In the past two years, you have had an at-fault accident and four moving violations. Your present coverage will cease January 31, 2005. A nonstandard auto insurance company is willing to provide you coverage.

Revised Version

Last week Hawkeye Casualty, your auto insurance carrier, discontinued all policies for "high-risk" drivers. Your at-fault accident and four moving violations put you in that category. As a result, Hawkeye Casualty has cancelled your auto insurance policy effective January 31, 2005. However, I have found another company that will cover you. While the cost of the new policy is somewhat higher than your present policy, the coverage is comparable, and the company is reliable.

The Closing

Did you pull your message together? Have you summarized points, recommended steps, and specified action to be taken?

Original Passage

This coverage issue needs attention soon.

Revised Version

Please call me at 555-1228 so that, together, we can determine how best to meet your auto insurance needs.

Revising for Voice

Fine-tune your writing voice as necessary so that it fits your purpose, your topic, and your audience. (See Chapter 12, "Trait 3: Conversational Voice.")

Attitude

Keep attention focused on the topic and the reader. In most business writing, avoid putting yourself in the spotlight.

> **Original Passage**
>
> We at Birks Cleaners aim to provide the best dry cleaning. We're in our third generation of quality-minded dry cleaners. Our quality speaks for itself.
>
> **Revised Version**
>
> At Birks Cleaners, you'll find services to meet all your dry cleaning needs. For three generations, our goal has been to treat your clothes as if they were our own.

Transparency

Create an invisible style that lets the reader focus on content, not fancy phrases. Rewrite passages that say, "Look at these words!"

> **Original Passage**
>
> From the starting pistol of this pivotal project, Janice worked with superhuman strength to maneuver the company ship through the turbulent transition waters.
>
> **Revised Version**
>
> From start to finish, Janice worked hard to help us manage during this often difficult transition.

Level

Maintain a consistent tone. Test your draft for a positive attitude, an appropriate seriousness, and the correct level of technical discussion.

> **Original Passage**
>
> Even though your inconvenience wasn't entirely our fault, we're obviously sorry.
>
> **Revised Version**
>
> We apologize that your confirmed room was unavailable last night, and we are sorry for the inconvenience this may have caused you.

Energy

If your voice drags, rewrite the passage to show that you care about the topic.

> **Original Passage**
>
> There is very little doubt that the multiple organizational changes designed to meet our business challenges create uncertainty for employees at all levels.
>
> **Revised Version**
>
> We know that change creates concern for all employees, but we are looking for creative ways to meet these challenges.

REFINING

To develop quality, polished writing that impresses your readers, pay close attention to details. Take time to make the line-by-line changes to style and grammar that ensure your document is clean, clear, and correct.

Tips and Techniques

Determine whether you're ready to refine. Before you dedicate time to editing and proofreading, make sure you've finished revising for overall meaning. Otherwise, you might waste time polishing material that still requires substantial change. If possible, let the draft sit for a while so that you can mentally prepare to check it sentence by sentence, word by word.

Gather your tools. Have writing aids at hand:

- Use the models and Part 9, "Proofreader's Guide," in this handbook.
- Keep within reach a dictionary and a thesaurus, whether print or electronic.
- If available, refer to your company's style sheet—the guidelines that your company sets for its documents (e.g., format expectations such as margin size, typeface, type size; preferred spelling of key words; proper uses of the company's name versus personal pronouns).
- Keep computer tools (spell checker, online thesaurus, grammar checker, templates, cut-and-paste, find-and-replace) a mouse click away.

Set limits. Decide how much time and effort to put into refining. Consider the overall importance of the document. The wider the distribution, the higher the reader's status, or the more complex the content—the more refining you should do.

Focus on details. Follow logical steps that address separate editing and proofreading issues.

- First, edit for effective word choice and sentence smoothness.
- Second, proofread for errors in grammar, usage, and mechanics.
- Third, check the document's format and design for correctness, readability, and consistency with similar documents.
- Fourth, address details related to distributing the document.

Ask for help. Find an objective, skilled reader.

- Turn to a co-worker with good editing skills.
- Share documents with managers, technical or content experts, and legal counsel as needed.

Use your computer wisely. Remember that computer tools aren't foolproof. Use a grammar checker and spell checker to catch basic errors, but print your document and proofread it carefully to catch those slips that these programs miss (e.g., typing *peace* when you meant *piece*, *it's* when you need *its*).

WRITE Connection

For tips on collaborative refining and team projects, see pages 60–67.

Editing for Word Choice

In effective business messages, each word is necessary, clear, and appropriate. To achieve these qualities, edit for the issues below. (See Chapter 13, "Trait 4: Clear Words.")

Wordiness

Be concise: Don't use many words to make your point when few will do.

> **Original Passage**
>
> The committee has been organized and set up to provide leadership in our effort to improve our abilities to communicate in and between departmental structures.
>
> **Revised Version**
>
> The committee's goal is to improve communication between all departments.

Unclear Words

Replace clichés, jargon, and flowery phrases with clear, simple words—plain English.

> **Original Passage**
>
> Pursuant to his request, I analyzed Bob's physiological constitution and determined that his physical state of being was unimpaired.
>
> **Revised Version**
>
> I checked Bob for injuries and determined that he was unharmed.

Negative Words

Replace negative or insensitive words with neutral or positive ones.

> **Original Passage**
>
> I read your memo requesting that all sales reps be given a whole day off for training in Cincinnati. While that stuff is important, Oscar, the bottom line is no less important! So I have no choice except to say, "No."
>
> **Revised Version**
>
> I reviewed your request to send all the sales reps to the training seminar in Cincinnati. Oscar, I agree that this training would help your staff be more productive. However, our budget for . . .

Inappropriate Level

Choose conversational words, but make fitting choices about contractions, personal pronouns, use of names, and forms of politeness.

> **Original Passage**
>
> Hey, Lois, I sure hope that you'll keep helping us get all adults up and reading. You can do that by writing us a check and sending it back ASAP.
>
> **Revised Version**
>
> Please continue to support our effort to get every adult reading. To send a check, make it out to National Campaign for Literacy and return it in the enclosed envelope.

Editing for Sentence Smoothness

To test sentence style, read your draft aloud. When you stumble over phrasing or get confused, it's time to edit. Here are three editing tips, along with three common sentence problems. (For more, see Chapter 14, "Trait 5: Smooth Sentences.")

1. **Highlight the first few words of each sentence.** Do several sentences start the same way? Vary your sentence openings.
2. **List word counts for each sentence.** Are many sentences similar in length? Try combining some sentences, or dividing long ones.
3. **Examine the verbs.** Are there strong action verbs, or weak linking verbs (*is, are, was*)? Rewrite sentences to replace weak verbs with strong ones.

Choppy Sentences

Do you have a series of short sentences with poor flow, simplistic thoughts, or poor connections? Combine sentences by relating ideas and adding transitions.

Original Passage

I am responding to your job advertisement. It appeared in the *Seattle Times*. The date was November 11, 2005. I am applying for the Software-Training position.

Revised Version

In response to your advertisement in the *Seattle Times* on November 11, 2005, I am applying for the position of Software-Training Specialist.

Tired Sentences

Are your sentences sluggish, overly negative, or repetitive? Energize them by using the active voice, making negative statements positive, and varying sentence openings.

Original Passage

Your getting back to me in a short time was greatly appreciated. Some people take a lot longer. Some of your other people may also want to meet me next Thursday, and that's fine with me.

Revised Version

Thank you for your quick response to my letter of application for the position of Software-Training Specialist. I look forward to meeting you and the staff at Evergreen Medical Center next Thursday.

Rambling Sentences

Are your sentences strung out, confusing, or packed with too many ideas? Then divide sentences, cut sentence parts, or use lists to order sentence parts.

Original Passage

I enjoyed touring your facilities and I would enjoy contributing to the work you and others do, for which I believe that my hospital training would be an asset.

Revised Version

After touring your facilities, I am certain that I would enjoy working at Evergreen Medical Center. I believe that my hospital training would be an asset.

Chapter 1 Using the Writing Process   35

Proofreading for Correctness

In addition to spelling errors and typos, the issues listed below commonly plague business documents. When refining your writing, use this list to find and correct the problems. For more information, consult Chapter 15, "Trait 6: Correct Copy," and Part 9, "Proofreader's Guide."

- Pronoun–Antecedent Agreement (page 280)

    To help a new employee learn our day-to-day procedures, please take ~~them~~ *him or her* through the orientation program.

- Shift in Person (page 283)

    When new employees go through this orientation, ~~you~~ *they* learn ~~your job~~ *their jobs* more quickly.

- Subject–Verb Agreement (pages 278–279)

    The procedure, as well as the attached checklist of steps and items, ~~cover~~ *covers* key orientation topics.

- Dangling Modifier (page 276)

    After filling in the review form for the new employee, ~~the form should be sent~~ *you should send the form* to Human Resources.

- Unparallel Construction (page 282)

    The form describes issues at each stage in the evaluation process: after 1 day, *after* 1 week ~~later,~~ *after* ~~then~~ 2 weeks ~~after starting,~~ and ~~when~~ *after* 30 days.

- Sentence Fragment (page 281)

    The form will streamline orientation for new employees, *and shorten the* ~~Much less~~ time it takes them to learn their assignments.

- Comma Splice (page 281)

    The new checklist includes more information than the old one*;* therefore, the new one will take more time to complete.

- Comma Omission After Introductory Phrases and Clauses (page 284)

    After the final review*,* the supervisor and the employee will sign the form.

- Comma Omission Between Independent Clauses (page 284)

    Reviewers should use the form for 30 days*,* and then they should forward the form to Human Resources.

- Comma Omission Around Nonrestrictive Modifiers (page 285)

    For each new employee, Human Resources*,* which is responsible for the initial orientation*,* will prepare a checklist with the employee's name on it.

Part 1 Challenges for Workplace Writers

Refining Document Design

Document design should follow company standards, make information accessible, and invite reading. Before sending your document, check and improve its appearance by reviewing the following issues. (For more on reader-friendly design, see Chapter 16, "Trait 7: Reader-Friendly Design.")

Format

- Have you correctly followed a format (e.g., memo, letter, report)?
- Is the document printed on paper of an appropriate size, color, and weight?
- Does your document include features that make information accessible: headings, page numbering, icons, and color?

Page Layout

- Are pages open, balanced, and readable? Are column choices sensible and attractive?
- Does white space frame the message and give readers rest from text?
- Do headings separate and highlight blocks of text?
- Are headings consistent in size and presentation, parallel in wording, and easy to see (**boldface** or underlining)?
- Do lists make details digestible through parallel phrasing, alignment of items, and proper use of bullets or numbers?
- Is the right margin ragged (uneven), unless full justification is required?
- Are headings, hyphenated words, and the first lines of paragraphs NOT placed at the bottom of a page?
- Are single words or lines NOT placed at the top of a page?
- Are lists, if possible, NOT split between pages?

Typography

- Does typography help your reader read, understand, and use the document?
- Is regular text 10–12 point type size, subheadings 14–16, and major headings and titles 18–20?
- Is the typeface a readable *serif* such as Times New Roman, Schoolbook, Garamond, or Bookman (like this type)?
- Do special text (e.g., headings) and on-screen documents use *sans serif* type (like this type)?
- Are special type styles used effectively for emphasis: underlining, *italics*, highlighting, boxes, shading, UPPERCASE, and color?

F.Y.I.

Word-processing and desktop publishing software offer powerful design tools, but be careful: Design should always be functional and consistent with your company's guidelines, not flashy and distracting.

Chapter 1 Using the Writing Process

ONE WRITER AT WORK

To see the writing process in action, follow Julia, a Human Resources Assistant, as she writes a welcome memo to a new employee as part of the company's orientation procedure.

Prewriting

Julia thinks about her assignment (the memo's purpose, reader, and context), jotting some notes as she does so. These notes function as her rhetorical analysis of the writing situation.

Goals
-- Introduce myself and the HR Dept.
-- Schedule an orientation meeting with Robert

Reader
-- Robert Pastorelli: new to Rankin's management team, but lots of experience
-- Already on the job but may have questions about how things work here
-- Needs to know about benefits and policies

Context/Info
-- Send memo by Wednesday
-- Follow orientation procedure
-- Supply an overview of key benefits and policies so that Robert can get the big picture and think of some questions he wants answered

---

Next, Julia gathers the information she needs:

- Robert's application file and employment agreement
- The orientation checklist for new employees
- Manuals, policy statements, and benefit descriptions used in her department
- A model welcome message in *The Business Writer*

Before drafting the memo, Julia develops a scratch outline growing out of the information she's gathered:

---

1. Explain who I am, and where I'm located
2. Tell what our Human Resources Department does
3. Describe Rankin's benefits and policies
4. Schedule meeting to review benefits, policies, procedures

Drafting

Julia clicks on her word-processing program, pops up a memo heading, fills in the necessary information, and drafts the message.

Date: Oct. 14/05

To: Rob Pastorelli

From: Julia Westmoreland

Subject: Meeting

The writer focuses on recording her ideas and supplying the necessary information, not on correctness.

Using her outline, she shapes the message into paragraphs.

She focuses on her goal of welcoming the reader and setting up a meeting.

She double-spaces the draft to make revising it easier.

The Human Resources Department at Rankin Manufacturing aims to help new employees settle into their new jobs and succeed. All the members of the department are here to serve you. We are on the second floor of the administrative office complex. We have opportunities for Rankin employees to grow in their professional lives.

Rankin Manufacturing provides its employees with a top-notch benefit package. Among other things, Rankin has: dental/optometry procedures, equity programs to ensure fair treatment of all employees and customers, Blue Cross medical plan, convention funding, profit sharing (as you may recall from your interview), and we also have counseling services and advancement procedures.

In Human Resources, we realize that their is a great deal of material to absorb. Therefore, we recommend an orientation meeting to discuss all the program, benefits, and procedures.

Contact me so that we can review this information. If you have questions right now, please phone me at extension 342 or contact juliaw@rankin.com.

Chapter 1 Using the Writing Process 39

Revising

Julia sets aside her draft for a while so that she can look at it more objectively. She then prints a hard copy and jots notes in the margin about changes to improve the ideas, organization, and voice. She revises the memo itself and makes the changes to her electronic copy.

Date: ~~Oct. 14/05~~ October 14, 2005

To: Rob*ert* Pastorelli
 Production Manager

From: Julia Westmoreland
 HR Assistant

Subject: Meeting *to Discuss Benefits and Policies*

Welcome to Rankin Manufacturing! With your background, I was glad to learn that you accepted the Production Manager position. ~~The Human Resources Department at Rankin Manufacturing aims to help new employees settle into their new jobs and succeed. All the members of the department are here to serve you. We are on the second floor of the administrative office complex.~~ We have opportunities for Rankin employees to grow in their professional lives.

The Human Resources Department is here to help you grow. I would like to explain Rankin's benefit package, policies, and procedures. I would like to share information on: ~~Rankin Manufacturing provides its employees with a top-notch benefit package. Among other things, Rankin has:~~ dental/optometry procedures, equity programs to ensure fair treatment of all employees and customers, Blue Cross medical plan, convention funding, profit sharing (as you may recall from your interview), and we also have counseling services and advancement procedures.

I will call on Friday to set up an orientation meeting. It will take about 45 minutes. ~~In Human Resources, we realize that their is a great deal of material to absorb. Therefore, we recommend an orientation meeting to discuss all the program, benefits, and procedures.~~

I look forward to ~~Contact me so that we can~~ review*ing* this information *with you*. If you have questions ~~right~~ *me at* now, please phone me at extension 342 or contact juliaw@rankin.com.

Fix items in heading and make subject more precise.

Focus on the reader, not the department.

Use "grow" idea to focus and strengthen message.

Avoid repetition and call for a meeting.

Refining

After typing in the changes, Julia prints out a hard copy and reviews her revised memo. First, she reads the memo aloud to see if words are clear and sentences are smooth. Then she proofreads for grammar, punctuation, spelling, and typos—using both word-processing tools and her own eyes. Finally, she strengthens appearance on the page.

Date: October 14, 2005

To: Robert Pastorelli
          Production Manager

From: Julia Westmoreland
          HR Assistant

Subject: Meeting to Discuss Benefits and Policies

Because of your 15 years of experience at Driscoll Corporation, Welcome to Rankin Manufacturing! ~~With your background,~~ I was glad to ~~learn~~ hear that you accepted the Production Manager position. ~~We have~~ I hope you will find many opportunities for growth with us. ~~Rankin employees to grow in their professional lives.~~

The Human Resources Department is here to help you grow. I would like to explain Rankin's benefit package, policies, and procedures. Specifically, I'd like to share information on the following:
- Medical plan benefits for you and your family
- Submission procedures for dental and optometry ~~dental/optometry procedures, equity programs to ensure fair treatment of all employees and customers, Blue Cross medical~~
- Profit sharing • Counseling services ~~plan, convention funding, profit sharing (as you may recall from your~~
- Advancement policies and procedures • Continuing education programs ~~interview), and we also have counseling services and advancement~~
- Workplace rules (from hazardous waste to equal opportunity) ~~procedures.~~

I will call ~~on Friday~~ this to set up ~~an~~ orientation meeting at a time convenient to you. ~~It~~ The meeting will take about 45 minutes.

I look forward to reviewing this information with you. If you have questions now, please phone me at extension 342 or contact me at juliaw@rankin.com.

Julia makes phrasing more specific.

She adds transition words and phrases.

She reformats information as a list.

She replaces vague words with precise phrasing.

She fixes a misplaced phrase.

The Final Product

By using the writing process efficiently, Julia sends a clear message to Robert. Moreover, the memo becomes a template for future messages to new employees, helping Julia and other HR assistants in her department work productively. Here is the finished memo, with elements of reader-friendly design included.

R T Rankin Technologies
401 South Manheim Road ❖ Albany, NY 12236 ❖ Phone 708.555.1980 ❖ Fax 708.555.0056

Date: October 14, 2005

To: Robert Pastorelli
           Production Manager

From: Julia Westmoreland J.W.
           Human Resources Assistant

Subject: Meeting to Discuss Benefits and Policies

Welcome to Rankin Technologies! I am pleased to hear that you have accepted the Production Manager position. I believe that you will find many opportunities for professional growth with us.

Our Human Resources Department is here to help you grow. To that end, I would like to discuss Rankin's benefit package, policies, and procedures.

Specifically, I'd like to share the following information:

- Profit-sharing plan
- Medical-plan benefits for you and your family
- Procedures for submitting dental and optometry receipts
- Counseling services
- Continuing-education programs
- Advancement policies and procedures
- Workplace policies (from hazardous-waste rules to equal opportunity)

On Friday, I will call to set up a convenient time for your orientation meeting.

I look forward to spending time with you reviewing this useful information. In the meantime, if you have questions, please contact me at extension 342 or juliaw@rankin.com.

BEATING WRITER'S BLOCK

A common problem faced by workplace writers is writer's block. Having writer's block is like being tongue-tied—except the knot is in your pen or keyboard. The bad news is that writer's block is both inconvenient and frustrating. The good news is that it's curable. To beat writer's block, first diagnose its causes. Which of these symptoms fit your case?

- You're trying for instant perfection, perhaps without enough preparation.
- You're unsure where and how to start.
- You're procrastinating, perhaps because you've hit a mental wall or because the message or the situation is difficult or unpleasant.

The Cure

Once you've diagnosed the causes, set about curing your writer's block by trying some of the strategies below.

> "The art of writing is the art of applying the seat of the pants to the seat of the chair."
> —Mary Heaton Vorse

Do more prewriting. Perhaps you're actually not ready to draft, so experiment with techniques that get you prepared—focusing on your purpose and reader, brainstorming, research, freewriting, more detailed outlining.

Turn off the censor. Repeat after me: "The first draft is just a rough draft." Relax, take a deep breath, glance at your outline, and get going. Try starting, "I'm just writing to . . ."

Switch drafting techniques. Try paper and pen rather than computer, or record the message on tape. Try a new location or a different time.

Start with an easy part. Try a section that you're confident you can tackle—something you can understand clearly.

Find a model. A template similar to your piece may provide a boost. Follow the model closely. You can always customize your work during revision.

Let the piece percolate. Take a walk, or discuss the topic with a colleague.

Print out what you do have. Read what you've managed to draft, jotting ideas for pushing the draft further.

Attack procrastination. Remind yourself that a quality process leads to a quality product; stalling doesn't. Get rid of distractions, and commit yourself to drafting for a set time, whether fifteen minutes or two hours.

F.Y.I.

To avoid future bouts of writer's block, write regularly. The more you do it, the more natural it becomes.

CHECKLIST for the Writing Process

When you are developing any workplace document, use this checklist to guide you toward successful completion of the project.

Prewriting

- ☐ **Ideas.** Have I analyzed my purpose and audience; developed a measurable goal; and collected information through techniques such as brainstorming, clustering, freewriting, consulting, completing planners, and doing research?
- ☐ **Organization.** Have I organized my thoughts by using graphic organizers or outlining?
- ☐ **Voice.** Have I determined what tone fits the audience and the situation?
- ☐ **Design.** Have I selected and mapped out a format? Have I consulted models and style guides?

Drafting

- ☐ **Ideas.** Have I recorded all my key points and essential information?
- ☐ **Organization.** Have I arranged my ideas and details into an opening, middle, and closing?
- ☐ **Voice.** Have I maintained a person-to-person tone from start to finish?

Revising

- ☐ **Ideas.** Have I fixed problems with the draft's overall focus, clarity, and content? If appropriate, have I gotten feedback from a colleague?
- ☐ **Organization.** Have I sharpened the opening to pull readers in, reworked the middle paragraphs to improve flow and coherence, and improved the closing to help readers take the next step?
- ☐ **Voice.** Have I fixed any lapses in attitude, level, or energy so that the topic and reader are in the foreground and I am in the background?

Refining

- ☐ **Words.** Have I eliminated clichés, jargon, and redundancy so that wording is concise, clear, and positive?
- ☐ **Sentences.** Have I edited choppy, tired, and rambling sentences by using combining, parallelism, transitions, active voice, and variety?
- ☐ **Copy.** Have I proofread for sentence, punctuation, mechanics, spelling, and usage errors (using but not relying solely on a grammar checker and a spell checker)? If appropriate, have I gotten editing help?
- ☐ **Design.** Have I fine-tuned the format, page layout, and typography to make the document professional, inviting, and readable?

CRITICAL THINKING Activities

1. Reflect on your use of the writing process in various assignments. What steps have you typically followed? With what success? How do you plan to use the strategies in this chapter to write business documents in your anticipated job or profession?
2. Identify a job for which you are currently training and list five communication situations common to that occupation. If helpful, draw upon your job, internship, or volunteer experiences. Review "To Speak, Write, or Do Both" on pages 14–15. Then describe the ten situations, your goal in each, the communication media that will help you achieve that goal, and the reasoning behind your choice.
3. For each of the following situations, analyze the purpose, audience, and context of the message. Use the "Tips and Techniques" on page 18 to develop a measurable goal, strong reader profile, and clear sense of context.
   - Instructions to staff about using the company's new e-mail system
   - A memo to employees about recent accounting scandals
   - A monthly report to your manager on production activities
   - A letter to current customers about a new service you are offering
   - Your résumé and application letter, sent to the HR department of a medium-sized manufacturing company
   - A report to your manager on improving company security
   - A news release on your company's establishment of a charitable foundation
4. Louis Brandeis has argued that "[t]here is no such thing as good writing, only good rewriting." Given what you have learned in this chapter about the writing process, support or dispute this statement.

WRITING Activities

1. Interview three business people or other professionals who write regularly. Ask them about their writing habits and practices with an eye toward learning what challenges they face and what process they follow. Write a report presenting and analyzing their real-world writing practices.
2. Work through the writing process to develop a report for classmates on one of the following topics: campus housing issues; entertainment and cultural opportunities in your city or region; your campus's appearance, architecture, or ergonomics; campus safety and security; computer software, labs, and services for students; or a topic of special interest on your campus. Submit both your final report and the writing-process materials that you generated.
3. The chair of the department in your major has asked you to write a letter to high school students who request information about the major's requirements and career opportunities. Develop the letter by using the full writing process.
4. Find a document that needs work—your own writing, a business document, or Web material. Use revising and refining strategies to fix the piece.

In this chapter

Strategies for Learning New Software 46
Word-Processing Software 47
Special Applications 49
Digital Resources: Databases and the Web 51
Checklist for Writing and Technology 52
Critical Thinking Activities 52
Writing Activities 52

CHAPTER
2

Writing and Technology

Most business writing today is generated by people typing on computer keyboards and using a wide variety of software applications, from simple text editors to complex content-management systems. Modern technological tools enable writers not only to create sophisticated print and electronic documents, but also to tailor and repurpose content for different rhetorical situations—new audiences, new projects, new markets, and so on.

Because of the complexities inherent in electronic publishing, writers are often intimidated by the tools of technology. New versions of software programs are frequently released—along with flashy functions and puzzling terminology. Unfortunately, software programs do not always perform as intended—another complication.

With these electronic tools, however, writers can publish content collaboratively, instantly, and globally. This chapter discusses several categories of software tools that professional writers use and provides some suggestions for learning more about them.

STRATEGIES FOR LEARNING NEW SOFTWARE

Your goal when using software to complete a writing project should be the same as if you were using a pen and paper. The technology—whether a ballpoint pen or a desktop publishing program—is merely a tool. Try not to get distracted by "bells and whistles" or deterred by frustrations you experience. And don't try to learn the entire program at once; break the process into manageable tasks. These tips will help.

Learn by Doing

Get familiar with a software program by using it to complete a simple project. For example, you might create a brochure for your class, business, or other organization. Try to use the software every day, and attempt to learn one new thing each time you use it.

Learn by Investigating

Before starting a project, learn the interface and functions of the software.

- Set a timer for 30 minutes, open the software, and click each menu bar option to view available functions.
- Look up any unfamiliar functions or terms in the program's Help file.
- At the end of 30 minutes, list the functions you understand, those that you don't, and those that are most important for you to master.

Learn by Reading

- Peruse chapters of the user's manual that explain functions you want to use. Don't read the whole manual—just what you need!
- Go to the library and check out books on the software program. Again, you don't have to read the entire book!
- Go on the Web and find user groups that discuss the software program. Use search functions to find the specific information you want.

Use the Built-in Educational Features

- Read through the Help file's table of contents. Look up any terms you don't understand in the index.
- Walk through any step-by-step tutorials that show you how to complete tasks.

Learn from Others

- Communicate with a colleague in your class or office who has already mastered the software.
- Work together with someone else who is also learning the software.
- Sign up for a course—in a classroom or online. Some courses enable you to become a certified user of the software—a fact you can add to your résumé.
- Join an online user group.

Keep a Log of What You've Learned

After listing the important functions you want to learn and scheduling dates for mastering each one, track your progress in a log.

WORD-PROCESSING SOFTWARE

Software programs for writing range from simple text editors, such as the Notepad built into Microsoft Windows, to very specialized programs used by lawyers, scientific researchers, screenwriters, software developers, and technical writers. Most business writers rely on one of several widely used programs—Microsoft Word, Corel WordPerfect, or Open Office Writer.

Word-processing software was so named because it was first designed to enable people to manipulate text electronically. Today, the software can manage much, much more: images, sounds, animations, comments written by reviewers, even information merged from other sources.

The versatility of these programs makes them highly useful to business writers. You may use word-processing software during any stage of the writing process—prewriting, drafting, revising, and refining. In addition, you may use these programs to work on a wide variety of content, from simple memos to complex product catalogs.

Editing Tools

All word-processing software programs share many functions for working with text and other elements. These functions are described below and on the next page.

Standard Functions

Standard functions for manipulating text include the following:

- **Insert and overtype modes:** Settings that allow you to insert text where the cursor is located simply by typing or to type over existing text with new text.
- **Highlight/select text:** Using the mouse or a combination of keys (such as Shift and an arrow key in Word) to specify text or other elements that will be edited.
- **Drag and drop:** Grabbing selected elements with the mouse and moving them to another location in the document.
- **Copy and paste:** Copying selected elements to an electronic clipboard where they are stored temporarily so that they may be inserted elsewhere in the document or into another document.
- **Cut and paste:** Moving selected elements to an electronic clipboard where they are stored temporarily so that they may be inserted elsewhere in the document or into another document.
- **Find and replace:** Find is a function that searches for all instances of a specific word or phrase within a document. Replace allows you to edit any or all instances of that word or phrase.

Formatting Tools

Formatting allows you to change the appearance and positioning of document elements. These characteristics include many fundamental textual elements: typefaces, text styles (e.g., bold, italic, underline), text and background colors, line spacing, paragraph spacing, tabs and other indentations, lists, and columns.

Templates

Templates or **Sheets,** the basic blueprints on which documents are built, establish the default settings for page layout, text formatting, and many editing tools. Companies often use a standard template to ensure that all of their business correspondence looks the same.

Spelling Tools

Spelling tools are effective at finding simple typos, but use such tools with care! They cannot distinguish between homonyms—for example, *cite, sight,* and *site*; *it's* and *its*; *they're, their,* and *there*. Nor can such tools catch typographical errors that create other real words—for example, typing "*form*" when you intended to type "*from*."

Grammar Tools

Grammar tools, like spelling tools, should be used with care. While they may alert you to improper use of punctuation, such tools may also advise you to change a perfectly good sentence because it contains too many words. Nothing compares to close editing by human eyes. (See Chapter 15, "Trait 6: Correct Copy.")

Thesaurus Tools

Thesaurus tools are helpful for writers who are having trouble coming up with the exact (precise, correct, or particular) word. (See Chapter 13, "Trait 4: Clear Words.") Because these synonym lists do not include definitions, you should consult a dictionary before using an unfamiliar word. Also, avoid the temptation to embellish the language in your document simply because a thesaurus is available.

Automated Functions

Automated functions may be used to correct errors—changing common typos like "teh" to "the"—or to complete phrases—converting "Dear S . . ." to "Dear Sir or Madam:"—automatically. These automated functions are often customizable, which can be very handy. For example, if you were a chemist writing about polyethylene terephthalate (PET) polymers or a lawyer writing about the Fair Debt Collections and Practices Act (FDCPA), you might add these phrases to the automated text entries to avoid typing the entire phrase each time you use it.

Track Changes Tools

Track Changes and **Merge/Compare** functions enable writers to keep track of changes made during editing or to compare subsequent versions of a document. These functions can be used by writers working alone or in collaboration with others. (See Chapter 3, "Teamwork on Writing Projects.")

Commenting Tools

Commenting and **Review** tools, usually designed for collaboration, may also be used by writers working alone. In a research paper, for example, you might insert comments where language seems unclear or a fact needs to be checked.

SPECIAL APPLICATIONS

While most routine business documents can be generated by word-processing software, more sophisticated documents often require specialized software: desktop publishing programs, image editing tools, groupware, and presentation tools.

Desktop Publishing Software

As word-processing software has become more complex, the line between it and desktop publishing software has become blurred. *Desktop publishing* typically refers to programs capable of creating dynamic documents with elaborate layouts that meet precise publishing standards.

Some of the most popular programs—QuarkXPress, Adobe InDesign, Corel Ventura, and Microsoft Publisher—are used by advertising agencies, catalog houses, graphic design firms, magazines, newspapers, printers, and publishers to create a wide spectrum of printed and electronic content.

Desktop publishing software was not created to replace word-processing or image editing programs, but rather to aggregate content created in those programs and prepare it for publication. Chapters of this textbook, for example, were written with word-processing software, but the textbook pages were designed and paginated with desktop publishing software.

Image Editing Tools

The availability of powerful, yet inexpensive digital cameras and image scanners has given rise to many image editing tools. If you will be creating documents that contain images, you should learn about image software, which is classified into four categories (some categories overlap):

- Computer-aided design (CAD) and computer-aided manufacturing (CAM) are used for commercial- and industrial-design applications. These specialized systems are geared toward architecture, engineering, product design, and manufacturing.
- Illustration software—CorelDraw, Adobe Illustrator, and Macromedia Freehand—may be used by graphic designers to create digital images, such as diagrams, icons, logos, and other visual elements.
- Image editing software—Adobe Photoshop, Jasc Paint Shop Pro, Microsoft Picture It, and Roxio PhotoSuite—is used to manipulate digital images such as artwork, photographs, and scans.
- Web graphics software, like illustration software, is used to create visual elements. Web graphics are dynamic objects or environments with which users can interact for education, entertainment, or reference. Such programs include Adobe Atmosphere and Macromedia Fireworks.

WRITE Connection

For more on using graphics in your business writing, see Chapter 5, "Using Graphics in Business Documents."

Groupware

You've probably been using groupware for some time without knowing it. Groupware, also known as collaborative software, includes Lotus Notes, Microsoft Outlook, and Novell Groupwise. These programs incorporate communication tools (contact lists, e-mail, instant messaging), scheduling tools (calendars), and work flow tools (document folders, task lists, project management).

Groupware gives a single computer user an almost seamless integration of commonly used computer functions. Groupware also enables multiple computer users to communicate, coordinate, and collaborate electronically.

As a business writer, you may need to coordinate your work with that of people from other departments—for example, marketing, sales, or research—on a project. Groupware would enable you to send e-mail, schedule meetings, hold online meetings, share files, and publish a project schedule.

Groupware continues to evolve to keep pace with the rapid developments in mobile computing. In the near future, business writers may collaborate via cellular phones and handheld wireless devices.

WRITE Connection

For more on group projects, see Chapter 3, "Teamwork on Writing Projects."

Presentation Tools

Presentation software enables business people, educators, and trainers to enhance their presentations with visual elements, to make their presentations using projectors or computer screens, and to publish or distribute their presentations electronically.

These programs are relatively simple to learn and use, and they come with a wealth of built-in visual features. Microsoft PowerPoint has been the industry standard since the 1990s, but Macromedia Flash, Open Office Impress, and Apple KeyNote have gained popularity as alternatives.

Some business people may find presentation software too limiting: Only one slide may be displayed at a time, and the slides must be displayed in linear order. An alternative is to create a presentation as Web content and use a Web browser (such as Microsoft Internet Explorer or Mozilla Firefox) to display the presentation.

Web browsers, like presentation software, can generate text in various sizes and colors, display images and animations, and play video and audio files. Unlike presentation software, however, Web browsers are not limited to linear presentation and can display multiple windows simultaneously.

WEB Link

Check your word-processing program or other special applications for readability tests such as the Gunning Fog Index and the Flesch Reading Ease Formula. But be careful. These formulas provide a general, limited measure of your writing—not a real measure of writing quality and clarity. Check http://college.hmco.com/english for more information.

DIGITAL RESOURCES: DATABASES AND THE WEB

Business and professional writers today rely not only on word-processing and specialized programs, but also on electronic databases and the Internet to obtain much of their information.

Databases

Databases are simply collections of information—such as the names, addresses, and numbers in a telephone book—organized and stored for easy retrieval by computer programs. They can be fantastically complex and store immense amounts of information.

However, databases store more than data, including much of the content you see displayed on Web sites. For example, a Web page may not exist as an actual page; rather, when you request the page, the computer generates it by assembling content elements within a template. For instance, the stories and photos displayed on the Yahoo! News Web page (http://news.yahoo.com/) are constantly updated in a database. Each time you visit that page, you request that the most recent stories and photos be displayed from the database.

As a business writer, perhaps you haven't considered breaking content down into small chunks of structured information that might be managed in a database and repurposed for print and online applications. However, structured documents and content databases are quickly becoming a valuable resource in many companies.

Content databases can be built with a variety of database programs, including Macromedia ColdFusion, Microsoft Access, FileMaker, dBase, and Lotus Approach. The primary business of some companies, such as Documentum, Vignette, and FileNet, is electronic content management.

The Web

As a business writer, you will probably take advantage of a broad array of Web-based information—customer demographics, financial data, historical archives, library catalogs, mailing lists, market data, news stories, stock photo repositories, and research reports. Some of this information is free, but some may require a subscription or fee.

From a business-writing perspective, Web-based information is valuable for several reasons:

- Because the Web usually offers multiple information sources on a given topic, the information can be considered comparatively.
- Online information resources can remain more up-to-date than any printed information source.
- Information in online databases is managed for you. In other words, someone else takes care of the filing.

WRITE Connection

For help with using the Internet to conduct research, see pages 145–150. For instruction on writing for the Web, see Chapter 36, "Writing for the Web."

CHECKLIST for Writing and Technology

Use the items below to check how efficiently and effectively you are using technology to support your workplace writing.

- ☐ When faced with new software, I learn by doing, investigating features, reading available resources, using built-in educational features, consulting others, and tracking progress with a log.
- ☐ For most business documents, I fully use the power of word-processing tools:
  - Standard functions for manipulating text
  - Tools to format text, paragraphs, lists, and columns
  - Templates or style sheets
  - Spelling tools, grammar tools, and the thesaurus
  - Automated functions that are available
  - Tracking, Compare, Commenting, and Review functions
- ☐ For sophisticated workplace documents, I rely on specialized software tools: desktop publishing, image editing, groupware, and presentation software. As needed, I rely on experts to ensure a positive outcome for such documents.
- ☐ I am familiar with the electronic databases and the Web resources that will be most helpful for my workplace writing.

CRITICAL THINKING Activities

1. What are key differences between word-processing and desktop publishing software? When should you use each, and why?
2. After reviewing the discussion of presentation software on page 50, explore the uses and abuses of these tools. If helpful, reflect on your own experience using such software or attending presentations created with such software.
3. Reflect on your anticipated profession: What writing tools will be important to your work? If necessary, interview a professional about these issues.

WRITING Activities

1. "The medium is the message." This famous statement by Marshall McLuhan implies that communication technologies are never neutral but rather are integral to the message conveyed. Doing research as needed, write a report exploring how communication tools influence business messages.
2. Write a brochure (two- or three-fold) using word-processing or desktop publishing software that you are only somewhat comfortable using. The brochure is intended to recruit new members to your class, club, or company.
3. Using word-processing or desktop publishing software, create a new template for a business letter, memo, fax, form, or Web page.

In this chapter

Using Teamwork to Strengthen Documents **54**
Using Peer Review for an Early Draft **56**
Using Peer Editing for a Later Draft **60**
Working on a Group-Writing Project **63**
Testing Documents with Readers **68**
Checklist for Peer Review and Editing **70**
Critical Thinking Activity **70**
Writing Activity **70**

CHAPTER

3

Teamwork on Writing Projects

In the workplace, writing is often a team effort. At a software company, for example, a marketing manager might ask a programmer to review a promotional brochure for accuracy and also ask a sales manager whether the brochure emphasizes the most important features of the product. Without the input of others, the writer of the brochure could overlook important aspects of the audience, purpose, and content. Similarly, many reports, proposals, and manuals are better developed as team projects.

In truth, even a writer working on a brief e-mail message can benefit from feedback—as long as both the writer and the reviewer understand the writing process and are committed to it. With that shared knowledge and commitment, two people working together will likely produce better writing than either one working alone.

Productive collaboration (whether it's two people or twenty) takes time, patience, and knowledge—and sometimes requires a thick skin. If done well, the benefits are great. Working together combines colleagues' different skills and perspectives, and it gives writers feedback before they get it from readers—when it may be too late. In fact, because the consequences of poor writing can be serious, no important document should be sent off without input from colleagues.

This chapter will help you develop the team-writing skills needed to offer such input. First, you'll learn how to work one on one in a peer-review situation—how to seek helpful feedback and how to provide it. Second, you'll learn how to work with a group of colleagues on a writing project that requires your shared knowledge and skills. Both lessons are based on this workplace-writing truth: Two heads are always better than one when both work to develop a strong document.

USING TEAMWORK TO STRENGTHEN DOCUMENTS

In the modern workplace, work gets done through project teams, collaboration, partnerships, committees, and quality circles. Such workplace writing depends on healthy teamwork. To strengthen your teamwork skills, adopt the strategies described here.

Foundations of Teamwork

For teamwork of any type to succeed, it should adhere to a few foundational principles.

Authority

In any team-writing situation, be clear about where the authority and responsibility lie for the document. When a co-worker gives you feedback, for example, you can use his or her suggestions as you see fit to improve the piece. If your supervisor supplies feedback, however, it takes on the quality of a directive to change the document.

Perspective

Any collaboration on a document should happen in the larger context of the company's good. Specifically, all team members should share the company's mission, values, policies, and goals as they seek to develop a document.

Direction

Teamwork must involve progress on the project. Collaboration fails when it becomes bogged down in details and loses sight of the goal, or when it gets mired in personalities and loses sight of what's good for the project. Strong leadership is essential.

Differences

To be successful, colleagues must rely on cooperation and professionalism during collaboration. Such productive behavior includes respecting and, in fact, embracing differences. Consider the value of the following differences:

- **Conflict.** Within certain bounds, conflict benefits collaborative work. By honoring divergent voices and considering contrasting opinions, you scrutinize multiple viewpoints, test reasoning, and tap into creative possibilities. The key is to manage the conflict through respecting, listening, probing, and negotiating.
- **Criticism.** Writing improvement depends on authentic, in-depth, constructive criticism—not wishy-washy generalities. Of course, criticism must always be directed toward the document, not toward the writer; toward fixing problems, not toward the problems themselves.

Audience

Whether you are the writer or the reviewer, put yourself in your reader's shoes during collaboration. Your own personal reactions, no matter how valid, are secondary to the potential responses of real readers.

WRITE Connection

For in-depth treatment of group issues, see Part 8, "Speaking, Listening, and Giving Presentations."

Types of Teamwork

Various forms of teamwork can help you write well. Review the methods described here and use them with your writing projects.

Informal Teamwork

Collaboration may be as simple as asking a co-worker for input or feedback at a key point in the writing process. Perhaps you need to brainstorm ideas early in the process or clarify your assignment with your manager. Maybe you need someone to look at your first draft to see whether you're on track or to look at your final draft to check for typos.

> **Media:** Informal teamwork can happen face to face, over the phone, or by e-mail.
>
> **Suitable projects:** Correspondence, short reports and proposals, presentations, job applications and résumés.
>
> **Examples:** An engineer reviews the technical content of your progress report to a client concerning work on a building project. A colleague in the next office reviews the tone of your e-mail response to a customer complaint about the late delivery of a product. An administrative assistant edits your monthly sales report.

Formal Review

Formal collaboration occurs when the primary writer shepherds an important document through systematic review by content experts, technical editors, managers, public-relations personnel, legal staff, graphic designers, and so on.

> **Media:** Formal reviews are usually structured processes that include face-to-face meetings and other correspondence.
>
> **Suitable projects:** Major reports and proposals; public-relations, marketing, and human resources documents.
>
> **Example:** The public-relations writer in the communication department of your company heads up the writing of the annual report—collecting data from various departments, soliciting or ghost-writing materials from the company's chief officers, and running the drafted report through a sequence of readers who check its content, style, and correctness.

Group Projects

Group collaboration relies on the contributions of many writers to accomplish the prewriting, drafting, revising, and refining tasks. The group may include a mix of communication specialists—content experts, Web site editors, and others—as well as professionals from different business functions—sales, human resources, and so on.

> **Media:** Group projects usually require multiple face-to-face meetings, constant correspondence, and use of file sharing and editing software.
>
> **Suitable projects:** Sales proposals, major reports, Web sites, product documentation, policy handbooks, mission statements, procedures, business plans.
>
> **Example:** New marketing materials for direct mailings, conventions, and your company's Web site are developed by a team made up of marketing, graphic design, finance, product development, and information management staff. Each colleague brings his or her expertise to the project and contributes appropriately to the documents generated.

USING PEER REVIEW FOR AN EARLY DRAFT

Sometimes, you may need help after completing a first or second draft of your document. To give and receive helpful feedback, both writers and peer reviewers need to understand their roles (outlined below) and to use reliable strategies and techniques (shown on the pages that follow). Essentially, writers and reviewers need to focus on the draft's global strengths and weaknesses—the ideas, the organization, the voice, and the overall format of the piece. (For more on revising techniques, see pages 28–31.)

Writer's Role

1. Share a hard copy or an electronic format that's easy to read, print out, and respond to orally or in writing. Clearly indicate that the document is a draft; print the word in capital letters and bold type, or use a stamp.
2. Introduce the context of your piece, without saying too much. Let the writing speak for itself.
3. Specify the kind of feedback you want and give a reasonable deadline for receiving it.
4. Remain open to reviewers' comments. Take brief notes about reactions, asking for clarification when necessary. Then reflect on those responses and develop a revision plan.
5. Raise any specific concerns.

Reviewer's Role

1. Consider the document's purpose and intended reader. Then listen to or read the draft attentively.
2. Take brief notes. Annotate the draft with comments, questions, and underlining.
3. Make specific, focused statements about what you hear in the writing. Balance constructive criticism with praise of good material.
4. Ask questions to help the writer think and talk about the draft.
5. In a group session, listen and add to other people's comments.
6. Focus comments on the success of the writer, the document, the project, and the group.
7. For a first-draft review, focus on improving ideas, organization, voice, and overall format. Avoid the following:
   - Commenting extensively on word choice and sentence style
   - Addressing grammar, punctuation, mechanics, usage, and spelling mistakes
   - Focusing on the writer, not on the writing
   - Praising or criticizing in ways that don't direct revision
   - Pressuring the writer's revision decisions more than your authority allows (A colleague should make suggestions, whereas a supervisor or manager can give directives.)

Peer Review Systems

Effective peer review of an early draft goes beyond "It looks okay." Instead, strong feedback offers specific suggestions aimed at strengthening content for intended readers. Two systems can help peer reviewers give such practical feedback.

Reacting to Writing

In *Writing Without Teachers,* Peter Elbow offers four strategies for reacting to a draft:

- **Pointing** refers to highlighting words, phrases, and ideas that make a positive or negative impression. This technique alerts writers to both strong and potentially dangerous passages in the draft.
- **Summarizing** refers to a general reaction to or understanding of the draft. List the main ideas, or focus on a key sentence. This strategy helps the writer sense whether the document communicates the necessary content.
- **Telling** refers to what happens as you read the piece. First, this happens; second, this; third, this—and so on. This technique helps the writer hear the draft from the reader's perspective.
- **Showing** refers to expressing feelings about a piece by using a comparison. For example, you might say, "Your introduction feels a bit muddy. It's hard to tell exactly why you are writing." A good comparison gives the writer a handle for understanding the positive or negative impact of something in a draft.

Practicing OAQS

A four-part reviewing scheme—Observe, Appreciate, Question, Suggest—helps you make productive comments about an early draft.

1. **Observe.** Note the document's purpose, and then note something about the content and design as it relates to that purpose. ("Your goal is to prove that the present cleaning methods are inefficient. Adding a cost analysis would strengthen your point.") Such a strategy helps writers refocus content around their goal for the document.
2. **Appreciate.** Praise something in the piece. ("Your sales letter opens with a real attention-grabber.") This strategy helps writers see what's already good.
3. **Question.** Offer questions that help writers think about the draft's content.
   - Reflect on purpose and audience: *Why are you writing, and what does your reader need?*
   - Focus their ideas: *What point are you making? How can you explain and support it?*
   - Consider organization: *What do you need to share in the opening? What's the best way of ordering the middle? What should happen at the end?*
   - Hear their voice: *What tone do you want to adopt, and why?*
4. **Suggest.** Give thoughtful, courteous advice about possible changes. ("Your discussion of the potential benefits of counseling services is strong. Consider summarizing those benefits up front.") This technique offers constructive, writing-focused feedback that writers can chew over, debate, and act on—without feeling bullied or forced into changing the draft.

Part 1 Challenges for Workplace Writers

Reviewer Response Sheet

Use or modify this response sheet for writing projects on which you frequently collaborate with co-workers.

Writer's Name: Reviewer's Name:

Date Document Submitted: Deadline for Response:

Project: Type of Document:

Form of Response (written and/or oral):

Purpose and Reader:

Ideas (content, information, details):

Organization (opening, middle, closing; pattern of development; paragraphs):

Voice (tone, attitude, style):

Overall Design (format, layout, typography):

Questions and Comments:

 WEB Link

For a printable version of this form, go to http://college.hmco.com/english.

Peer Review in Action

Lawrence King, the director of the Hancock County Arts Council, drafted a letter to area business people to encourage them to participate in the Purchase Awards Program. To improve this important letter, he turned to a colleague at the council, Ardith Brooks, asking for her feedback on what worked and didn't work with respect to the overall content—the ideas, the organization, the voice.

Dear [name]:

Try different opening, a shorter para. You need to give ArtBurst background first, then talk about P.A. Prog. Switch first and second paragraphs?

Do you like art? Then come to ArtBurst! Many Bar Harbor businesses see it as their duty to participate in the Purchase Awards Program. The program works because business people agree to attend ArtBurst and also agree and promise to purchase artwork (at a designated dollar amount), hence attracting artists and visitors. Everyone's a winner with the Purchase Awards Program! *Why does everyone win? Explain.*

Get the focus on the reader, on "you," not on "I."

I am the director of the Hancock County Arts Council. This council sponsors each year ArtBurst—a fair where artists display and sell their work. Well-known artists like William Drummond and Leslie Blass and many local artists like Susanna Reese show their wares: beautiful stained glass, classic landscapes in oil and watercolor, glass sculptures, wood work, pots, and much much more. ArtBurst is a real art feast for the community. This year, it will be held in Central Park on Saturday, May 4. Last year, ArtBurst brought in many artists and thousands of visitors who were good for the local economy. *Extra details about artists—are they needed at this point?*

Add concrete numbers and move from last year to this year—more natural order.

The tone seems to switch here; it becomes more formal.

Completion of the enclosed application form will ensure your commitment and participation in this grand event! Therefore, please give this request due consideration. *Add an action P.S.?*

Yours sincerely,

Lawrence King, Director

USING PEER EDITING FOR A LATER DRAFT

With some documents, you need editing help with a later draft—one that has gone through revision and is close to completion. Before distributing the document, you want a colleague with an eye for detail to strengthen your word choice and sentence smoothness, as well as to catch any errors in grammar, punctuation, and spelling that you missed. Consider two methods of refining your document through peer editing:

- Print out a hard copy of the document, preferably double-spaced, and ask your colleague to make changes to the text with a pen or pencil.
- Share an electronic copy of the document and encourage the reviewer to make needed changes to the file itself, perhaps using word-processing tools that track changes.

To give and receive helpful peer editing, study the editing symbols and model that follow.

Proofreading and Correction Symbols

Peer editing is more efficient when both writer and editor use simple editing symbols. Table 3.1 shows the most common symbols and how to use them. For more detail on these issues, see Part 9, "Proofreader's Guide." Note the specific cross-references added to items in the table.

Table 3.1 Common Proofreading Symbols

Symbol	Action Necessary	Example
Punctuation		
⊙	Add a period (p. 723)	all the hardware we need⊙
⋀	Add a comma (p. 726)	Once you've read the data⋀ I believe that . . .
⊙	Add a colon (p. 731)	Here are the three steps we need to take⊙
⋀;	Add a semicolon (p. 732)	Those are the positives⋀; however, here are the negatives.
=	Add a hyphen (p. 736)	Melissa demonstrates great self⁼control under stress.
⋁	Add an apostrophe (p. 729)	The customer⋁s needs come first.
Mechanics		
(cap)≡	Capitalize a lowercase letter (pp. 743–746)	the Human <u>r</u>esources <u>d</u>epartment
(lc)/	Lowercase a capital letter (pp. 743–746)	the /Loan was approved
/	Change a letter.	the total expend/ature
○	Close up a space.	the company's well⌒ness program
#	Add a space.	the fund#raising letter
gram	Fix a grammatical error.	The customer canceled ~~their~~ ʰᵉʳ order.
(ital)	Italicize material underlined (p. 740)	Check out <u>Chicken Soup for the Soul.</u>
sp.	Fix a spelling error (p. 754)	Our department hosted the discus⋀ˢjon.

Table 3.1 (Continued)

Symbol	Action Necessary	Example
Wording		
∧	Insert a missing word or letter.	this month's figures (*sales*)
⌐	Change a word.	the investment report (*investigative*)
ℓ	Leave out a word, letter, or punctuation mark.	once in a ~~long~~ while
tr. ↶	Change the order of letters or words.	the quality/total approach
ambig	Fix ambiguous wording.	implications of the decision are (curious)
awk	Fix awkward expression.	the bottom line's (brutal expression)
W	Fix wordiness (p. 244)	the basic, essential thinking on this week's events and happenings
Sentence		
¶	Begin a new paragraph (p. 225)	... is my first conclusion.¶The second conclusion follows from the first....
choppy	Fix choppy sentence rhythms (p. 262)	We should act now. We should also change the policy. The policy should now read...
logic	Fix faulty logic (p. 175)	Because our blender retails for $9.95, it's the best on the market.
∥	Fix faulty parallelism (p. 282)	The rotomaster breaks the hardened crust, and the top ten inches of earth is churned too.
ramb	Control rambling sentence (p. 270)	That's only one way to go to fix this problem, which we know is serious, but another, alternative way, something more desirable, should still be sought after for the foreseeable future.
trans	Fix weak transition (pp. 230–231)	First, we must investigate the CO problem. Conclusions about the problem should lead us to investigate viable solutions.
TS	Fix or add a topic sentence (pp. 227–228)	Maybe a power washer would be a good idea. First, it would ...

Peer Editing in Action

What does peer editing look like? After drafting and revising the persuasive request below, Lawrence King had his assistant edit and proofread it. With fresh eyes, she suggested important changes that made the message clear, polished, and professional.

Dear [name]:

Does your store have a wall that needs a painting, photograph, or sculpture to fill it? *(insert: or office, or a corner)*

For 14 years, your Hancock County Arts Council has sponsored ArtBurst—a fair where artists display and sell their work. Last year, ArtBurst ~~turned into a real art feast, an extravaganza that~~ attracted more than 90 artists and 15,000 visitors. This year, ArtBurst will be held in Central Park on Saturday, May 4.

The Purchase Awards Program, supported by local businesses, is key to ArtBurst's success. Business people like you join the program by agreeing to purchase artwork. Your commitment attracts *better* artists and *more* visitors. You, the local economy, and the arts all win. ¶ You win ~~big~~ *two ways*. First, you get a ~~nice~~ *beautiful* print, painting, drawing, photograph, or sculpture to decorate your business. *Second,* ArtBurst materials ~~also~~ advertise your support ~~in addition~~.

So please join the Purchase Awards Program! Just complete the enclosed form and return it in the postage-free envelope by April 12. A few weeks later, [name], you and your customers will be *enjoying beautiful* ~~covered with nice~~ art!

Yours sincerely,

Lawrence King, Director

p.s. An Arts Council member can help you ch*o*se the right artwork for your business. Call for details.

WORKING ON A GROUP-WRITING PROJECT

Teams—project teams, committees, panels, task forces, and so on—are a fact of life in today's business world. Such teams often develop important documents as part of their work—major reports, proposals, policies, and more. Successful teamwork will help ensure a successful document. To be a good team member, do the following:

- Stay focused on the goals of your group and the purpose of the document.
- Maximize your contribution by considering the expertise and writing strengths you bring to the group. Share your skills and abilities liberally; honor the skills and abilities of other members of the group.
- Respect the group's deadlines and your assignments. To ensure smooth development of the document, get your work done right and on time.
- Work efficiently. Help keep meetings productively on track. Use e-mail and other technologies to share ideas, research leads, and drafts.

Managing a Group-Writing Project

The many tasks associated with most business-writing projects can overwhelm even experienced writers. In a busy workplace, people are usually working on different phases of several projects every day; moreover, projects may be added, expedited, delayed, or cancelled, so writers' schedules are often in a state of flux. While collaboration helps lighten each writer's workload, it does require paying extra attention to organization and communication.

Like other business enterprises, writing projects may be judged on cost, time, and quality. *Did the project come in on budget, on schedule, with positive results?* From complex Web sites to simple brochures, business-writing projects that succeed do so because their managers mastered two skills: **planning** the project and **controlling** it once work began. Project plans may include more steps than are outlined here, so when you lead a team project adapt the following lists to your situation. The goal of such a step-by-step approach is to break down the project into manageable pieces.

Planning steps	Questions to ask
1. Assess needs	- *Why (purpose)—and for whom (audience)—are we doing this project?*
2. Define the project goals	- *What specific goals does this project need to achieve?* - *What related goals are beyond the scope of this project? (Those can be addressed at another time.)*
3. Create a cohesive team	- *What expertise and contributions are required?* - *What are the main assignments? How can the work be fairly and effectively shared?*
4. Select appropriate delivery media	- *Which media—print, audio, video, Web—are best suited for our purpose, audience, and topic?* - *How much will each medium cost?* - *How long will it take to develop content for each selected medium?* - *What style guidelines will we follow?*

5. Schedule and allocate resources
   - *How much can we spend?*
   - *Who's available to help?*
   - *Who will do what?*
   - *Who's in charge?*
   - *What are the technical requirements—software, hardware, and personnel?*
   - *How much time do we have before the deadline?*
   - *When—and how often—should we schedule meetings?*
   - *How will members of the project communicate?*

Control steps

1. Create and develop the content
   - *Do we have all the information we need? Is more research needed?*
   - *Have we addressed software requirements and electronic file format issues?*
   - *Have we addressed specific elements of document design—layout, headings, text formats, the use of images, and so on?*
   - *What are the specific writing guidelines? Does our topic require a style guide for grammar, punctuation, acronyms, abbreviations, or special terminology?*
   - *Do we need to investigate any legal or corporate policy issues related to this project?*
   - *What about possible workflow bottlenecks? Is there content that can't be created until other tasks are complete?*

2. Review and revise
   - *Are we missing anything?*
   - *What could be condensed or cut?*
   - *Does the content support the project's goals?*
   - *Are we communicating the right ideas?*
   - *Is the voice appropriate?*
   - *How do the visual aspects of the content support our goals?*

3. Publish
   - *How is the content being delivered?*
   - *Are companion pieces—mailings, flyers, and so on—being coordinated to accompany this content?*
   - *How long does the printer—or Web developer—need to publish the content?*
   - *Are we controlling quality and costs?*

4. Evaluate
   - *Were we successful?*
   - *Were we on budget and on schedule?*
   - *Can we be proud of the quality of this work?*
   - *What have we learned from this project that we can apply to other projects?*
   - *Have we celebrated completion and success?*

Teamwork During the Writing Process

Like an individual writer, your team needs to develop the document by working through the writing process. However, different steps in that process will require different team strategies—as do different projects. Work may be divided based on team members' skills or on elements of the project itself. Assigning tasks based on the expertise of team members builds on the particular strengths of individuals. For example, a marketing manager may oversee promotional sections of a document, or a Web developer may be paired with a customer service representative to ensure that a Web interface meets customer needs. Work may also be assigned according to the organizational or topical elements of the document itself.

Several strategies can be used to divvy up the writing process and to carry out the process—whether in conference room meetings, teleconferences, or online discussions. Explore below how to work together each step of the way.

Prewriting

Work with others to bring a writing project into focus and to generate ideas.

- Discuss the project's purpose, the document's readers, and the context (format, deadlines, resources) so that everyone is "on the same page."
- Using white boards, poster paper, or computer projection, brainstorm initial content for the document. Encourage full participation, recording all points uncritically. Focus discussion on potential ideas and possible organization. After reflection, separate gold from dross and "clump" points together.
- Review model documents, working toward completion of an outline. With the outline settled, divide drafting assignments.
- Generally, group members should do their own research and flesh out their own section outlines. However, to keep everyone honed in on the same target, share tips, finds, outlines, and progress reports before drafting.

Drafting

Generally, having individuals draft separate sections works best. This approach enables each writer to focus on a specific topic, to do a thorough job within this boundary, and to create content quickly. The danger is that the resulting pieces of the document may not fit together well. Writing styles may clash and information may be presented differently so that extensive revision is needed. To minimize such problems, each writer should do the following:

- Remember the big picture while working on his or her section.
- Follow the outline settled on by the group, but still exercise some creativity.
- Concentrate on getting down the ideas needed in the section.
- Adhere to whatever style guidelines the team has adopted.

F.Y.I.

Other approaches to dividing drafting duties are possible—from having all team members contribute to any aspect of the document to having one main author who is supported by other members supplying raw information. Generally, these methods are less efficient than the one described above.

Revising

First-draft feedback is vital. Honest, constructive criticism that focuses on ideas, organization, and voice helps the team move the draft forward. (See "Using Peer Review for an Early Draft," pages 56–59.) However, circulating the draft raises several logistical problems. The group must agree on routing procedures before the review process begins. These questions will help you establish a procedure:

- Will we route one copy of the draft, photocopies, or an electronic file?
- If we use electronic files, how will we safeguard the original draft?
- Will comments be written on a single copy of the document, attached as comments to an electronic file, or submitted on separate copies?
- How will we make sure that everyone is reading the most current draft?
- How will we compile comments to review them?

Two strategies for circulating documents during the review process are commonly used, and each may be suitable for paper or electronic distribution:

- **Round-robin routing** is a traditional form of routing in which a single copy of the document is passed from one team member to the next. This method is the slower of the two routing strategies, as each person must wait his or her turn to comment, but such routing has the advantage of keeping everyone's comments in one place. Problems with this method can include delays as the document sits in one person's inbox, the possibility of losing the document and the reviewer comments, and the tendency of reviewers to comment on others' comments rather than on the document itself.
- **Centralized routing** means that one team member distributes copies of the document to the rest of the group. Everyone reviews the document at the same time and has the same deadline for submitting comments. The quicker of the two methods, this strategy also ensures that reviewers don't become distracted by other reviewers' comments. Compiling reviewers' comments can be labor-intensive, albeit less so when the routing is done electronically. However, the group must follow strict file-naming conventions to avoid confusion over subsequent drafts of the document.

Refining

Once the team has revised the document, one or more editors should check the master document for the following:

- Accuracy, consistency, and clarity of information
- Appropriate tone and smooth flow throughout
- Consistent formatting (e.g., page layout, heading system, numbering)
- Correct and consistent spelling, punctuation, grammar, and mechanics

F.Y.I.

Word-processing software, desktop publishing, and e-mail can make collaborating interactive and fast—even across great distances. Explore the possibilities for your group project.

Checklist for a Team Project

During a writing project, all team members should be working toward the same goal while focusing on their own assignments. Use the checklist below to ensure that your group works effectively and efficiently.

The team
- ☐ The right people constitute the core group: appropriate experts in technical matters, finance, management, law, and so on.
- ☐ If possible, the team includes people affected by the document: employees, clients, or community members, for example.
- ☐ People outside the core group are routinely consulted.

The schedule
- ☐ Realistic goals and deadlines have been set.
- ☐ A Gantt chart (page 101) maps out the project's stages and milestones.

The plan
- ☐ Work procedures have been established: meeting times and places; methods and media for sharing information (e.g., e-mail group alias); and a style guide.
- ☐ Each group member understands the project: its goal, its relation to the organization's mission, and its expected outcome.
- ☐ Project assignments capitalize on group members' strengths.

The process
- ☐ Team leaders regularly share feedback and information with the team.
- ☐ During prewriting, the whole group works together on the document by
  - Brainstorming about the topic.
  - Establishing seven-traits benchmarks for the piece (page 186).
  - Developing an outline, template, and/or prototype.
  - Making drafting assignments.
- ☐ During drafting, each group member develops a portion of the document, conducting research and requesting input as needed.
- ☐ During revising, the group uses peer review (page 56) through either round-robin or centralized routing to test the overall content, shape, and voice of the draft.
- ☐ During refining, team editors fix stylistic inconsistencies, correct errors, and sharpen the design.
- ☐ If appropriate, the document goes through systematic review or testing.

The outcome
- ☐ Results are measured; follow-up is traced and pursued.
- ☐ The team celebrates its work, and has its work celebrated by others.

TESTING DOCUMENTS WITH READERS

Testing a document with real readers can be an excellent strategy for improving the piece before publishing it. Such testing may ensure that a document will achieve its goals with hundreds, thousands, or even millions of readers from different educational, cultural, and socioeconomic backgrounds. Documents and Web sites are often tested for **usability,** the degree to which a reader (or user) can successfully learn and use the information presented. By testing, you can determine—through scientific observation—how valuable a document is to your intended reader.

Although it may sound like a major undertaking, document testing should not be limited to large projects with huge budgets. Instead, it can be accomplished with fifteen, five, or even one reader. In other words, as important as team members are to creating the document, the insights of prospective readers can be even more invaluable. Readers who are completely uninvolved with a project may uncover serious errors, illogical assumptions, confusing elements, insensitive language, or other problems that would undermine the success of a document or Web site. Effective testing, then, will improve reader understanding, reduce costly follow-up, and ensure user safety.

What You Can Test

Your testing can address one or more of the following issues:

- **Comprehension.** If you wish to know how clear and logical the document's content is, ask the reader to restate its main points by summarizing the document or developing an outline.
- **Accessibility.** If you wish to determine whether a document's design and shape are reader-friendly, ask readers to predict content by simply reading introductions or to find specific things in the document.
- **Tasks.** For instructions, manuals, and procedures, you may want to test the readers' ability to actually perform a process. As they do so, note any problems that arise.
- **Comparisons.** Which version of a document or a section is better? Check by observing readers interacting with different versions.

Types of Tests

Documents can be tested for qualitative data (e.g., experiential, emotional, verbal, thoughtful) and/or quantitative data (e.g., numbers, statistics, trends).

Qualitative Testing

Much like the reviews performed within the project team, such testing focuses on readers' opinions. *What do they think about the document?* Qualitative testing can be accomplished two ways, via focus groups and/or protocol testing:

> **WRITE Connection**
>
> For a discussion of comprehension-related issues that might guide your testing, see Chapter 10 on strong ideas and Chapter 11 on logical organization. For a review of concepts and practices related to accessibility, see Chapter 16 on reader-friendly design.

- **Focus groups.** Usually no more than a dozen potential readers are paid to meet together to read and discuss a document. Focus group meetings are overseen by a moderator who initiates and facilitates the discussion, which is often guided by a carefully scripted questionnaire. Focus groups can provide valuable insights, but the discussions may lead to biased conclusions, as outgoing participants may contribute more than others, and shy participants may not share valuable information.
- **Protocol testing.** Interviewing readers one by one ensures that each reader's opinion is explored. Protocol testing helps determine what the reader actually thinks the document says, and helps writers gauge whether the intended message is being communicated. These interviews—like focus group discussions—are guided by a carefully scripted questionnaire and may be conducted either while the reader reads through the document or afterward.

Quantitative Testing

Quantitative testing is more involved than qualitative testing and, much like laboratory experiments, requires the comparison of data from a control group with those from an experimental group. Before sending a sales letter to 45,000 names on a mailing list, for example, a marketing department might conduct a **control study** that tests two versions of the letter on subsets of the mailing list. This study would allow marketing managers to compare customer responses to each letter before incurring the large expense of printing and mailing 45,000 sales letters. Quantitative tests are typically conducted after qualitative testing is completed. (*Note:* For control testing to produce scientifically valid results, the test sample must conform to established statistical methodologies, which are beyond the scope of this chapter.)

Testing Principles

As much as possible, you want your reader-focused testing to produce reliable, valid, useful results. To get such results, follow these principles:

Test early. Testing late in the document's development can help, but testing early (e.g., by using a document prototype) can save weeks of misdirected work.

Prepare well. Follow these steps:
- Identify your readers and find representative participants.
- Clarify the test's purpose: What aspects of the document require feedback?
- Design the test to effectively collect the needed information.
- If you are working with a team, assign responsibilities to team members for preparing test materials, interacting with readers, and observing.

Be objective. During the actual testing, you may be tempted to interfere or interject. Do your best to stand back and just collect observations and data.

Analyze results rigorously. Collate test results and examine them or discuss them thoroughly with the team. What can you conclude about the strengths and weaknesses of the document? Address weaknesses by mapping out changes.

CHECKLIST for Peer Review and Editing

Writer

- ☐ I used the reader's time well by developing the writing as far as I could, arriving on time, providing a clear paper or electronic copy, and leaving on time.
- ☐ I treated the reader professionally by introducing my writing (purpose, topic, and audience) and describing the kind of help I wanted.
- ☐ I listened carefully to the reader's response, taking notes and asking appropriate questions for clarification.
- ☐ I showed no defensive body language, made no argumentative responses, and thanked the reviewer.

Reader/Listener

- ☐ I made the reader feel welcome and offered my help freely.
- ☐ I listened to the reader's request and limited my comments to those issues.
- ☐ I evaluated whether the message, style, and voice matched the reader's purpose, topic, occasion, and audience.
- ☐ I annotated the writing correctly using the symbols on pages 60–61.
- ☐ For early drafts, I focused on issues related to content (ideas, organization, and voice); for later drafts, I also focused on sentence fluency, word choice, correctness, and document design.
- ☐ I balanced deserved praise with constructive criticism.
- ☐ My criticism was specific and helpful: "I noticed that three sentences on this page begin with 'Moreover.'" versus "The style is repetitive."
- ☐ My praise was specific and helpful: "Your opening effectively forecasts the arguments that follow." versus "Great opening!"

CRITICAL THINKING Activity

Reflect on group projects that you have participated in during the course of your college work. Assess (a) the success or failure of these projects, along with reasons for it, and (b) the quality of your own contribution to the projects. If you have done team projects at work, compare and contrast the school and work situations.

WRITING Activity

Work with three classmates to do the following: (a) Choose a business-related issue in the news about which you could write an eight- to ten-page report; (b) use the guidelines in this chapter to write the report; (c) use the checklist on page 67 to identify both strengths and weaknesses in your work process and product; and (d) present your report to the class, along with your assessment of your work process and product.

In this chapter

Strategies for Intercultural Communication **72**
Writing to an Intercultural Audience **74**
Showing Respect for Diversity **78**
Effective Attention to Diversity: A Model **83**
Checklist for Intercultural Communication **84**
Critical Thinking Activities **84**
Writing Activities **84**

CHAPTER

4

Writing for Diversity

We've all seen people commit *intended* cultural offenses. In response, we feel hurt for the victim, shame for the act, and disgust for the perpetrator. But we've also seen people—including ourselves—commit *unintended* cultural blunders. While we meant no harm by such gaffes, the actions often caused harm, regardless of whether we acted out of ignorance, carelessness, or misunderstanding.

Because our global work world includes people from many cultures who have a broad range of language skills, learning about others' social and communication practices is crucial. For example, we must understand that while many people in this diverse workplace write or speak English fluently, many others do not. For some, English is our first language, and we're so familiar with it that we don't give it a second thought. For others, English is our second or third language, requiring varying degrees of effort to use for communication. With this cultural and linguistic complexity, we must all strive to communicate so clearly and politely that each listener or reader feels welcome to join the dialogue. But how can we communicate freely while offending no one—particularly when our messages zip electronically around the world?

Clear, polite communication is not easy, but it is possible. Whether you are writing an e-mail to a co-worker with an ethnic background different from yours, a report that will be read in both U.S. and Japanese offices, or sales literature that will be read by people from diverse cultures, start by knowing yourself and knowing your readers. Understand your own cultural position, research your reader's perspective, and then connect on common ground. In addition, practice the tips and strategies in this chapter. Here you'll find sound advice for communicating across a whole range of cultural differences that impact business at home and abroad.

STRATEGIES FOR INTERCULTURAL COMMUNICATION

To communicate successfully with a diverse audience, you need to follow some basic principles. Specifically, you should be aware of some basic intercultural concepts and know how to do intercultural research.

Basic Intercultural Concepts

What a group of people share—language, beliefs, experiences, values, religion, education, technologies, laws, institutions, and attitudes—constitutes their *culture*. You and your readers are all products of complex cultural forces. *Intercultural communication* refers to exchanges between people from distinct cultural groups, whether these groups live in the same city or different countries. Such communication succeeds when it is based on positive practices like those described here.

Be Open

Much in your writing depends on your attitude. You want to avoid a closed mind and cultivate an open one.

Avoid	Cultivate
Ethnocentrism	Global awareness
Assumptions, ignorance	Knowledge, experience
Stereotypes	Sensitivity, empathy
Prejudice	Appreciation of difference

Ethnocentrism refers to the practice or habit of seeing one's own culture as natural, right, and superior, and of measuring all other cultures against that norm.

Assumptions are conclusions that are taken for granted or accepted without question; they discourage inquiry about and appreciation of cultural differences.

Stereotypes are generalizations that characterize or label all members of a cultural group without leaving room for individual differences.

Prejudices are ignorant assessments of a cultural group's nature and value.

The point? These closed-minded attitudes demean other people, alienate readers, and short-circuit communication.

Value Differences

Differences can expand your perspective, insights, and creative thinking. Tapping into differences with co-workers and customers also empowers you to communicate respectfully and build honest, substantive relationships.

Value Relationships

U.S. business people generally stress financial concerns over substantive connections. In contrast, many other cultures look at relationships as foundational for doing business. Don't bypass people; instead, work to establish and maintain good working relationships through goodwill. In the end, business is always about people.

Basic Intercultural Research

As a writer, you must understand your own cultural perspective: where you come from, what you value, and what principles guide your actions. But it's equally important to understand your reader's culture. Using the Web, library resources, nonprofit organizations, government agencies, personal contacts, and travel, investigate these aspects of your reader's culture: values, government system, natural resources, laws, gender roles, regions, language(s), traditions, ethnic groups, history, religions, and business institutions.

In addition, study *the role of context in communication*. In low-context cultures communication is direct: People state their points in a straightforward, forceful manner; words carry the weight of meaning. In high-context cultures, readers must read between the lines because the social context determines what the words mean. A simple "yes" or "no" is not simple at all. For example, a "yes" response during a negotiation may mean that the listener or reader understands your point or position, not that he or she agrees with and accepts it. Study the scale below; then review the attitudinal differences between low- and high-context cultures mapped out in Table 4.1. Note that these are generalities. Treat them with caution.

Low Context ◄─────────────────────────► **High Context**
Germany, Scandinavia, U.S., Canada, France, U.K., Italy, Spain, Mexico, Saudi Arabia, Japan, China

Table 4.1 Low-Context Versus High-Context Cultures

Attitude Toward	Low-Context Cultures	High-Context Cultures
Writing	Considered part of the process of doing business and making decisions. Written documents often seen as legally binding records. Directness preferred.	Often used only at the end of discussion, negotiation, and decision making. Written documents may point to ideal intention, not actual outcome. Indirectness preferred.
Relationships	Stress on individual responsibility, initiative, success, independence. Relationships seen as contractual, based on mutual benefit.	Stress on group, corporate decision making, social obligations, social connections. Issues of "face"—respect and status—deeply significant.
Gender	Women tend to be equal participants in business.	Women tend to be outside positions of influence.
Roles	Democratic and horizontal: gap between group members relatively small.	Authority-based and hierarchical: strong differences in status, power, and privilege.
Time	Seen as linear quantity that must be managed productively. Focus is often on immediate results.	Time fluid or circular. May be more past oriented (tradition) or future oriented (long-term viability).
Work	Central to individual's identity and self-worth.	Secondary to family and community obligations.
Risk and uncertainty	Favor and reward risk taking.	Prefer and practice caution.

WRITING TO AN INTERCULTURAL AUDIENCE

When writing to a diverse audience, your goals are to communicate clearly and avoid unnecessary conflict and follow-up. What strategies and techniques will help you achieve these goals?

Choosing a Language

A key first decision is whether to use English, the reader's native language, or both. Obviously, if your reader belongs to a culture that uses English (British, Canadian, Australian), use English. In other situations, the decision is tougher. Even though English is generally used for global business, translating documents or producing them in another language may make sense. Weigh these factors:

How strong is your reader's grasp of English? More specifically, how familiar is your reader with standard written English? Some individuals' spoken versus written facility with English may differ sharply.

How long-term and important is the business relationship? For brief, one-time contacts, use plain English, but encourage the reader to seek help with the message. For long-term relationships, consider learning the language yourself or hiring people who are fluent in it. Another potential resource is translation software, which can, for example, convert e-mail messages from one language to another.

What advantages could be gained by using English, the reader's language, or both? For example, will translation open new markets or meet government regulations?

How important, detailed, complex, or public is the document? Is it a legal document, a technical description, health and safety information, sales material, a Web page? The more public, the more serious, or the more dangerous the topic, the better it is to present the message in your reader's most familiar language.

How will the reader use the document? How often? Under what conditions? The more difficult the reading use, the more complex the conditions, or the more critical the document for the reader, the more likely you should translate it.

What does the law require? What is the ethically right choice? For example, Canada requires product labeling in both English and French. Similarly, a product's health risks should be stated in the user's language.

Translation tips

- Use a qualified translator—someone familiar with the reader's language and culture.
- Make sure that the translator knows the message's context, topic, and purpose.
- Get feedback from co-workers who are familiar with that language and culture.
- Follow a double process—translating from English to the other language, and then having someone translate from the other language back into English.
- To avoid embarrassing slips and costly misunderstandings, test the document by sharing it with a test group of people who speak the language.

Using Plain English for Multilingual Readers

If you choose to use English with a multilingual audience (i.e., an ELL—English language learner—audience using English as their second or third language), use straightforward, grammatically correct language. Different readers will have differing levels of facility with English. Although your reader may be thoroughly trained in English, it may be stiff, textbook English. Or your reader may be struggling at a basic level with grammar and diction. Given this diversity, follow these tips:

Avoid humor. Normally, humor doesn't translate well from one culture to another. Moreover, the reader may find a joke insulting.

Avoid cultural references. Be careful with references to people, places, and events specific to your culture. These may confuse or alienate your reader. Specifically, avoid sports, pop culture, religious, and military references and metaphors. Similarly, be careful with references to the reader's culture. Thoughtful, knowledgeable references help connect with readers; uninformed references may divide you from them and create embarrassment for both parties.

Avoid jargon, slang, acronyms, and abbreviations. Such shorthand has a restricted use that will trip up multilingual readers. Use plain English instead.

Use simple, objective words. Avoid words that have a lot of emotional baggage or that create ambiguity. Use straightforward words—nouns with clear meanings, and verbs that express clear action. However, don't confuse simplicity with a condescending tone: Don't write as if your reader were a child.

Use clear, obvious transitions. At the beginnings of paragraphs and sentences, use obvious transitions such as *however, in addition, first,* and so on. These transitions highlight the relationships between statements and help multilingual readers follow your thoughts.

Be grammatically correct. Spelling errors, typos, misplaced modifiers, sentence fragments, and faulty comma usage—all these errors create confusion for multilingual readers. Be especially careful to spell words correctly and to use accents or other diacritical marks properly.

Keep sentences and paragraphs short. Avoid long, complex sentences (more than fifteen words) and big, intimidating paragraphs (eight lines or longer). Multilingual readers may find such sentences and paragraphs difficult to read and digest.

F.Y.I.

- Not all English is the same. For example, U.S., Canadian, British, and Australian English differ in vocabulary and other elements. Even when writing to other cultural groups that use English fluently, use standard English.
- Be especially careful with the content and design of Web documents. The Web is not a homogenous environment. See page 577 for more.

WRITE Connection

If English is your second or third language and you want help improving your writing skills, see Chapter 50, "Addressing ESL Issues."

Connecting with Multilingual Readers

Communicating well with multilingual readers requires more than plain English. It requires that you adopt strategies that take into account cultural issues and practices that impact communication. Essentially, tailor your writing style and approach to the needs of your multilingual readers. Follow these tips:

Work through designated channels. Know the system of communication that your reader prefers and expects. Observe social hierarchies. If you're not sure, ask: Tactfully inquire about the proper practices and procedures for writing within the culture.

Be polite—almost to a fault. Because courtesy is extremely important in most cultures, pay special attention to these elements in your writing. Learn and observe the expected etiquette to avoid embarrassment and the perception of rudeness. For example, learn the culture's conventions concerning proper forms of addressing men versus women.

Be formal—but not impersonal. Avoid contractions of words and maintain an objective tone. Switching tones within a message will prove confusing (and perhaps insulting) to many intercultural readers. Accommodate your writing style to your reader's expectations (without compromising your own values). For instruction on achieving and creating a formal tone, see pages 236–237.

Be fittingly direct or indirect. Based on your understanding of your reader's culture and expectations, state your main point directly and forcefully or couch it in careful, indirect phrasing. For more on direct versus indirect approaches, see page 219. Be especially careful about how you say no to readers who value indirectness, harmony, and "face."

Be specific, not vague. Readers will have a difficult time digesting a string of generalizations and abstract ideas. Balance these with concrete, precise details. See page 248 for more on precise wording.

Use visual elements to communicate, when possible. First, make words accessible through design strategies—white space, attractive typography, and bulleted or numbered lists, for example. Second, integrate graphics into your document—pie graphs, flowcharts, line drawings, photographs, and so on. For more, see Chapter 16, "Trait 7: Reader-Friendly Design," and Chapter 5, "Using Graphics in Business Documents."

Convert information for your reader's convenience. Think about the information systems that your readers use: measurements, currency, dates, time references, and so on. When necessary, convert information from the system you prefer to that which your reader prefers. Table 4.2 gives some key considerations.

Table 4.2 Information Conversion

System	U.S. Practices	International Practices
Dates	August 15, 2005	15 August 2005
Times	3:45 p.m.	15.45
Numbers	4,567,344.73	4.567.344,73
Measurements	60 miles	100 km

Corresponding with Multilingual Readers

Be especially careful when corresponding with a diverse audience. Because letters, memos, and e-mail messages are the most common (and sometimes least polished) forms of business writing, they can create big problems in an intercultural setting. To craft your correspondence effectively for intercultural readers, follow these tips:

Make e-mail messages more formal. Avoid the typical informality used when writing e-mails to colleagues. Be careful with grammar and spelling, format the message for easy reading, and avoid emoticons (smileys) and shorthand (see page 373). In fact, treat the e-mail message as a formal, electronic letter with appropriate greeting, formal tone, and complimentary closing.

Use a conservative format for letters. Avoid the simplified and full-block methods (see page 384). Use the semiblock format or even a more traditional format with indented paragraphs. Better yet, learn the format used in the reader's culture and follow that in your letter. Specifically, learn the correct address format for your reader's country and his or her postal system.

Build feedback mechanisms into your writing. To ensure that readers have understood your message, directly request feedback. Rely on quick clarification through e-mail and faxed exchanges—not snail mail. When possible, follow up a written document with a phone call or an in-person discussion.

Show respect. Honor your reader through these practices:

- Use the reader's full and correct title, a proper form of address, polite salutations, and respectful closings. (See pages 391–395.)
- For most cultures, start your message with greetings, goodwill, and background.
- Avoid using your reader's first name unless you are certain that doing so will not offend him or her.
- Learn a few words of greeting, politeness, and parting in your reader's native language. Then use those words correctly in your correspondence. Be particularly careful to use diacritical marks (accents, etc.) with such words.

Use your reader's correspondence style. Learn the approach, tone, organization pattern, and other conventions favored in your reader's culture. Write your message by respecting that style while remaining true to your own cultural style.

WEB Link

Go to http://college.hmco.com/english for

- examples of address formats for Canadian, Mexican, Russian, Japanese, and other international correspondence
- examples of some basic courtesy terms in a variety of languages
- specific details and models of different cultural approaches

SHOWING RESPECT FOR DIVERSITY

In your writing, you want to show respect. When you are writing to a diverse audience, respect boils down to being consistently polite, treating others as you would like to be treated, and imagining yourself as the person you are writing about or to. It means paying attention to *what* you say, *how* you say it, and what you leave *unsaid*. Show respect for the members of your diverse audience by observing the guidelines that follow.

> **F.Y.I.**
> Unless absolutely necessary, avoid identifying a person's sex, age, race, religion, sexual orientation, ethnicity, or disability. Treat people as individuals, not group representatives.

Respecting Disabilities or Impairments

When writing to or about people with disabilities or impairments show respect by putting the person first and avoiding negative labels.

Put People First

Avoid referring to people as if they *are* their condition rather than people who *have* that condition. Put the person first, the disability second.

Not recommended (emphasis on disability)	Preferred
The disabled	People with disabilities
The retarded	People with mental retardation
Dyslexics	Students with dyslexia
Neurotics	Patients with neuroses
Quadriplegics	People who are quadriplegic
Wheelchair users	People who use wheelchairs

Avoid Negative Labels

First, avoid degrading labels such as *crippled, deformed, maimed, idiot,* and *invalid*. Second, avoid terms with negative connotations, such as *victims of, suffering from,* or *confined to*.

Not recommended	Preferred
Handicapped	Disabled
Birth defect	Congenital disability
Stutter, stammer, lisp	Speech impairment
Harelip	Cleft palate
Subjects, cases	Participants, patients
AIDS victim	Person with AIDS
Suffering from cancer	Person who has cancer
Epileptic fit	Epileptic seizure
Confined/bound to a wheelchair	Using a wheelchair

Respecting Ethnicity and Race

Obviously, you want to avoid racial and ethnic slurs in your writing, as well as words that create negative associations about and for specific ethnic and racial groups. In addition, you want to avoid clichés, stereotypes, and extraneous references to cultural characteristics. By contrast, you want to use words that show your familiarity with and respect for various cultures.

Use acceptable general and specific terms. Follow these practices to respect racial and ethnic diversity within the United States.

- For citizens of African descent, use *African American* or *black*. The first term is now widely accepted, although some people prefer the second.
- To avoid the notion that *American* by itself means *white*, use *Anglo-Americans* (for English ancestry) or *European Americans*. More specific terms would be *Irish Americans* or *Ukrainian Americans*.
- Avoid using the term *Orientals*. Instead, use *Asian*, and when appropriate, *Asian Americans*, and the specific terms *Japanese Americans* and *Chinese Americans*.
- As a general term, use *Hispanic Americans* or *Hispanics*. For specific terms, use *Mexican Americans, Cuban Americans,* and so on.
- As a general term, use *Native Indians* or *Native Americans*. Note that some native peoples prefer to be called by their tribe: *Cherokee people, Inuit,* and so on.

POOR: I recommend that Donna Kao, a petite Oriental woman, be hired for the HR Assistant position.

ANALYSIS: The sentence includes an unacceptable term, focuses on the person's appearance, offers a stereotype, and includes irrelevant information.

BETTER: I recommend that Donna Kao be hired for the HR Assistant position.

Avoid dated or insulting terms. Replace negative terms with more neutral, current ones.

Instead of	Use
Eurasian, mulatto	Person of mixed ancestry
Nonwhite	People of color
Caucasian	White
Foreigner	Immigrant

Use the term *American* with care. When writing to other U.S. citizens, the terms *America* and *American* are perfectly acceptable. However, the reality is that anyone living in North or South America is also an American. Therefore, when communicating with people from outside the United States, use the terms *United States* and *U.S. citizen*.

POOR: American politicians remain concerned about your foreign cows coming across the American border with Mad Cow disease.

BETTER: U.S. politicians remain concerned about the potential of Mad Cow disease coming across the Canadian–U.S. border through beef imports.

Respecting Gender

Sexist language stereotypes males and females or slants your discussion toward one gender. The guidelines below will help you avoid such language.

Take care with occupations and titles. Some words imply that one sex "need not apply." Instead, use the preferred forms listed.

Poor	Strong
Salesman	Sales representative, salesperson
Mailman	Mail carrier, postal worker, letter carrier
Insurance man	Insurance agent
Fireman	Firefighter
Businessman	Executive, manager, businessperson
Congressman	Member of Congress, representative, senator
Steward/stewardess	Flight attendant
Policeman/policewoman	Police officer
Chairman	Chair, chairperson, moderator
Foreman	Supervisor
Newsman	Reporter/journalist
Waiter/waitress	Server
Delivery man	Delivery person
Clergyman	Pastor, minister

F.Y.I.

Avoid using *man, woman, male,* or *female* as an adjective modifying an occupation.

NOT: male nurse, woman doctor, female Webmaster, female minister
BUT: nurse, doctor, Webmaster, minister

Take care with salutations and courtesy titles. In correspondence, usually avoid reference to a woman's marital status in a courtesy title. Use *Ms.* unless you know that the reader prefers *Mrs.* or *Miss.* When you don't know the reader's name, avoid gender-specific salutations.

Poor	Strong
Dear Sir:	Dear Sir or Madam:
Dear Gentlemen:	Dear Marketing Manager:
Dear Ladies:	Attention: Human Resources Department
Dear Housewife:	Dear Customer:

WRITE Connection

For more on courtesy titles and salutations, see "Forms of Address," page 391.

Use parallel language for both sexes. Avoid phrasing that implies inequality. Look especially for subtle differences in the way you treat men and women—attention to appearance, attitudes toward their work, and so on.

Poor	Strong
The men and their wives rebuilt the school.	The men and women rebuilt the school.
Harry Hilt and Ms. Susan Deptford	Harold Hilt and Susan Deptford

Use neutral terms and expressions rather than gendered ones. When a term implies that only one gender is included, use a more inclusive term.

Poor	Strong
Man/men	Person/people
Mankind	Humanity
Man-made	Synthetic
Man the office	Staff the office
Manpower	Human resources, workforce, personnel
My girl/boy	My assistant
Housewife	Homemaker
Forefathers	Ancestors

Avoid masculine-only pronouns (*he, his, him*) when referring to people in general. At one time, this practice was acceptable. Such pronoun use now is unequal and outdated.

Problem

A good *manager* seeks input from all people affected before *he* implements a change.

Solutions

1. Make pronouns and antecedents (the nouns to which they refer) plural. Change verbs as needed, and check that pronoun references remain clear.

   Before good *managers* implement changes, *they* seek input from all people affected.

2. Use "he or she" with care. (Overuse sounds awkward and actually draws attention to gender.)

   A good manager seeks input from all people affected before *he or she* implements a change.

3. If possible, rewrite the sentence so that it addresses the reader directly.

   To be a good manager, *you* should seek input from all people affected before *you* implement a change.

4. Rewrite the sentence to eliminate pronouns altogether.

   A good manager seeks input from all people affected before implementing a change.

Respecting Age

In your writing, respect generation gaps by showing respect to people of all ages. Use acceptable, neutral terms and avoid terms with negative connotations.

Use the following terms for general age groups:

Up to age 13 or 14	Boys, girls, child, children
Between 13 and 19	Youth, young people
Late teens and early 20s	Young adults, young women, young men
20s to age 60	Adults, men, women
60 and older	Older adults, older people
65 and older	Seniors, senior citizens

Avoid age-related terms that are too informal or that carry negative connotations. For example, avoid terms such as *kids, juveniles, yuppies, old man,* or *old woman.*

> **POOR:** The magazine's target market will be the older crowd, especially those silver citizens interested in travel, old folks' health, and careful management of their fattened retirement accounts.
>
> **ANALYSIS:** The sentence uses negative terms and implies stereotypes.
>
> **BETTER:** The magazine's target audience will be older adults interested in travel, health, and sound financial management.

Resources for Diversity

Information on diversity, cross-cultural issues, workplace sensitivity, global etiquette, and business manners can be found in many places—from your library to government agencies to nonprofit groups to Internet resources. Examples follow.

Print Books or E-Books

Here is a sampling of titles related to conducting business in different cultures.

- *The Global Etiquette Guide to Mexico and Latin America: Everything You Need to Know for Business and Travel Success* (part of a series that includes Africa, Asia, and Europe)
- *Harvard Business Review on Managing Diversity*
- *Creating the Multicultural Organization: A Strategy for Capturing the Power of Diversity*

Government Resources

The broadest electronic entry point into government information is www.firstgov.gov. From that page, you can access services for businesses and nonprofit organizations. A more specific resource for global information is the U.S. Department of State Web site at www.state.gov.

Web Resources

Internet resources abound concerning business etiquette, diversity, and global awareness. For an excellent starting point, visit www.executiveplanet.com for a guide to business etiquette, cultural guides, and links to specific countries.

Chapter 4 Writing for Diversity

EFFECTIVE ATTENTION TO DIVERSITY: A MODEL

Writing that disregards diversity among readers disrespects those readers. That disrespect can cause misunderstanding, pain, anger, and loss of business. For example, compare the two passages below.

Original Company Profile

> Providing tech fixes for wastewater problems, Abix Technologies, Inc., is an international player the speciality of which is the designing of applications for the disinfection of wastewater systems that offer environmentally safe, penny-pinching alternatives to chlorination. Since it rocketed onto the high-tech scene in 1981, Abix Technologies has been 100% energized for a full-court press against environmental problems. Our major league market is, of course, America. However, Canada, South America, Europe, Australia, New Zealand, the Middle East, and the Far East have proven worthy of Abix's attention. With offices in Sydney, Australia; London, England; Flint, Michigan; and Toronto, Ontario; foreigners are able to benefit from our American know-how.

Words: slang, jargon, unfamiliar comparisons

Long, complex sentences

Weak transitions

Informality and mixed tone

Cultural assumptions

Revised Company Profile

> Abix Technologies, Inc., is an international company that provides technological solutions to wastewater problems. Abix specializes in designing applications for disinfecting wastewater systems using environmentally safe, cost-effective alternatives to chlorination.
>
> Since its beginning in 1981, Abix Technologies has been committed to its goal of creating lasting solutions to environmental problems. Our markets include the United States, Canada, South America, Europe, Australia, New Zealand, the Middle East, and the Far East. With offices in Sydney, Australia; London, England; Flint, Michigan; and Toronto, Ontario; we are able to respond quickly and effectively to our customers' needs.

Clear, simple word choice

Straightforward sentences

Manageable paragraphs with clear transitions

Formal, polite, consistent style

Cultural sensitivity

WEB Links

- Visit our Web site http://college.hmco.com/english for information about specific cultural practices, links to intercultural resources, and advice about tailoring Web sites for intercultural readers.

- To learn more about another country's cultures and business practices, find quality Web sites from that country. For example, information on Canada can be accessed by starting with the federal government's Web site at www.canada.gc.ca.

CHECKLIST for Intercultural Communication

- ☐ I choose reading materials, TV programs, films, cultural events, and travel opportunities that broaden my understanding of other cultures.
- ☐ I choose words, forms of address, forms of writing, and organizational patterns that are used in my readers' cultures to show respect for others.
- ☐ My word choice, sentence structure, and paragraph structure can be read and understood by all my readers.
- ☐ I refer to people with disabilities politely, naming the person before the condition and avoiding negative labels.
- ☐ My references to ethnicity and race avoid clichés and stereotypes.
- ☐ My references to occupations and titles are fair with respect to gender.
- ☐ My references to age and generational differences are polite.
- ☐ For multilingual audiences, my communication avoids humor, cultural references, jargon, slang, acronyms, and abbreviations.
- ☐ For multilingual audiences, I make an extra effort to communicate through proper social channels, and to be formal, specific, and clear.

CRITICAL THINKING Activities

1. Reflect on your background, including where you have lived, what the composition of your communities was, where you went to school, what languages you speak, and where you have traveled. Then think about how these experiences and others have or have not prepared you to participate in a diverse workplace.
2. Analyze five business documents (including a letter, brochure, and newsletter) to determine whether the documents are accessible to multilingual readers.
3. Find a business news story that relates to a failed or successful intercultural workplace event (e.g., a merger, a marketing campaign, an expansion, a training program). Analyze the causes and effects of success or failure.

WRITING Activities

1. Working with a classmate, review this chapter's guidelines regarding respectful treatment of all people. Then exchange three pieces of your recent writing and analyze whether the writing follows the chapter's guidelines. Revise any problematic passages.
2. Find a business document that does a poor job of addressing diversity issues. Revise the document and submit the following to your instructor: (a) the original document, (b) your revision, and (c) a memo in which you explain why your revisions are needed.

In this chapter

Guidelines for Designing Graphics **86**
Parts of Graphics **87**
Using the Computer to Develop Graphics **88**
Integrating Graphics into Text **90**
Choosing the Right Graphics **92**
Tables **93**
Graphs **94**
Charts **99**
Visuals **102**
Checklist for Using Graphics in Business Documents **106**
Critical Thinking Activities **106**
Writing Activities **106**

CHAPTER

5

Using Graphics in Business Documents

Graphics portray visually what words alone can't communicate—or at least can't communicate as clearly, quickly, or dramatically. In other words, graphics "picture" information so we can see relationships (e.g., percentages of data), qualities (e.g., colors or shapes), and expressions (e.g., a smile or a grimace).

Because many of us are visual learners (we learn better by seeing *plus* reading or hearing versus reading or hearing alone), using graphics is a smart choice for workplace writing. In fact, workplace documents benefit from graphics for several reasons:

- They grab readers' attention and persuade them.
- They make information attractive and readable.
- They simplify complex ideas and dramatize important points.
- They save words by condensing information.
- They help transform a document into a presentation.

For these reasons, learning to read, analyze, and design graphics is a must for workplace writers. This chapter will help you understand what graphics are, how to read them, and how to design and use them. At some point while learning about graphics, you may need someone to help you understand digital scanners, desktop publishing programs, or graphic design concepts. However, you need not wait for help: You can begin right now by reading this chapter and completing the activities.

GUIDELINES FOR DESIGNING GRAPHICS

As you design graphics, your goal is to create clear, interesting, and helpful visuals to complement your words.

1 Prewrite

Rhetorical Analysis: Consider your purpose, your audience, and the context.
- ☐ What key information or ideas should your graphic display?
- ☐ Who are your readers, and how will a graphic help them?
- ☐ In what environment will the graphic appear—paper, electronic, both? In what type of document—a brochure, a report?

Choose the right graphic and gather necessary details.
- ☐ Tables show precise numbers or words in a grid.
- ☐ Line graphs show numerical trends, bar graphs give numerical differences, and pie graphs indicate proportions or percentages.
- ☐ Organizational charts show structure, flowcharts outline a process, and Gantt charts show scheduled activities.
- ☐ Line drawings, maps, and photographs provide informative images.

2 Draft

Sketch out your graphic, using shapes, patterns, and colors that clarify information.
- ☐ On a computer or on paper, experiment with your design.
- ☐ Work on the title, captions, legends, labels, keys, and notes.
- ☐ To prevent confusion, use a limited number of patterns and colors; avoid flashy elements that serve no purpose—elements called *chartjunk*.
- ☐ Use contrasting colors to show differences and shades of a color to show similarities.
- ☐ Use bright colors for accents and pale colors for backgrounds.

3 Revise

Check for clear ideas, correct details, and strong "pictures."
- ☐ Are all your data correct? Is the graphic's message clear and accurate?
- ☐ Is the picture you created realistic, or is it distorted?

4 Refine

Check your graphic for these characteristics:
- ☐ Precise wording in titles, labels, and legends
- ☐ Design elements that confuse, distract, or distort (chartjunk)
- ☐ Smooth, crisp sentences or phrases in notes and captions
- ☐ Correctness, clarity, and overall effectiveness

> "You never get a second chance to make a good first impression."
> —Dad

F.Y.I.
- Shapes, colors, and icons may have different meanings in different cultures.
- Color graphics may lose sharpness when photocopied.

Chapter 5 Using Graphics in Business Documents

PARTS OF GRAPHICS

While each type of graphic has its own guidelines and features, all graphics share some common elements. Use Figure 5.1 to identify these components.

Title

Each graphic needs a descriptive title that clearly indicates its topic and purpose. If a document contains more than one graphic, each graphic should be numbered as well (e.g., Table 3, Figure 5). Organize tables in a sequence called "tables," and place graphs, charts, and visuals in a second set called "figures."

Numbers and titles may be placed above or below the graphic. Simply be consistent by using one placement style throughout a document.

Body

Provide what is needed for the type of graphic: columns, rows, bars, lines, arrows, boxes, and so on. For clarity, include labels (written horizontally whenever possible), numbers and units of measurement, and a legend for symbols and colors.

Note or Caption

To explain information in the graphic, provide a footnote or caption. If you requested permission to use the information or graphic, be sure to acknowledge your source below the graphic with *Source:*, *Adapted from:*, or *Used with permission of.*

Figure 5.1 Sample Graphic

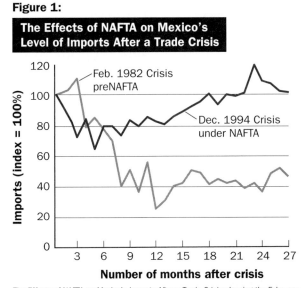

The Effects of NAFTA on Mexico's Imports After a Trade Crisis showing the February 1982 crisis without NAFTA (gray line) vs. the December 1994 Crisis with NAFTA (red line). **Adapted from:** "The Economic Effects of NAFTA," 1995, *Supervisory Management*, 40, p. 1.

USING THE COMPUTER TO DEVELOP GRAPHICS

Computer software puts informative, eye-catching graphics within the reach of most computer users. Today, creating great graphics doesn't require an art department, painstaking hand drawings, expensive software, or even much time or skill. In addition, computer-developed graphics can easily be saved, shared with others, edited, and altered to reflect updated information.

Which Graphics Program Should You Use?

While the choices for graphics programs can seem overwhelming, remember that most programs are designed for a specific task. The first step is to determine your graphic needs. Once you have decided whether you want a diagram for a business report, photos to enhance your marketing materials, or clip art for a newsletter, you can choose which type of software will best meet that need.

For many graphics, you may not need a specialized program. Most word-processing programs offer an "insert" option on their toolbar. From here, you can select a variety of options (see Figure 5.2), ranging from creating charts and diagrams to importing clip art. For example, when you create a graph, you select the type: pie, bar, line, and so on. All data (numbers and/or percentages) are entered into a table. You can name the various areas of your chart and create a legend, which helps others quickly understand your graphic. You can also choose the font and colors and adjust the graphic's size to best fit your document. Moreover, by creating and saving your graphic in a new file, you can design, save, retrieve, and edit the graphic without affecting your text document.

You can create more impressive graphics using specialized software programs.

Figure 5.2 Creating a Chart on the Computer

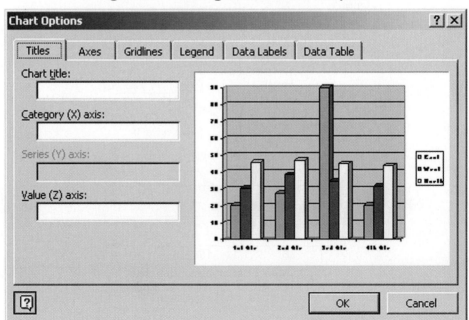

Photo Editing Programs

Photo editing programs allow you to alter photographs on your computer (either digital photos or those you have scanned in). You can crop pictures to include only the key subject and adjust both the shape and the size of the photo. For example, a color group photo of top executives could be cropped and edited to a black-and-white close-up of one person's face in a circle with blurred edges. Most photo editing programs allow you to easily correct issues such as lighting, red-eye, and out-of-focus subjects. Some programs let you transform a photograph using special-effect filters, such as charcoal drawing, blur, and sketch.

Drawing and Paint Programs

Drawing and paint programs allow you to paint a watercolor, sketch a pencil drawing, or create an oil painting—all using the computer and software as your supplies. These highly detailed programs produce realistic art effects as if done by hand. Options include the size and shape of your brush and brushstroke, colors, medium (oil or watercolor; pencil or pastel), and more.

Drafting Programs

Drafting programs are used to create architectural blueprints and plans of buildings, homes, parks and gardens, streetscapes, and designed products, such as furniture. Anything that is designed and built can be outlined in a visual plan that clearly illustrates the layout, overall view, and construction plan for the project. Many drafting programs offer 3-D options that generate 360-degree views. Some will even incorporate shadows, based on the location, date, and time.

Chart and Graph Software

Chart and graph software produce many types of graphs and charts, ranging from line graphs to flowcharts. Typically, data are entered into a table or spreadsheet, and then the program produces the chart or graph of your choice using the information provided.

Clip Art

Clip art is ready-made art or images that can be effortlessly imported into documents. Plenty of free clip art can be found on the Internet, and even more is available if you are willing to pay for it. Most word-processing programs come with their own cache of clip art, including shapes, lines, and other images, which can be accessed from the toolbar.

Desktop Publishing Programs

Desktop publishing software, which boasts formatted pages, is often used for publications such as magazines, newspapers, and newsletters. To lay out your publication, you can set up templates that allow you to swap in new text, photos, and art each time you publish it. Using multimedia software, you can view, listen to, edit, and save both visual and audio clips, which can then be imported into a presentation.

F.Y.I.

For documents that require truly professional-looking graphics, seek help from a trained graphic designer.

INTEGRATING GRAPHICS INTO TEXT

Graphics should enhance your document, whether it's a report, a proposal, or a brochure. Integrate graphics using the tips below and on the facing page.

Select an appropriate page design for the document as a whole. Consider the amount of text and the number of graphics you plan to use, and select a layout to balance the two. Here are examples:

- A single column with a large graphic (Figure 5.3a): text and graphic create a balanced page; table is placed at bottom after reference in text; table set off by box and shading.
- Two even columns with a small graphic (Figure 5.3b): even text columns create balance; small graphic allows text more weight; graphic is properly contained within column margins; graphic follows first in-text reference.
- Two even columns, one text and one graphics (Figure 5.3c): text and graphics have equal weight; visual and textual "stories" run side by side; text refers to graphics on same page.
- Two uneven columns, one text and one graphics (Figure 5.3d): imbalance gives weight to text; visual and textual stories run side by side.
- Multiple columns, multiple graphics (Figure 5.3e): weight given primarily to visual story; text supports and supplements graphics.

Position graphics logically. Insert a graphic close to the first reference to it—preferably after the reference.

- Place a small graphic on the same page as the reference.
- Place a large graphic on a facing page in a double-sided document, or on the page following the reference in a single-sided document.
- Turn a secondary or large graphic into an attachment or an appendix.

Follow good page-layout principles. Use white space around and within graphics as necessary.

- Position a graphic vertically on the page whenever possible.
- Adjust the size and shape of graphics to fit with the written text.
- Place partial-page graphics in boxes at the top or bottom of a page.
- Keep all graphics within the text margins.
- Use white space around graphics so that they don't appear crowded.

Direct readers to graphics. Unless the graphic is primarily decorative (e.g., a photograph in a brochure), discuss the graphic by following this pattern:

- Make points based on key ideas that the graphic shows.
- Refer to the graphic by noting its number, title, or location.
- Comment on the graphic to help readers see and understand the information.
- Use the same terminology in the written text and the graphic.

Figure 5.3 Sample Page Designs

a.

b.

c.

d.

e.

F.Y.I.

As you plan your document's graphics, consider the following questions. The answers will help you shape the graphical "story" of the document.

- What mix of writing and graphics should you use? Will written text support graphics, will graphics support text, or should the two be balanced?

- In what order should the graphics appear? Will later graphics build on earlier ones? For example, should you present a table of numerical data and then break it out into line, bar, or pie graphs?

- Will the graphics be used in an oral presentation? If so, how will they look when photocopied or on screen? What impact will they create with an audience?

CHOOSING THE RIGHT GRAPHICS

The pages that follow address specific types of graphics. The list below offers an overview of these types so that you can quickly review your options.

Tables

Arrange data in a grid of rows and columns to show the intersection of two factors.

- **Numerical tables:** provide amounts, dollars, percentages, and so on.
- **Text tables:** use words, not numbers.

Table

Figures

Shape information into a visual "story"; interpret data into 2-D images.

Graphs: show relationships between numbers.

- **Line graphs:** reveal trends by showing changes in quantity over time.
- **Bar graphs:** compare amounts by using a series of vertical or horizontal bars.
- **Pie graphs:** divide a whole quantity into parts to show proportions.

Graph

Charts: show relationships between parts or stages (not numbers).

- **Organizational charts:** picture structures and relationships within a group.
- **Flowcharts:** map out steps in a process.
- **Gantt charts:** provide an overview of project tasks and phases.

Chart

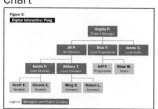

Visuals: provide images of some portion of reality.

- **Maps:** present information oriented according to geography.
- **Photographs:** show detailed reality in a range of formats.
- **Line drawings:** present simplified images of objects.

Visual

TABLES

Tables arrange numbers and words in a grid of rows and columns. Each slot in the grid contains data where two factors intersect (e.g., the type of trip and the number of days taken as in Figure 5.4). Use tables for these purposes:

- To categorize data for easy comparison of several factors
- To provide many exact figures in a compact, readable format
- To present raw data that are the foundation of later line, bar, or pie graphs

Design Tips

Set up rows and columns in a logical order. Make tables easy to read by using patterns of organization: category, time, place, alphabet, ascending, or descending order. Make the columns or rows containing totals clear and prominent.

Label information. Identify columns at the top and rows at the left. Use short, clear headings and set them off with color, screens (light shades of color), or rules (lines).

Present data correctly. In a numerical table, round off numbers to the nearest whole number (if appropriate) and align them at their right edge. Otherwise, align numbers on the decimal point. Indicate a gap in data with a dash or n.a. (not available). In a text table, use parallel wording.

Figure 5.4   Sample Numerical Table

Table 1:

Types of Personal Weekend Trips in the United States: 2004			
	Length of Trip		
Type of Trip	**1–2 Nights**	**3–5 Nights**	**Total**
Number			
Total personal trips	252,581	188,804	**441,382**
Business	32,358	33,172	**65,530**
Pleasure	186,219	134,659	**320,878**
Visit friends and relatives	104,438	74,151	**178,589**
Leisure	81,781	60,508	**142,289**
Personal business	34,004	20,970	**54,974**
Percent			
Total personal trips	100.0	100.0	**100.0**
Business	12.8	17.6	**14.8**
Pleasure	73.7	71.3	**72.7**
Visit friends and relatives	41.3	39.3	**40.5**
Leisure	32.4	32.0	**32.2**
Personal business	13.5	11.1	**12.5**

WRITE Connection

For a sample text table, see Table 7.2, "Information Sites," on page 121.

GRAPHS

Graphs show relationships between numbers—differences, proportions, or trends. When properly designed (without distortion), graphs can clearly portray complex ideas.

Line Graphs

A line graph, like that in Figure 5.5, reveals trends by showing changes in quantity over time—increases and decreases in sales revenue, stock prices, fuel consumption, and so on. Typically, the horizontal axis measures time (days, months, years), and the vertical axis measures a quantity (costs, products sold). The quantity for each time period creates a "data point." When joined, these points produce the lines that show trends.

Design Tips

Start the vertical axis at zero. If it's impractical to show all increments, show a break on the axis.

Avoid distortion. Use all available data points (no skipping). Make sure vertical and horizontal units match well (roughly equal in scale or ratio).

Identify each axis. Label the units of measurement (years, dollars), and use consistent increments. Print all words horizontally, if possible.

Vary line weights. Make data lines heavy, axis lines light, and background graph lines even lighter.

Use patterns or colors to distinguish multiple lines.

Use a legend if necessary.

Figure 5.5   Sample Line Graph

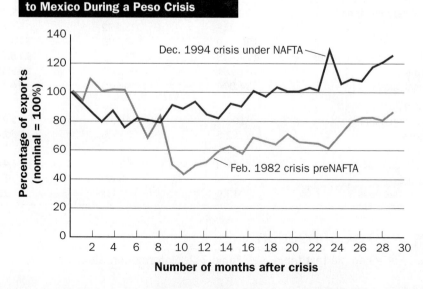

Bar Graphs

A bar graph compares amounts by using a series of vertical or horizontal bars. The height or length of each bar represents a quantity at a specific time or in a particular place. Moreover, different designs show different types of information:

- **Single-bar graphs** show quantity differences for one item (see Figure 5.6).
- **Multiple-bar graphs** present groups of bars so that readers can (1) compare bars within each group and (2) compare one group to the next (see Figure 5.7).
- **Segmented-bar graphs** divide individual bars to show which parts make up their wholes (see Figure 5.8).
- **Bilateral-bar graphs** show amounts both above and below a zero line (see Figure 5.9).
- **Pictographs** build bars with images representing the bar content (e.g., stacked cars to represent auto sales; see Figure 5.10).

Design Tips

Develop bars that are accurate and informative.

- Choose a scale that doesn't exaggerate or minimize differences between bars. Keep increments accurate and consistent.
- Maintain a consistent bar width and spacing between bars.
- Use 2-D bars; 3-D bars can blur differences and amounts.
- For pictographs, build bars by stacking images, not by enlarging one image. Simply enlarging the icon distorts differences.

Design bar graphs that are easy to read.

- Label bars and axis units clearly, and add a legend if necessary.
- Whenever possible, print words and numbers horizontally.
- Within multiple-bar and segmented graphs, limit the number of bars in a group or segments in a bar to five.
- Use patterns or colors to distinguish between different bars or segments.
- Present items in a logical order within a group or single-bar segments.

Figure 5.6 Sample Single-Bar Graph

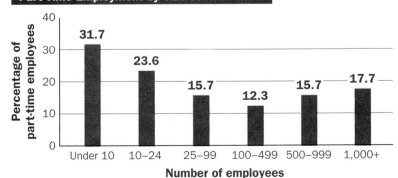

Figure 5.7 Sample Multiple-Bar Graph

Figure 4:

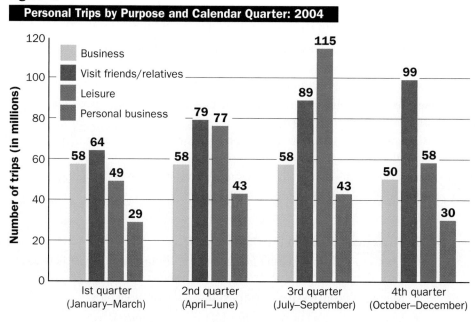

Figure 5.8 Sample Segmented-Bar Graph

Figure 5:

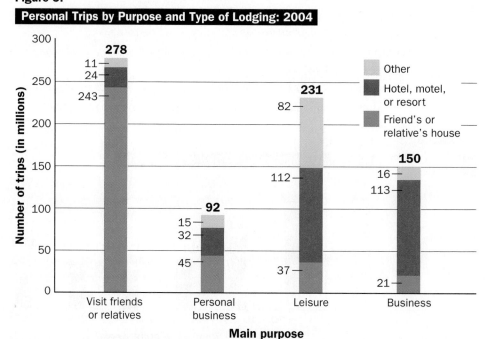

Chapter 5 Using Graphics in Business Documents

Figure 5.9   Sample Bilateral-Bar Graph

Figure 6:

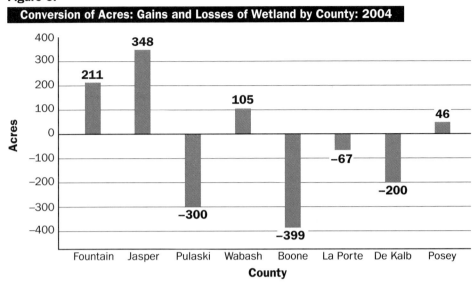

Figure 5.10   Sample Pictograph

Figure 7:

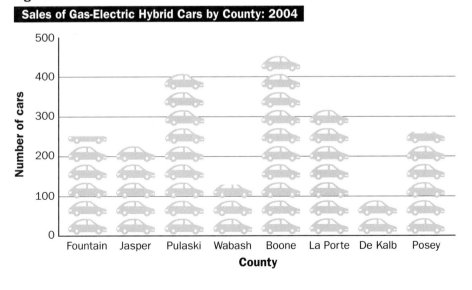

Pie Graphs

Pie or circle graphs divide a whole quantity into parts (see Figure 5.11). They show how individual parts relate to the whole and to each other. Use pie graphs to show proportions, give the big picture, and add visual impact.

Design Tips

Keep your graph simple. Divide the circle into six or fewer slices. If necessary, combine smaller slices into a "miscellaneous" category, and explain its contents in a note or in the text.

Make your graph clear and realistic. Avoid confusion and distortion.

- Use a moderate-sized circle and avoid 3-D effects.
- Distinguish between slices through shading, patterns, or colors.
- Use a legend if necessary.
- Measure slices (number of degrees) to assure accuracy. Use graphics software or the formula below.

To Calculate Degrees for Each Slice:
Amount of part ÷ Whole quantity × 100 =
Percentage × 3.6 = Number of degrees

Slices	$1,936	100%	360.0	degrees
Part 1	$775	40%	140.0	degrees
Part 2	$484	25%	90.0	degrees
Part 3	$415	21.4%	77.0	degrees
Part 4	$262	13.5%	48.6	degrees

Figure 5.11 Sample Pie Graph

Figure 8:
Office Expenditures: 2004

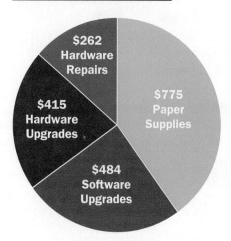

CHARTS

Whereas graphs show numerical relationships, charts show relationships between parts, steps, or stages. Use a chart to diagram spatial or temporal elements.

Organizational Charts

An organizational chart "pictures" a company, division, department, or other group (see Figure 5.12). Boxes show parts, and lines show relationships. Such a chart can provide a variety of details: overall structure, key people, positions, units, functions, responsibilities, authority, and communication channels.

Design Tips

Work downward. Place people and positions of highest authority at the top, and then branch down through each level.

Supply an appropriate level of detail. Do readers want an overview of the company, a sense of all the positions, or the names of actual people?

Be accurate. For nonhierarchical or "flattened" organizations that rely on teams, be creative with your chart design.

Be clear. Use boxes for people, positions, and units. Space boxes evenly, make them a good size, and put accurate information in each. Connect boxes with solid lines to show direct reporting relationships, and use dotted lines to show indirect relationships (communication or coordination).

Figure 5.12 Sample Organizational Chart

Figure 9:
Digital Interactive: Pong

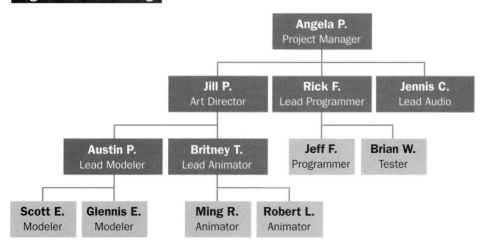

Legend: Managers and Project Leaders

Flowcharts

Using shapes, lines, and arrows, a flowchart shows the steps in a process (see Figure 5.13). In your document, a flowchart might describe a procedure, map out a solution, or clarify steps required to manufacture a product.

Design Tips

Establish direction. Design charts to flow from top to bottom or from left to right, whichever works best for the reader and your document.

Focus on the big picture. Keep charts uncluttered to give an overview of the process. Provide the minor details elsewhere—in the text or a secondary flowchart.

Strive for clarity.

- Use shapes to show different activities. Use ovals or circles for starting and ending points, rectangles for actions, and diamonds for decisions.
- Add a legend if necessary.
- Label each shape with a word or phrase.
- Use lines and arrows to show pathways through the process.

Figure 5.13   **Sample Flowchart**

Figure 10: The Hiring Process

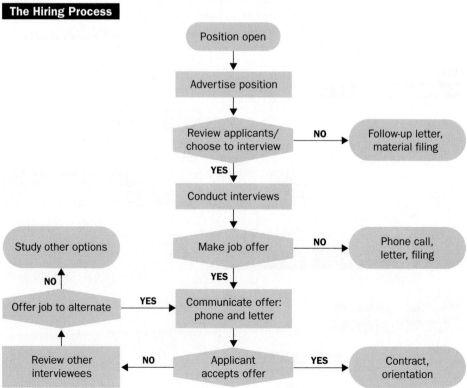

Gantt or Milestone Charts

A Gantt or milestone chart presents readers with a project overview: which tasks must be completed, when each task should start, and when each should finish. (See Figure 5.14.) Such a time line accomplishes three goals:

- It shows planned and scheduled activities at a glance.
- It clarifies how these activities are related.
- It helps readers manage a project's diverse tasks by giving time frames for many activities.

Design Tips

Establish two axes. On the vertical axis, list major project activities, with the beginning activity at the top and the other activities listed below in sequence according to their starting times. On the horizontal axis, list time units (days, weeks, months, years), including dates. Make sure the time increments are accurate and consistent.

Use bars to indicate start and finish times for activities. To keep the chart clear and uncluttered, show only *major* activities and discuss the details elsewhere. Indicate milestones with dots, triangles, or other icons. Provide a legend, if necessary.

Figure 5.14   Sample Gantt Chart

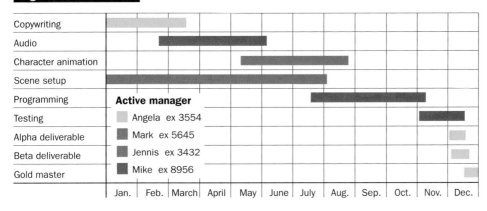

F.Y.I.

Establishing a realistic schedule for a project and its various stages can be tricky. Some time management experts recommend that you make your best estimate of the time required for a task—and then double it! As a general rule, build in some extra time as you establish deadlines. You'll be happy that you did when you are in the middle of a project.

VISUALS

Graphics like maps, photographs, and drawings provide visual images of places, objects, or people. By stressing certain details, these images help readers understand ideas or visualize concepts.

Maps

By representing geographically related information (usually related to the earth's surface), a map can present a wide range of useful details: communication and transportation data, distances, directions, regions and zones, natural and urban features, market segments, and so on. (See Figure 5.15.)

Design Tips

Make the geographic area clear. While your map need not be geographically accurate to be highly effective, it should clarify the area in question. If necessary, orient the reader by providing a directional arrow (usually north).

Provide useful content. Select details carefully, making sure that features, markings, and symbols are distinct and easy to understand. If necessary, use labels and a legend. Add a note or caption to help the reader interpret the map.

Indicate differences for regions. Use any of these techniques:
- Coloring, shading, cross-hatching, or dotting
- Creating bar or pie graphs in each area on the map
- Labeling areas with actual numbers

Figure 5.15   Sample Map

WEB Link

To make a map, try an online resource such as The National Atlas http://nationalatlas.gov.

Photographs

Photographs and other visual images can provide a wealth of vivid, informative details. Traditional photographs, whether black-and-white or color, show the surface appearance of reality—from people to objects to landscapes.

The very latest photographic technologies can be used to show external and internal images of a different nature: infrared pictures, X-rays, ultrasounds, CAT scans, 2-D and 3-D computer-generated images, and so on. Moreover, the development of digital photography processes gives you tremendous flexibility and power over these images. Not only can you create your own digital images to import into your documents, but you can also download images from a wealth of resources available online. But create and choose images wisely, always keeping in mind your communication goals, your audience's needs, and the type of document.

Design Tips

Be creative and ethical. Software such as Adobe Photoshop and Paint Shop Pro allows you to modify images, but make reasonable changes that do not distort or distract.

- Removing blemishes and cropping images are standard practice.
- Adding, subtracting, and distorting should be done cautiously. Avoid such editing if it will mislead or confuse; document or note it if necessary.

Take, develop, and print quality images.

- Use good equipment and seek expert help, if needed.
- Select distances and angles to clarify size and details.
- Enlarge and crop images to focus on key objects. (See Figure 5.16.)
- Effectively mount images on the page or import them digitally into the document.

Obtain permission to use any recognizable photo of a person or place.

Clarify images. Use labels, arrows, size references, and so on to help the reader get an accurate picture. (See Figure 5.17.)

Figure 5.16 Use of Enlargement

Figure 5.17 Use of Arrows

Line Drawings

Because line drawings screen out unnecessary details, they provide clear, simplified images of objects. Moreover, different types of line drawings focus on different key features:

- Surface drawings present objects as they appear externally (Figures 5.18 and 5.19).
- Exploded drawings "pull apart" an object, showing what individual parts look like and how they fit together (Figure 5.20).
- Cutaway drawings or cross sections show an object's internal parts (Figure 5.21).
- Diagrams, schematics, and blueprints provide detailed design plans and specifications (Figure 5.22).

Design Tips

Use the right tools. Specifically, rely on drawing instruments or a computer-aided design (CAD) program. Get expert help, if necessary.

Select the viewpoint and angle needed to show key details: surface, cutaway, or exploded; front, side, diagonal, back, top, or bottom. Indicate the angle and perspective in the drawing's title.

Pay attention to scale and proportion to represent the object accurately. Add measurements, if helpful. If the drawing reflects the actual size of the object, say so.

Include the right level of detail for your specific readers. Avoid clutter, and use colors to highlight and clarify information, not simply for flash (see Figure 5.23).

Use labels, or "callouts," to identify key parts of the object. Whenever possible, write callouts horizontally, and clearly link the callout to the part with a line. Avoid crisscrossing lines; instead, have labels radiate around the drawing to avoid interference.

Figure 5.18
Sample Surface Drawing: Front View

Figure 5.19
Sample Surface Drawing: 3/4 View

Figure 13:
Front View

Figure 14:

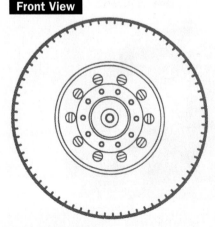
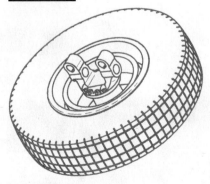

Chapter 5   Using Graphics in Business Documents

Figure 5.20   **Sample Exploded Drawing**

Figure 15:

Front-Wheel Assembly

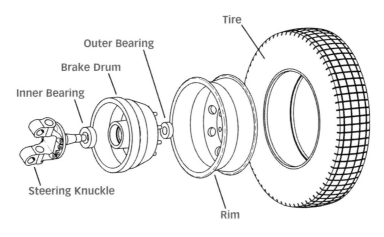

Figure 5.21   **Sample Cross Section**

Figure 16:
Cross Section

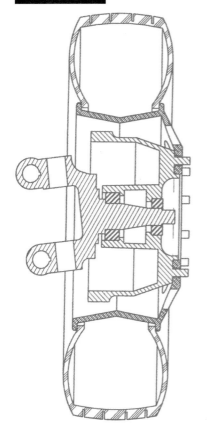

Figure 5.22
Sample Blueprint

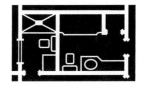

Figure 5.23   **Using Color in Line Drawings**

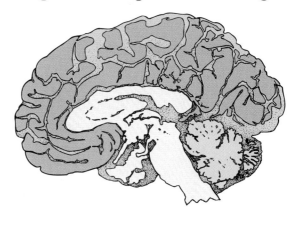

CHECKLIST for Using Graphics in Business Documents

Use these seven traits to check your graphics and then revise as needed.

- ☐ **Ideas.** Graphics include important information presented clearly:
  - Without distortion or deception
  - Appropriate for the content, audience, purpose, and document type
  - With accuracy that helps tell the document's "story"
- ☐ **Organization.** Graphics have all needed parts, including labels, legends, notes, and captions.
- ☐ **Visual tone.** The combination of words and images is professional, attractive, informative, and helpful.
- ☐ **Words.** Titles, labels, legends, and notes are precise and consistent with words used in the text.
- ☐ **Sentences.** Graphics include brief phrases or complete sentences as needed, along with clear and smooth captions or notes.
- ☐ **Copy.** Graphics include no errors in grammar, spelling, and punctuation.
- ☐ **Design.** Graphics make information easily accessible:
  - Avoid irrelevant material ("chartjunk").
  - Use colors, shapes, patterns, and white space to clarify information.
  - Use boxes or white space to set off or feature information.

CRITICAL THINKING Activities

1. Review journals related to your major for graphics. Analyze how writers in this field use graphics. Next, analyze the quality of each graphic's content and design. Finally, present a report to the class in which you display each graphic, and share your analysis of its quality.
2. Review your textbooks to find a table, a graph, a chart, and a visual. Analyze the design features of each graphic and describe why the graphic is or is not effective.

WRITING Activities

1. Select one of your past assignments that could be enhanced by two or three graphics. Create and incorporate the graphics into the text, along with necessary references. Submit the new document to your instructor, along with a memo in which you explain why the graphics enhance the writing.
2. Relying on an appropriate range of graphics, develop a report that tells the story of your education, training, and career up to this point.

In this chapter

Getting Technical: An Overview **108**

Ineffective Versus Effective Technical Communication **109**

Strategies for Technical Communication **110**

Features of an Effective Technical Style **112**

Checklist for Technical Communication **116**

Critical Thinking Activities **116**

Writing Activities **116**

CHAPTER

6

Communicating Technical Information

Technical writing relays specialized or complex information (often related to an applied science) that readers need to understand a topic or complete a task. Your challenge when doing such writing is to clearly, precisely, and accurately convey the information so that readers can understand it and use it. Conveying technical information is a challenge shared by individuals in nearly all occupations.

While some workers (e.g., chemists, engineers) do more technical writing than others, all workplace writers act as technical writers when they transmit specialized or complex information using language or a format not commonly understood by nonspecialists. For example, every specialized area of knowledge—accounting, human resources, engineering, social work, and so on—has its own way of knowing, its own content, and its own vocabulary or technical jargon. Whenever a writer in one of these professions shares information using the language of that field, he or she is communicating technical information.

Are you, then, a technical writer? In all likelihood, yes. Whenever you are called upon to share the specialized knowledge, insight, and methods of your profession, you are involved in technical communication. This chapter will help you succeed in that important task.

GETTING TECHNICAL: AN OVERVIEW

Technical—what does it mean, and why is the word important to workplace communication? Consider these points and principles.

Technical Goals

In the strictest sense, technical communication refers to knowledge shared within and by the applied sciences—engineering, health fields, and so on. Such communication extends scientific thinking, principles, and patterns into the workplace. More broadly, technical communication refers to the sphere of workplace communication in which experts share their knowledge with a wide-ranging audience, from fellow experts to general readers. In other words, the writer's knowledge *of the given topic* is equal to or, more likely, greater than the audience's knowledge. An expert, then, can be an accountant, an HR director, a computer programmer, a psychiatrist—virtually anyone with specialized training and knowledge to share.

Technical communication, whether spoken or written, has two goals:

- To serve an organization with *technique*—expert knowledge that defines, describes, classifies, analyzes, synthesizes, or evaluates
- To get something done safely and efficiently by sharing information, interpreting data, making projections, or mapping out plans

Technical Features

While following the guidelines for all workplace communication, technical writing should also have some special features:

- Accurate, objectively presented information supported by sound scientific research and reasoning
- A strong combination of words and graphics to clarify concepts—relevant graphics such as tables, charts, and photographs that "picture" the message
- Content presented in patterns consistent with scientific thinking (deductive and inductive logic, data interpretation, principled practice of definition, classification, and so on)
- Methodical organization with a consistent heading system that makes information readily accessible
- Word choice tailored to the reader's familiarity with the technical vocabulary

Technical Documents

Virtually any workplace document may contain technical content. However, the following are common forms of technical writing: instructions, user manuals, product documentation, policies and procedures, specifications, journal articles reporting research, field and lab reports, analytical/investigative reports, project progress reports, impact statements, justification/troubleshooting proposals, project bids, research and grant proposals, and feasibility studies.

INEFFECTIVE VERSUS EFFECTIVE TECHNICAL COMMUNICATION

Strong technical communication is accurate, objective, complete, and clear. It supplies specific readers exactly what they need to understand, act, or decide. Contrast the two examples below addressing ultraviolet-radiation methods of purifying wastewater.

Ineffective Technical Writing

> The process of UV disinfection is the process of making alterations to microorganisms that are single-celled by using UV-C and altering them through genetic means. LI lamps emit UV at 254 nm. All of the UV-C is used by the HI lamps, so they are more effective and it would be foolish for big wastewater plants and those that have poor quality not to use them. Flow pacing is another issue that should be considered. Meaning that the HI ballast allows for changing the setting of the lamps to account for flow, unlike the LI system.

Problems
- Unclear definitions and acronyms
- Poor explanations and incomplete information
- No paragraphing, headings, lists, or graphics
- Wordy, complex sentences

Effective Technical Writing

> The ultraviolet (UV) disinfection process targets single-celled microorganisms by altering their genetic material. The UV spectrum found to be most effective in altering the genetic material of organisms such as bacteria and viruses is UV-C (wavelengths in the 200–300 nm range).
>
> Approximately 90 percent of the UV radiation produced by low-intensity (LI) lamps is at germicidal wavelength of 254 nm (see Figure 1). In contrast, high-intensity (HI) lamps use the entire UV-C spectrum (see Figure 2). When one compares the biocidal doses delivered by the two lamp technologies, HI requires fewer lamps. For large wastewater treatment plants and plants with poor water quality, the HI option offers dramatic capital cost savings.
>
> With HI systems, operation and maintenance costs are further reduced by flow pacing. Typically, UV systems are designed for peak flow conditions. Thus, under average flow conditions, only some of the lamps are needed. The HI electronic ballast provides 16 intensity set points for each lamp (much like a dimmer switch). If an online UVT monitor is installed, this intensity-rate feature allows the wastewater plant to pace the lamps continuously according to flow rates and UV transmittance. As a result, the HI unit uses less power and wears more slowly.

Strengths
- Clear definitions and acronyms
- Complete information and precise details
- Use of graphics along with words
- Crisp, energetic sentences
- Effective comparisons to clarify complex technical matters
- Clear cause/effect thinking

STRATEGIES FOR TECHNICAL COMMUNICATION

Effective technical communication follows the strategies shared by other forms of workplace writing and presented throughout this book—seven-traits benchmarking, three-part structure, "you attitude," and so on. In addition, the particular strategies that follow apply especially to technical situations.

Know Your Audience

Profile Your Readers

With technical communication, it's crucial that you anticipate your readers' abilities and limits. To reach your audience, profile them by analyzing these issues:

- **Their knowledge:** Education level, professional expertise, reading skills, and familiarity with the topic and its vocabulary
- **Their positions:** Jobs, professions, roles within or outside your organization, levels in the company, responsibilities, powers
- **Their attitudes:** Motivations, values, needs, worries, concerns
- **Their reasons for reading:** To understand, to appease curiosity, to decide, to act

Connect with Your Readers

Based on the content of your message and your purpose, your audience will generally fall into one of the three categories shown in Table 6.1.

Table 6.1 The Three Types of Audiences

Audience	Reasons for Reading	Strategies for Connecting
Technical: Those familiar with the specialized content • Technicians • Operators • Specialists/researchers • Engineers	Some combination of understanding, researching, doing, operating, fixing, and implementing	• If appropriate, include theory discussion. • Use precise technical terms (jargon) as shorthand. • Keep discussion practical, how-to.
Managerial: Decision makers probably unfamiliar with the technical content • Department heads • HR, Public Relations, Finance staff • Upper management (VPs, CEOs)	Some combination of understanding, supervising, approving, troubleshooting, direction setting, and problem solving	• Use plain English and define technical terms that must be used. • Focus on decision-related concerns, not technical details. • Provide context and background, but get to the point. • Supply clear summaries, conclusions, and recommendations.
General: Laypeople (e.g., customers, the public) • Customers and clients • Some co-workers • Civic officials • The public	Some combination of satisfying curiosity, deciding to purchase, assembling, cleaning, repairing, reviewing, supporting, criticizing	• Use plain English and define technical terms. • Focus on what readers need to know, not everything you know. • Supply helpful background. • Be factual, but engaging. • Keep sentences short but varied.

Consider Audience Complexity

Frequently, your technical documents will have *multiple audiences* (a complex range of readers) and *secondary readers* (people who may see the document, even though they are not the primary readers). Here are three strategies for reaching a complex audience:

- **Develop different documents for different audiences.** Although potentially time-consuming and costly, this approach ensures that each category of readers gets exactly the information needed in understandable terms without having to wade through irrelevant material.
- **Designate different sections for different readers.** While creating one document, use clear headings to indicate which sections are relevant to technical, managerial, or general audiences. Define technical terms in a glossary.
- **Attach an audience-specific cover letter to the document.** In different cover letters for different types of readers, interpret the document's importance for the specific audience, point readers to the most relevant information, and clarify any potentially confusing information.

Use Building-Block Organization

In your document, each section, paragraph, and sentence should build on the preceding section, paragraph, or sentence. Start with clear purpose and forecasting statements, and then present information selectively and sequentially. In this known–new pattern, you begin with what the reader knows and follow with related new information. Note how this building-block approach works in the following paragraph where the known is in italics and the new is in boldface:

> *Deep ozone losses over both the Arctic and Antarctic* **are the result of special conditions that occur over polar regions in winter and early spring**. *As winter arrives in each hemisphere,* **a vortex of winds develops around the pole and isolates the polar stratosphere**. *Without milder air flowing in from the lower latitudes and in the absence of sunlight,* **air within the vortex becomes very cold**. *At temperatures of -80°C or less,* **clouds made up of ice, nitric acid, and sulfuric acid begin to form in the stratosphere**. *These* **are called polar stratospheric clouds (PSCs), and they give rise to a series of chemical reactions that destroy ozone far more effectively than the reactions that take place in warmer air**.
> Source: Adapted from Environment Canada "Arctic Ozone" Web site.

Make Content Accessible

Reading a dense series of long paragraphs filled with technical details is daunting for any reader. Avoid density by using graphics and an airy design that help the reader navigate the document and understand the information. Try these strategies: recognizable, standard formats; a system of headings and subheadings; page headers or footers; section tabs; numbered or bulleted lists; graphics (tables, graphs, charts, visuals); color-coding; columns; generous margins; and white space.

FEATURES OF AN EFFECTIVE TECHNICAL STYLE

Technical writing should be clear, factual, and honest—not flashy or writer focused in its style. The tone should be objective, but communicate an attitude of helpfulness. To develop such a style and tone, pay attention to your word choice and sentence structure.

Word Choice Issues
Be Clear, Accurate, Precise, and Objective

Technical writing calls for scientific precision and objectivity. Use words that are straightforward and factual—not vague, general, slanted, or subjective. In particular, avoid these troublemakers:

- **Gobbledygook:** Dense, difficult diction chosen more to impress than clarify
- **Buzz words:** Popular but often meaningless catch phrases
- **Ambiguous words:** Ones with multiple meanings that create confusion
- **Loaded words:** Ones that leave readers defensive, angry, or hurt
- **Incorrect or incorrectly used signal words:** Terms and graphics such as Warning! or Caution!

UNCLEAR: Watch OUT! For the accomplishment of the cleansification of the quartz sleeves, citric acids are in no way to be recommended because of their high-cost benchmark and low-bar cleaning capability.

CLEAR: Caution! To clean the quartz sleeves, avoid using citric acids, which are more expensive and less effective than other cleansers.

Use Technical Jargon Judiciously

Technical jargon is the specialized vocabulary of a field, profession, or occupation—engineering, law, and so on. Practice caution by adhering to these rules:

- Use technical jargon only when (1) technical terms are needed for correctness or precision or (2) your readers are experts and the terms are a kind of communication shorthand.
- Avoid technical jargon with laypeople (managers and general readers unfamiliar with the topic) because such jargon will function as distancing gibberish.
- If you must use technical terms with a lay audience, offer definitions (see page 202).
- When you use a technical term in a document, be consistent with its meaning.

    POOR: Destabilization of the QS unit element occurs when N/P ions magnetize to the QS's skin.

    STRONG: Fouling of the quartz sleeves occurs when charged particles adhere to the sleeve surface.

Cut Wordiness

Be as concise as clarity allows. Replace long phrases and clauses with single words; eliminate unnecessary repetition. (For more help, see page 244.)

> **WORDY:** As the coating of particles that are charged continues to extensively and intensively build up, the level of intensity of the light in the UV wavelengths continues to diminish and lessen to the point where it becomes imperative that all of the build-up that has occurred must be necessarily lessened and removed to ensure that the system remains effective at its task of purifying the water.
>
> **CONCISE:** As the coating builds, the UV light intensity lessens to the point that the coating must be removed to ensure that the unit works effectively.

Be Correct in Mechanics

Because of its scientific nature, technical writing relies on accurate terms, precise measurements, standard abbreviations, and recognized acronyms.

> **INCORRECT:** Clean your sleeves with a cleaning solution having a 2 or three Ph balance, such as a 55% solution of phosphorescent acids or Limeaway.
>
> **CORRECT:** Clean the quartz sleeves using a cleaning solution with a pH value between 2 and 3, such as a 5% solution of phosphoric acid or a commercially available solution such as Lime-A-Way.

Show Sensitivity to Diversity

Many technical documents go to diverse readers—both men and women; people of different races, colors, and nationalities; those of different age groups, generations, and educations; and readers with varied levels of English-language skills. Choose words that include rather than exclude specific readers. (For more help, see page 71.)

> **INSENSITIVE:** White folks, especially really pale-skinned and red-haired (carrot-topped) old folks, are more likely to be skin-cancer victims. UV radiation may suppress man's immune response, increasing the risk of skin cancer and infections for the careless, including reactivating herpes lesions among the sexually promiscuous.
>
> **SENSITIVE:** People with light skin and red hair are more vulnerable to skin cancer. UV radiation may suppress the human immune response, increasing the risk of skin cancer and infections, including reactivating herpes lesions.

Use Positive Expressions

While you want to be objective, you also want to be positive. Negative statements tend to be wordy, difficult to understand, and less productive. Whenever possible, use positive words. (See also page 255.)

> **NEGATIVE:** Acute, long-term, and chronic eye conditions may not become less common if UV nonprotection does not decrease.
>
> **POSITIVE:** Acute, long-term, and chronic eye conditions may become more common if UV exposure increases.

Sentence-Style Issues
Use Active- and Passive-Voice Constructions in Fitting Ways

While the passive voice is generally accepted and often expected in scientific writing, it does have limits, especially when you are communicating with a general audience. (When the verb is active, the subject of the sentence is performing the action; when the verb is passive, the subject is being acted upon. For a detailed discussion, see page 266.) Follow these guidelines.

- Use the active voice to keep sentences focused on actors and actions, and to cut unneeded words.

    PASSIVE: Operating and maintenance costs are further reduced by flow pacing.

    ACTIVE: Flow pacing further reduces operating and maintenance costs.

- Use the passive voice to emphasize the action over the actor or to stress the action's receiver.

    Bleached pulp and paper products made by inexpensive processes are discolored by UV radiation. Nonplastic building materials such as roofing membranes are currently being studied for their resistance to UV.

Rely on Parallel Structure

Information becomes clear when it is ordered through parallelism—the presentation of similar material in the same grammatical format (words, phrases, clauses). For detailed help, see page 263.

UNPARALLEL: Effective wastewater treatment that is primary and combining it with UV disinfection will lead to reductions in cost per sea outfall of 90 percent, protecting marine life against bacterial contamination, and the meeting of European Union directives for bathing water.

PARALLEL: Combining effective primary treatment of wastewater with UV disinfection will reduce the cost per sea outfall by 90 percent, protect marine life against bacterial contamination, and meet the European Union directives for bathing water.

Use Positive Repetition

When key concepts are restated in new ways, the reader absorbs the information. Repeat key nouns and verbs so that readers can follow the flow of information.

Over the *Antarctic,* these processes commonly lead to the formation of a massive *ozone* hole. Over the *Arctic,* however, *ozone* amounts have not yet fallen to the very low levels observed in *Antarctica.* This difference occurs partly because the *Arctic* has more *ozone* to start with, but it also results from the more variable atmospheric circulation of the Northern Hemisphere, which makes the *Arctic* vortex less stable.

Avoid Strings of Modifiers

When several adjectives and adverbs are piled in front of a noun, or several adverbs in front of a verb, the result is a dense, difficult-to-digest sentence. Use only those modifiers that are necessary; replace general nouns and verbs with precise ones that require no modifiers; or move some modifiers into phrases that follow the word being modified.

> **DIFFICULT:** Parties to the international multilateral environmental ozone-depleting substances Montreal Protocol agreement, which was signed and ratified by more than approximately 180 participating countries, have largely and specifically agreed to quickly and effectively develop CFC and Halon life-cycle management phase-out strategies.
>
> **CLEAR:** Parties to the Montreal Protocol—an international, multilateral agreement on ozone-depleting substances that was ratified by more than 180 nations—agreed to develop strategies to manage and phase out CFCs and Halon.

Keep Sentences a Moderate Length

Long sentences filled with complex technical content kill comprehension, slow reading speed, and lead to frustrating and time-consuming rereading. For detailed help, see page 270. Generally, follow these guidelines:

- Provide ideas in manageable chunks—sentences of 12–20 words.
- Use a simple but varied sentence style, following natural syntax: subject–verb–object.
- Offer clear transition words.
- Use action verbs (e.g., *mix*) more often than linking verbs (e.g., *is*).

> **LONG SENTENCE:** The future of the Arctic ozone layer is primarily dependant upon our success in ridding the atmosphere of ozone-depleting chemicals, and on our ability to control greenhouse gases, which is also important, because the linkages between these issues means that we cannot treat either of them in isolation, meaning that it is indicative of the importance of developing a comprehensive strategy for moderating the human impact on the atmosphere.
>
> **MODERATE, VARIED SENTENCES:** The future of the Arctic ozone layer will depend primarily on our success in ridding the atmosphere of ozone-depleting chemicals. However, controlling greenhouse gases will also be important because the gases and chemicals are linked problems. In other words, we must develop a comprehensive strategy for moderating human impact on the atmosphere.

F.Y.I.

Some word-processing programs allow you to assess the readability of your writing—the difficulty level of both your word choice and sentence styles. Although such tests have their limits, consider using readability formulas to check the reading difficulty of your technical documents. Check http://college.hmco.com/english for more information. Alternatively, read aloud your technical writing to test its difficulty, or have appropriate readers check the document for you.

CHECKLIST for Technical Communication

Use this checklist to review, revise, and refine documents that contain technical content.

- ☐ **Ideas.** The thinking is scientifically sound and objective; details are precise, accurate, and complete.
- ☐ **Organization.** The document has a transparent three-part structure:
  - An effective opening that provides an overview and forecasts content
  - A middle that follows building-block organization
  - A closing that focuses on conclusions and next steps
- ☐ **Voice.** The tone is objective but helpful throughout the document.
- ☐ **Words.** Phrasing is concise, unclear terms are defined, and jargon is avoided when possible.
- ☐ **Sentences.** The document reads well aloud, contains varied sentences, and uses passive and active voices effectively.
- ☐ **Copy.** The document contains
  - No confusing or distracting grammar, punctuation, usage, and spelling errors.
  - Accurate mechanics—numbers, equations, acronyms, abbreviations.
- ☐ **Design.** The document offers a fitting, recognizable format, as well as features that make content accessible—a heading system, graphics, white space, and so on.

CRITICAL THINKING Activities

1. Find a technical document (e.g., instructions, a field report, or a policy statement). Analyze and assess its attention to audience, its organization strategies, its use of graphics, its design, its word choice, and its sentence style. Attach your analysis to the document.
2. Write a memo to your classmates and instructors in which you (a) identify your major or profession; (b) explain what is *technical* about its knowledge, content, processes, and language; and (c) discuss how that technical material may best be communicated to a lay audience.

WRITING Activities

1. Identify a key concept or term in your major or profession. Write (a) a memo explaining the term to other members of your profession or major and (b) a letter explaining the term to a layperson.
2. In a technical document such as those identified on page 108 or in Critical Thinking activity 1, locate a poorly written paragraph. Edit the paragraph for problems of presentation, word choice, and sentence style.

WEB Link

For more activities, check our Web site at http://college.hmco.com/english.

In this chapter

Research Overview: A Flowchart **118**
Planning Your Research **119**
Managing Your Project: Note-Taking Strategies **123**
Doing Primary Research **130**
Doing Library Research **137**
Doing Internet Research **145**
Organizing Your Findings **151**
Using and Integrating Sources **152**
Avoiding Plagiarism **156**
Following APA Documentation Rules **158**
APA References List **161**
Checklist for Research **167**
Checklist for a Research Report **167**
Critical Thinking Activities **168**
Writing Activities **168**

CHAPTER

7

Conducting Research for Business Writing

In your business writing, readers are counting on you to give them all the facts—the correct facts. Complete, accurate information (as opposed to misinformation) provides a solid foundation for making business decisions. For this reason, you need to develop sound research practices and skills.

Conducting sound research today is both easy and difficult. It's easy because research technology is powerful. It's difficult because that technology makes so much information available—both good and bad. Such research may involve searching online databases and conducting surveys, checking company records and locating reliable Web sites, studying journal articles and gathering statistics. In this sense, good business research is challenging but rewarding detective work that helps you both expand and share your knowledge. Research may be as simple as quickly checking a fact and sharing it in an e-mail. It may also be as complicated as spending months digging for information and developing insights to present formally in a major report or proposal that will shape the future of your organization.

Good business research, then, focuses on three facets: *curiosity* to find answers to important questions; *discovery* through uncovering and analyzing information; and *dialogue* with sources, experts, yourself, and readers. This chapter will guide you toward this curiosity, discovery, and dialogue in doing and writing up research. In particular, it offers a refresher on sound research strategies and extends these into the workplace.

> "The use of information for strategic purposes largely determines whether the firm anticipates change, or is controlled by it."
> —Michael Lavin

Part 1   Challenges for Workplace Writers

RESEARCH OVERVIEW: A FLOWCHART

Business research works best when you follow the dynamic process outlined in Figure 7.1. Broadly speaking, the process includes three phases: planning your research, conducting the research, and organizing and drafting your report. Note that many research activities can happen simultaneously even as you push the process forward. Use the flowchart to help you map out appropriate activities for any research project.

Figure 7.1   The Research Process

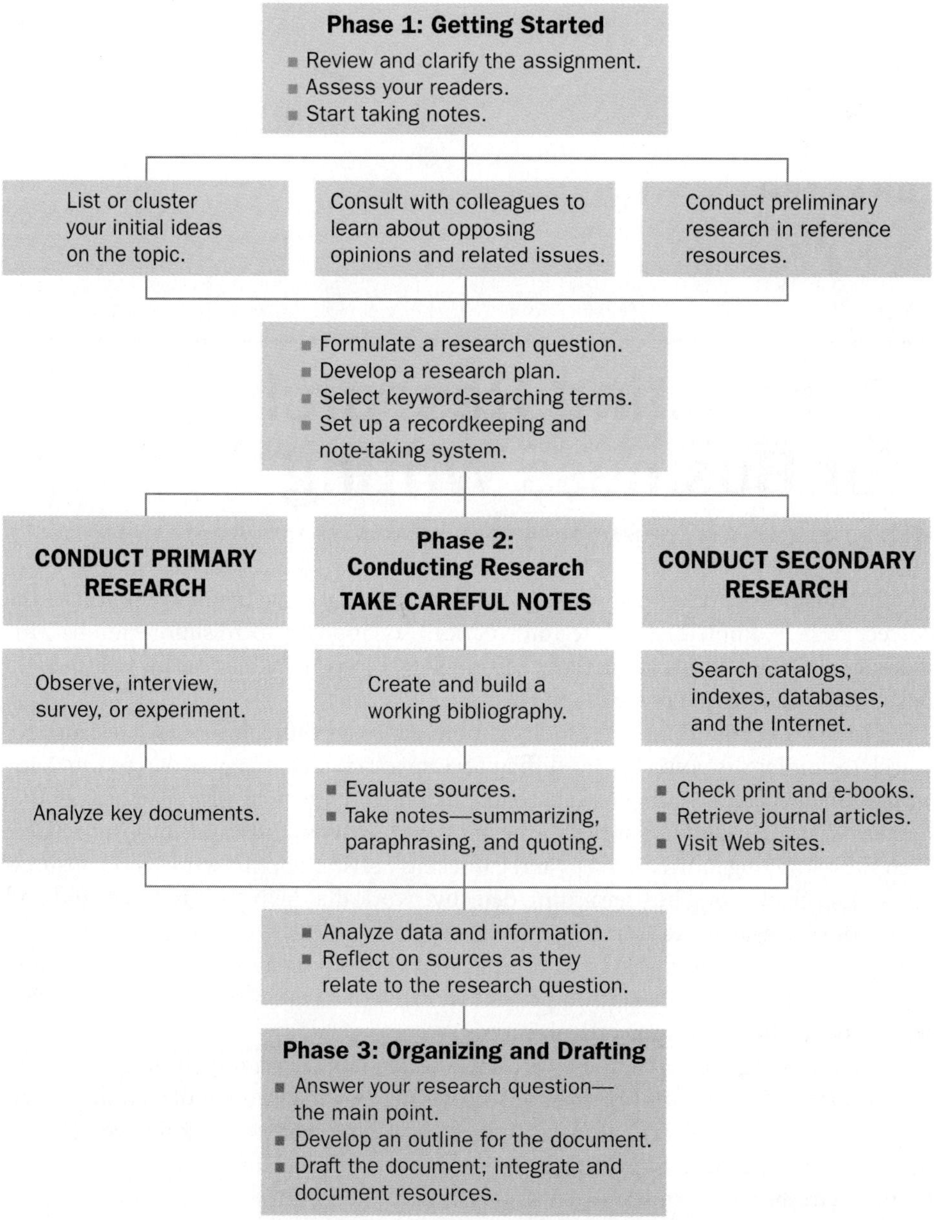

Phase 1: Getting Started
- Review and clarify the assignment.
- Assess your readers.
- Start taking notes.

- List or cluster your initial ideas on the topic.
- Consult with colleagues to learn about opposing opinions and related issues.
- Conduct preliminary research in reference resources.

- Formulate a research question.
- Develop a research plan.
- Select keyword-searching terms.
- Set up a recordkeeping and note-taking system.

Phase 2: Conducting Research
TAKE CAREFUL NOTES

CONDUCT PRIMARY RESEARCH
- Observe, interview, survey, or experiment.
- Analyze key documents.

- Create and build a working bibliography.
- Evaluate sources.
- Take notes—summarizing, paraphrasing, and quoting.

CONDUCT SECONDARY RESEARCH
- Search catalogs, indexes, databases, and the Internet.
- Check print and e-books.
- Retrieve journal articles.
- Visit Web sites.

- Analyze data and information.
- Reflect on sources as they relate to the research question.

Phase 3: Organizing and Drafting
- Answer your research question—the main point.
- Develop an outline for the document.
- Draft the document; integrate and document resources.

PLANNING YOUR RESEARCH

Planning research can save hours *doing* research. Planning involves understanding the assignment, considering resources, and distinguishing primary and secondary sources.

Understand the Assignment

Your assignment should direct your research. To understand the task required and the outcome desired, think through these four issues:

Purpose

What are you being asked to research? Why? By whom? What weight does the project carry in your work and in your organization? Consider these factors:

- **Topic.** What *should* you research, and what should you *avoid*?
- **Actions.** Are you expected to inform, persuade, analyze, compare and contrast, predict, evaluate, or recommend?
- **Key questions.** What questions must your research answer?
- **Scope.** What boundaries, benchmarks, and priorities do you need to follow?

Product

What should result from your research—a report, proposal, presentation, or Web site? Consider these issues as well:

- **Length:** The size of the document planned or needed
- **Format:** Margins, line spacing, type size, pagination, and so on.
- **Presentation of findings:** When and how the research will be shared; through what medium (electronic copy, paper, both); with what restrictions
- **Use of findings:** Decisions that will be made, changes that will result

People

Consider research participants, the audience, and those affected by results:

- Will specific people serve as resources, assistants, or research subjects? What needs to be done to involve participants productively and ethically?
- Who will read, view, or listen to the research results? What does this audience already know about the topic? What do they want or need to know?
- Who will be affected by the research, and how?

Process

How should the project unfold? Consider these issues:

- **Schedule:** What are due dates for project stages and completion?
- **Resources:** What equipment, personnel, and money are available?
- **Expectations:** What types of resources must you use? Are secondary sources sufficient, or should you conduct primary research such as interviewing?

Consider Different Information Resources

The information you need may be in a variety of resources. Consider the range of resources shown in Table 7.1. Which ones will give you the best data for your project?

Table 7.1 Types of Resources

Type of Resource	Examples	Case Study: Hybrid Cars
Personal, direct resources	Memories, journals, logs Experiments, tests, observations Interviews, surveys	Test drive of Toyota Prius Meeting with sales rep Survey of hybrid owners Interview of auto engineer
Workplace documents (e.g., business, trade, nonprofit)	Correspondence, reports Pamphlets, brochures, ads Instructions, manuals Policies and procedures Seminar and training materials Displays, presentations, demos Catalogs, newsletters, news releases	Manufacturer reports and brochures Hybrid model handbook or manual Manufacturer presentations promoting hybrid vehicles Company news releases about hybrid models
Audiovisual, digital, and multimedia resources	Graphics (tables, graphs, charts, maps, drawings, photos) Audiotapes, CDs, videos, DVDs Web pages, online databases, computer software, CD-ROMs	Diagrams of hybrid engines Manufacturer's Web pages (e.g., www.honda.com) Tessa Combs, *Travelling Light* (MediaMatters video)
Books (print and electronic)	Nonfiction, how-to, biographies Fiction, trade books, scholarly and scientific studies	Jim Motavalli, *Forward Drive: The Race to Build "Clean" Cars for the Future* (San Francisco: Sierra Club books, 2001)
Reference works (print and electronic)	Dictionaries, thesauruses Encyclopedias Almanacs, yearbooks, atlases Directories, guides, handbooks Indexes, abstracts, bibliographies	Bureau of Transportation Statistics (www.bts.gov) *The Beaulieu Encyclopedia of the Automobile*
Periodicals and news sources	Print newspapers, magazines, and journals Broadcast news and news magazines Online magazines, news sources, and discussion groups	Kerry Dolan, "Emissions-Free Investing." *Forbes* Dec. 22, 2003; Vol. 172, issue 13, p. 218.
Government publications	Guides, programs, forms Legislation, regulations Reports, records, statistics	Zero Emissions Vehicle Mandate (California Air Resources Board)

Consider Different Information Sites

Where do you go to find the resources that you need? Consider the information "sites" listed in Table 7.2.

Table 7.2 Information Sites

Information Location	Specific "Sites"	Case Study: Hybrid Cars
People	Experts (knowledge area, skill, occupation)	Engineers, sales staff, mechanics
	Population segments or individuals (with representative or unusual experiences)	Hybrid owners
		Staff presently using company cars
Workplace	Computer databases, company files	Company records of vehicles: purchase, maintenance, fuel costs, depreciation
	Desktop reference materials	
	Bulletin boards (physical and electronic)	www.honda.com (Honda site—links to Civic Hybrid)
	Company and department Web sites	
	Departments and offices (e.g., customer service, public relations)	www.gm.com (General Motors site—links to AUTOnomy project)
	Associations, professional organizations	U.S. Fuel Cell Council (www.usfcc.com)
	Consulting, training, and business information services (e.g., Better Business Bureau, Chamber of Commerce)	Society for Automotive Engineers (www.sae.org)
Computer resources	Computers: software, disks, CD-ROMs	What's New @ IEEE Online
	Networks: Internet and other online services (e-mail, limited-access databases, discussion groups, MUD, chat rooms, Web sites, Weblogs), intranets	Lii.org
		LexisNexis (business databases)
		www.howstuffworks.com
Libraries	General: public, college, online	WorldCat/FirstSearch
	Specialized: legal, medical, government, business	Library of Congress
		VERA (Virtual Electronic Resource Access): MIT libraries
Municipal, state, and federal government offices	Elected officials, representatives	Environmental Protection Agency, Department of Energy, Department of Commerce
	Offices and agencies	
	Government Printing Office (GPO, www.gpoaccess.gov)	Census Data Online (www.census.gov)
	Web sites	FirstGov (www.firstgov.gov)
Media	Radio (AM and FM)	*CNNfn*
	Television (network, public, cable, satellite)	*Scientific American*
	Print (newspapers, magazines, journals)	*Journal of Automobile Engineering*
Testing, training, meeting, and observation sites	Plants, facilities, field sites	Automotive and Transportation Technology Congress and Exhibition
	Lab, research centers, universities, think tanks	Hybrid Vehicles Symposium
	Conventions, conferences, seminars	Argonne National Laboratory (Argonne, IL)
	Museums, galleries, historical sites	

Distinguish Primary and Secondary Resources

Information sources for your project can be either primary or secondary. Depending on your assignment, you may be expected to use one or both kinds of sources. Here's a helpful rule of thumb: With your deadline and budget in mind, gather the information that is most relevant and most reliable for your project.

Primary Sources

A primary source is an original source, one that gives firsthand information on a topic. This source (such as a log, person, or event) informs you directly about the topic, not through another person's explanation. Common forms of primary research are observations, interviews, surveys, experiments, and analyses of documents and artifacts.

Upside: Produces information precisely tailored to your research needs; gives you direct, hands-on access to your topic.

Downside: Can take much time and many resources, as well as specialized skills (designing surveys, analyzing statistics, conducting experiments).

Secondary Sources

Secondary sources present information on your topic at least once removed from the original. The information has been compiled, summarized, analyzed, synthesized, interpreted, and evaluated by someone studying primary sources. Journal articles, encyclopedia entries, documentaries, and nonfiction books are typical examples.

Upside: Can save much research labor and provide extensive data; offers expert perspective on and analysis of topic.

Downside: Can require extensive digging to find precisely relevant data; information may be filtered through a bias; original research may be faulty.

If you were researching the potential of hybrid-car technology, the following would be examples of primary and secondary sources.

Primary sources	Secondary sources
E-mail interview with automotive engineer	Journal article discussing the development of hybrid cars
Fuel-efficiency legislation	Newspaper editorial on fossil fuels
Visit and test drive at Honda dealership	TV news magazine roundtable discussion of hybrid-car advantages and disadvantages
Published statistics about hybrid car sales	Automobile manufacturer promotional literature for a specific hybrid car

F.Y.I.

Whether a source is primary or secondary depends on what you are studying. For example, if you were studying U.S. attitudes toward hybrid cars (and not hybrid-car technology itself), then the newspaper editorial and TV roundtable discussion would be primary sources.

MANAGING YOUR PROJECT: NOTE-TAKING STRATEGIES

For a research project to proceed smoothly, you need to manage tasks effectively. The next few pages cover management strategies from keyword searching to taking notes.

Conduct Effective Keyword Searches

With the vast range of research tools now available (e.g., library catalogs, subscription databases, the Web, e-books), what you find depends on which keywords you select and how well you design your search. Keywords give you "compass points" for navigating a sea of information, and the right keywords will generate quality results. Consider these tips:

1. **Use a shotgun approach.** Brainstorm likely keywords for your project and choose the best term for your initial search. If you get no "hits," choose a related term. Once you get some hits, check the citations (especially subject terms) for clues about which words to use.
2. **Use Boolean operators to refine your search.** Often, combining keywords by using Boolean logic leads to better research results (see Table 7.3).

Table 7.3 Boolean Logic and Keyword Searching

Refinement	Example	Explanation
Narrowing a Search with And, +, not, −		
Use when one term gives you too many hits, especially irrelevant ones.	*hybrid and car* Or *+hybrid +car*	Searches for citations containing both keywords
	hybrid not corn or *+hybrid -corn*	Searches for citations containing "hybrid" but excludes those with "corn"
Expanding a Search with Or		
Combine a term providing few hits with a related word.	*hybrid or electric*	Searches for citations containing either term
Specifying a Search Phrase with Quotation Marks		
Indicate that you wish to search for an exact phrase.	*"fuel cells"*	Searches for the exact phrase "fuel cells"
Sequencing Operations with Parentheses		
Indicate that the operation should be performed first before other operations in the search string.	*(hybrid or electric) and (car or automobile)*	Searches first for citations containing either "hybrid" or "electric," then checks results for either "car" or "automobile"
Finding Variations of a Word with Truncation and Wild Card Symbols		
Depending on the database, symbols such as $, ?, or # will find variations of a word.	*electri#*	Searches for terms like *electric, electrical, electricity, electrify, electrifying*

Develop a Working Bibliography

A working bibliography lists sources that you have used and that you intend to use. It helps you track research and develop your final bibliography. Here's what to do:

Choose an orderly method. Consider topics like these:
- **Paper note cards.** Use 3- × 5-inch cards, and record one source per card.
- **Paper notebook.** Use a spiral-bound book to gather source information.
- **Computer program.** Record source information electronically, either by capturing citation details from online searches or by recording bibliographic information using word-processing software or research software such as TakeNote, EndNote Plus, or Bookends Pro.

Include complete identifying information for each source. Consider recording this information in the format required by the documentation system you are using (e.g., APA). Doing so now will save time later.
- **Books:** Author, title and subtitle, publication details (place, publisher, date)
- **Periodicals:** Author, article title, journal name, publication information (volume, issue, date), page numbers
- **Online sources:** Author (if available), document title, sponsor, database name, publication or posting date, access date, other publication information
- **Primary or field research:** Date conducted; name and/or descriptive title of person interviewed, place observed, survey conducted, document analyzed

Add locating information. For books, include the call number; for articles, where and how you accessed them (stacks, current periodicals, microfilm, database); for field research, a telephone number or e-mail address; for Web pages, the address.

Annotate the source. Add a note about the source's content, perspective, focus, and usefulness.

Sample Bibliography Note

Code	#5
Source information	Ogden, J. M., Williams, R., & Larson, E. (2004, January). "Societal lifecycle costs of cars with alternative fuels/engines." *Energy Policy, 32*(1), 7–28.
Location	Interlibrary loan
Annotation	Effectively analyzes through compare/contrast "lifecycle costs" of different types of transportation: traditional internal-combustion, hybrids, diesel, etc. Reliable source—affiliated with Princeton Environmental Institute

Evaluate Sources

As you begin working with sources, you need to test their reliability. Using trustworthy sources strengthens the credibility of your research and writing.

Is the author an expert? An expert is an authority—someone who has mastered a subject area. Is the author an expert *on this topic*?

> **WEB TEST:** Does a subject directory recommend this site? Who sponsors it? Are the author's name, credentials, and contact information available?

Is the source current? A five-year-old book on computers may be ancient history, but a book on Abraham Lincoln could be forty years old and still be the best source available.

> **WEB TEST:** Is the site up-to-date? When was it created and last updated? Are links current?

Is the source accurate and logical? Check for obvious factual errors, analysis that doesn't make sense, conclusions that don't add up, and poor grammar.

> **WEB TEST:** Is material easy to understand, with useful rather than decorative multimedia elements? How are claims backed up? Can you trace the sources used? Is language used well or poorly?

Does the source seem complete? Does the discussion seem full, or do you sense major gaps? Have you been given both sides of the argument?

> **WEB TEST:** Does the site seem thin? Review the topics covered, their depth, and the site's internal and external links.

Is the source unbiased? *Bias* means "a tilt to one side." While all sources come from a specific perspective, a *biased* source may be pushing an agenda. Watch for bias toward a region, country, political party, industry, gender, race, ethnic group, religion, or philosophy. Ask why this source was created.

> **WEB TEST:** Is the site a nonprofit (.org), government (.gov), commercial (.com), educational (.edu), business (.biz), informational (.info), network-related (.net), or military (.mil) site? Is it a U.S. or international site? In other words, what type of organization sponsors the site? What are its mission, agenda, and interests? Is it pushing a cause, product, service, or belief? Can you "follow the money"?

How does the source compare? Does it conflict with other sources? Do sources disagree on the facts themselves or on how to interpret the facts?

> **WEB TEST:** Is the site easy to get in and move around? Is the site's information logically consistent with print sources?

WRITE Connection

For more help with evaluating sources, check out Chapter 8, "Business Writing Ethics," especially "Avoiding Logical Fallacies" (pages 175–177).

Develop a Note-Taking System

Accurate, thoughtful notes create a foundation for your research writing. To take good notes, you must be selective—recording only information that is related to your research question. The key is to develop an efficient note-taking system. A good system, like the four outlined here, will help you to keep accurate records, distinguish between sources, record source material correctly, and explore your own ideas.

Copy and Annotate

This method involves working with photocopies or print versions of sources.

1. Selectively photocopy or print articles, chapters, and Web pages. Copy carefully, making sure that you have the full page, including page number.
2. Add identifying information right on the copy—author, publication details, and date. Each page should be easy to identify.
3. As you read, highlight key statements. In the margins, record your ideas by asking questions, making connections, adding your own response, defining key terms, and creating an index of topics.

Note Cards

Using paper note cards is the traditional method of note taking; however, note-taking software is now available with most word-processing programs and special programs like TakeNote, EndNote Plus, and Bookends Pro. After establishing one set of cards for your working bibliography (see page 124), take notes on a second set.

1. Record one point from one source per card.
2. Clarify the source: List the author's last name, a shortened title, or a code from the matching bibliography card. Include a page number.
3. Provide a topic or heading: Called a *slug*, the topic helps you categorize and order information.
4. Label the note as a summary, paraphrase, or quotation of the original.
5. Distinguish between the source's information and your own thoughts, perhaps by using your initials beside your responses.

Slug	PROBLEMS WITH INTERNAL-COMBUSTION CARS
Quotation Page number	"In one year, the average gas-powered car produces five tons of carbon dioxide, which as it slowly builds up in the atmosphere causes global warming." (p. 43) (Quote)
Comments	AM: Helpful fact about the extent of pollution caused by the traditional i-c engine
	AM: How does this number compare with what a hybrid produces?
Source	#7

Computer Notebook or Research Log

This method involves note taking on a computer or on sheets of paper.

1. Establish a central location for your notes—a notebook, a file folder, a binder, or an electronic folder.
2. Take notes on one source at a time, making sure first to identify the source fully. Number your note pages.
3. When recording your own thoughts, distinguish them from source material by using your initials or some other symbol.
4. When it's time to outline your paper, use codes to go through your notes and identify what information relates to which topic in your outline. Under each topic in your outline, write the page number in your notes where that information is recorded.

Double-Entry Notebook

Using a double-entry notebook or the columns feature of your word-processing program, do the following:

1. Divide notebook pages in half vertically.
2. In the left column, record bibliographic information and take notes on sources.
3. In the right column, write your responses. Think about what the source is saying, why the point is important, whether you agree with it, and how the point relates to other ideas.

Motavalli, Jim. (2001). *Forward Drive: The Race to Build "Clean" Cars for the Future.* New York: Sierra Club Books.

Ch. 2, "A Dizzying Ride: Internal Combustion's Rapid Rise and Coming Decline"	Motavelli's perspective interesting—obviously loves cars, but now concerned about impact and future. What's his perspective on the automobile industry—fair or not?
Author's love affair with car before environmental concerns, traffic issues became clear—cars and freedom (18)	
1996: 100th anniversary of the car (19)	Good to realize that the industry has constantly been faced with change, that it's an industry that is barely a century old!
"We are literally choking to death on our enduring love affair with the gasoline-powered car." (19)	Here's the issue—our connection with the gas car is more than practical; it's personal and cultural. Change is needed—culturally and in terms of business practice—but change is tough without vision, entrepreneurship, and the promise of practical results.
Stats: —Since 1969, car population has grown 6 × faster than human pop in US —US has 5% of world's pop but 34% of world's cars —US drivers travel 2 trillion miles annually (19)	Stats are helpful for getting a perspective on the size of the problem.

Summarize, Paraphrase, and Quote Source Material

As you work with sources, you need to determine what to put in your notes and how to record it—as a summary, paraphrase, or quotation. The passage below comes from an article on GM's development of hybrid-car and fuel-cell technology. After reading the passage, note how the researcher summarizes, paraphrases, and quotes from the source.

> **From Burns, L. D., McCormick, J. B., & Borroni-Bird, C. E. (2002, October). Vehicle of change.** *Scientific American, 287*(4), 10 pp.
>
> Now another revolution could be sparked by automotive technology: one fueled by hydrogen rather than petroleum. Fuel cells—which cleave hydrogen atoms into protons and electrons that drive electric motors while emitting nothing worse than water vapor—could make the automobile much more environmentally friendly. Not only could cars become cleaner, but they could also become safer, more comfortable, more personalized—and even perhaps less expensive. Further, these fuel-cell vehicles could be instrumental in motivating a shift toward a "greener" energy economy based on hydrogen. As that occurs, energy use and production could change significantly. Thus, hydrogen fuel-cell cars and trucks could help ensure a future in which personal mobility—the freedom to travel independently—could be sustained indefinitely, without compromising the environment or depleting the earth's natural resources.

Summarizing condenses in your own words the main points in a passage. Summarize when the source provides relevant ideas and information on your topic.

1. Reread the passage, jotting down a few key words.
2. Without looking back to the original, state the main point in your own words. Add key supporting points, leaving out examples, details, and long explanations.
3. Check your summary against the original, making sure that you use quotation marks around any exact phrases that you use from the original. Also, distinguish anything that is your opinion on or response to the original.

> A dramatic social change is now taking place in the shift from gas engines to hydrogen technologies. Fuel cells may make the car "greener," and perhaps even more safe, comfortable, unique, and cheap. These automotive changes will affect the energy industry by making it more environmentally friendly; as a result, people will continue to enjoy mobility while transportation moves to renewable energy.

WEB Link

For help with writing summaries and abstracts of whole documents, go to http://college.hmco.com/english.

Paraphrasing puts a whole passage *in your own words*. Paraphrase when you need to clarify a particularly important passage.

1. Quickly review the passage to get a sense of the whole.
2. Go through the passage carefully, sentence by sentence.
   - State ideas in your own words, simplifying and defining words as needed.
   - If necessary, expand the text to make it clearer, but don't change the meaning.
   - Do not borrow phrases directly unless you put them in quotation marks.
3. Check your paraphrase against the original for accurate tone and meaning.

> Automobile technology may lead to another radical economic and social change through the shift from gasoline to hydrogen fuel. By breaking hydrogen into protons and electrons so that the electrons run an electric motor with only the by-product of water vapor, fuel cells could make the car a "green" machine. This technology could also increase the automobile's safety, comfort, personal tailoring, and affordability. Moreover, this shift to fuel-cell engines in automobiles could lead to drastic, environmentally friendly changes in the broader energy industry, one that will be now tied to hydrogen rather than fossil fuels. The result of this shift will be radical changes in the way we use and produce energy. In other words, the shift to hydrogen-powered vehicles could maintain our society's valued mobility, while the clean technology would preserve the environment and its natural resources.

Quoting records the original source word for word, within quotation marks. Quote nuggets—statements that are well phrased, to the point, or authoritative.

1. Note the quotation's context—how it fits into the author's overall discussion.
2. Copy the phrase, sentence, or passage word for word—carefully. Enclose the original in clearly visible quotation marks.
3. If you leave out any words, note that omission with an ellipsis (. . .).
4. Check your quotation for accuracy.

> "[H]ydrogen fuel-cell cars and trucks could help ensure a future in which personal mobility . . . could be sustained indefinitely, without compromising the environment or depleting the earth's natural resources."

Note: This sentence is a good quotation because it captures the authors' main claim about benefits and their projection about where fuel-cell technology is going.

F.Y.I.

Whenever possible, include a page number, paragraph number, or other locating detail with your paraphrase, summary, or quotation.

DOING PRIMARY RESEARCH

Primary research gathers firsthand information. Consider doing such research if this information is central to your project, or if secondary information is incomplete, out-of-date, or difficult to find. (Check page 122 for more on the distinction between primary and secondary sources.) Check the options below, select forms useful for your project, and review the guidelines on the pages that follow.

Observations, inspections, and field research require systematic examination and analysis of places, spaces, scenes, equipment, work, events, and so on. Such research provides precise data on the state of materials and processes.

Interviews involve consulting people who are experts on your topic, or people whose experiences with the topic can supply useful information. Such research can lend authority and perspective to your research report.

Surveys, forms, and inquiry letters gather information as written responses you can review, tabulate, and analyze. Such research can conveniently pull together varied information, from simple facts to personal opinions and attitudes.

Analyses of documents, records, and artifacts involve studying original correspondence, reports, statistics, legislation, and so on. Such research often provides direct evidence, key facts, and in-depth insights into the topic.

Experiments test hypotheses—predictions about why things happen as they do—so as to arrive at conclusions that can be accepted and acted upon. Such testing often explains cause/effect relationships for varied projects, including problem–solution proposals and product research and development.

WEB Links

For an introduction to experimentation and experiment reports, go to http://college.hmco.com/english. For a sample test report related to hybrid cars, check EPA report 420R98006, "Evaluation of a Toyota Prius Hybrid System (THS)," available online at www.epa.gov.

Make Observations

Observing people, situations, processes, things, and places is a common method of workplace research. Such research, which involves watching, listening, recording, counting, or analyzing, may be especially useful when you are writing reports about site inspections, incidents, and trips. Here are some guidelines for making observations.

Before Observation

1. Know what you want to accomplish. Do you need to understand a place or a process? Solve a problem? Answer a question?
2. Consider your vantage point or perspective.
   - Should you observe the site passively or interact with elements of it?
   - Should you remain objective, or include analysis and impressions?
3. Do your homework before you go. For example, if you are visiting a fuel-cell laboratory, read up on the principles behind fuel-cell technology.

4. Make observation productive through careful planning:
   - Create a list of specific things to look for.
   - Ask permission if the site isn't public.
   - Ensure safety if the site involves dangerous conditions.
   - Get the timing right (choosing a good time to visit, setting aside enough time, and planning multiple visits, if needed).
   - Arrange appropriate transportation, clothing, equipment, and assistance (considering the site, weather, data-collection activities, and danger).
5. Gather reliable observation tools that will help you collect the data you need. Besides your eyes and ears, consider bringing these items:
   - Note-taking tools (pens, pencils, paper; field journal; laptop)
   - Recording equipment (camera, tape recorder, sketch book)
   - Measuring equipment (maps, counter, tape measure, watch, scale)
   - Other tools (containers for samples, flashlight, geologist's hammer)

During Observation

1. Follow your plan, but also let the situation shape your observing so that you remain open to surprises.
2. Identify your position. Where are you in the site? What is your angle? More broadly, what is your stance here, and does your presence change the site?
3. Pay attention to the big picture (context, time, and surroundings), but keep focused on the subject by filtering out unnecessary details.
4. Take notes on details and impressions—the sounds a machine makes, its motions, and the way people interact with it, for example. If appropriate, focus on your five senses (sights, sounds, smells, textures, tastes).
5. Gather other forms of evidence. Take measurements, use recording technology, gather samples, interview people, ask for brochures—do whatever you can to collect material for further study and analysis.

After Observation

1. Soon after your observing, look over your notes and evidence for key ideas and patterns in the information you've gathered.
2. List conclusions about what you've observed, along with supporting details from the data that your observations generated.
3. Relate your observations to other research. Explore how your observations confirm, contradict, complement, or build on other sources of information about your topic. Assess the overall value of the observation for your research and ways that the evidence gathered can be used.

WEB Link

For a sample field report rooted in observation, go to
http://college.hmco.com/english.

Conduct Interviews

The idea of an interview is simple. You talk with someone who has expert knowledge or who has had important experiences with your topic. For example, to learn about hybrid vehicles, you might talk with an automotive engineer or a hybrid owner. Such interviews may involve one question or dozens; they may take place in person, by phone, or via e-mail. To conduct productive interviews, follow these guidelines:

Preparing for the Interview

1. Think about the topic. Consider what you already know and what you need to know. Aim to arrive at the interview well informed so that you can build on that knowledge.
2. Consider the person. Think about your relationship and the person's expertise or experiences. How can you encourage the individual to share useful, vital information and insights?
3. When you arrange the interview, be thoughtful. Consider the interviewee's convenience in terms of time, place, and medium (person-to-person, telephone, e-mail). Identify yourself, your purpose, the process, and the topics you would like to cover.
4. Develop a set of questions. This exercise will give the interview some structure and help you get the information and ideas you need.
   - Brainstorm questions using: *who, what, when, where, why,* and *how.*
   - Understand open-ended and closed questions. Closed questions ask for simple, factual answers; open-ended questions ask for an explanation.

     CLOSED: How did you vote on the recent budget proposal?

     OPEN-ENDED: What were the key issues that members of Congress debated during the recent budget sessions?
   - Avoid slanted or loaded questions that suggest you want a specific answer.

     SLANTED: Don't you agree that liberal politicians are overspending?

     BETTER: What are some key differences between liberal and conservative political positions on economic issues?
   - Think about the specific topics you want to cover in the interview, and draft related questions for each topic.
   - Sequence questions so that the interview moves smoothly from one subject to the next. Start with a question that will establish rapport.
   - Write the questions on the left side of the page. Leave room for quotations, information, and impressions on the right side.
   - Highlight key questions—ones that you *must* ask.
5. Rehearse the questions, visualizing how the interview might go.
6. Be prepared. Take pens and paper. If you plan on taping the interview, get permission ahead of time and take along a tape recorder and camera. In addition, give yourself enough time to find the location if you are unfamiliar with the area.

Doing the Interview

1. From the start, be natural, sincere, and professional.
   - Respect the interviewee's time by being on time, efficient, and courteous.
   - Relax so that both you and the interviewee can be open.
   - Introduce yourself, remind the interviewee why you've come, thank the person for the interview, and provide helpful background.
   - Listen much more than you talk, and avoid impolite interruptions.
2. Create a record of the interview.
   - If the person gives permission, tape the interview and take pictures. Use a recorder with a counter so that when the interviewee makes an interesting or important point, you can jot down the number.
   - Write down key facts, quotations, and impressions.
3. Listen actively, using encouraging body language (nods, smiles, and eye contact). Pay attention not only to what the person says but also to how he or she says it—word choice, connotations, context, and body language.
4. Be tactful and flexible. If the interviewee avoids a difficult question, politely rephrase it or ask for clarification. If the person looks puzzled by a question, rephrase it or ask another question. If the discussion gets off track, gently redirect it. But don't limit yourself to just your planned questions. Instead, ask pertinent follow-up questions, like these:
   - **Clarifying:** Do you mean . . . ?
   - **Explanatory:** What do you mean by . . . ?
   - **Detailing:** What happened exactly? Can you describe that?
   - **Analytical:** Did that happen in stages? What were the causes?
   - **Probing:** What do you think that meant?
   - **Comparative:** How was that similar to or different from . . . ?
   - **Contextual:** What else was going on? Who else was involved?
   - **Summarizing:** Overall, what was the net effect?
5. End positively. Conclude by asking if the interviewee wants to add or emphasize anything. (*Note:* Important points may come up late in the interview.) Thank the person and part with a handshake.

Following Up on the Interview

1. As soon as possible, review your notes and fill in responses you remember but couldn't record at the time. If you recorded the interview, replay it.
2. Analyze the results. Study the information, insights, and quotations you gathered. What do they reveal about the topic? How does the interview confirm, complement, or contradict other sources on the topic?
3. Contact the interviewee by note, e-mail, or phone call to offer thanks, to confirm statements and details, and to offer to share the outcome of the interview—the research report, for example.

Conduct Surveys

Surveys are useful tools for collecting primary data—ranging from simple facts to opinions to statistics—on virtually any topic. An effective survey gathers this information efficiently because it persuades respondents that the survey is important, asks clear questions, and uses an easy-to-complete format. Follow the guidelines below.

F.Y.I.
Gathering scientifically reliable data from surveys requires special skills in methods and statistics. If you need statistically significant data, seek expert help. The material that follows simply allows you to do informal surveys.

1. **Do some planning.** Sort out your initial thinking on these issues:
   - **Purpose.** What is the limited, measurable goal of your survey? How will a survey be helpful for your project?
   - **Types of information.** Think about the data your survey might collect. Can you gather opinions, simple and clear facts, hard statistics?
   - **Respondents.** What target audience can provide the needed information? What information can they offer that you can't find elsewhere? How can you obtain an appropriate, reliable sampling from this group?
   - **Context.** What would be the best way to distribute and collect the survey—regular mail, e-mail, in person? How soon do you need responses? What conditions would encourage objective responses from respondents? What methods would encourage a high rate of return?

2. **Ask relevant, reliable, and valid questions.** *Relevant* questions clearly relate to the survey's topic. *Reliable* questions are completely clear in meaning. *Valid* questions give you exactly the type of data you want. To write such questions, do the following:
   - Phrase questions in language that readers understand, with answer options that are both complete and without overlap.
   - Keep questions objective and manageable for readers—not broad, slanted, time-consuming, or complicated.
   - Don't ask the same question twice, and cover one item at a time.

3. **Choose questions that fit your purpose.** Your survey can include open-ended questions and closed questions.
   - **Open-ended questions** ask respondents to provide extensive written responses. While such questions can collect in-depth and varied information, responses take time to complete and can be difficult to summarize and code during tabulation.
   - **Closed questions** give respondents a limited number of clearly distinguished answer options. They are easy to answer and tabulate although they can take a variety of forms: two-choice questions (yes or no, true or false), multiple-choice questions, rating scale and ranking questions, and short-answer questions.

4. **Organize your survey with the respondent in mind.** The following features make a survey easy to complete:
   - The title suggests the survey's purpose.
   - The introduction persuades and instructs by (1) stating who you are and who has authorized the survey; (2) indicating why you need the information; (3) offering benefits for completing the survey; and (4) explaining how to complete and return it, by when.
   - Numbers, instructions, and format guide readers through the survey.
   - Questions are grouped logically—moving smoothly from one topic to the next, and often beginning with basic, closed questions and ending with more complex, open-ended questions.
   - The survey is a reasonable length for these readers.
5. **Test your survey.** Ask colleagues to provide feedback to the drafted survey. To debug it, do a trial run with a small group representing the target group.
6. **Distribute your survey using an appropriate method.** Think about your target audience and the information you want. For example, do you need responses from an equal number of males and females? Depending on such factors, choose one of the following distribution methods:
   - **Systematic random sampling.** Choose a cross section of people from the target audience (e.g., by picking every fifth colleague from your company directory).
   - **Stratified random sampling.** If your target audience includes a number of subgroups, pick a proportionate cross section of people from each group (e.g., a proportionate number of colleagues from specific ethnic groups).
   - **Quota sampling.** Choose a number of representative individuals from the group, without using any particular system (e.g., the first 25 customers who call the service department).
7. **Tabulate the results and draw conclusions.** When responses come in, tally results manually or by computer. If tallying manually, try this method:
   - Set up a separate tally sheet for each question on your survey. Write the question at the top, and set up separate columns for (a) the respondent (either the person's name or a number code for that completed survey), (b) each answer option, and (c) a column for "no response."
   - Listing your respondents by code and/or name in the left column, check off responses. When you have checked that question on each returned survey, total the responses and calculate percentages.
   - Study the numbers: What do they mean? What are the trends?
   - With any open-ended questions, study the range of responses: What insights do they offer? What patterns emerge?

WEB Link

For a sample survey, go to http://college.hmco.com/english.

Analyze Documents, Records, and Artifacts

An original document or record is one that directly grew out of the phenomenon that you are researching. Examining original documents can involve studying letters, e-mail exchanges, case notes, customer files, project reports, sales records, legislation, and much more. As you analyze such records, you examine evidence in an effort to understand a topic, arrive at a coherent conclusion about it, and support that judgment. Here are some guidelines for working with such diverse documents and records.

Frame Your Examination with Questions

To make sense of the document, record, or artifact, understand what you are exploring and why. List pertinent questions that relate to the main question of your research project. For example, to study the legislative background behind the development of cleaner cars, such as the hybrid vehicle, you could access various documents on the Clean Air Act of 1990 (e.g., *The Plain English Guide to the Clean Air Act,* an EPA publication). You could frame your reading with these questions: What are the requirements of the Clean Air Act? How do those requirements impact automotive technology? Are schedules for change or deadlines written into the act?

Put the Document or Artifact in Context

So that the material takes on meaning, clarify the external and internal nature of the document. First, consider its external context—the five W's and H: What exactly is it? Who made it, when, where, why, and how? Second, consider its "internal criticism"— what the document means, based on what it can and cannot show you. What is the meaning of the language and to what does it refer? What is the structure of the document? What are its main points or ideas? Put your study of the Clean Air Act in context with questions such as the following.

> **EXTERNAL CONTEXT:** What type of legislation is this? Who initiated the bill, and why? What were the issues that the bill sought to address in the 1980s? What has happened to the act since it was passed? Who does the act affect, and how? What are the polities behind the act?
>
> **INTERNAL CONTEXT:** What are the different parts of the act? What does the legal language mean in plain English? What does the act require of businesses, industries, and citizens? Do these requirements make sense? Do they have "teeth"? Are they fair and sound?

Draw Coherent Conclusions About Meaning

Based on an appropriate analysis of the source, make sense of its meaning in relation to your research topic and questions. What connections does the source reveal? What important changes or developments? What cause/effect relationships? What themes? With what voice does it speak? (Arriving at logical conclusions involves using complex thinking skills. Check pages 151 and 194 for more on developing logical ideas.)

Your study of the Clean Air Act might lead you to a whole range of conclusions about the relationship between environmental legislation and the development of hybrid technology—for example, that both the automobile industry and U.S. citizens must turn to clean cars to ensure acceptable air quality and deter global warming.

DOING LIBRARY RESEARCH

A good library (whether public, college, or specialized) is your doorway to print and electronic information. Inside its doors, the library provides a wide range of research resources, from books to reference librarians to Internet access. To get the most out of a library, learn to use its catalogs, tap reference works, and find periodical articles.

Search the Library Catalog

In most college libraries, books, videos, and other holdings are catalogued in an electronic database organized by the Library of Congress classification system. To find material, learn to do basic and advanced searches of the online catalog. As you search, practice the keyword-searching strategies found on page 123.

Study the Start-up Screen

Figure 7.2 shows a typical start-up screen for a Web-based catalog. Notice the different features that allow you to develop a successful search.

Build on Your Initial Search Results

An initial keyword search will typically lead to a list of brief citations. This list will offer you a summary of your search results (e.g., the number of "hits"), as well as print and capture choices and view features. Each brief citation might contain the resource's locating call number, author, title, and library status (available or checked out). Review this information to decide on the relevance of the results for your project, to expand or narrow your search through Boolean operators, to explore alternative keywords if the results seem unrelated to your topic, and to mark citations to explore further.

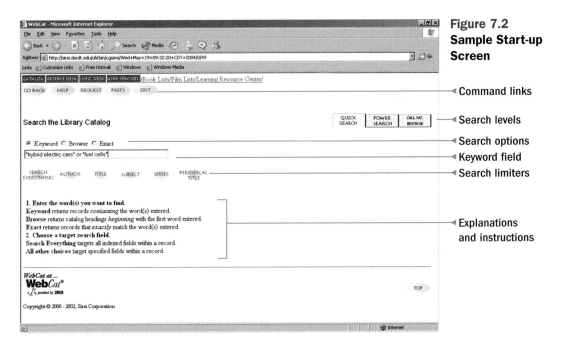

**Figure 7.2
Sample Start-up Screen**

Study Full Citations

When you find a citation for a book or other resource, some or all of the information shown below will be provided. Use that information (1) to determine whether the resource is worth exploring further, (2) to locate the resource in the library, and (3) to discover other research pathways.

❶ Author's or editor's name

❷ Title and subtitle

❸ Publisher and copyright date

❹ Descriptive information

❺ Subject headings (crucial list of topics)

❻ Call number

❼ Format/Medium

❽ Location

TL221.15 .M68 2001
Forward drive : the race to build "clean" cars for the future
Motavalli, Jim.

Personal author:	Motavalli, Jim. ❶
Title:	Forward drive : the race to build "clean" cars for the future / Jim Motavalli. ❷
Publication info:	San Francisco : Sierra Club Books, c2001. ❸
Physical descrip:	xxiv, 280 p. ; 21 cm. ❹
Subject term:	Hybrid-electric cars.
Subject term:	Fuel cells.
Subject term:	Automobiles, electric.
Subject term:	Automobiles—Environmental aspects. ❺
Subject term:	Automobiles—Technological innovations.
Subject term:	Automobiles—Fuel systems.

Call Numbers		Copy	Material ❼	Location ❽
1)	TL221.15 .M68 2001 ❻	1	BOOK	On-shelf

Locate Resources Using Call Numbers

Library of Congress call numbers combine letters and numbers to specify a resource's broad subject area, topic, and authorship or title. Therefore, finding a resource involves combining the alphabetical and numerical orders. Here is a sample call number for *Forward Drive: The Race to Build "Clean" Cars for the Future:*

> TL221.15.M68 2001
> Subject area (TL); topic number (221); subtopic number (15); cutter number (M68)

To find this resource in the library, follow the core elements one at a time:

1. Find the "TL" section (motor vehicles, aeronautics, and astronautics).
2. Follow the numbers until you reach "221."
3. Within the "221" items, find those with the subtopic "15."
4. Use the cutter "M68" to locate the resource alphabetically with "M" and numerically with "68."

In the Library of Congress system, pay careful attention to subject area letters, topic numbers, and subtopic numbers: T416 comes before TL221; TL221 before TL1784; TL221.M98 before TL221.15.M68.

Search Other Catalogs

Your library may offer access to other online catalogs that allow you to search for resources that are not available in your local library, but that might be available electronically or through interlibrary loan. Check for links to the following.

NetLibrary

Your library may link you with this e-book collection. With more than 7,500 professional, scholarly, and reference e-books and more than 3,400 publicly accessible e-books, NetLibrary offers fully searchable books to "browse" or "sign out."

State Libraries

Your library may provide links to a statewide system of libraries, giving you access to state, county, and city information.

National, Global, and Specialized Libraries

Your library may connect you with catalogs such as these:

- The Library of Congress catalog (www.loc.gov).
- Advanced research libraries specializing in science and technology, such as the MIT Libraries (http://libraries.mit.edu).
- Business libraries, such as the Harvard Business School's Baker Library (www.library.hbs.edu/).
- OCLC's FirstSearch WorldCat, a catalog of materials in libraries worldwide and on the Internet. The search screen in Figure 7.3, illustrates the power and flexibility made available by searching WorldCat.

Figure 7.3 WorldCat Search Screen

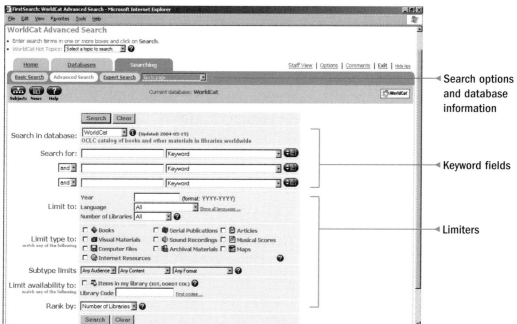

Use Reference Works

Reference works can be handy sources of information or useful research tools. They may include books, CD-ROMs, subscription databases, or Web sites. They can be general resources, such as general encyclopedias, or highly specialized. Reference works can be crucial resources for research:

- They can provide an overview of your topic.
- They can offer basic facts—definitions, statistics, dates, and names.
- They can indicate what is common knowledge about the topic.
- They can present clues for further research, including lists of resources to consult.

In general, reference works can be classified into two categories: (1) those that supply information and (2) those that function as research tools.

Reference Works That Supply Information

From encyclopedias to atlases, many reference works supply facts about your topic, along with clues for further research.

Almanacs, yearbooks, and statistical resources, which are typically published annually, contain diverse facts.

- *The World Almanac and Book of Facts* presents information on politics, history, religion, business, social programs, education, and sports.
- *Statistical Abstract of the United States* provides data on population, geography, social trends, politics, employment, business, science, and industry.

Encyclopedias supply facts and overviews for topics arranged alphabetically.

- General encyclopedias cover many fields of knowledge: *Collier's Encyclopedia*.
- Specialized encyclopedias focus on a single area of knowledge: *Encyclopedia of Busine$$ and Finance; The Blackwell Encyclopedic Dictionary of Business Ethics; McGraw-Hill Encyclopedia of Science and Technology*.

Vocabulary resources supply information on languages.

- General dictionaries, such as *Webster's Dictionary* and *Oxford English Dictionary,* supply definitions, pronunciations, and histories for a whole range of words.
- Specialized dictionaries and glossaries define words common to a field, topic, or group of people: *Dictionary of Marketing Terms* (e-book), *Dictionary of Engineering and Technology, Dictionary of Environment and Sustainable Development*.
- Bilingual dictionaries translate words from one language to another: *Diccionario Bilingüe de Negocios (Bilingual Business Dictionary)*.
- A thesaurus presents synonyms—words with similar meanings.

Biographical resources supply information about people. General biographies cover a broad range of people. Other biographies focus on the deceased, the living, or people from a specific group (e.g., an industry, vocation, race). Examples: *Who's Who in America, Who's Who in Economics, Dictionary of Scientific Biography.*

Atlases contain maps, charts, and diagrams presenting data on geography, climate, and human activities. Generally, they present two types of map information:

- General atlases portray physical boundaries (land/water, lakes, rivers, mountains) and/or political boundaries (nations, states, counties).
- Specialized atlases offer thematic maps that represent topics such as natural resources, climate, or population density. For example, *The Rand McNally Commercial Atlas and Marketing Guide* includes U.S. maps as well as transportation, communication, economic, and population information.

Directories list and supply contact information for people, groups, and organizations. Examples: *The National Directory of Addresses and Telephone Numbers, USPS ZIP Code Lookup and Address Information* (online), *Official Congressional Directory.*

Reference Works That Are Research Tools

From guides to bibliographies, these resources explain how to research your topic and find specific resources.

Guides and handbooks help readers explore a knowledge area or work in a discipline. Such guides can be print or electronic.

- **Guides to Fields or Disciplines:** *Hoover's Handbook of American Business; Standard Handbook for Electrical Engineers; Treasurer's and Controller's Desk Book* (e-book).
- **Topical Guides:** *Write for Business: A Compact Guide to Writing and Communicating in the Workplace; How to Find Chemical Information; Business Information Desk Reference; The Manager's Desk Reference; Microsoft Office 2000: Introductory Concepts and Techniques; Tests: A Comprehensive Reference for Assessments in Psychology, Education, and Business; Peterson's Graduate and Professional Programs; The Global Etiquette Guide to Mexico and Latin America.*

Indexes point you toward useful resources. Some indexes are general, such as *Readers' Guide to Periodical Literature*. Others specialize on a topic, such as *Environment Index, Business Periodicals Index, Engineering Index,* and *Index to Legal Periodicals*. Print indexes are often available electronically through subscription databases. See "Find Periodical Articles," pages 142–144.

Bibliographies list resources on a specific topic or by a specific author. A good, current bibliography can save you a lot of the work compiling your own bibliography on a topic. Example: *The Basic Business Library: Core Resources.*

Abstracts, like indexes, direct you toward articles on a topic. But abstracts also summarize those materials so you know whether a resource is relevant before you invest time in locating and reading it. Abstracts are usually organized into subject areas: *Computer Abstracts, Work-Related Abstracts, Environmental Abstracts, Social Work Abstracts.* They are often integrated into online database citations.

Find Periodical Articles

Periodicals are publications or broadcasts produced at regular intervals (daily, weekly, monthly, quarterly). As a rule, periodicals focus on a narrow range of topics geared toward a particular audience.

- **Daily newspapers and newscasts** focus on current events, opinions, and trends (*The Wall Street Journal, USA Today, The Newshour*).
- **Weekly and monthly magazines** generally provide more in-depth information on a wide range of topics (*Time, Business Week, 60 Minutes*).
- **Journals,** generally published quarterly, provide specialized, scholarly information for a narrowly focused audience (*Human Resource Development Quarterly, Energy Policy*).

To find useful articles, learn what search tools your library offers, what periodicals it has available in what forms, and how to gain access to those periodicals.

Search Online Databases to Find Promising Articles

If your library subscribes to EBSCOhost, LexisNexis, or a similar database service, use keyword searching (see page 123) to find citations on your topic. You might start with the general versions of such databases. A more focused research strategy would involve turning to specialized databases, such as EBSCOhost's Business Source Elite or LexisNexis's Business, Legal, or Medical databases.

The search screen in Figure 7.4 uses EBSCOhost's Business Source Elite as the chosen database for an article search focused on hybrid electric cars. Notice how limiters, expanders, and other advanced features help you find quality materials.

Figure 7.4 EBSCOhost Search Screen

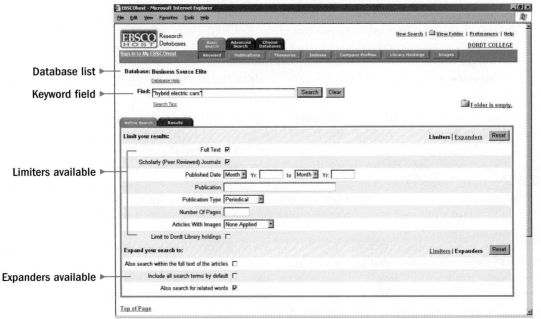

Build on the Initial Results

The search in Figure 7.4 would result in a list of citations to specific articles that address hybrid electric cars. At this point, you can study the results and do the following: refine the search by narrowing or expanding it, mark specific citations for "capture" or further study, re-sort the results, and follow links in a specific citation to further information.

Study Promising Citations and "Capture" Identifying Information

Once your search of databases generates descriptions of promising periodical articles, you need to study citations carefully, especially the abstracts (summaries). By scrutinizing them, you can determine three things: Does the article look relevant? Is an electronic, full-text version of the article available? Does the library have the periodical in its print or microform collection? In addition, you should "capture" identifying details for the periodical article by using the save, print, or e-mail function, or by writing down key details: the periodical's title, the issue and date, and the article's title and page numbers. These functions are all shown in the EBSCOhost citation in Figure 7.5.

Figure 7.5 EBSCOhost Citation

Find, Retrieve, and Evaluate Articles

When citations indicate promising articles, then access those articles as efficiently as possible.

1. Access an electronic copy. The citation may provide a link to the full text. From there, you can print, save, copy and paste, or e-mail the article.
2. If the article is not available electronically, find a print version:
    - Check if the online citation indicates that the article is available in your library. If the citation doesn't offer this detail, check your library's list of periodicals held, a list available online or in print. To get the article, follow your library's procedure.
    - If an article is unavailable online or in your library, use interlibrary loan. (Interlibrary forms may be available in paper or online.)
3. Evaluate each article, considering especially the periodical's quality, its goals, its target audience, and the article's currency. (See page 125.)

Specialized Databases for Different Fields

Most libraries offer access to databases from a wide range of fields, disciplines, and professions. Check your library's Web site for links to databases like these:

- **Communication & Mass Media Complete** offers access to resources related to a wide range of issues—from public speaking to TV broadcasting.
- **Engineering E-journal Search Engine** offers free, full-text access to more than 150 online engineering journals.
- **ERIC** (Educational Resource Information Center) offers citations, abstracts, and digests for more than 980 journals in the education field.
- **GPO** (Government Printing Office) offers access to more than 450,000 citations or records for U.S. government documents (e.g., reports, hearings, judicial rules, executive orders, addresses).
- **Health Source** (in Nursing and Consumer editions) offers access to almost 600 medical journals through abstracts, indexing, and full text. Consumers can find information on numerous health-related topics, from nutrition to sports medicine, through a range of health care publications.
- **JSTOR** offers full-text access to scholarly articles covering a broad range of disciplines, including finding articles going back in time or "recovering" electronically articles that were once available only in print.
- **National Environmental Publications Internet Site (NEPIS)** offers access to more than 6,000 EPA documents (full text, online).
- **PsycINFO,** a database offered by the American Psychological Association, offers access to materials in psychology and psychology-related fields (e.g., social work, criminology, organizational behavior).
- **Vocation and Career Collection** offers full-text access to more than 400 periodicals related to different trades and industries.

DOING INTERNET RESEARCH

The Internet can be a great research resource—or a great waste of time. Consider these benefits and drawbacks:

Benefits

- The Internet contains a wealth of current and specific information in verbal, oral, and visual formats.
- Because the information is digital, it can be searched quickly and conveniently; it can also be downloaded, saved, and sent.
- The Internet is always open.

Drawbacks

- Because of the vast quantity of material and its relative disorganization, finding relevant, in-depth, reliable information can be difficult.
- "Surfing" can encourage "shallow" research practices.
- The Net changes rapidly—what's here today may be gone, changed, or outdated tomorrow.

Web Sites for Business: A Short List

The Internet, of course, is filled with business-related Web sites—from individual companies' sites to e-commerce sites to training sites. Here are some recommended Web sites for workplace researchers:

- Advertising Council (public service advertising): www.adcouncil.org/
- American Accounting Association: http://aaahq.org/index.cfm
- American Marketing Association: www.marketingpower.com/
- AuditNet: www.auditnet.org/
- CNNfn The Financial Network: http://money.cnn.com/
- Current Industrial Reports: www.census.gov/cir/www/
- GlobalEDGE International Business Resources: http://globaledge.msu.edu/ibrd/ibrd.asp
- Hoover's Online: www.hoovers.com/free/
- Industry Research Desk: http://virtualpet.com/industry/
- Internet NonProfit Center: www.nonprofits.org/
- Marketing Virtual Library: www.knowthis.com/
- The Internet Public Library, Business and Economics: http://ipl.sils.umich.edu/div/subject/browse/bus00.00.00/
- National Federation of Independent Business: www.nfib.com/cgi-bin/NFIB.dll/Public/SiteNavigation/home.jsp
- WWW Virtual Library: Business and Economics: http://vlib.org/BusinessEconomics.html

Locate Information on the Internet

Because the Internet contains so much information of varying reliability, you need to become familiar with search tools that can locate information you can trust.

Use Your Library's Web Site

This tool gives you access to quality Internet resources by providing the following:

- Tutorials on using the Internet
- Guides to Internet resources in different disciplines
- Links to online document collections (e.g., Project Gutenberg, Etext Archives, New Bartleby Digital Library)
- Connections to virtual libraries, subscription databases, search engines, directories, government documents, periodicals, and reference works

Your library may give you access to WorldCat (see page 139). If you click on the "Internet" limiter, you will be able to search specifically for Web sites and other Internet information recommended by librarians.

Use a URL

Finding useful Internet resources can be as easy as typing in a URL:

- If you have a promising Web address, type or cut-and-splice it into the location field of your Web browser. Then hit "Enter" to go to that page.
- If you don't have the exact URL, sometimes you can guess it, especially for an organization (company, government agency, or nonprofit group). Try the organization's name or a logical abbreviation to get the home page.
  Formula: www.organization-name-or-abbreviation.domainname
  Examples: www.nasa.gov, www.honda.com, www.habitat.org, www.ucla.edu
- If you find yourself buried deep in a Web site and you want to find the site's home page, backspace on the URL shown in the address bar.

Follow Helpful Links

Locating information on the Net can mean "surfing" for leads:

- If you come across a helpful link (often highlighted in blue), click on it to go to the new page. This link may be internal to the site, or it may be external.
- Your browser keeps a record of the sites that you visit during a search session. Click the back arrow on your browser's toolbar to go back one site, or the forward arrow to move ahead again. Click the right mouse button on these arrows to show a list of recently visited sites.
- If you get lost while on the Net, click the "home" symbol of your browser's toolbar to return your browser to its starting place.

F.Y.I.

"Sponsored links" listed at a Web site are essentially a form of advertising.

Follow the Branches of a Subject "Tree"

A *subject tree,* sometimes called a *subject guide* or *directory,* lists Web sites organized into categories by experts who have reviewed those sites. Use subject trees or directories when you need to narrow a broad topic, you want evaluated sites, or you desire quality over quantity.

How does a subject tree work? Essentially, it allows you to select from a broad range of subjects or "branches." With each topic choice, you narrow your selection through subtopics until you arrive at a list of Web sites. Conversely, a subject tree may also allow you to keyword-search selected Web sites.

You might want to check whether your library subscribes to a service such as NetFirst, a database in which subject experts have catalogued Internet resources by topic. Here are some other subject directories that you can likely access at your library:

WWW Virtual Library	http://vlib.org/Overview.html
Argus	www.clearinghouse.net
Lii	www.lii.org
Magellan	http://magellan.excite.com/
LookSmart	http://looksmart.com

Case Study

Study the subject tree in Figure 7.6 provided by the Librarians' Index to the Internet (Lii). To find reviewed Web sites containing information on hybrid electric cars, you could select from one of the categories listed, depending on the angle that you want to explore: *Business, Finance, & Jobs; Government & Law;* or *Science, Technology, & Computers.* Each of these starting points will lead to a different listing of relevant sites. Another option would be to use the keyword search feature shown in Figure 7.6.

Figure 7.6 Lii Subject Tree

If you chose *Science, Technology, & Computers*, the subcategories shown in Figure 7.7 would appear. At this point, you could (1) select *Environment,* (2) follow *Transportation,* or (3) do a keyword search of this more limited grouping of Web sites.

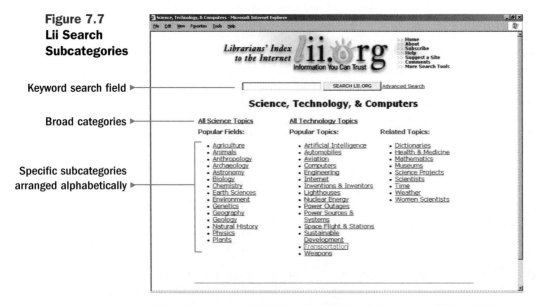

Figure 7.7 Lii Search Subcategories

Keyword search field ▶

Broad categories ▶

Specific subcategories arranged alphabetically ▶

As you work down through increasingly narrower branches of the subject tree, you will arrive at a listing of relevant Web sites. For example, the site in Figure 7.8 is listed under the topic *Electric vehicles.* The sites found in this way have all been reviewed in terms of quality, but you still need to evaluate what you find. In the citation for a site, study the site title, the description of information available, the site sponsorship, and the Web address (particularly the domain name).

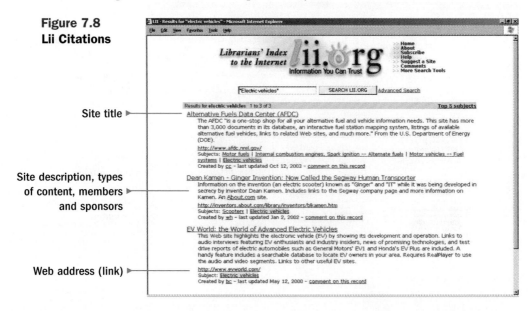

Figure 7.8 Lii Citations

Site title ▶

Site description, types of content, members and sponsors ▶

Web address (link) ▶

Use Search Engines and Metasearch Tools

Unlike a subject directory, which is constructed with human input, a search engine is a computer program that automatically scours a large amount of Internet material for keywords that you submit. Search engines are useful when you (1) have a very narrow topic in mind, (2) have a specific word or phrase to use in your search, (3) want a large number of results, (4) are looking for a specific type of Internet file, or (5) have the time to sort through the material looking for reliability.

Be aware that not all search engines are the same. Some search *citations* of Internet materials, while others conduct *full-text* searches. Choose a search engine that covers a large portion of the Internet, offers high-quality indexing, and offers high-powered search capabilities. Here's an overview of the most common search engines:

- **Basic search engines.** Search millions of Web pages gathered automatically.

Alta Vista	www.altavista.com
AllTheWeb	www.alltheweb.com
Google	www.google.com
HotBot	www.hotbot.com
Vivísimo	www.vivisimo.com

- **Metasearch tools.** Search several basic search engines at once, saving you the time and effort of checking more than one search engine.

Ask Jeeves	www.ask.com
Dog Pile	www.dogpile.com
Ixquick	www.ixquick.com
Northern Light	www.northernlight.com

- **"Deep Web" tools.** Check Internet databases and other sources not accessible to basic search engines.

Complete Planet	www.completeplanet.com
Invisible Web	www.invisibleweb.com

One key to using search engines successfully lies in effective keyword searching (see page 123). Essentially, it's best to become familiar with a few search engines—which areas of the Internet they search, whether they produce full-text results, what rules you must follow to ensure successful searches, and how they order or prioritize results. Here are some helpful tips:

- Enter a single word to find sites that contain that word or a derivative of it. Example: The term *apple* covers "apple," "apples," "applet," and so on.
- Enter more than one word to find sites containing any of those words.
- Use quotation marks to find an exact phrase. Example: The term *"apple pie"* (together in quotation marks) yields only sites with that phrase.
- Use Boolean operators (*and, or, not*) or appropriate symbols such as +, −, !, and ? to narrow or broaden your search.

Part 1 Challenges for Workplace Writers

Case Study

If you were interested in information on only Toyota's hybrid electric vehicles, you could conduct a search using the metasearch engine Ask Jeeves, a search that might lead to the initial results shown in Figure 7.9. At this point, you have several choices:

- Narrow or broaden your search by using the drop-down menus in the box.
- Follow the links on the right to related topics.
- Search the "Sponsored Web Results"—essentially advertising sites.
- Search "Web Results"—sites that generally offer more depth although the site sponsorship is still important.
- Explore a related search of other topics and of other Internet resources.

Figure 7.9   Ask Jeeves Search Results

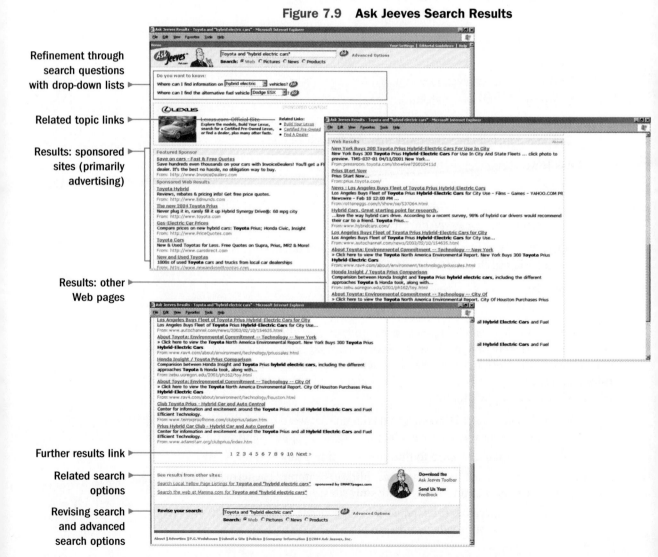

ORGANIZING YOUR FINDINGS

Depending on the size of your project, your research may generate a mass of notes, printouts, photocopies, electronic files, and so on. The challenge is to move from this unwieldy mass to a coherent structure or outline for the research document that you need to write. The tips given here will help you make sense of your research.

Develop and Order Your Ideas

Good thinking is foundational to good research writing and sound business decisions. However, good thinking must be well organized for readers to access and make sense of it. To develop ideas and order your information, follow these tips:

Refocus on your research questions. Review the questions you are trying to answer, as well as the context—your assignment, your goal, your readers.

Study the evidence. Review your materials for as long as it takes for ideas to percolate and information to make sense. Consider these questions:

- Is the information complete or at least sufficient for the project?
- Does the information seem reliable and accurate?
- How do the data and the ideas relate to the research topic?
- What connections link different pieces of evidence?

Develop sound conclusions. Because logical analysis of research findings leads to strong ideas (conclusions, recommendations), practice these strategies:

- Work against personal biases that create blind spots toward the evidence.
- Practice using logic in your analysis, but also tap into your intuition, creativity, and imagination.
- Interpret statistical data carefully and correctly.
- Logically distinguish between causes and effects; carefully link them.
- Avoid either-or, black-and-white thinking; circular arguments; and slippery-slope claims. Make fine distinctions rather than sweeping generalizations.
- Check your conclusions against counterarguments, your experience, and common sense.
- Develop recommendations that are realistic, precise, and complete.

Follow a template. An organizational template may be available through your word-processing program, in company files, or in this book.

Clump and split material. Using your key ideas or conclusions as main headings, arrange support, evidence, and information under the most appropriate heading. After placing information in these categories, decide how best to sequence the topics in an outline.

Rely on tried-and-true patterns. Many traditional organizational patterns offer sound methods for developing your thinking, answering your research question, and making sense out of a large body of information. See pages 195–213.

USING AND INTEGRATING SOURCES

After you've found strong sources and taken good notes on them, you want to select researched facts and ideas to develop and support your own ideas. Making the best choices requires that you clearly show (1) what information you are borrowing and (2) where you got it. By making this distinction, you create credibility. To be fair to your sources, follow the strategies below and on the next pages. Note: The examples shown here follow the American Psychological Association (APA) format (see pages 158–166).

Document All Borrowed Material

Generally, credit any information that you have summarized, paraphrased, or quoted from any source, whether that information consists of statistics, facts, graphics, phrases, or ideas. Your reader can then see what's borrowed and what's yours, understand your support, and do appropriate follow-up research.

Common knowledge (information that is generally known to your readers or easily found in several sources) need not be cited. For example, the fact that automakers are developing hybrid electric cars is common knowledge, whereas the precise nature and details of GM's AUTOnomy project are not. Still not sure about the difference? **When in doubt, document.**

Integrate Sources Thoughtfully

Fold source material into your discussion by carefully relating it to your own thinking. Use this pattern:

1. State and explain your idea, creating a context for the source.
2. Identify the source and link it to your discussion.
3. Summarize, paraphrase, or quote the source, providing a citation in an appropriate spot.
4. Use the source by explaining, expanding, or refuting it.

Writer's ideas Attributive phrase Paraphrase, quotation, or summary Citation Commentary Conclusion	The motivation and urgency to create and improve hybrid-electric technology come from a range of complex forces. Some of these forces are economic, others environmental, and still others social. In "Societal Lifestyle Costs of Cars with Alternative Fuels/Engines," Joan Ogden, Robert Williams, and Eric Larson argue that "continued reliance on current transportation fuels and technologies poses serious oil supply insecurity, climate change, and urban air pollution risks" (2004, p. 7). Because of the nonrenewable nature of fossil fuels and their negative side effects, the transportation industry is confronted with making the most radical changes since the introduction of the internal-combustion automobile more than 100 years ago. Hybrid-electric vehicles represent one response to this pressure.

Identify Clearly Where Source Material Begins

Your discussion must offer a smooth transition to source material. For first references to a specific source, use an attributive statement that indicates some of the following: the author's name and credentials, the title of the source, the nature of the study or research, and helpful background.

> *Joan Ogden, Robert Williams, and Eric Larson, members of the Princeton Environmental Institute, explain* that modest improvements in energy efficiency and emissions reductions will not be enough over the next century because of anticipated transportation increases (2004, p. 7).

For subsequent references to a source, use a simplified attributive phrase, such as the author's last name or a shortened version of the title.

> *Ogden, Williams, and Larson go on to argue* that "effectively addressing environmental and oil supply concerns will probably require radical changes in automotive engine/fuel technologies" (2004, p. 7).

In some situations, such as quoting straightforward facts or providing an overview of research, simply skip the attributive phrase. The parenthetical citation supplies sufficient attribution.

> Various types of transportation are by far the main consumers of oil (three-fourths of world oil imports); moreover, these same technologies are responsible for one-fourth of all greenhouse gas sources (Ogden, Williams, & Larson, 2004, p. 7).

The verb that you use to introduce source material is a key part of the attribution. Use fitting verbs, such as those in the list below.

Attributive verbs: synonyms for "says"

Accepts	Confirms	Enumerates	Points out	Stresses
Acknowledges	Considers	Emphasizes	Praises	Studies
Adds	Contradicts	Explains	Proposes	Suggests
Affirms	Contrasts	Highlights	Proves	Summarizes
Argues	Criticizes	Hypothesizes	Refutes	Supports
Asserts	Declares	Identifies	Rejects	Urges
Believes	Defends	Insists	Responds	Verifies
Cautions	Denies	Interprets	Shares	Warns
Claims	Describes	Lists	Shows	
Compares	Discusses	Maintains	Speculates	
Concludes	Disagrees	Outlines	States	

Normally, use attributive verbs in the present tense. Use the past tense only when you want to stress the "pastness" of a source.

> In their 2004 study, "Societal Lifecycle Costs of Cars with Alternative Fuels/Engines," Ogden, Williams, and Larson present a method for comparing and contrasting alternatives to the traditional internal-combustion engine. In an earlier study (2001), these authors *had made* preliminary steps. . . .

Indicate Clearly Where Source Material Ends

Obviously, closing quotation marks and a citation indicate the end of a source quotation. With summaries and paraphrases, the situation is more challenging.

- Generally, place the citation immediately after a quotation, paraphrase, or summary. You may also place the citation early in the sentence or at the end.
- When you discuss several facts or quotations from a page in a source, use an attributive phrase at the beginning of your discussion and a single citation at the end.
- Provide a citation each time that you use additional information from a previously cited source.

> As the "Lifestyle Costs" study concludes, when greenhouse gases, air pollution, and oil insecurity are factored into the analysis, alternative-fuel vehicles "offer lower LCCs than typical new cars" (Ogden, Williams, & Larson, 2004, p. 25).

Smoothly Integrate Quotations

Because quotations can interrupt the flow of your discussion, you need to pay attention to style, punctuation, and syntax. Use enough of the quotation to make your point without changing the meaning of the original. All key phrases that appear in a source must be enclosed in quotation marks.

> Ogden, Williams, and Larson (2004) also conclude that the hydrogen fuel-cell vehicle is "a strong candidate for becoming the Car of the Future," given the trend toward "tighter environmental constraints" and the "intense efforts under way" by automakers to develop commercially viable versions of such vehicles (p. 25).

If a quotation is longer than forty words, set it off from the main text:

- Introduce the quotation with a complete sentence and a colon.
- Indent the whole quotation one-half inch (five spaces) and double-space it, but don't put quotation marks around it.
- Put the parenthetical citation outside the final punctuation mark.

> Toward the end of the study, Ogden, Williams, and Larson (2004) argue that changes to the fuel delivery and filling system must be factored into planning:
>
>> In charting a course to the Car of the Future, societal LCC comparisons should be complemented by considerations of fuel infrastructure requirements. Because fuel infrastructure changes are costly, the number of major changes made over time should be minimized. The bifurcated strategy advanced here—of focusing on the H_2 FCV for the long term and advanced liquid hydrocarbon-fueled ICEVs and ICE/HEVs for the near term—would reduce the number of such infrastructure changes to *one* (an eventual shift to H_2). (p. 25)

Indicate Changes to Quotations with Brackets and Ellipses

Normally, you should maintain the wording, capitalization, and punctuation of the original. However, you may shorten or change a quotation so that it fits more smoothly into your sentence—but don't alter the original meaning.

Use an ellipsis to indicate that you have omitted words from the original. An ellipsis is three periods with a space before and after each.

> In their projections of where fuel-cell vehicles are heading, Ogden, Williams, and Larson (2004) discuss GM's AUTOnomy vehicle, with its "radical redesign of the entire car. . . . In these cars steering, braking, and other vehicle systems are controlled electronically rather than mechanically" (p. 24).

Use brackets to indicate a clarification, or to change a pronoun or verb tense.

> As Ogden, Williams, and Larson (2004) explain, "even if such barriers [the high cost of fuel cells and the lack of an H_2 fuel infrastructure] can be overcome, decades would be required before this embryonic technology could make major contributions in reducing the major externalities that characterize today's cars" (p. 25).

F.Y.I.

Here are four additional tips about using research in your writing to help you develop a clear, useful, reliable document.

1. Don't cram everything that you gathered into the research document. In fact, if you've done good research, you probably have four or five times more material than you could possibly fit in the document. That's good, because having the material gives you choices. Sift through it for the information most relevant to your research question and to your reader's needs. You can always put secondary data in appendices.

2. Make sure that each claim you make is backed up with solid evidence—not with an argument that the point is self-evident. See Chapter 10, "Trait 1: Strong Ideas."

3. Understand readers' perspectives to make sense of the information for them, but resist the temptation to simply say what they want to hear. For example, if your supervisor gave you the research assignment, you may naturally want to supply pleasing information, but accurate information is much more important. See Chapter 8, "Business Writing Ethics."

4. Use quotations sparingly. In most research documents, restrict your quoting to nuggets—key statements by authorities, well-phrased claims and conclusions, passages where a careful word-by-word analysis and interpretation is important for your argument.

AVOIDING PLAGIARISM

Plagiarism is borrowing someone's words, ideas, or images (what's called "intellectual property") and using the material so that it appears to be yours. In school, such theft can have dire consequences—failing an assignment or a course, putting a blemish on your transcript, or even being expelled from college.

In the workplace, the situation is parallel but different. Plagiarism does not refer to typical in-house recycling of written documents, which are essentially company property, not the property of an individual employee. However, plagiarism does occur when workplace writers steal someone else's material and pass it off as their own work. Such theft violates a workplace ethic of honesty and may also violate copyright laws. Moreover, such plagiarism destroys your credibility and may prompt punishment such as a reprimand, dismissal, or legal action. Because such plagiarism reflects on your company, it, too, is punished.

Along with basic honesty, effectively integrating and documenting source material will help you avoid most forms of plagiarism and other source abuses. The following tips offer more specific help.

Forms of Plagiarism

Plagiarism refers to a whole range of source abuses—some obvious, some not. Read the passage below from an article related to hybrid-electric vehicles. Then review the five types of plagiarism, noting how each misuses the source.

> **From Joan M. Ogden, Robert H. Williams, & Eric D. Larson. (2004). "Societal lifecycle costs of cars with alternative fuels/engines."** *Energy Policy, 32,* **7–27.**
>
> [S]peculation that FCVs [fuel-cell vehicles] might be cheaper to manufacture is based on the vision that substantial scale economies might be realized by mass-producing the compact "skateboard" chassis in truly huge centralized facilities (much larger than what is involved with ICEV [internal-combustion engine vehicle] manufacture) and shipping these compact chassis to many traditionally sized auto factories around the world that would specialize in making car bodies tailored to the needs of the local economy and where the chassis would be attached to the car bodies. (pp. 24–25)

Submitting Another Writer's Work

The most blatant plagiarism is taking an entire piece of writing and claiming it as your own. Some examples include

- Downloading an article, reformatting it, and distributing it as your own work
- Buying a paper from a "paper mill"
- Taking a "free" paper or a publicly available report off the Internet
- Turning in another student's or colleague's work as your own
- Submitting a workplace report without crediting a colleague who co-authored it

Using Copy-and-Paste

Avoid taking chunks of material from a source and splicing them into your paper without acknowledgment. In the example below, the writer pastes in part of the original passage without using quotation marks or a citation. Even if the writer changes some words, this practice would still be plagiarism.

> Because of potential design modifications, fuel-cell vehicles may turn out to be cheaper than conventional vehicles. The basis for the speculation that FCVs might be cheaper to manufacture is based on the vision that scale economies might be realized by mass-producing the compact "skateboard" chassis.

Failing to Cite a Source

Even if you use information accurately and fairly, don't neglect to cite the source. Here the writer correctly summarizes the passage but offers no citation.

> Fuel-cell vehicles may turn out to be highly economical to produce because the compact chassis can be mass-produced at enormous factories and then distributed to smaller auto plants that manufacture the auto bodies preferred in that geographical area.

Neglecting Necessary Quotation Marks

Whether it's a paragraph or a phrase, if you use a source's exact wording, that material must be enclosed in quotation marks. Here the writer cites the source, but doesn't use quotation marks around a borrowed phrase (italicized).

> The key to the fuel-cell vehicle's success just might be *mass-producing the compact "skateboard" chassis* (Ogden, Williams, & Larson, 2004, p. 25).

Confusing Borrowed Material with Your Own Ideas

Through carelessness (often in note taking), you may confuse source material with your own thinking. In the passage below, the writer indicates that he borrowed material in the first sentence, but fails to indicate that he also borrowed the next sentence.

> The FCV's flexible design will allow for vehicles "tailored to the needs of the local economy" (Ogden, Williams, & Larson, 2004, p. 25). Shipped from large, centralized manufacturing plants, chassis could be distributed to auto plants that make and attach the body.

F.Y.I.

Besides plagiarism, other source abuses can derail your research writing with readers. Avoid these pitfalls:

- Using sources inaccurately (e.g., an incorrect quotation or statistic).
- Taking source material out of its context and forcing it into your own.
- Overusing source material, especially quotations.
- Using "blanket" citations at the ends of paragraphs, leaving readers unclear about what is and isn't taken from the source.
- Relying heavily on just one source or one perspective.

FOLLOWING APA DOCUMENTATION RULES

The documentation system that you may be required to use will differ from document to document—depending on the situation, the profession, the industry, and the publication method. We recommend the American Psychological Association (APA) system because (1) it is based on in-text citations combined with a references page, making it relatively straightforward; (2) it is the main system used in the social sciences, and business studies are traditionally part of this division; and (3) its use of publication dates within in-text citations quickly highlights information currency.

APA Paper Format

Title Page

On the first page, include your paper's title, your name, and your institution's name on three separate lines. Double-space and center the lines beginning approximately one-third of the way down from the top of the page.

Abstract

On the second page, include an abstract: a 100- to 150-word paragraph summarizing your paper. Center the title "Abstract" one inch from the top of the page.

Body

Format the body (which begins on the third page) of your paper as follows:

- **Margins.** Leave a one-inch margin on all four sides of each page.
- **Line spacing.** Double-space your entire paper, unless your instructor allows single spacing for tables, titles, captions, and so on.
- **Headings.** Main headings should be centered, using standard uppercase and lowercase text.
- **Page numbers.** Place your short title (the first two or three words) and the page number at the upper-right margin. The title should be either just above or five spaces to the left of the page number.

Citations

Within your paper, give credit for others' ideas by including the author and year in a citation. For specific information, add the page number(s).

References

Place full citations from all sources in an alphabetized list at the end of your paper. Place the title "References" approximately one inch from the top of the page and center it.

Appendix

If your instructor requires it, place your charts, tables, and graphs in an appendix. Otherwise, include them within the body of your paper.

WEB Link

For a sample APA research report and for information on other documentation systems, see http://college.hmco.com/english.

Model APA In-Text Citations

In APA style, you cite your source in the text (in parentheses) each time you borrow from it. Each of these parenthetical citations must be matched to an entry in an alphabetized list called "References" at the end of your paper. Each item in the "References" list should, in turn, be cited in the text.

The APA documentation style is sometimes called the "author-date" system because both the author and the date of the publication must be mentioned in the text when citing a source. Both might appear in the flow of the sentence, like this:

> Only South Africa has more people infected with AIDS than India, according to a 2001 article by Mike Specter.

If either the name or the date does not appear in the sentence, it must be mentioned in parentheses at the most convenient place, like this:

> According to an article by Mike Specter (2001), only South Africa . . .
>
> According to a recent article (Specter, 2001), only South Africa . . .

One Author: Citing a Complete Work

The correct form for a citation for a single source by a single author is parenthesis, last name, comma, space, year of publication, parenthesis. Note that final punctuation should be placed outside the parentheses.

> . . . and the great majority of Venezuelans live near the Caribbean coast (Anderson, 2001).

One Author: Citing Part of a Work

When you cite a specific part of a source, give the page number, chapter, or section, using the appropriate abbreviations (p. or pp., chap., or sec.). Always give the page number for a direct quotation.

> . . . Bush's 2002 budget, passed by Congress, was based on revenue estimates that "now appear to have been far too optimistic" (Lemann, 2001, p. 48).

Two to Five Authors

In APA style, all authors—up to as many as five—must be mentioned in the citation, like this:

> Love changes not just who we are, but who we can become (Lewis, Amini, & Lannon, 2000).

The last two authors' names are always separated by a comma and an ampersand (&) when enclosed in parentheses.

For works with more than two but fewer than six authors, list all the authors the first time; after that, use only the name of the first author followed by "et al." (the Latin abbreviation for *et alii*, meaning "and others"), like this:

> These discoveries lead to the hypothesis that love actually alters the brain's structure (Lewis et al., 2000).

Six or More Authors

If your source has six or more authors, refer to the work by using the first author's name followed by "et al.," both for the first reference in the text and all references after that. Be sure to list all of the authors (up to six) in your references list.

Anonymous Book (Work)

If your source lists no author, treat the first two or three words of the title (capitalized normally) as you would an author's last name. A title of an article or a chapter belongs in quotation marks, whereas the titles of books or reports should be italicized:

> . . . including a guide to low-stress postures ("How to Do It," 2001).

Group Author

A group author is an organization, association, or agency that claims authorship of a document. Treat the name of the group as if it were the last name of the author. If the name is long and easily abbreviated, provide the abbreviation in square brackets. Use the abbreviation without brackets in subsequent references, as follows:

> First text citation: (National Institute of Mental Health [NIMH], 2002)
> Subsequent citations: (NIMH, 2002)

A Source Referred to or Quoted in Another Source

If you need to cite a source that you have found referred to in another source (a "secondary" source), mention the original source in your text. Then, in your parenthetical citation, cite the secondary source, using the words "as cited in," like this:

> . . . theorem given by Ullman (as cited in Hoffman, 1998).

Note: In your references list at the end of the paper, you would write out a full citation for Hoffman (not Ullman).

Two or More Works in a Parenthetical Reference

Sometimes it is necessary to lump several citations into one citation. In that case, cite the sources as you usually would, separating the citations with semicolons. Place the citations in alphabetical order, just as they would be ordered in the references list:

> These near-death experiences are reported with conviction (Rommer, 2000; Sabom, 1998).

Informal, Unpublished Communications

In business research, informal communications may have provided you with knowledge (e.g., personal letters, phone calls, e-mail messages, and memos). Because such documents are not published in a permanent form, APA style does not require you to list them in your references. Instead, cite them only in the text of your paper in parentheses, like this:

> . . . according to M. T. Cann (personal communication, April 1, 1999).
> . . . by today's standard (M. T. Cann, personal communication, April 1, 1999).

APA REFERENCES LIST

Quick Guide

The references section at the end of your paper lists all the sources you have cited in your text. Begin your list on a new page (the next page after the text) and number each page, continuing the numbering from the text. Follow this format:

1. Type the short title and page number in the upper-right corner, approximately one-half inch from the top of the page.
2. Center the title, "References," approximately one inch from the top; then double-space before the first entry.
3. Begin each entry flush with the left margin. If the entry runs more than one line, indent additional lines approximately one-half inch (five to seven spaces).
4. Double-space between all lines on the references page.
5. List each entry alphabetically by the last name of the author or, if no author is given, by the title (disregarding "A," "An," or "The").
6. Leave one space following each word and punctuation mark in an entry.
7. For both book titles and article titles, capitalize only the first letter of the title (and subtitle) and any proper nouns.
8. Use italics for titles of books and periodicals rather than underlining.
9. Include the state, province, and/or country in the publisher's location only if the city is not well known for publishing.

Reference Entries: Books

The general form for a book or brochure entry is this:

   Author, A. (year). *Title.* Location: Publisher.

For a single chapter of a book, follow this form:

   Author, A., & Author, B. (year). Title of chapter. In *Title of book* (pp. xx–xx). Location: Publisher.

The entries that follow illustrate the information needed to cite books, sections of a book, brochures, and government publications.

One Author

   Guttman, J. (1999). *The gift wrapped in sorrow: A mother's quest for healing.* Palm Springs, CA: JMJ Publishing.

Two or More Authors

   Lynn, J., & Harrold, J. (1999). *Handbook for mortals: Guidance for people facing serious illness.* New York: Oxford University Press.

Note: Follow the first author's name with a comma; then join the two authors' names with an ampersand (&) rather than with the word "and." List up to six authors; abbreviate subsequent authors as "et al."

Anonymous Book

If an author is listed as "Anonymous," treat it as the author's name. Otherwise, follow this format:

> *Publication manual of the American Psychological Association* (5th ed.). (2001). Washington, DC: American Psychological Association.

Chapter from a Book

> Tattersall, I. (2002). How did we achieve humanity? In *The monkey in the mirror* (pp. 138–168). New York: Harcourt.

Single Work from an Anthology

> Marshall, P. G. (2002). The impact of the cold war on Asia. In T. O'Neill (Ed.), *The nuclear age* (pp. 162–166). San Diego: Greenhaven Press.

Note: When editors' names appear in the middle of an entry, follow the usual order: initial first, surname last.

Group Author as Publisher

> Amnesty International. (2000). *Hidden scandal, secret shame: Torture and ill-treatment of children*. New York: Author.

Note: If the publication is a brochure, identify it as such in brackets after the title.

Two or More Books by the Same Author

> Dershowitz, A. (2000). *The Genesis of justice: Ten stories of biblical injustice that led to the Ten Commandments and modern law*. New York: Warner Books.
>
> Dershowitz, A. (2002). *Shouting fire: Civil liberties—past, present, and future*. Boston: Little, Brown.

Article in a Reference Book

> Lewer, N. (1999). Non-lethal weapons. In *World encyclopedia of peace* (pp. 279–280). Oxford: Pergamon Press.

Note: If no author is listed, begin the entry with the title of the article.

Technical or Research Report

> Taylor, B. G., Fitzgerald, N., Hunt, D., Reardon, J. A., & Brownstein, H. H. (2001). *ADAM preliminary 2000 findings on drug use and drug markets: Adult male arrestees*. Washington, DC: National Institute of Justice.

Government Publication

> National Institute on Drug Abuse. (2000). *Inhalant abuse* (NIH Publication No. 00-3818). Rockville, MD: National Clearinghouse on Alcohol and Drug Information.

Note: If the document is not available from the Government Printing Office (GPO), the publisher would be either "Author" or the government department.

Reference Entries: Periodicals

The general form for a periodical entry is this:

> Author, A. (year). Article title. *Periodical Title, Volume Number,* page numbers.

Include some other designation with the year (such as a month or season) if a periodical does not use volume numbers.

Article in a Scholarly Journal, Consecutively Paginated

> Epstein, R., & Hundert, E. (2002). Defining and assessing professional competence. *JAMA, 287,* 226–235.

Note: Pay attention to the features of this basic reference to a scholarly journal: (1) last name and initial(s) as for a book reference; (2) year of publication; (3) title of article in lowercase, except for the first word; title not italicized or in quotations; (4) title and volume number of journal italicized; and (5) inclusive page numbers.

Journal Article, Paginated by Issue

> Lewer, N. (1999, Summer). Nonlethal weapons. *Forum, 14*(2), 39–45.

Note: When the page numbering of the issue starts with page 1, the issue number (not italicized) is placed in parentheses after the volume number. (Some journals number pages consecutively, from issue to issue, through their whole volume year.)

Journal Article, More Than Six Authors

> Wang, X., Zuckerman, B., Pearson, C., Kaufman, G., Chen, C., Wang, G., et al. (2002, January 9). Maternal cigarette smoking, metabolic gene polymorphism, and infant birth weight. *JAMA, 287,* 195–202.

Magazine Article

> Silberman, S. (2001, December). The geek syndrome. *Wired, 9*(12), 174–183.

Note: If the article is unsigned, begin the entry with the title of the article.

> Tomatoes target toughest cancer. (2002, February). *Prevention, 54*(2), 53.

Newspaper Article

> Stolberg, S. C. (2002, January 4). Breakthrough in pig cloning could aid organ transplants. *The New York Times,* pp. 1A, 17A.

Note: If the article is a letter to the editor, identify it as such in brackets following the title. For newspapers, use "p." or "pp." before the page numbers; if the article is not on continuous pages, give all the page numbers, separated by commas.

Newsletter Article

Newsletter article entries are very similar to newspaper article entries; only a volume number is added.

> Teaching mainstreamed special education students. (2002 February). *The Council Chronicle, 11,* 6–8.

Reference Entries: Electronic Sources

APA style prefers a reference to the print form of a source, even if the source is available on the Internet. If you have read *only* the electronic form of an article's print version, add "Electronic version" in brackets after the title of the article. If an online article has been changed from the print version or has additional information, follow the same general format for the author, date, and title elements of print sources, but follow it with a "retrieved from" statement, citing the date of retrieval and the electronic address.

Periodical, Identical to Print Version

>Author, A., & Author, B. (year, month day). Title of article, chapter, or Web page [Electronic version]. *Title of Periodical, volume number or other designation,* inclusive page numbers (if available).

>Ashley, S. (2001, May). Warp drive underwater [Electronic version]. *Scientific American (2001, May).*

Periodical, Different from Print Version or Online Only

>Author, A., & Author, B. (year, month day). Title of article, chapter, or Web page. *Title of Periodical, volume number,* inclusive page numbers if available. Retrieved Month day, year, from electronic address

>Nicholas, D., Huntington, P., & Williams, P. (2001, May 23). Comparing web and touch screen transaction log files. *Journal of Medical Internet Research, 3.* Retrieved November 15, 2001, from www.jmir.org/2001/2/e18/index.htm

Note: Include an issue number in parentheses following the volume number if each issue of a journal begins on page 1. Use "pp." (page numbers) for newspapers. Page numbers are often not relevant for online sources. End the citation with a period unless it ends with the electronic address.

Multipage Document Created by Private Organization

>National Multiple Sclerosis Society. (n.d.). *About MS: For the newly diagnosed.* Retrieved May 20, 2002, from www.national/mssociety.org

Note: Use "n.d." (no date) when a publication date is unavailable. Provide the URL of the home page for an Internet document when its pages have different URLs.

Document from an Online Database

>Author, A., & Author, B. (year). Title of article or Web page. *Title of Periodical, volume number,* inclusive page numbers. Retrieved Month day, year, from name of database.

>Belsie, L. (1999). Progress or peril? *Christian Science Monitor, 91*(85), 15. Retrieved September 15, 1999, from DIALOG online database (#97, IAC Business A.R.T.S., Item 07254533).

Note: If the document cited is an abstract, include "Abstract" before the "retrieved" statement. The item or accession numbers may be included but are not required.

Other Nonperiodical Online Document

> Author, A., & Author, B. (year, month day). Title of work. Retrieved Month day, year, from electronic address
>
> Boyles, S. (2001, November 14). World diabetes day has people pondering their risk. Retrieved November 16, 2001, from http://my.webmd.com/content/article/1667.51328

Note: Break URLs after a slash or before a period; do not insert a hyphen.

> Catholic Near East Welfare Association. (2002). Threats to personal security. In *Report on Christian emigration: Palestine* (sect. 5). Retrieved May 20, 2002, from www.cnewa.org/news-christemigrat-part1.htm

Note: To cite only a chapter or section of an online document, follow the title of the chapter with "In *Title of document* (chap. number)." If the author is not identified, begin with the title of the document. If a date is not identified, put "n.d." in parentheses following the title.

U.S. Government Report Available on Government Agency Web Site

> Name of government agency. (Year, month day). *Title of report*. Retrieved Month day, year, from electronic address
>
> United States Department of Commerce, Office of the Inspector General. (2001, March). *Internal controls over bankcard program need improvement.* Retrieved July 23, 2001, from www.oig.doc.gov/e-library/reports/recent/recent.html

E-mail

E-mail is cited only in the text of the paper, not in the references list. See "Informal, Unpublished Communications," on page 160.

Reference Entries: Other Print and Nonprint Sources

The following citation entries are examples of audiovisual media sources and sources available electronically.

Specialized Computer Software with Limited Distribution

Standard, nonspecialized computer software does not require a reference entry. Treat software as an unauthored work unless an individual has property rights to it.

> Carreau, S. (2001). Champfoot (Version 3.3) [Computer software]. Saint Mandé, France: Author.

Television or Radio Broadcast

> Crystal, L. (Executive Producer). (2002, February 11). *The newshour with Jim Lehrer* [Television broadcast]. New York and Washington, DC: Public Broadcasting Service.

Television or Radio Series

> Bloch, A. (Producer). (2002). *Thinking allowed* [Television series]. Berkeley: Public Broadcasting Service.

Television or Radio Program (Episode in a Series)

> Berger, C. (Writer). (2001, December 19). Feederwatch [Radio series program]. In D. Byrd & J. Block (Producers), *Earth & sky*. Austin, TX: The Production Block.

Audio Recording

> Kim, E. (Author, speaker). (2000). *Ten thousand sorrows* [CD]. New York: Random House.

Music Recording

> ARS Femina Ensemble. (Performers). (1998). *Musica de la puebla de los angeles: Music by women of baroque Mexico, Cuba, & Europe* [CD]. Louisville, KY: Nannerl Recordings.

Note: Give the name and function of the originators or primary contributors. Indicate the recording medium (CD, record, cassette, and so on) in brackets, immediately following the title.

Motion Picture

Give the name and function of the director, producer, or both. If its circulation was limited, provide the distributor's name and complete address in parentheses.

> Jackson, P. (Director). (2001). *The lord of the rings: The fellowship of the ring* [Motion picture]. United States: New Line Productions, Inc.

Published Interview, Titled, No Author

> Stephen Harper: *The Report* interview. (2002, January 7). *The Report* (Alberta, BC), *29,* 10–11.

Published Interview, Titled, Single Author

> Fussman, C. (2002, January). What I've learned [Interview with Robert McNamara]. *Esquire, 137*(1), 85.

Unpublished Paper Presented at a Meeting

> Lycan, W. (2002, June). *The plurality of consciousness*. Paper presented at the meeting of the Society for Philosophy and Psychology, New York, NY.

Unpublished Doctoral Dissertation

> Roberts, W. (2001). *Youth crime amidst suburban wealth*. Unpublished doctoral dissertation, Bowling Green State University, Bowling Green, OH.

CHECKLIST for Research

Use this checklist to keep your research projects on track.

- ☐ I understand my research assignment: the purpose, expected products, people involved, and process (schedule, resources, expectations).
- ☐ My research is clearly focused on a well-defined set of research questions.
- ☐ I have followed a strong research plan that includes, when appropriate, background research, field (or primary) research, library research, and Internet research—from a good range of information sources and sites.
- ☐ I have developed a systematic note-taking system.
- ☐ I have evaluated sources for accuracy, currency, and other factors.
- ☐ With any primary research (e.g., observations, interviews), I have carefully and ethically gathered reliable data.
- ☐ I have effectively used my library's collections and research tools.
- ☐ I have found a range of reliable resources: a balance and blend appropriate for the project.
- ☐ I have used refined techniques to search the Internet for reliable information.

CHECKLIST for a Research Report

Use this checklist to review your research report before submitting it.

- ☐ **Ideas.** Researched data are accurate and complete. Conclusions follow logically from the data. All ideas and information are properly credited.
- ☐ **Organization.** Information is delivered in a structured format that allows the reader to make sense of it.
- ☐ **Voice.** The tone is confident but also sincere, measured, and objective. Where uncertainty exists, it is noted objectively.
- ☐ **Words.** Precise, clear phrasing is used—language that readers will understand, as opposed to obscure jargon. Terms are defined as needed.
- ☐ **Sentences.** Constructions flow smoothly, especially with respect to integrating source material (whether summary, paraphrase, or quotation).
- ☐ **Copy.** Grammar, punctuation, mechanics, usage, and spelling are all correct—especially punctuation used with source material and parenthetical documentation. Documentation of research is complete and correct, following APA style or another appropriate system.
- ☐ **Design.** The document's format, page layout, and typography are all reader-friendly. Data are effectively presented in discussion, lists, tables, charts, graphs, and other formats.

CRITICAL THINKING Activities

1. Reflect on research projects that you have completed in your college courses. Have those experiences been positive or negative? Why? Review a research paper that you recently wrote for another class. List the current strengths and weaknesses of your research practices, and sketch out how you would change your research habits for conducting workplace research.

2. Find two articles on the same workplace-related topic: one from a popular magazine, and one from an academic journal. Compare and contrast the articles in terms of their treatment of the topic and approach to research.

3. Become familiar with some information resources central to your major or career. With the help of a reference librarian, if necessary, track down the following: key reference resources (print and electronic); professional associations; the names of prominent authors in your field; the titles of two or three quality journals; and three or four rich and reliable Web sites.

4. Interview a professional in a field related to your career concerning research. What information skills are needed in this profession? What types of resources does he or she work with regularly? How and why is research important to his or her work?

WRITING Activities

1. As directed by your instructor, develop a research topic from one of the broad options listed below. Then work through the research process outlined in this chapter to develop a carefully documented research report for your instructor and classmates.

   - A workplace-related challenge that you and your classmates may face (e.g., health and safety, affirmative action, current labor laws, global trends, conducting business on the Web)

   - A workplace communication issue (e.g., shifts and developments in communication technology, the nature and role of nonverbal communication, the shift from paper to electronic communication, challenges of intercultural communication)

   - Current developments and trends in a business sector or industry (e.g., the trend toward hybrid and hydrogen cars in the auto industry, outsourcing of human resources work)

   - Challenges for consumers (e.g., "green" shopping, identity theft, credit-card debt)

2. As directed by your instructor, complete Writing Activity 1 as a team project. For help with this task, see Chapter 3, "Teamwork on Writing Projects."

In this chapter

Strategies for Ethical Writing 170
Information Ethics 171
Persuasion Ethics 174
Checklist for Business Writing Ethics 178
Critical Thinking Activities 178
Writing Activities 178

CHAPTER

8

Business Writing Ethics

Doing business involves transactions between real people. Ethical transactions are principle-based: They treat people according to sound standards spelled out in company policies, professional codes, societal rules, government laws, and religious beliefs.

As a powerful tool for conducting transactions, business writing always has an ethical dimension. For example, writing based on sound principles seeks to treat all people fairly and respectfully. But as you can imagine, writing ethically at work is challenging. Business writers must ethically work through their own—and their organization's—competing interests. In addition, writers must ethically deal with many complex or murky issues—from responsibility for reader safety to proper documentation of research to persuading readers to buy, believe, donate, or act.

Despite these challenges, writing ethically pays off in terms of responsibility to others, compliance with regulations, long-term goodwill, lasting respect, personal integrity, and fulfillment of social and faith-related commitments. In addition, ethical writing honors all parties involved in defining and demanding ethical behavior: the writer, the reader, and interested groups ranging from the local community to the global community.

In a real sense, attention to ethics is present throughout this handbook. (See, for example, the business writer's code of ethics on page 11 in the introduction.) This chapter simply offers focused help on the topic.

WEB Link

For more information on the broader topic of business ethics, check our Web site at http://college.hmco.com/english.

STRATEGIES FOR ETHICAL WRITING

Too often, ethics gets only lip service. These guidelines will help you "walk the talk" when it comes to sound ethics in your business writing.

Measure Impacts

Who will be affected by this message, how, and when? Who benefits, and who doesn't? Analyze consequences of projections, conclusions, and recommendations.

Aim for Clarity

Your main goal should be your reader's clear understanding. Avoid distortions that might create confusion, waste, and danger. Instead, use plain English.

Be Honest and Objective

Present all sides and facts fairly; consider cons as well as pros. Distinguish verifiable facts from opinions, and resist biases that distort the issue.

Follow Standards

Check that the message is consistent with your company's mission and policies, as well as with societal codes, government regulations, and religious principles.

Consult and Confide

Go to a trusted colleague for advice. For serious situations, take the document through a systematic review, including legal counsel.

Avoid Destructive Behavior

These verbal abuses short-circuit your writing:

- **Negative criticism:** Slamming co-workers or your company.
- **Swearing:** Using profane or vulgar language that offends readers.
- **Defamation or libel:** Printing attacks that harm a person's reputation.
- **Discrimination:** Labeling and/or excluding a person based on a stereotype.
- **Cultural insensitivity:** Sharing assumptions, jokes, or offensive terms focused on race, ethnicity, sexual orientation, or other characteristics.
- **Harassment:** Sending messages with sexual innuendo or threats; creating a poisonous, disrespectful work environment.
- **Correspondence warfare:** Sending angry messages that prompt harmful exchanges (e.g., e-mail flaming) as opposed to healthy debate or disagreement.
- **Grapevine gossip:** Spreading juicy news, rumors, and secrets.

Show Integrity

Make commitment and credibility your priorities by avoiding shoddy research, dishonest application writing, illegal interview questions, distorted graphics, deceptive sales pitches, and false advertising. Be open about your point of view, give credit to sources, honor copyright, and guard confidentiality.

> "Ethics is about how we treat each other, every day, person to person. If you want to know a company's ethics, look at how it treats people—customers, suppliers, and employees. Business is about people. And business ethics is about how customers and employees are treated."
> —R. Edward Freeman

INFORMATION ETHICS

Information is power. For that reason, using information requires paying attention to ethical issues. In your writing, address these information problems.

Sharing Inaccurate Information

Problem

When information is vague, distorted, or just plain wrong, your readers may make bad decisions that have wide-ranging harmful results. Sometimes the problem is a matter of emphasis: deemphasizing crucial information or overemphasizing minor details.

Solution

Get and keep your facts straight. In the long run, accuracy pays.

- If necessary, verify your data through a second (or third) source.
- Ask a colleague to review your numbers and calculations.
- While striving for simplicity, honor the complexity of the issue and the data.
- Avoid "tweaking" the data to produce the outcome that you or readers want.
- Give proper weight and perspective to the information that you share.
- Seek feedback to ensure that readers understand the information.

Suppressing Vital Information

Problem

When you don't share vital information openly and fully, you fail to meet your responsibilities to the company, clients, and others. For example, if your product does not include clear instructions, a customer could be frustrated, or harmed. Keeping information scarce, making it inaccessible, and covering it up are the main violations.

Solution

Regularly share information in a timely fashion with those who need it.

- Distribute information to readers based on your answers to these questions: Who needs this information and why? By when is it needed? What form or format will make the information accessible?
- Make products and services safe by providing readily accessible documents. Accurately describe the product, its purpose, and its proper uses. Provide instructions on assembling, installing, using, and disposing of the product; use understandable language, and offer clear cautions, warnings, and graphics. (See Chapter 35, "Writing Instructions.")
- When you know of wrong-doing involving harm, danger, or violation, raise the alarm. If a cover-up happens, "blow the whistle" to the proper authorities.

WEB Link

For more on whistle-blowing, check our Web site at
http://college.hmco.com/english.

Abusing Information Sources

Problem

When you fail to credit information sources or if you copy them without permission, you abuse those sources. Here are the chief abuses:

- **Plagiarism:** Using published material without properly acknowledging the debt.
- **Misrepresentation:** Distorting a source by using a quotation or statistic out of context to say what you want it to say, not what the original means.
- **Unauthorized adaptation:** Changing a source (e.g., a photo) without permission from the creator or copyright holder.
- **Copyright infringement:** Without permission or payment, copying and distributing books, articles, documents, CDs, or other "intellectual property."

Solution

Give credit where credit is due and payment where payment is due.

- Properly document all sources by using quotation marks, offering citations (even for paraphrases and summaries), and (if appropriate) including a references list. See pages 156–166 for more on plagiarism and documentation.
- Honor the information source. Offer enough of the source to accurately convey the original meaning. In addition, don't change the source (e.g., a graphic) without permission.
- Understand what's *common knowledge*. When a fact is well known or available as basic information in several sources, that information is considered common knowledge and you need not document your source. However, if in doubt, document.
- Observe *fair use* guidelines. You may be able to quote small portions of a document for limited purposes, such as education or research. Copying large portions for your own gain is not fair use. You must seek permission and perhaps pay to use the material.
- Understand what's *public domain*. While you need not obtain permission to copy and use public domain materials—primarily documents created by the U.S. government—you should document the source.
- Observe intellectual property and copyright laws. First, know your company's policies on copying documents. Second, realize that copyright protects the expression of ideas in a whole range of formats—writings, videos, songs, photographs, drawings, computer software, and so on. Always obtain permission to copy and distribute copyrighted materials. Third, respect trademarked words by using the trademark symbol (®) when you first use the word. Example: Coca-Cola®.

WEB Link

Add the copyright symbol to important documents that you generate. In fact, have your company register with the U.S. Copyright Office any document with commercial value. Visit www.copyright.gov for more information.

Abusing Confidential Information

Problem

If you dig for or share confidential information, you may violate someone's rights and cause harm. These practices are wrong:

- Stealing secrets through eavesdropping on conversations or reading confidential documents
- Recording conversations without permission from the people involved
- Reading private information in files unrelated to your own work
- Sharing business secrets or private information with inappropriate people or without permission

Solution

Respect privacy and confidences by guarding sensitive information as if it were about yourself or a family member.

- Don't snoop for confidential information—period.
- Restrict others' access to private information you have. For example, when you're away from your desk, don't leave private files on your computer screen or sensitive papers on your desktop. Watch what you say in conversations, what you send to a shared printer, and what you send via e-mail.
- If sharing private information is necessary, follow your company's policies, record your rationale for doing so, and get permission from relevant people.

Mismanaging Information

Problem

If information is managed in a haphazard fashion, the result is chaos. For example, poor financial records may lead to tax problems (or worse), poor product design records could lead to liability lawsuits; poor human resources records could discredit personnel decisions; and faulty environmental records could result in accusations and fines.

Solution

Maintain good paper and electronic records, and keep them well organized.

- For yourself and future readers, make the paper trail and e-trail easy to follow. Information should be simple to retrieve quickly.
- Establish a document retention and destruction policy that distinguishes important from unimportant pieces of writing. In the policy, specify which documents to keep, how to store them, which to discard, and how to do so.
- Develop a sound filing system (paper and electronic) that includes logical names, labels, and categories. See http://college.hmco.com/english for instruction on filing systems and practices.

F.Y.I.

Many of these information-ethics issues relate to legal concerns—copyright, trademark, contract, and liability laws. When appropriate, seek legal counsel.

PERSUASION ETHICS

Business requires persuasion. To handle persuasion ethically in your business writing, use the practices that follow.

Adopting Ethical Habits

Ethical persuasion begins with a positive attitude, proper practices, and sound habits.

- **Honesty.** Be truthful about the cause that you are promoting. Avoid blurring the facts, exaggerating a point, downplaying cons, leaving something out, or telling people only what they want to hear.
- **Responsibility.** Use your influence productively by aiming to motivate, not manipulate. Avoid abusing your power, influence, or position to bully, flatter, trick, or guilt-trip the reader. Also avoid fostering prejudice or groupthink, arguing that the ends justify the means, or adopting a win-at-all-costs mentality.
- **Sensitivity.** Know who your readers are—their identities, objectives, needs, and values. Such awareness will help you consult with and serve them, avoiding assumptions and neglect.
- **Commitment.** Ethical persuasion takes resources—time and energy, for example. Invest these resources in contact, follow-up, and long-term goodwill.

Treating Opposition Fairly

When you face counterarguments, alternatives, competition, and doubts that oppose your cause, treat such opposition—both the people and their arguments—fairly.

Don't trash-talk the opposition. Show respect and understanding for alternatives and competition. Focus on issues, not personalities.

Make concessions. When you acknowledge "points" scored by the opposition, you admit your argument's limits and the truth of other positions. Paradoxically, such concessions strengthen your persuasive power by making your overall argument more credible. Concede points graciously, using words like these:

Admittedly	Granted	I agree that	I cannot argue with
It is true that	You're right	I accept	No doubt
Of course	I concede that	Perhaps	Certainly it's the case

Develop rebuttals. Even when you concede a point, you can often respond with a rebuttal—a small, tactful argument responding to a flaw in the opposition.

- By examining the larger context, show that the opposition leaves something important out of the picture.
- Tell the other side of the story. Offer your own valid interpretation of the evidence; counter with stronger, more reliable evidence; or stress your side's greater benefits.
- Address logical fallacies in the opposition's thinking.

Avoiding Logical Fallacies

The heart of ethical persuasion is a logical argument—a line of reasoning that carries the reader to an acceptable conclusion. Start by making good claims and effectively supporting them (see pages 181–191). In addition, you want to properly combine logical and emotional appeals by engaging readers positively, avoiding emotional excess, and addressing readers' needs (see pages 452–453). However, it's equally important to avoid logical fallacies, unethical mental maneuvers based on weak reasoning. Study the fallacies below, and then test your own writing against this list.

Distorting the Issue

Circular Reasoning. Also known as begging the question, this pattern of thought tries to prove a point by simply rephrasing it.

> The present method of undercoating is *dangerous* because *it isn't safe*.

Either/Or Thinking. Also known as black-and-white thinking, in this fallacy the writer considers only two possible solutions or choices (usually opposites or extremes).

> We've got only two choices: immediately get rid of all internal-combustion forklifts or install a $400,000 ventilation-humidifier system.

Oversimplification. Check your writing for phrases such as "It all boils down to . . ." and "It's a simple question of . . . ." Few things are that simple.

> As far as eliminating cockroaches goes, it's simply a matter of having professionals thoroughly spray all units. (*Never mind common areas, nontoxic methods, ongoing monitoring, or other factors.*)

Bare Assertion. The most basic way to distort the issue is to deny that it exists. This fallacy claims, "That's just how it is."

> High CO levels in the factory must be accepted as an unchangeable by-product of manufacturing.

Straw Man. In this fallacy, the writer argues against a claim that is easily refuted. Typically, such a claim exaggerates or misrepresents the opponents' actual arguments.

> Those who oppose drilling in the ANWR must not care about energy shortages.

Sabotaging the Argument

The Bandwagon Mentality. Groupthink plays on the desire to belong or be accepted. Bandwagon claims tend to avoid the real question by suggesting that if a group of people believe something, it must be true.

> It's obvious to intelligent people that cockroaches are present only in the units of people whose cleanliness leaves something to be desired. If these people clean up their act, the problem will be solved.

Appeal to Emotion. Emotions have a place in persuasion. However, appealing only to pity, fear, sentimentality, negative humor, guilt, or happiness is poor logic.

> Picture Bob Irving struggling frantically beneath the truck, saying his last prayers. Do you want his blood on your hands? Then buy us some new oxygen suits!

Appeal to Popular Sentiment. This fallacy sidesteps critical thought by associating your position with something commonly loved: the American flag, baseball, apple pie.

> It would be totally un-American to consider layoffs while our troops are fighting overseas.

Red Herring. This strange term comes from the practice of dragging a stinky fish across a trail to throw tracking dogs off the scent. With this fallacy, the writer uses a volatile idea to pull readers away from the real issue.

> In 1989, the infamous *Exxon Valdez* oil spill led to widespread animal deaths and enormous environmental degradation. Therefore, oil drilling in the ANWR should not proceed.

Drawing Faulty Conclusions from the Evidence

Hasty Conclusion. When the logical link is missing between the facts and the conclusion, you've leapt without looking.

> Because the Cattle Processing Room is often dirty, we must purchase a power washer.

Broad Generalization. Sweeping statements take in everything and everybody at once, without exceptions or subtleties.

> Automobile manufacturers are not producing fuel-efficient vehicles.

Appeal to Ignorance. This fallacy unfairly shifts the burden of proving a point from the writer to someone else. Since no one has *proven* a particular claim, says the writer, it must be false; or, since no one has *disproved* a claim, it must be true.

> No study has proven that chemicals don't control cockroaches. Therefore, we should spray all apartments.

False Cause. This fallacy confuses sequence with causation: If *A* comes before *B*, *A* must have caused *B*. However, *A* may be one of several causes, or *A* and *B* may be only loosely related, or the connection between *A* and *B* may be entirely coincidental.

> Since the new plant opened, employee absenteeism has sky-rocketed. The plant should not have been built.

Slippery Slope. This fallacy argues that a single step will start an unstoppable slide. While such a slide may occur, the prediction lacks real evidence.

> If we implement affirmative action, it's only a matter of time before our workforce makes us uncompetitive.

Misusing Evidence or Language

Weak or Misleading Comparison. When an idea compares two or more things, test the comparison for clarity and compatibility. Is the comparison full and fair?

> Because a fire hose sprays a greater volume of water, the hose is clearly superior to a power washer.

Half-Truth. Using some data but not all of it, oversimplifying information, presenting details out of context, or burying readers in a blizzard of data—all of these practices can distort the facts at the service of a faulty conclusion.

> At 35 ppm, CO levels factory-wide are only 10 ppm above OSHA recommendations. The problem, therefore, is minor. (*What's left unsaid: The 10 ppm difference is significant, and some areas of the factory have experienced levels between 40 and 80 ppm.*)

Unreliable Testimonial. If your claim relies on a recognized expert on the topic, great. If you use someone without that expertise, your claim loses credibility.

> Nine out of ten dentists fly Royal Tasmania—shouldn't you?

Attack Against the Person. This fallacy directs attention to a person's character, lifestyle, or beliefs rather than the issue.

> Jane Erdrich's wild hobby of rock climbing makes her an unsuitable candidate for this promotion. Do we want a dangerous risk-taker at the helm?

Slanted Language. Words that carry strong positive or negative feelings (connotations) can pull the reader's attention away from a valid argument.

> Before you accept the landscaping bid from Green Thumb, look at how *the slobs* maintain their own property.

Obfuscation. Also known as gobbledygook, this fallacy involves using fuzzy terms to muddy the issue. These words may make simple ideas sound more profound than they really are or false ideas sound true.

> Through the fully functional developmental process of a streamlined target refractory system, the U.S. military will successfully reprioritize its data throughputs.

F.Y.I.

Your persuasive powers are nothing if you lack credibility. Here's how to build trust:

- Create polished documents that are well designed and free of basic errors.
- Demonstrate that you are honest, knowledgeable, and respectful.
- Show sound thinking by treating the topic objectively and logically.
- Develop and guard your reputation through training, experience, and recommendations.
- Keep your promises, avoiding pie-in-the-sky projections and undeliverable deals.

CHECKLIST for Business Writing Ethics

Integrity. My message
- Is honest, true, and clear—and will look positive in the long run.
- Advocates only those actions that are legal and moral.
- Creates an effective electronic and paper trail concerning the issue.
- Avoids manipulating or misleading readers, as well as violating anyone's rights.

Logic. My message
- Looks at the issue from all sides and answers the opposition.
- Avoids rationalizations, excuses, self-deception, and groupthink.
- Offers sound thinking rather than logical fallacies (faulty reasoning).

Loyalty. My message
- Is consistent with my deeply held values and beliefs.
- Guards the best interests of my company, clients, community, and myself.
- Avoids gossip and guards confidentiality.
- Properly "blows the whistle" if the situation involves serious wrongdoing.

Credibility. My message
- Shows that I am fair, consistent, and objective in my thinking and behavior.
- Properly gives credit to contributions from individuals and organizations.
- Is based on advice and feedback from colleagues, managers, and experts.
- Avoids verbal abuse.

CRITICAL THINKING Activities

1. Review "Information Ethics" (pages 171–173). Then analyze three incidents related to violations of information ethics recently reported in the media.
2. Using the principles cited in "Persuasion Ethics" (pages 174–177), analyze how three business documents do or do not follow these principles.

WRITING Activities

1. Working with two classmates, analyze samples of one another's writing—assessing the ethical challenges of each piece, addressing any information or persuasion problems, and looking for logical fallacies. Then revise and refine your document.
2. Research a work-related ethical challenge relevant to your major or profession. In particular, explore how communication issues relate to this ethical challenge.

PART 2

Benchmarking Writing with the Seven Traits

9	The Seven Traits at Work
10	Trait 1: Strong Ideas
11	Trait 2: Logical Organization
12	Trait 3: Conversational Voice
13	Trait 4: Clear Words
14	Trait 5: Smooth Sentences
15	Trait 6: Correct Copy
16	Trait 7: Reader-Friendly Design

In this chapter	The seven traits
Traits of Ineffective Writing **182**	1. Strong Ideas
Assessing an Ineffective Document **183**	2. Logical Organization
Traits of Effective Writing **184**	3. Conversational Voice
Assessing an Effective Document **185**	4. Clear Words
Checklist for Seven-Traits Benchmarking **186**	5. Smooth Sentences
Critical Thinking Activity **186**	6. Correct Copy
Writing Activities **186**	7. Reader-Friendly Design

CHAPTER

9

The Seven Traits at Work

What qualities make for effective business writing? How do you know that your writing is doing what it should, given the dynamic and shifting nature of the workplace and of work itself?

Workplace writers need criteria to assess (or *benchmark*) the quality of writing—both their own writing and that done by others. The seven traits of good writing provide such a benchmarking or quality-control system. These qualities should be present in all workplace writing: ideas, organization, voice, word choice, sentence smoothness, correctness, and document design. For any given document, the question is whether these traits are effective or ineffective. When defined carefully and addressed consistently, the seven traits provide a common vocabulary and set of criteria (a yardstick) for everyone within an organization (classroom or workplace) to measure, revise, and refine writing.

Part 2 of this book presents this seven-traits benchmarking system. Chapter 9 begins that discussion by defining all the traits and showing how they can be used to assess a piece of writing. Chapters 10–16 then take an in-depth look at each of the seven traits, showing in detail what that trait is and how to strengthen it in your writing. Parts 3–6 of the book then explain how you can develop—and measure—the seven traits in all workplace forms, from memos to mission statements.

You can use this set of eight chapters in two ways. First, you can use them in class to study, discuss, and practice specific writing strategies. Second, you can use the chapters out of class to work through your own self-directed writing tutorials.

Part 2 Benchmarking Writing with the Seven Traits

TRAITS OF INEFFECTIVE WRITING

Ineffective writing has one or more of the traits that are described below and illustrated in the letter that follows. Study each trait, along with the sample letter. As you do so, reflect on which problem areas give you the most challenge in your own writing.

Weak ideas
- The main point is unclear.
- Discussion is fuzzy, sketchy, disjointed, incomplete, or poorly researched.
- Thinking is simplistic or illogical.

Disorganization
- The reader gets lost because of rambling or confusing order.
- The opening lacks background, context, or a purpose statement.
- Ideas are unconnected.
- The closing fails to conclude or to clarify action.
- The patterns used (e.g., cause/effect, chronological order, question and answer, problem/solution, alphabetical order) don't fit the form or company guidelines.

Awkward voice
- The tone is bored, pompous, defensive, negative, wishy-washy, or phony.
- Expression is too familiar or too formal for the situation.
- The attitude seems self-focused rather than topic- or reader-focused.

Unclear words
- The prose is wordy, repetitive, and vague.
- The piece contains clichés, slang, jargon, and negative or insensitive language.
- Key words and technical terms are used incorrectly or without definition.

Rough sentences
- The prose is hard to read and it requires rereading to make sense.
- The sentence rhythm is choppy, rambling, or tired.
- Parallelism, coordination, and subordination need work.
- Active and passive voice are poorly used.

Incorrect copy
- The piece contains grammar, punctuation, spelling, and mechanics mistakes.
- Errors trip up the reader and make a bad impression.

Unprofessional design
- The format is incorrect, incomplete, or inconsistent with the form.
- The page layout and typography are unattractive, unclear, and unreadable.
- The absence of lists and headings makes information difficult to access.

Chapter 9 The Seven Traits at Work

ASSESSING AN INEFFECTIVE DOCUMENT

The annotations to the right of this letter show the problems cited by the writer's peer editor. Study both the letter and the annotations to understand the problems and their likely effects on the reader.

HANCOCK COUNTY ARTS COUNCIL

327 South Road • Bar Harbor, ME 04609-4216 • (207) 555-1456

March 28, 2005

Ms. April Wadsworth,
Belles Lettres Books,
The Harbor Mall:
Bar Harbor, MA 046093427

Dear ms. Watsworth,

    Do you like art? Many Bar Harbor businesses see it as their duty to participate in The Purchase Awards Program. The program works because business people agree to attend ArtBurst and also agree and promise to purchase artwork (at a designated dollar amount), hence attracting artists and visitors. Everyone's a winner with the Purchase Awards Program!

I am the director of the Hancock County Arts Council. This council sponsors each year ArtBurst—a fair where artists display and sell their work. Well-known artists like William Drummond and Leslie Blass and many local artists like Susanna Reese show their wares: beautiful stained glass, classic landscapes in oil and watercolor, glass sculptures, wood work, pots, and much much more. ArtBurst is a real art feast for the community. Artbust will be help this year in Central Park on Saturday, May 7. ArtBurst brought in last year many artists and thousands of visitors that were good for the local economy, I am proud to say.

    Completion of the enclosed application form will ensure commitment and participation in this grand event! Therefore, I implore you to please give this request due and proper consideration.

Yours Sincerely;

Lawrence King

Lawrence King, Director

[ideas]
- Content incomplete
- Argument lacks power

[organization]
- Opening unclear
- Order weak
- Transitions missing

[voice]
- "I" focused in middle
- Too formal at end

[word choice]
- Some business jargon at end
- Some vagueness about action

[sentence smoothness]
- Some rambling and choppy sentences

[correctness]
- Many distracting errors, including misspelling the reader's name

[design]
- Several errors in letter format

Part 2 Benchmarking Writing with the Seven Traits

TRAITS OF EFFECTIVE WRITING

Effective writing has the traits described below and illustrated in the letter that follows. As you study each trait and the sample letter, reflect on why these qualities make writing effective. In addition, consider which qualities are strong in your own writing.

Strong ideas
- The piece focuses on a main point.
- Supporting points are logically developed and well explained.
- Information is accurate, precise, complete, current, and well researched.

Logical organization
- A strong three-part structure (opening, middle, closing) guides the reader.
- The piece is direct or indirect, as appropriate.
- Details are ordered sensibly (e.g., categories, problem/solution, order of importance, chronological order); the patterns used fit the form and company requirements.
- Transitions link sentences, paragraphs, and sections.

Conversational voice
- The tone is positive, polite, knowledgeable, confident, sincere, and convincing.
- The piece shows "you attitude"—paying attention to the reader's perspective.
- The voice helps connect with and encourage readers.

Clear words
- Words are fresh, natural, and understandable.
- Key words and technical terms are precise—and defined, if necessary.
- Language is sensitive to gender, race, ethnicity, and international cultures.

Smooth sentences
- Sentences are concise and easy to read.
- Lengths and patterns are varied through parallelism, coordination, and subordination.
- Active and passive voice are used effectively.

Correct copy
- Grammar, punctuation, spelling, and mechanics are correct.
- Correctness creates logic, guides the reader, and makes a positive impression.

Reader-friendly design
- The format is correct, complete, and consistent with the form and company guidelines.
- The page layout and typography make the document attractive and easy to read.
- Lists and headings make information accessible.

Chapter 9 The Seven Traits at Work

ASSESSING AN EFFECTIVE DOCUMENT

The annotations to the right of this letter show the strengths cited by the writer's peer editor. Study both the letter and the annotations to understand why the writing will likely have a positive impact on the reader.

HANCOCK COUNTY ARTS COUNCIL
327 South Road • Bar Harbor, ME 04609-4216 • (207) 555-1456

March 28, 2005

Ms. April Wadsworth
Belles Lettres Books
The Harbor Mall
Bar Harbor, ME 04609-3427

Dear Ms. Wadsworth:

Does your store or office have a bare wall or corner that needs a painting, photograph, or sculpture to fill it?

For fourteen years, your Hancock County Arts Council has sponsored ArtBurst—a fair where artists display and sell their work. Last year, ArtBurst attracted more than ninety artists and 15,000 visitors. This year, ArtBurst will be held in Central Park on Saturday, May 7.

The Purchase Awards Program, supported by local businesses, is key to ArtBurst's success. Business people like you join the program by agreeing to purchase artwork. Your commitment attracts better artists and more visitors. You, the local economy, and the arts all win.

You win two ways. First, you get a beautiful print, painting, drawing, photograph, or sculpture to decorate your business. Second, ArtBurst publicity materials advertise your support.

So please join the Purchase Awards Program! Just complete the enclosed form and return it in the postage-free envelope by April 12. A few weeks later, Ms. Wadsworth, you and your customers will be enjoying beautiful art!

Yours sincerely,

Lawrence King

Lawrence King, Director

P.S. An Arts Council member can help you choose the right artwork for your business. Call for details.

[ideas]
- Main point is clear (boldface)
- Argument is persuasively developed
- Information is complete

[organization]
- Tidy three-part structure
- Effective indirect organization
- Clear sequencing and transitions
- Tidy paragraphing
- Closing focuses on action

[voice]
- Tone positive, polite
- "You attitude" throughout

[word choice]
- Fresh, natural, plain English

[sentence smoothness]
- Easy to read
- Short to middle length
- Nice variety

[correctness]
- No distracting errors

[design]
- Proper letter format
- Attractive layout and typography

CHECKLIST for Seven-Traits Benchmarking

Use the seven traits to check your document and then revise as needed.

Strong ideas
- ☐ The piece focuses on a main point.
- ☐ Supporting points are developed logically and explained well.
- ☐ Information is accurate, complete, current, and well researched.

Logical organization
- ☐ A strong three-part structure (opening, middle, closing) guides the reader.
- ☐ The organization is direct or indirect, as appropriate.
- ☐ Details are ordered sensibly (e.g., categories, order of importance).
- ☐ Transitions link sentences, paragraphs, and sections.

Conversational voice
- ☐ The tone is positive, polite, knowledgeable, and natural.
- ☐ The piece shows "you attitude"—paying attention to the reader's perspective.

Clear words
- ☐ Words are fresh and natural; technical terms are defined, if necessary.
- ☐ Language is sensitive to gender, race, ethnicity, and international cultures.

Smooth sentences
- ☐ Sentences are concise and easy to read.
- ☐ Lengths and patterns are varied through parallelism, coordination, and subordination.

Correct copy
- ☐ Grammar, punctuation, spelling, and mechanics are correct.
- ☐ Correctness creates logic, guides the reader, and makes a positive impression.

Reader-friendly design
- ☐ The format is correct, complete, and consistent with the form.
- ☐ The page layout and typography make the document attractive and easy to read.

CRITICAL THINKING Activity

Use the seven traits to evaluate the quality of three business documents that you find. Based on your assessment, grade each document as an A, B, C, D, or F. Report your findings to the class.

WRITING Activities

1. Working with a classmate, exchange a piece of your writing and use the checklist above to assess both pieces. Discuss strengths and weaknesses.
2. Find a piece of workplace writing that you recently composed. Using the seven traits, assess and revise the piece.

In this chapter	The seven traits
Stating Ideas Clearly **188**	1. **Strong Ideas**
Supporting Ideas Effectively **190**	2. Logical Organization
Thinking Creatively **192**	3. Conversational Voice
Thinking Logically **194**	4. Clear Words
Using Thinking Patterns (from Describing to Evaluating) **195**	5. Smooth Sentences
	6. Correct Copy
Checklist for Ideas **214**	7. Reader-Friendly Design
Critical Thinking Activities **214**	
Writing Activities **214**	

CHAPTER

10

Trait 1: Strong Ideas

Without quality content, writing is just a lot of hot air. In fact, nothing deflates a document faster than weak ideas and thin support. Conversely, nothing commands the reader's respect more convincingly than powerful ideas. For this reason, ideas are the substance of any business document, regardless of its form—report, proposal, or even brief e-mail. While the other traits of effective writing are important, they depend upon the ideas. Ideas always provide the substance of any workplace document, and the other traits are merely strategies for shaping that substance.

How do you develop powerful ideas? First, you strengthen your thinking skills, such as creativity and logic. These skills help you generate ideas, test their quality, and polish their content. Second, you deliver those ideas. For example, you state an idea effectively, support it well, draw logical conclusions, and shape recommendations. Third, you use reliable idea-development methods such as thinking patterns. These patterns (such as defining, comparing, and evaluating) help you develop and deliver your ideas.

Developing powerful ideas takes time and work, involvement and investment, creativity and logic. However, while developing ideas is complex, it's a *skill*—not a *mystery*. The purpose of this chapter is to help you strengthen that skill.

STATING IDEAS CLEARLY

Some workplace writing simply presents facts. However, most business writing requires that you work with information by stating ideas or making claims about it: analyzing data, drawing conclusions, and developing recommendations. A good idea or claim has these characteristics:

- It requires explaining or arguing to make sense to readers.
- It can be developed or defended with sound logic and sufficient evidence.
- It offers an ethically responsible statement.
- It is precise and clear, not general or vague.
- It offers interest and practical insight for readers.

To state ideas effectively in your writing, use the four guidelines that follow.

Obey the Fact–Idea Distinction

A *fact* is information that can be checked and proven correct. An *idea* is a thought about facts—an opinion, conclusion, prediction, or summary. A good idea is based on facts, but it is not a fact itself. Build your writing with strong ideas supported by facts.

Idea

In the factory, CO levels between 35 and 80 ppm are a serious problem.

Supporting Facts

- Those levels are 10–55 ppm above OSHA recommendations and could lead to a fine.
- High CO levels may potentially lead to employee illness.
- High CO levels force us to use exhaust fans that reduce relative humidity, shrinking the wood we use for manufacturing.

Phrase Claims Precisely

Unless a good idea is well phrased, the reader can't understand it and can't use it. State your claims effectively by following this pattern:

> A specific subject (*in italics below*)
> + a specific thought, conclusion, opinion (**in boldface below**)
> = a good claim statement

A power washer **will keep the Cattle Processing Room cleaner than our present method.**

The intangible benefits of your gift **include helping millions of Americans get better jobs and personal dignity.**

Earth-Scape's natural, sustainable landscape **will require less long-term care, put less stress on the Nooksack watershed, and offer educational value.**

Use Claims That Match Your Purpose

Claims fall into three categories: statements of truth, value, or policy. Whether you're making a point, offering a conclusion, arguing for a recommendation, or stating a principle, choose the best type of claim for the job.

Use **claims of truth** to argue that something is or is not accurate or correct. For readers to accept your claim as trustworthy, you must support it carefully. Avoid truth claims that are exaggerated, impossible to prove, harmful, or libelous.

> **REASONABLE TRUTH CLAIM:** The current clean-up procedure in Cattle Processing falls short of what's needed by being labor-intensive, costly, wasteful, and unsanitary.

Use **claims of value** to argue that something does or does not have worth. For readers to accept your judgment, you must support it by referring to a known standard or by establishing one. Avoid value claims based solely on personal taste, individual preference, or emotional bias.

> **VALUE CLAIM WITH CLEAR STANDARDS** (customer satisfaction and savings): Replacing our current cleaning equipment with a power washer would have distinct client-related and economic benefits.

Use **claims of policy** to argue that something should or should not be done. For readers to be convinced enough to approve your course of action, your policy claim must be practical and desirable. Avoid claims that are simply complaints or wishful thinking.

> **PRACTICAL POLICY CLAIM:** After checking several retailers, I recommend that HVS purchase a Douser washer.

F.Y.I.

As the sequence of sample claims shows, workplace documents often move from claims of truth and value to claims of policy. Follow this pattern in your own writing, relying on a solid foundation of truth claims when you construct conclusions and recommendations.

Qualify Your Ideas

While you want to state your ideas clearly and boldly, you also want to be reasonable. Statements including words that are strongly positive or negative (such as *all, best, every, never, none,* or *worst*) may sound unconvincing because they are tough to support. That's where qualifiers come in. They make your statements more moderate: *almost, usually, maybe, probably, possibly, often, some, most, might,* and *in most cases*. Note the difference between the two statements below. The first is an all-or-nothing claim, whereas qualifiers make the second more reasonable.

> As an independent consultant, I **always** regulate my own work. With your proposal, I would **certainly** lose that flexibility and **never** get it back.

> As an independent consultant, I **can** regulate my own work. **Although** your proposal is financially attractive, I would lose **some** of that flexibility **in order to** do a proper job for you and your clients.

SUPPORTING IDEAS EFFECTIVELY

If you want readers to accept an idea, it must be well supported with reasoning and evidence. Effective support meets four criteria:

- **Accurate.** Each detail is correct, not vague or confusing.
- **Complete.** Full, pertinent facts are provided.
- **Concrete.** General statements are clarified with specific details.
- **Focused.** All the information relates to the stated idea.

To develop such support, follow these four guidelines.

Use Provable Facts

Any details you present should be clearly provable. For example, compare the supporting facts for this statement: "We decided not to use protective air systems."

> **PROVABLE FACT:** The gas leak was minor—detectable only with an agent.
> **NOT PROVABLE FACT:** We knew nothing dangerous would happen with the leak.

Answer Your Readers' Questions

When you make a point, readers generally have three questions that you need to answer.

[what do you mean?]
Clarify, expand, restate the idea

[can you prove it?]
Add supporting details, such as statistics, facts, examples, and analysis
Interpret the data so that their meaning is clear

[why is it important?]
Summarize impacts, outcomes, costs, benefits, and/or effects

Problem: Emissions from Lift Trucks

Since November 2004, we have been recording high levels of carbon monoxide (CO) in the factory, particularly in Area 3, during the winter months (November–March). General CO levels in the factory have exceeded 35 ppm, and many office spaces have experienced levels of 40–80 ppm, when the OSHA recommendation is 25 ppm.

These CO levels are a concern for three reasons:

- They range from 10 to 55 ppm above recommendations and could bring a substantial fine.
- High CO levels make people ill. The results are sick leave, lower productivity, and dangerous work conditions.
- Using summer exhaust fans in winter reduces relative humidity, shrinking wood used for manufacturing.

Acknowledge Alternatives and Opposition

As you support your idea, show awareness of other options and perspectives. Especially in persuasive writing, concede points and offer counterarguments. For more help, see "Persuasion Ethics," pages 174–177.

Use Fitting Types of Support

Aim to provide the type and amount of evidence that leads the reader to understand and agree with your idea. Avoid both too little and too much support—measured by how simple or complex your document is, and by what your reader needs.

Observations and anecdotes share what you or others have sensed or experienced. Such evidence offers an "eyewitness" perspective, which can be powerful.

> I then checked all the lift trucks to see whether they were producing more CO than normal. All lifts were in good condition and were being used properly.

Illustrations, examples, and demonstrations support general claims with specific instances, making such statements concrete and observable.

> Replacing our current cleaning equipment with a power washer would have distinct benefits. The intense pressure quickly removes even the driest material from both flat and hard-to-reach places. While brushes leave streaks and dirt behind, a power washer does not.

Numbers and statistics offer quantitative data to back up a claim. However, numbers don't speak for themselves. They need to be interpreted properly, not slanted or taken out of context. They also need to be up-to-date, relevant, and accurate.

> General CO levels in the factory have exceeded 35 ppm, and many office spaces have levels of 40–80 ppm, when the OSHA recommendation is 25 ppm.

Expert testimony offers insights from an authority on the topic. But be aware that such testimony has limits: Experts don't know it all, and they may disagree.

> After checking several retailers and consulting *Consumer Reports*, I recommend that HVS purchase a Douser power washer.

Predictions offer insights by forecasting what might happen under certain conditions. To be plausible, a prediction must be rooted in a logical analysis of present facts.

> The shipping area shows the best potential for using electric lift trucks. Shipping lift trucks do not travel long distances or use ramps.

Analysis examines parts of a topic through thought patterns—cause/effect, compare/contrast, classification, process, or definition. Such analysis helps make sense of a topic's complexity, but muddles the topic when poorly done.

> High CO levels force us to use summer exhaust fans during the winter. However, using these fans reduces relative humidity, shrinking wood used for manufacturing.

Tests and experiments provide hard data developed through the scientific method, although these data must be carefully studied and properly interpreted.

> Through testing, I discovered that our current cleaning method costs $53.15 per day, whereas a power washer reduces that figure to $13.05.

Analogies compare two things, creating clarity by drawing parallels. However, every analogy breaks down if pushed too far.

> A power washer cleans like a vigorous scrub brush but without the abrasion.

THINKING CREATIVELY

Workplace writing is often writing-on-the-run, no-nonsense writing to get results. At other times, getting results means getting creative. To think creatively, learn to explore *what if* questions, tap innovation, and develop comparisons.

Explore *What If* Questions

Creative thinkers see things differently than other people. They see challenges rather than problems. They suspend the rules and imagine by asking "what if" questions. To develop fresh ideas, use these questions with your topic:

- *What if* the topic (person, place, thing, problem, idea) existed in another time or place?
- *What if* one key element of the topic were changed? What would happen?
- *What if* object A were brought together with object Z? What would the reaction be?
- *What if* the topic could talk? What would it say?
- *What if* an object were made of completely different material?
- *What if* the situation were the opposite of the present?

Consider the thinking behind the development of the Internet and the World Wide Web. *What if* computers that are far apart could be networked together? *What if* Internet information could be accessed through hyperlinks?

Tap Innovation

Innovation makes space for positive change to happen. By finding or developing creative solutions, innovation seeks to improve things. Tap your innovative potential with the following techniques.

Get Inspired

Take a lesson from art and religion by seeking out inspiration. Be open to new challenges, commitments, and callings. Look out for accidental discoveries and the teaching power of mistakes. Read books, magazines, and newspapers; watch documentaries; join discussion groups.

Develop Synergies

Synergy refers to combined effects that are greater than the sum of individual parts. For instance, two prescribed drugs taken together may interact to more potently combat illness than the same drugs taken separately. To develop synergistic thinking, pull ideas and information from a diverse range of fields to spark potent results. Forge and nurture partnerships with people and ideas.

Shift the Paradigm

Models and structures are tremendously helpful for creating productive frameworks and processes. However, sometimes you need to break the mold by asking, "Why must we do it this way? What principle or rule tells us that this is the way to go?" Once you've isolated that fundamental principle, test it with new knowledge. Does current thinking encourage you to shift or change the paradigm?

Consider the development of the personal computer. That revolutionary change required a paradigm shift away from seeing computers as large devices that filled rooms.

Develop Fresh Thoughts Through Dynamic Techniques

Encourage the free flow of your ideas by using some of these techniques: freewriting (page 22), clustering (page 22), and graphic organizers (pages 20–21). Better yet, encourage idea development by collaborating with others (page 53).

Develop Comparisons

Perhaps you remember comparisons (metaphors and similes) from a literature class. While business writing isn't poetry, a strong comparison can deepen and extend your idea by connecting it to something unexpected, or by relating it to something familiar. In the next example, notice how the basic idea comes to life through comparison. Then create your own comparisons by following the four guidelines below.

> **BASIC IDEA:** A power washer cleans effectively.
>
> **COMPARISON:** A power washer cleans like a vigorous scrub brush but without the abrasion.

Create Original Comparisons

Avoid clichés—tired metaphors that no longer convey original ideas. If you find yourself asserting that a problem *is a fine kettle of fish* that's taking the company *out of the frying pan into the fire,* it's time to give your writing *a sound thrashing.*

> **CLICHÉ:** A power washer cleans like magic.

Make Comparisons Clear and Consistent

Avoid confusing and mixed comparisons.

> **CONFUSING:** A power washer is a vigorous *scrub brush* that can *lick the plate clean* for any dirty job.

Extend Comparisons Effectively

A good metaphor or simile can often unify a paragraph, a section, or even a whole document.

> **EFFECTIVE:** A power washer cleans like *a vigorous scrub brush* but without the abrasion. The intense pressure quickly *breaks up and dissolves* even the driest material from both flat and hard-to-reach places. But the water *brushes away* the dirt without scratching painted and stainless steel surfaces. While *brushes* leave streaks and dirt behind even after tiring *scrubbing,* a power washer quickly and effortlessly rinses all surfaces spotless.

Avoid Stretching Comparisons Too Far

Don't overdo or overuse metaphors to the point that your writing sounds forced.

> **STRETCHED:** A power washer is the flame thrower of the cleaning industry— incredibly powerful and deadly to dirt.

THINKING LOGICALLY

For thinking to be logical, it has to be **reasonable** (supported with good explanations), **reliable** (backed up with solid evidence), and **believable** (convincing to readers). To develop logical ideas that pass these tests, practice sound thinking methods, develop a solid line of reasoning, and use the thinking patterns on pages 195–213.

Practice Sound Thinking Methods

When you need to make a decision, solve a problem, or investigate a topic, follow the steps outlined below:

Decision-making method

Imagine the outcome or result desired.
Define your goal as a thinker and writer.
Explore (list) your options.
Choose two or three best options.
Investigate the chosen options; compare and test them.
Decide on the best option.
Evaluate your choice.

Problem-solving method

Identify the problem.
Describe the problem (e.g., parts, causes, effects, history).
Explore possible solutions or responses.
Adopt and defend the best plan or solution.

Scientific method

Identify the topic, issue, or question.
Make observations directed at deeper understanding.
Advance a hypothesis: an educated guess, explanation, or prediction.
Gather data and test it against your hypothesis.
Investigate further, observing and collecting more data.
Note data and draw possible conclusions.
Establish a single conclusion.

Develop a Line of Reasoning

Develop a clear line of reasoning in which each point you make clearly supports your conclusions. This line of reasoning might develop naturally as you study the topic, or you could adopt the following formal approach:

1. Introduce the question or concern; make your claim.
2. Make, discuss, and support a strong argument for the claim.
3. Present other arguments supporting the claim.
4. Acknowledge objections and offer concessions and rebuttals.
5. Make, discuss, and support your strongest argument for the claim.
6. Conclude by consolidating, modifying, and amplifying the claim, as needed.

USING THINKING PATTERNS (FROM DESCRIBING TO EVALUATING)

A clear thought pattern (sometimes called a method of development) helps you develop an idea quickly and effectively. Check out the thought patterns briefly introduced in Table 10.1, study "Combining Thinking Patterns" on the next page, and review, as needed, the specific thinking patterns discussed on the pages that follow.

Table 10.1 Thinking Patterns

Whenever You Need to		Then
Inform		**Share What You Know**
List	Describe	• Collect and organize information
Name	Label	• Clarify main points
Share	Forward	• Supply details
Summarize	Reply	
Explain		**Answer Questions and Apply Knowledge**
Define	Clarify	• Discuss principles
Restate	Illustrate	• Clarify the meanings of terms
Demonstrate	Show	• Give examples
Model		• Rephrase ideas and restate information
		• Provide steps in a process
		• Show how something works
Analyze		**Break Down a Topic**
Break down	Rank	• Carefully examine a topic
Conclude	Classify	• Partition it into parts and subparts
Compare	Contrast	• Create categories of information
Tell why	Give results	• Show connections and cause/effect links
Interpret	Characterize	• Discuss similarities and differences
		• Prioritize points
Synthesize		**Reshape Knowledge into New Forms**
Combine	Connect	• Invent a better way of doing something
Speculate	Design	• Blend the old with the new
Compose	Create	• Predict or hypothesize
Predict	Develop	• Mix concepts and practices from different fields
Invent	Imagine	
Innovate	Assemble	
Persuade		**Convince Others to Believe and Act**
Recommend	Judge	• Evaluate a subject's quality, value, or worth
Criticize	Argue	• Explain what needs to be done and why
Evaluate	Rate	• Point out strengths and weaknesses, benefits and costs
Convince	Assess	
Propose	Choose	
Measure	Anticipate	

WRITE Connection

To effectively practice these thinking patterns, you must avoid logical fallacies—soft spots in your reasoning. For help, see pages 175–177.

Combining Thinking Patterns

The thinking patterns outlined on the following pages isolate specific thinking strategies and show them at work. In some workplace writing, a single thinking pattern dominates the document. For example, instructions are shaped by the thinking pattern of process description.

However, most workplace writing seeks to smoothly combine thinking patterns. In such writing, the thinking patterns are put at the service of communicating with real readers about a specific topic. While seeking to demonstrate sound thinking, writers aim to get something done with their writing—to inform, to solicit feedback, to prompt a decision or action. Various thinking patterns are simply tools used to that end. To combine thinking patterns productively, consider the following tips.

Blend Naturally

When you are familiar with the different thinking patterns and practice them effectively, blending patterns in your documents can become second nature. But don't force the issue or overthink it while you are drafting. Instead, when you are revising the document, spend some time examining the thinking embedded in your writing: Is the direction of your thinking solid? Are specific passages of cause/effect, compare/contrast, or classification properly developed? Does a section require an extended definition to clarify a key concept?

Signal Thought Patterns Structurally

Sometimes, the situation or document may call for a more structured and systematic presentation of thinking. For example, a troubleshooting proposal requires (1) an objective description of the problem, (2) a clear cause/effect analysis of its origins and seriousness, (3) a synthesis and evaluation of alternative solutions, and (4) an implementation plan (process description) for the recommended solution. In this case, the dominant patterns of thought for each section can be indicated in headings. In fact, the pattern of thought can be clearly signaled in each paragraph's topic sentence, and paragraph structure can be used as a way of ordering thinking patterns. (For more help with paragraphing, see pages 225–231.)

Give Patterns Appropriate Space or Weight

For each document you write, think about its purpose and its audience. With these two issues in mind, review the range of thinking skills and patterns that workplace writers use. Then research, plan, draft, and revise your document with these questions in mind:

- What types of thinking are foundational for this document—description, definition, partitioning, classification, compare/contrast, cause/effect, synthesis, evaluation, inductive/deductive?
- Should the foundational types of thinking receive equal weight in the document, or should one or another pattern dominate the whole, with other patterns put at the service of that overarching pattern?
- If you use multiple thinking patterns, what headings and transitions would signal a smooth shift from one pattern to another?

Chapter 10 Trait 1: Strong Ideas

Example: Passage Combining Thinking Patterns

In the article excerpt below, notice how the writer combines definition, compare/contrast, partitioning, and more to explain hybrid-vehicle technology. Later sections of the article use process description to clarify the evolution of the hybrid, as well as evaluation to discuss performance and efficiency. (Note that graphics indicated are not included.) The article is taken from the "How Stuff Works" Web site (www.howstuffworks.com).

What Makes it a "Hybrid"?

Any vehicle is a hybrid when it combines two or more sources of power. In fact, many people have probably owned a hybrid vehicle at some point. For example, a **mo-ped** (a motorized pedal bike) is a type of hybrid because it combines the power of a gasoline engine with the pedal power of its rider. ⎫ *Informal definitions*

Hybrid vehicles are all around us. Most of the locomotives we see pulling trains are **diesel–electric hybrids.** Cities like Seattle have diesel–electric **buses**—these can draw electric power from overhead wires or run on diesel when they are away from the wires. Giant **mining trucks** are often diesel–electric hybrids. **Submarines** are also hybrid vehicles—some are **nuclear–electric** and some are **diesel–electric.** Any vehicle that combines two or more sources of power that can directly or indirectly provide propulsion power is a hybrid. ⎫ *Extended definition by example* / *Formal definition*

The **gasoline–electric hybrid car** is just that—a cross between a gasoline-powered car and an electric car. Let's start with a few diagrams to explain the differences. . . . ⎫ *Compare/contrast*

Hybrid Structure

You can combine the two power sources found in a hybrid car in different ways. One way, known as a **parallel hybrid,** has a fuel tank, which supplies gasoline to the engine. But it also has a set of batteries that supplies power to an electric motor. Both the engine and the electric motor can turn the transmission at the same time, and the transmission then turns the wheels.

By contrast, in a **series hybrid** (**Figure 4** below) the gasoline engine turns a generator, and the generator can either charge the batteries or power an electric motor that drives the transmission. Thus, the gasoline engine never directly powers the vehicle.

Take a look at the diagram of the series hybrid, starting with the fuel tank, and you'll see that all of the components form a line that eventually connects with the transmission. ⎫ *Classification, partition, contrast, process description*

Hybrid Components

Hybrid cars contain the following parts:
- **Gasoline engine:** The hybrid car has a gasoline engine much like the one you will find on most cars. However, the engine on a hybrid is smaller and uses advanced technologies to reduce emissions and increase efficiency.
- **Fuel tank:** The fuel tank in a hybrid is the energy storage device for the gasoline engine. Gasoline has a much higher energy density than batteries do. For example, it takes about 1,000 pounds of batteries to store as much energy as 1 gallon (7 pounds) of gasoline. . . .

⎫ *Partition, physical description, contrast*

Part 2 Benchmarking Writing with the Seven Traits

Thinking Pattern 1: Describing People, Places, or Things

When describing things that exist spatially (people, places, or things), focus on physical details—shapes, sizes, parts, distances, composition, colors, textures, and so on. Spatial description is especially important in trip reports (page 504), incident reports (page 486), and instructions (page 568). Whether you're describing a microscopic cell, a simple thermos, or a factory site, follow these guidelines.

Base the Description on Good Observation Strategies

For a discussion of effective observation, see pages 130–131. Follow these tips:

Create an accurate record by using a field book and recording device.

Use your five senses. Record impressions and details available:

- **Sights.** Record colors, shapes, and appearance. See the big picture and the little details.
- **Sounds.** Listen for loud and subtle, harsh and pleasant, natural and mechanical sounds. If people are present, record relevant conversations.
- **Smells.** Check out both pleasant and unpleasant odors: what's sweet, spicy, sweaty, pungent, sour, rancid, and so on.
- **Textures.** Safely test things for temperature, smoothness, roughness, thickness, and so on.
- **Tastes.** If the site permits it, taste for sweet, sour, bitter, and so on. Note the relationship between smells and tastes.

Take measurements (e.g., distances, sizes, weights) to gain precise physical data.

Partition the site or object into physical or functional parts (see page 204). For example, a simple fork is made of one physical part (one metal whole), but it has at least two distinct functional parts (the handle and the tines).

Determine the object's composition and material. Out of what substances is it made—wood, plastic, a metal alloy?

Build Your Description so That Readers "See" the Object

A successful physical description makes sense of the site, object, or person by presenting in a sensible order those details that the reader needs.

Provide a "big picture" overview that defines, orients, and locates the object. Preview the parts, and use signposts (e.g., *next to, above, below*) to show links between those parts. For example, a thermos description might begin with this overview:

> The wide-mouth thermos is a cylindrical container used to store both hot and cold liquids. Also called a vacuum flask, it maintains food temperature by creating "dead air" insulation. The 9-inch-high thermos sits on a four-inch-diameter, disc-like base. This container has six parts that screw together: (1) the cup, (2) the lid, (3) the outer plastic casing, (4) the middle glass cylinder, (5) the inner plastic food compartment, and (6) the base.

Chapter 10   Trait 1: Strong Ideas

Organize Details to Meet Readers' Needs

Organize the description into parts, moving from left to right, from top to bottom, from outside to inside, from far to near (whatever order makes sense). For example, a thermos description could move from top and outside to bottom and inside.

Include as much detail as your reader needs, explaining features and their significance. For example, if your reader needs a basic understanding of a thermos, a paragraph might do; otherwise, your description might dedicate a full paragraph or more to each thermos part. In addition, to stress the significance of the three-inch-diameter thermos mouth, you could add that this feature allows for easy filling and pouring.

Provide graphics such as drawings, photographs, and maps. (See pages 102–105.) For example, a thermos description could be accompanied by a cut-away or exploded line drawing.

Clarify shapes with geometrical terms familiar to your reader. Here are both two-dimensional and three-dimensional options:

Two-dimensional shapes	Three-dimensional shapes
Angles (right, acute, obtuse)	Cubes
Triangles (equilateral, isosceles, right angle)	Cones
Squares, rectangles	Cylinders
Pentagons, hexagons	Spheres (full or hemi)
Circles (full, half or hemi), ovals	Ellipsoids

Compare the object to something familiar (see page 209). Such comparisons can cover a broad range of objects; specific comparisons might be alphabet-based (A-frame house, C-clamp, O-ring) or based on body parts (saw teeth, needle eye).

Example: Physical Description

The following passage is taken from a trip report (pages 506–507). Note how the writer uses the guidelines above to describe the site of a gas leak.

> The Eugene crew had exposed the leak area on both sides and the top of the pipe. As a first step, we assessed damage and conditions:
> - Approximately 6–12 inches of mud covered the work area, including the ramps into the ditch.
> - The ditch contained water up to the pipe's bottom.
> - A steady rain was falling.
> - The leak came from two quarter-inch cracks on the seam at 10 o'clock looking south.
> - We had three hours of daylight left, with a Light Plant on site.
> - We had limited space for a hydro crane and emergency trailer.

F.Y.I.
Note in the example above how the writer's details help the reader picture both the problem (cracked pipe) and the conditions affecting the crew's repair work.

Thinking Pattern 2: Describing Processes (Chronological Order)

Workplace writing is often time logical. Time logic is particularly important in collection letters (page 466); chronological résumés (page 321); periodic, incident, progress, and trip reports (page 485); meeting minutes (page 673); research and troubleshooting proposals (page 521); and procedures (page 618) and instructions (page 565). In such documents, you may have to lay out sequences, steps, or events to describe

- How to do something (expecting the reader to perform the actions).
- How something is done, works, or happens (expecting the reader simply to understand the process).

Whatever the process, follow these guidelines.

Thoroughly Understand the Process

To describe a process, you must study it closely. Most useful is first-hand experience through observing or participating. You must know and show how each step leads logically to the next and how all the steps work together to complete the process.

Be Clear and Complete

Shape the description based on how readers will use it and how much they already know about the process. Consider addressing your description to the reader who knows the least about the process. Here are some options:

- **Process description.** To inform readers how something happens naturally, describe how the process unfolds (e.g., how cancer cells multiply).
- **Process explanation.** To help readers who want to know how something is made or done, explain how someone would complete each step (e.g., how microchips are manufactured).
- **Instructions.** To help readers perform the process themselves, supply step-by-step, how-to directions (e.g., how to sterilize your hands for hospital work—see page 571).

Give the Big Picture First

As with a physical description, provide an overview and necessary background before the details. Explain the process's principle, purpose, or outcome, for example. Alternatively, relate the process to a larger set of phenomena.

Divide the Process into Stages and Sub-stages

Given the whole sweep of the process, how does it break down into discernible stages? How is each stage composed of a grouping of related events or actions?

Make the Process Familiar

Use precise terms, well-chosen adjectives, and clear action verbs so that readers can see the process unfold. Add graphics (e.g., flowcharts or time lines) as needed.

Use Time Cues

Send time signals to readers with words such as *step, phase, stage*; transitions like *first, second, next, finally*; or an actual numbering system (*1, 2, 3, 4*).

Chapter 10 Trait 1: Strong Ideas

Example: Process Description

In the passage below, the writer practices these guidelines in describing for nonscientists how hair grows.

Hair Today, Gone Tomorrow

Imagine a pitcher's mound covered with two layers of soil: first a layer of clay, and on top of that a layer of rich, black dirt. Then imagine that 100,000 little holes have been poked through the black dirt and into the clay, and at the bottom of each hole lies one grass seed.

Slowly each seed produces a stem that grows up through the clay, out of the dirt, and up toward the sky above. Now and then every stem stops for awhile, rests, and then starts growing again. At any time about 90 percent of the stems are growing and the others are resting. Because the mound gets shaggy, sometimes a gardener comes along and cuts the grass.

Your skull is like that pitcher's mound, and your scalp (common skin) is like the two layers of soil. The top layer of the scalp is the epidermis, and the bottom layer is the dermis. About 100,000 tiny holes (called follicles) extend through the epidermis into the dermis.

At the base of each follicle lies a seed-like thing called a papilla. At the bottom of the papilla, a small blood vessel drops like a root into the dermis. This vessel carries food through the dermis into the papilla, which works like a little factory using the food to build hair cells. As the papilla makes cells, a hair strand grows up through the dermis past an oil gland. The oil gland greases the strand with a coating that keeps the hair soft and moist.

When the strand reaches the top of the dermis, it continues up through the epidermis into the open air above. Now and then the papilla stops making new cells, rests awhile, and then goes back to work again.

Most of the hairs on your scalp grow about one-half inch each month. If a strand stays healthy, doesn't break off, and no barber snips it, the hair will grow about 25 inches in four years. At that point hair strands turn brittle and fall out. Every day between 25 and 250 hairs fall out of your follicles, but nearly every follicle grows a new one.

Around the clock, day after day, this process goes on . . . unless your papillas decide to retire. In that case you reach the stage in your life—let's call it "maturity"—that others call "baldness."

- The writer uses an analogy to introduce the process and distinguish its steps.
- An illustration shows parts of a hair stem.
- He explains the analogy.
- The writer explains steps analogous to those in the illustration.
- He closes with a brief summary and humor.

Thinking Pattern 3: Defining Terms

A definition clarifies meaning through an equation: *term x = explanation y*. Your job is to show that the explanation on the right amounts to the same thing as the term (or referent) on the left. In workplace writing, defining may be important in any document, but it is especially relevant in technical reports and policy statements. Sometimes, such definitions are collected in a glossary or appendix for a longer document. Generally, however, such definitions are handled in the body of the text, as the need arises. Definitions can be formal, informal, or extended.

Formal Definitions

Write a formal, single-sentence definition when readers need a term defined precisely. Follow this formula:

Term	= Larger class	+ Distinguishing features/characteristics
A brand name	is a design, symbol, or term	that identifies the goods or services of a specific company and distinguishes them from those of competitors.
A general ledger	is an accounting journal or computer program	where all business financial transactions are recorded and the accounts are balanced.
To swerve	is to diverge	from a regular line of motion.

In the larger class and features, use words that readers will clearly understand. Don't confuse your definition by introducing terms that themselves require definition. For example, when defining a thermos for children, avoid using the term *vacuum*.

Make the larger class fairly narrow. If it's too broad, you have to add many distinguishing features. For example, when defining a thermos, the larger class *food-storage container* is better than *object*.

Distinguishing features should focus on the term's key or unique properties, its action or function, its appearance or substance. For example, a definition of a personal digital assistant (PDA) could focus on its compactness and portability, its ability to store and manage information, or the technological elements that make it work.

Avoid repeating the term in the definition.

> **INCORRECT:** *Breakeven* is the point at which sales income and production costs break even.
>
> **CORRECT:** *Breakeven* is the point at which sales income equals production costs.

Avoid using *is when*, which is an ungrammatical structure.

> **INCORRECT:** Importing *is when* you purchase goods and services from other countries.
>
> **CORRECT:** *Importing* is the process of purchasing goods and services from other countries.
>
> **OR:** *To import* is to purchase goods and services from other countries.

Informal Definitions

Write an informal definition when (1) readers will easily understand a synonym or phrase placed within parentheses, commas, or dashes, or (2) you want to specify a term's meaning for the particular document.

> An oxygen suit (a suit that supplies air and protects its wearer from fumes and skin contact) must be worn during undercoating.
>
> Successful affirmative action, defined for this report as a 15 percent increase in minorities and women within the company, was the goal set by the 2001 task force.

Extended Definitions

Write an extended definition when readers need an expanded discussion of a term. Start with a formal definition of the term, follow up with explanatory sentences, and conclude by relating the term to the rest of the document. To expand a definition, select strategies that fit your purpose and the reader's needs:

- Explain the origin of the word—its history and background.
- Explain specific terms in the formal definition—clarify each distinguishing feature.
- Provide examples that make the basic definition concrete.
- Explain the word with comparisons familiar to the reader.
- Use graphics to help the reader "see" the term.
- Explain the term's basic principle and provide applications.
- Explain causes and effects related to the term.
- Explain what the term is not.

Example: Extended Definition

In the following example, the writer offers a formal definition of "brand name" and then extends the definition by explaining what makes for a good brand name and offering three examples (including graphics or logos).

> A brand name is a design, symbol, or term that identifies the goods or services of a specific company and distinguishes them from those of competitors. From the company name and logo mounted on a building to the product label, a brand name is key to a company's identity. Therefore, a company should develop its brand name by considering (1) product name, sound, and lettering;
> (2) typographical style, color, and size; and (3) distinguishing symbols and logos. In particular, the brand name should be timeless and distinct—memorable to the public and different from that of competing companies. So give brand names serious thought to come up with something attractive and memorable, such as the names and logos shown here.

Thinking Pattern 4: Partitioning Objects or Processes

Partitioning takes one object or process and divides it into parts or stages. That is, partitioning logic is parts/whole logic. Partitioning is particularly helpful in any document describing an object, place, or process (e.g., incident reports, trip reports, and instructions). To partition effectively, follow these guidelines.

Choose a Logical Principle for Partitioning the Object, Event, or Process

Make sure that your reader understands this principle. If necessary, explain the principle and offer a rationale early in your writing. Consider these options:

Physical parts. Break up the physical object based on its actual physical components. For example, a simple pencil could be divided into (1) the graphite "lead," (2) the wood surrounding the lead, (3) the eraser, and (4) the metal collar around the wood and eraser.

Functional parts. Focus on specific parts in terms of their purpose or role. A single physical part may have several functional parts. For example, a simple thermos cup has these functional parts: a base for sitting level on a surface, sides for containing food or drink, a lip for pouring, a handle for lifting, and threads for attaching the cup to the thermos.

Phases or stages. Divide a process according to (1) set time periods, such as minutes or hours, or (2) completion of key actions or developments. For example, the process of developing and marketing a new product might be broken down by monthly benchmarks or by key achievements (e.g., development of a prototype, testing of a prototype, development of an initial marketing plan).

Give Readers an Overview of the Whole Before Treating Each Part

Individual parts of an object or specific phases of a process make sense only if understood in the context of the whole object or entire process. Therefore, provide the big picture: Define the object or process, explain its nature, relate it to similar objects and processes, list its main parts or phases, and explain the organizing principle behind it. For example, the partitioning of a thermos could be prefaced by a brief explanation of the vacuum principle behind it, a statement of its purpose, and a preview list of its main parts.

Divide the Material Systematically

In terms of levels, depth, and balance, be guided by the reader's need for detail and order. For example, in a document intended for consumers, the partitioning of a DVD player would focus on supplying practical, user-focused information. In a document intended for technicians, the DVD partitioning would require greater detail related to maintenance and repair of components.

Use Graphics

A good graphic helps readers visualize parts or phases. For object partitioning, use line drawings such as exploded drawings and cross sections (page 105). For process partitioning, use a flowchart (page 100).

Example: Partitioning of an Object

The description of an ultraviolet water disinfection system shows parts/whole logic at work.

> The Rankin System UV2000 combines the benefits of efficient electronic ballast technology with the advantages of our open-channel modular design.
>
> **UV Modules**
>
> UV modules are the basic building blocks of the UV Package Treatment System. Disinfection takes place within an array of lamps, with the UV lamps submerged in the liquid parallel to the flow.
>
> Power supply to the UV modules is provided through standard outdoor ground-fault protected receptacles. The UV lamps are enclosed in quartz sleeves attached to the module inside a waterproof connector. All connecting wires are protected inside the waterproof frame.
>
> Electronic ballasts are enclosed within a waterproof housing, which eliminates the need for separate ballast panels and simplifies installation and maintenance.
>
> **UV Channels**
>
> UV modules are mounted in stainless steel channels. An overflow weir maintains correct water depth over the complete range of flow rates to be treated, and a built-in drain facilitates channel cleaning . . .

Example: Partitioning of a Process

In the implementation section of a troubleshooting proposal, the writer maps out a process of switching from internal-combustion forklifts to electric forklifts.

> Rankin buys an average of four lift trucks annually. The switch to electric lifts in Shipping can take place through purchasing only electric ones for the next four years. Here's how the process would work:
>
> 1. An area requests replacement of an existing lift truck.
> 2. Management approves the request and purchases a new electric lift truck.
> 3. The new electric lift truck goes to the Shipping Department.
> 4. The newest internal-combustion lift truck in Shipping is transferred to the area that requested the new lift truck.
>
> Through this process, all lift trucks in Shipping would be electric by December 31, 2007. To make this process work, we would need to create a charging location near Shipping.
>
> Therefore, I recommend that Rankin management approve this plan, and phase it in over the next four years as outlined. Improved safety, increased product quality, and lower total long-term operating costs outweigh higher initial costs.

Thinking Pattern 5: Classifying Information

When you classify information, you place items in distinct categories. Classification is particularly useful in résumés (page 320), major reports (page 509), job descriptions (page 626), company profiles (page 620), brochures (page 552), and newsletters (page 560). To classify, clump material under headings that identify the information.

Establish a Principle for Grouping

Given your purpose for presenting the information, find a basis or principle for categorizing items. This principle becomes the "common denominator" of your classifying work. For example, you could group or divide trees based on principles like these:

- **Size:** Types of trees grouped by height categories.
- **Geography:** Trees common to different areas, zones, or elevations.
- **Structure** or **composition:** Division by leaf type (deciduous versus coniferous).
- **Purpose:** Wind-break trees, shade trees, flowering trees, fruit trees.

Develop the Classification Scheme Sensibly

To help the reader, develop categories by following these guidelines:

- Limit the number of main categories. For example, two to five main categories make for a manageable structure.
- Make sure that the categories you designate cover the topic or body of information completely, but without overlap.
- Create subcategories to show depth of structure. Work down two, three, or four levels—depending on how much detail you must share with readers.
- Order the categories and subcategories according to a sensible principle that will help readers digest the classification scheme.
- Complement your discussion with graphics such as tables, charts, diagrams, drawings, and photographs—whatever can help your reader understand both the overall structure of the classification scheme and the individual categories.

Example: Classification Passage

Below, the writer clumps the company's products based on the types of applications and services. She can then discuss each category in greater depth (not shown).

> Since 1979, Rankin Technologies has been leading the switch from chemical disinfection to safe, efficient, ultraviolet, water disinfection. Rankin's research and development team brings together international experts in microbiology, chemistry, and engineering to create some of today's most successful innovations in UV technology. In particular, Rankin provides these water purification technologies:
>
> - Wastewater disinfection (medium- and low-pressure applications)
> - Community drinking water, industrial process, and ultrapure applications
> - Residential lower-flow purification
> - Recreational waters disinfection

Thinking Pattern 6: Ranking by Importance

When you rank information, you are using order-of-importance logic. Ranking is especially helpful in informative messages (page 400), résumés (page 320), investigative reports (page 490), major reports (page 509), proposals (page 521), job descriptions (page 626), and news releases (page 548). Whether your ranking scale is based on numbers (e.g., quantities, sizes) or judgment (e.g., a list of recommendations), follow these guidelines.

Establish the Logic of Your Scale

Early in your writing, state the principle behind your ranking scale. If necessary, explain and support that principle. What rule, value, or benchmark allows you to position items relative to each other, and why is that principle valid and valuable? For example, in the marketing report excerpt below, market segments are prioritized based on their profit potential—the benchmark is economic.

Choose Between Ascending and Descending Order

Most often, use descending order—from most important to least important—because readers tend to want the most important information first. Moving from least to most important may make sense in some writing when you wish to build climactically or persuasively to the best item by first presenting and ruling out other choices.

Signal Each Item's Position in the Scale

Let readers know where they are in the ranking with numbers, letters, or words such as *greatest, least, highest,* and *lowest*.

Example: Ranking Passage

Check the ranking logic at work in the following passage, an excerpt from a major report (see page 512). The writers rank potential markets for their company's herbicides.

> The North American Small Grains market contains these four main segments, listed in priority: the North Central States, Western Canada, the Pacific Northwest, and the South Central States (Figure 1). Order of priority is based on grain types planted, total acreage planted, major broadleaf weeds and grasses found in the segment as they match with Rankin's product portfolio, and regional economic conditions.
>
> **A. North Central States**
>
> The North Central segment of the United States provides the greatest market potential for Rankin. In this area, growers plant . . .
>
> **B. Western Canada**
>
> The Western Canada market segment contains 43,100,000 acres . . .
>
> **C. Pacific Northwest**
>
> Rankin's third market priority, the Pacific Northwest, offers potential sales of . . .
>
> **D. South Central States**
>
> The South Central States constitute Rankin's fourth Small Grains market . . .

Thinking Pattern 7: Following Deductive and Inductive Logic

When you analyze information to arrive at conclusions, you are using your powers of deduction and induction (See Figure 10.1). Deduction and induction are especially useful in writing claims (page 446), adjustments (page 416), special requests (page 454), incident reports (page 486), investigative reports (page 490), and proposals (page 521). To use these lines of reasoning in your writing, follow these patterns.

Deduction

Move from general to specific by first making a general claim or stating a principle, then supporting the claim or applying the principle to develop conclusions.

Induction

Move from specific to general by first presenting and exploring specific facts, then developing generalized principles or conclusions from those facts.

Figure 10.1 Induction Versus Deduction

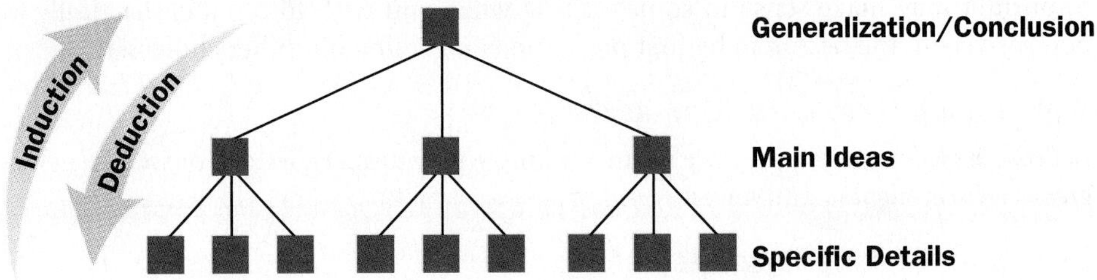

However you use or combine these methods, develop a logical train of thought and use transitions to highlight the logical links.

Example: Induction Passage

The writer uses induction to explain decisions that a gas-line repair crew made. After describing the site conditions (not shown, but see page 506), he answers three questions.

> **Repair Plan and Decisions**
>
> With these conditions in mind, we discussed three repair issues: the possibility of more leaks, clamp selection, and safety precautions.
>
> *Were there any more leaks?* Eugene crew members probed and sniffed a 3-foot section of pipe on each side of the leak and found no evidence of more leaks.
>
> *Which clamp would be best?* Klem indicated that the clamp would not need to be welded. We decided to use the 24-inch bolt-on clamp so that we could clamp past the girth weld next to the leak.
>
> *What safety precautions were necessary?* We decided not to use the air systems for these reasons:
>
> - The gas from the leak was minor and detectable only with a detecting agent.
> - Using air systems was necessary and time consuming.

Thinking Pattern 8: Comparing and Contrasting

Compare/contrast thinking examines options—with comparison focusing on similarities, and contrast focusing on differences. Essentially, this writing holds up side by side two or more objects, events, plans, or ideas so that they can be related and distinguished for the purpose of understanding and evaluation. Compare/contrast thinking is especially important in sales messages (page 460), proposals (page 521), and most reports (page 481). To compare and contrast effectively, follow these guidelines.

Know Your Purpose
Do you need to supply just the facts, or do you need to recommend one item over the other? Is your writing informative or evaluative?

Have a Solid Basis for Comparison
Make sure that all the items can be compared fairly. Comparable items are types of the same thing (e.g., two photocopiers) and of the same order—one cannot simply be an example of the other. For instance, IBM-compatible PCs and Macs are of the same order, whereas a Dell is simply an example of a PC.

Choose Criteria on Which to Base the Comparison
To develop the comparison, select features that will focus the comparison. Then apply these criteria consistently, perhaps using a Venn diagram to map out areas of similarity and difference (see page 20). For example, if you were comparing trees for a landscaping project, you could use these features: height, canopy, shade density, leaf type, leaf color, flowers and fruit, growing rate, and longevity. You would prioritize these features and establish preferences based on your purpose or need.

Develop Your Comparison
Using one of the following patterns as a framework, flesh out the comparison with sufficient detail.

- Whole versus whole discusses items separately, giving a strong overview of each. Use this pattern with short, simple comparisons.
- Topic by topic holds items up side by side, criterion by criterion. This pattern stresses fine distinctions, and is helpful in long, complex comparisons.
- Similarities versus differences highlights shared and distinct traits. Consider what relative weight you want to give the similarities and differences.

Example: Comparison and Contrast Passage
The writer uses purity, taste, and odor as criteria to compare and contrast two methods of disinfecting water. She employs a whole versus whole pattern.

> Although widely used for disinfection, chlorine has disadvantages. It reacts with organic materials found in water to produce unpleasant tastes and odors. In addition, chlorine produces some by-products, such as chloroform, that . . .
>   UV, by comparison, naturally disinfects without chemicals—no tastes, odors, or by-products. UV destroys pathogenic microbes by delivering concentrated . . .

Thinking Pattern 9: Analyzing Cause and Effect

Cause and effect are particularly important in claims (page 446) and adjustments (page 416); incident, investigative, and periodic reports (pages 486, 490, & 496); troubleshooting, justification, and sales proposals (page 522); and policies (page 614) and procedures (page 618). Cause/effect thinking can move in two directions. First, it can explore the effects of a particular event, action, or phenomenon—the logical results. Second, it can trace backward from a particular result to those forces that created that result—the causes. With this thinking pattern, your job is to establish and explain cause/effect links.

Explore the Cause/Effect Evidence

Initially, your job is to test all possible explanations for a given phenomenon's causes or effects. Consider these options:

- **Causes.** What forces can be designated as primary causes? What forces are secondary or contributing causes? What relations are simply coincidental? What evidence—measurements, testimony, and so on—supports or disproves the causal links? Can the links be tested through experiment?
- **Effects.** What are the primary, secondary, and ripple effects? Which are immediate, and which are long-term? What is the seriousness of each effect?

Supply Reliable Details That Support the Logical Connection

When you present your cause/effect analysis, readers must be able to believe your logic.

- Avoid relying on circumstantial evidence.
- Do not draw conclusions without adequate support.
- Avoid mistaking sequence for a cause/effect link. (If A came before B, A didn't necessarily cause B. See the false-cause fallacy, page 176.)

Consider Alternative Explanations

Address other possible explanations and outside forces that might have affected the cause/effect links.

Example: Cause/Effect Passage

Here the writer works out logical cause/effect relationships by explaining the effects of water quality on ultraviolet disinfection.

> The quality of water to be treated with UV radiation can have a large effect on an ultraviolet unit's performance. First, suspended solids can shield bacteria from ultraviolet light and should be removed by pre-filtration. Second, chemicals such as iron, manganese, and sulfides can lead to coating of the integral lamp/sleeve. This coating, which blocks UV light, can be addressed with a UV monitor plus regular cleaning of the integral lamp/sleeve. Third, optical clarity below 254 nm interferes with how deeply UV light penetrates. If water has a faint yellow color when viewed to a reasonable depth, such as in a white bucket, then organic UV-absorbing substances such as humic acids are present and should be neutralized.

Thinking Pattern 10: Evaluating Options

When you evaluate, you judge a topic by examining all its angles. Evaluation is particularly important in claims (page 446) and adjustments (page 416), responses to proposals (page 412), incident reports (page 486), proposals (page 521), and employee evaluations (page 636). Rooted in strong understanding and analysis, evaluation typically wrestles with two issues: judgments of quality or judgments of responsibility.

Judgments of Quality

Does the service, product, or program match the need?

- Establish evaluation criteria—standards or benchmarks for judging the topic. If necessary, explain and justify your criteria.
- Apply the criteria fairly, objectively, and systematically to the topic. How does it measure up against your benchmarks?
- Once you've drawn conclusions, state them clearly, along with details.

Judgments of Responsibility

Who or what is responsible for an event? (Note that this question can apply to either assigning fault or giving credit.)

- Establish definitions needed to develop your judgment. What are the key concepts surrounding the issue of responsibility in this case?
- Construct a cause/effect chain of reasoning that is accurate, logical, and supportable. Avoid the false-cause fallacy (page 176).
- Objectively assign the fault and/or credit; include supporting details.

Example: Evaluation Passage

In the passage below, the writer establishes criteria for measuring solutions to the problem (high CO levels in the factory) and then advocates for the strongest solution.

> Ideally, any solution implemented should accomplish the following in a timely and cost-effective manner:
>
> - Bring down CO levels so that they are consistently at or below 25 ppm
> - Maintain the relative humidity of the factory air to ensure product quality
>
> Given these criteria, we could continue using the exhaust fans and install humidifying equipment, but such a renovation would cost $32,000. (See attached estimate.) Alternatively, we could immediately replace all internal-combustion lift trucks in Area 3 with electric lift trucks for $175,000.
>
> Instead, I propose that we gradually replace the internal-combustion lifts in the Shipping Department with electric lift trucks, for these reasons:
>
> - The Shipping Department currently has 14 lift trucks that operate almost 24 hours per day. These lifts are clearly the major source of CO in Area 3.
> - The Shipping area also shows the best potential for using electric lift trucks. Shipping lift trucks do not travel long distances or use ramps.
> - The pros of purchasing electric lift trucks outweigh the cons.

Thinking Pattern 11: Synthesizing Material

When you combine information and ideas creatively, you are synthesizing. Synthesis is especially important in promotional writing (page 547), sales messages (page 460), investigative reports (page 490), proposals (page 521), mission statements (page 610), and brochures (page 552). While analysis (partition, compare/contrast, cause/effect) breaks down a topic into parts to see how it works, synthesis pulls material together to see what will happen. (See "Thinking Creatively," page 192.) Synthesis takes knowledge that you have, reshapes it or combines it with other ideas, and explores new possibilities. To synthesize effectively, follow these tips.

Think "Sideways" Rather Than Straight Ahead

While logic generally moves linearly forward, creative thinking moves sideways to connect things that are not normally related. Stretch your imagination by looking closely at what you have for surprises, twists, and odd relationships. Then combine your information with other material, reshape it into another form, or look at it through a different "window." For example, to improve the hiring process that your HR Department has in place, you could synthesize multiple forms of feedback from department managers, job candidates who've been through the process, HR assistants who take care of the nuts and bolts, and outside consultants who review your process.

Ask Questions of Your Topic

Synthesis works when you ask pointed questions and seek openly for answers. Here are some options:

- **Combining.** How can things be linked, associated, or blended?
- **Predicting.** Under these conditions, what might happen?
- **Inventing.** What new thing can be made?
- **Designing.** How can material be pulled together and shaped?
- **Proposing.** What fresh changes can you suggest?
- **Redesigning.** How can it be done better?
- **Imagining.** How can the topic be looked at differently?

Bring Order to Your Synthesis

The process of synthesizing material, because it's creative in nature, is often dynamic but messy. When you present the results of your synthesis, give the material a logical shape. For example, a working group that has spent months researching and brainstorming a vision for your organization's future might synthesize its findings in a strategic plan that, in an orderly but inspiring way, maps out a mission, a set of goals and objectives, action steps, and a time line for implementing the plan.

Example: Synthesis Passage

The passage that follows is taken from an investigative report (see pages 490–495). The writers have synthesized their research, problem analysis, and solution brainstorming to design a cockroach-control program. Like a lot of synthesis, the writing seeks to be comprehensive and visionary.

We recommend that Sommerville Development adopt an Integrated Program of Cockroach Prevention and Control for its 5690 Cherryhill Blvd. building. Management would assign the following tasks to appropriate personnel:

Education

1. Compile a list of home remedies.
2. Prepare fliers and/or newsletters providing tenants with information on sanitation, prevention, and home remedies.
3. Hold tenant meetings to answer questions.

Cooperation

To ensure 100% compliance, revise the current lease to include program requirements and management's access to all units. Provide tenants with advance notice and accurate information. Plan treatments to minimize inconvenience and ensure safety.

Habitat Modification

Revise the maintenance program and renovation schedule to give priority to the following:

1. Caulk cracks and crevices where roaches hide or gain access to water and food sources (baseboards, cupboards, pipes, sinks). Insert steel wool in large cavities (plumbing, electrical columns).
2. Apply residual insecticides before sealing cracks and entry points.
3. Repair leaking pipes and faucets. Wrap pipes with insulation to eliminate condensation.
4. Ensure regular cleaning of common garbage areas.

Treatment

In addition to improving sanitation and prevention through education, attack the roach population through these methods:

1. Use home remedies, traps, and hotels.
2. Place diatomaceous earth, a nontoxic, safe, yet residual pesticide, in dead spaces to repel and kill cockroaches.
3. Use borax or boric acid powder formulations as residual, relatively nontoxic pesticides.
4. Use chemical controls as a last resort—only on an emergency basis, no more than once per year.
5. Ensure safety by arranging for a Health Department representative to make unannounced visits to the building.

Monitoring and Evaluation

Monitor the cockroach population in the following ways:

1. Every six months, use sticky traps to collect data on cockroach activity.
2. Keep good records on the degree of occurrence, population density, and . . .

CHECKLIST for Ideas

Use the following directives to strengthen the ideas in your writing.

- ☐ **Be patient and thorough.** Do you examine your topic carefully? Do you research the topic thoroughly, test the quality of sources, and take accurate notes?
- ☐ **Ask questions.** Do you ask pointed questions that challenge easy answers?
- ☐ **Be objective.** Do you examine the topic in an unbiased way, exploring all possibilities, testing all conclusions, and seeking ideas from others?
- ☐ **Think creatively.** Are you open to inspiration, discovery, new sources, fresh angles, and new solutions?
- ☐ **Think logically.** Do you follow a logical pattern of thought such as cause/effect, comparison/contrast, or process? Do you test for logical fallacies?
- ☐ **Make connections.** Do you explore relationships between details by looking for patterns and using comparisons and analogies to clarify your thinking?
- ☐ **Write down ideas.** Do you explore your ideas through writing strategies such as listing, clustering, and freewriting? Do you reread and revise?

CRITICAL THINKING Activities

1. Working with two classmates, exchange a piece of your writing. Read each piece together, looking for each of the following: (a) a clear thesis or main idea, (b) solid supporting ideas, (c) relevant supporting details, and (d) logical thinking patterns (e.g., cause/effect, comparison/contrast, definition). Discuss the effectiveness of these strategies.
2. Find a sample of workplace writing composed of at least three or four paragraphs. Study the claims that the piece makes and the support it offers. In a memo to your classmates and instructor, assess the effectiveness of the thinking in the document.

WRITING Activities

1. With two classmates, select one of these campus-related topics: housing, food services, off-campus programs, distance learning, general-education requirements, computer services, or your preferred issue. Develop claims of truth, value, and policy for your claim. Test each claim by seeking to support it with reliable evidence. Brainstorm types of research that would be required to develop such support.
2. Select a common but fairly simple workplace or household object (e.g., staple remover, nail clippers, coffee grinder). Using techniques of description, definition, partitioning, classification, and comparison/contrast, develop a document introducing the object to readers who are unfamiliar with it.

In this chapter	The seven traits
Strategies for Getting Organized 216	1. Strong Ideas
Foolproof Organization Strategies 218	**2. Logical Organization**
Structuring Documents Through Paragraphing 225	3. Conversational Voice
Checklist for Organization 232	4. Clear Words
Critical Thinking Activities 232	5. Smooth Sentences
Writing Activities 232	6. Correct Copy
	7. Reader-Friendly Design

CHAPTER
11

Trait 2: Logical Organization

In the workplace, time is money, and reading takes time. Poorly organized writing wastes this valuable time, whereas well-organized writing saves it—thanks to qualities like strong titles, descriptive headings, purposeful openings, clear topic sentences, succinct lists, and tidy closings. Such devices help readers quickly identify the topic, learn the purpose, grasp the ideas supporting the purpose, and respond to the message.

Imagine a reader tackling your e-mail, report, or proposal. Maybe she's preoccupied with other work, perhaps she's wondering how your writing is relevant, and undoubtedly she's busy. This reader starts by skimming the piece—looking for the "bottom line." If she doesn't find your topic, main idea, and overall argument *quickly*, she will probably discard the writing without reading it. Even if she does struggle through the poorly organized piece to get its point, she will likely think less of both you and your message because the reading process took longer than necessary.

How do you organize your writing to help such busy readers? You start by learning how effective organization strengthens any workplace document. Then you practice strategies that make your document's structure transparent, simple, and logical. Finally, you use those strategies whenever you draft and revise your writing. While this learning may take you some time, the outcome will be worth it. You will be able to produce a well-organized document that gets your reader's attention, delivers your message efficiently, and earns her respect!

STRATEGIES FOR GETTING ORGANIZED

Any document you write—from a simple e-mail to a major report—should have a reader-friendly, efficient, and transparent organization. To achieve this goal, choose structures that make sense and control the flow of information.

Choose Structures That Make Sense

The structure of your memo, letter, or report should logically fit the context, create order in a common-sense way, and show awareness of the reader's reading of the document. To discover what's sensible for your document, consider this advice.

Follow the Rules

If your company expects your document to follow a certain structure, then use that pattern. For example, all periodic reports might be expected to follow the same format. Modify such structures only with good reason—and after obtaining permission.

Think About Your Readers

Decide how to organize your writing for readers by answering these questions:

- What order, structure, and format do readers expect—a brief, direct memo; chronologically ordered instructions; a cause/effect impact statement?
- What do readers need first, second, third, and so on? In other words, how should information be sequenced so that readers can access the information and build understanding? Consider, for example, how you would supply five W's information (who, what, when, where, why) in the lead to a news release.
- In the whole document and individual paragraphs, how should you answer these questions: *What's your point? Can you prove it? Why is it important?*

Follow the Internal Logic of Your Content

Organize your document, in part or whole, around a specific train of thought. For specific help, check the patterns of thought outlined in Chapter 10, "Trait 1: Strong Ideas."

Use Guidelines and Models

If the piece you're writing is a common document type (e.g., a proposal or a complaint letter), then find a template or guidelines:

- Review the guidelines, outlines, and models elsewhere in this handbook. For example, use the SEA formula for good-news messages (page 399), the BEBE formula for bad-news messages (page 427), and the AIDA formula for persuasive messages (page 453). Or check the outlines provided for different reports and proposals.
- Use one of the graphic organizers shown on pages 20–21.
- Follow one of the standard outlines provided later in this chapter (pages 222–223).

Control the Flow of Information

If information comes too slowly or with too much verbiage, readers get impatient. If information comes too fast, readers feel buried in details. Control the flow of information by using the techniques described here.

Put First Things First

Start with the most important information—your reason for writing, your main point, background and context, a polite greeting, conclusions, recommendations, or a summary of your findings. Follow this flow:

Main points → Supporting points → Details

Clump Like with Like

Keep related points together, not sprinkled around, so that the reader sees their relationship. If you find yourself frequently writing, "As I mentioned earlier," or coming back to the same details, related ideas are probably too scattered. Especially if your document is long, divide the content into clear sections and unified paragraphs.

Break up Long Sections of Text

Avoid long, wordy paragraphs—dense blocks of prose that are difficult to decipher and digest. Also, use headings and lists to "break out" information for readers and to indicate relationships between sections and pieces of information. Such techniques make your document's organization transparent.

Stress Beginnings and Endings

Openings and closings of documents, paragraphs, and sentences are positions of greatest emphasis with readers. This material is what they see first and read most carefully. Place key points and words in those spots.

Forecast and Summarize

Use introductions to documents and sections to forecast which topics and ideas will be discussed. (See page 498 for a sample forecasting statement.) Then sprinkle your documents with well-placed summaries that remind readers what they've just read. (For help with writing summaries, see page 128.)

Use Transitions

Transitional or "linking" words, sentences, and paragraphs help readers connect information and follow your train of thought. (See "Paragraph Coherence," page 230, and "Transition Paragraphs," page 226.)

Follow the Known/New Principle

Because readers "build" meaning as they proceed, shape your documents (from sections to paragraphs) so that new information is introduced gradually and is linked to information that the reader already knows.

FOOLPROOF ORGANIZATION STRATEGIES

With workplace writing, you don't have to invent a new structure for each document. The three-part structure and direct versus indirect patterns provide well-tested blueprints. Outlines, summaries, subject lines, heading systems, and lists provide additional order. These organizational strategies help readers quickly discern levels of ideas, patterns of thought, and relationships between points.

Three-Part Structure

Most workplace writing involves creating variations on an opening, middle, closing pattern—whether the message is short and simple or long and complex. This pattern is so useful that we have indicated that structure in the sidenotes for virtually every model in this book. Review the structure below, noting the options available for each part.

[opening goals]
To establish the purpose of the piece.
To get the reader "tuned in."

Strategic Choices
State your reason for writing.
Greet the reader; start with goodwill.
Get the reader's attention with a dramatic statement.
State your main point.
Summarize your conclusions and recommendations.
Give background details and context.
Define key terms.
Preview your document's contents; map out where you're going.

[middle goals]
To advance the message's purpose by providing the full explanations and details the reader needs.

Strategic Choices
Develop a structure that fits the document type, supports your purpose, and builds to the closing.
Arrange details by drawing on one or more of these patterns (see "Using Thinking Patterns," page 195, for help):

Questions and answers   General to specific
Order of importance   Specific to general
Alphabetical order   Compare/contrast
Categories   Problem/solution
Partitioning   Cause/effect
Spatial order   Claims and counterarguments
Chronological order

[closing goals]
To focus on outcomes, action, the future—fulfilling the purpose.

Strategic Choices
Stress the importance of the information.
State conclusions and recommendations.
Explain how to use the information.
State what you will do next.
Request or clarify action the reader needs to take: what to do, when, and how.
Look forward to future contact.
Offer further help, communication, or steps.

Direct Versus Indirect Organization

Knowing your reader's likely response to your message helps you effectively organize the message. To forecast that response, do the following:

Consider your reader's position. What is his or her role and authority? What is your relationship with the reader?

Measure the reader's knowledge of the topic and its context. What does your reader already know about the situation? What will be new or surprising?

Think about the reader's likely attitudes toward your topic and main point. What will he or she be thinking while digesting your message?

Based on your conclusion, choose either a direct or an indirect approach to your topic. If the reader will be open to your message, be direct by getting to the point quickly. If the reader might be closed to your message, be indirect by building up to the point with an explanation or argument. See Table 11.1.

Table 11.1 Direct Versus Indirect Organization

Direct	Indirect
Be direct when the reader's response will be neutral or positive.	Be indirect when the reader's response will be negative, indifferent, or resistant.
Opening	**Opening**
Supply context. State the main point (e.g., conclusion, request, information).	Get the reader's attention. Indicate the topic. Supply necessary background.
Middle	**Middle**
Discuss the main point and supporting points. Supply necessary information (information the reader needs or wants, arranged logically).	Supply information, proof, reasons, and analysis to build a persuasive argument. State your main point. If useful, address counterarguments or offer other options.
Closing	**Closing**
Perhaps restate the main point. Look forward. End positively and politely.	Focus on action, next steps, or the future.

WRITE Connection

Three-part structure and direct versus indirect patterns are discussed and shown in action throughout this handbook in the guidelines and models for each type of writing. In addition, outlines for reports and proposals provide helpful templates. Check out the following:

- "The Art of Being Direct" (pages 398–399)
- "The Art of Being Tactful" (pages 426–427)
- "The Art of Persuasion" (pages 452–453)

Outlines

An outline can be formal or informal. It can list topics, phrases, or complete sentences. It can guide your drafting, and then become a polished outline, a heading system, or a table of contents. (See, e.g., the major report, page 513.) To develop effective outlines, especially formal outlines, follow these guidelines:

- Set up your outline using either an alphanumeric or a decimal numbering system. Properly indent as you divide and subdivide.

Alphanumeric System		Decimal Numbering System	
I.	First level	**1.**	First level
A.	Second level	1.1	Second level
1.	Third level	1.2	
a.	Fourth level	1.2.1	Third level
b.		1.2.2	
2.		**2.**	First level
B.			
II.	First level		

- Move logically from main points to subpoints. Each division should contain at least two items. In other words, where there's an A, there should be a B.
- Make sure that all the items within a subdivision are of the same nature and level (properly coordinate and subordinate).
- Order points within each level logically.
- Use parallel wording and phrasing within each level of the outline.
- Aim for two to four levels within your outline.

Example: Outline

The following outline is for part of the investigative report on pages 492–495.

I. Introduction: context of cockroach infestation study
II. Summary: conclusions and recommendations
III. Research Methods and Findings
    A. Overview of Research: reading, consulting, monitoring
    B. The Nature of Infestation
        1. The Cockroach Population
        2. The German Cockroach's Biology
    C. Methods of Control

    A. Education
        1. Information for Tenants
        2. Tenant Meetings Concerning Program
    B. Cooperation
    C. Habitat Modification
        1. Alterations to Regular Maintenance Program
        2. Renovation Priorities
    D. Treatment
        1. Home Remedies
        2. Chemical Controls in Emergency

Subject Lines, Titles, and Headings

The subject line for an e-mail or a memo, the title of a document, and the headings and subheadings that divide the document all signal structure to readers. When well phrased, these elements indicate the idea, content, and order of what lies below them—helping your reader "get the point" at a glance. To generate high-information subject lines, titles, and headings, follow these guidelines.

Precise Wording

Focus on precise and clear wording; worry less about length. Readers should be able to follow your document's logic simply by scanning the subject line, title, or headings. For a bad-news message, your subject line should be informative but neutral—don't give away the bad news when you need to build up to it in the message. Also, avoid subject lines that "scream" at the reader. Consider these strategies as well:

Avoid say-nothing words or phrases.	**NOT:**	Trash
Avoid mystifying labels.	**NOT:**	Implementation Policy Objectives re: Reusification of Wood-Based Fibrous Products
Use strong action words and precise nouns. Try this pattern:	**BUT:**	Policy [broad topic] on Recycling Waste Office Paper [specific issue]
Subject line, title, or heading = broad topic + specific angle, issue, or concern		

Parallel Structure

Make phrasing within headings parallel (e.g., questions, how-to statements).

NOT: What Happened to the Air Hose? Conclusion: Probable Causes Implementing New Safety Measures	**BUT:** The Incident: Tangled Air Hose Conclusions: Probable Causes Recommendations: New Safety Measures

Typographical Techniques

Signal heading levels in a document either with an outlining system or with typographical techniques such as typeface, type size, indenting, and boldface. Most models in your handbook show these principles at work, but check the subject line and headings below from the troubleshooting proposal (see page 525).

> Subject: Reducing Levels of Carbon Monoxide in the Factory
> Problem: Emissions from Lift Trucks
> Proposal: Phase Out Shipping Department Internal-Combustion Lifts
> Implementation: Phase in Electric Lifts Through Regular Purchases
> Conclusion

F.Y.I.

Because subject lines and document titles are used for filing and searching (both paper and electronic), be careful to include key words in them.

Traditional Organizational Patterns

Often, the information you are working with suggests a natural structure. At other times, consider some tried-and-true patterns that are commonly used in both academic writing and in workplace writing. See examples in Table 11.2.

Table 11.2 Organizational Patterns

Object or Process Description

Introduction: definition of term; background, context, and overview of whole; materials needed
1. Initial part or step
   - Subparts or substages presented in detail
   - Use of measurements, comparisons, graphics
2. Next part or step
3. Next part or step
4. Next part or step
5. Last part or step

Conclusion: summary of object or process; review of the whole; statement of importance

Cause to Effect

Introduction: present the cause—the phenomenon, force, event; supply context
1. Effect 1: support of cause/effect link; description, details, discussion of importance
2. Effect 2
3. Effect 3
4. Effect 4
5. Effect 5

Conclusion: summary of effects and their importance

Effect to Cause

Introduction: present the effect observed—the event, phenomenon; provide context; explain the effect's importance
1. Contributing cause 1: cause/effect link analyzed; cause described in detail
2. Contributing cause 2
3. Contributing cause 3
4. Contributing cause 4
5. Contributing cause 5

Conclusion: summary of probable causes; weighing and balancing of causes

F.Y.I.

Think about organizational patterns as both writing and reading strategies. First, a well-chosen pattern helps you as a writer analyze your topic, understand it, and present it clearly in writing. But the pattern also helps the reader follow your line of reasoning from one concept or step to the next.

Table 11.2 (continued)

Comparison/Contrast

Introduction: objects, processes, or ideas to be compared/contrasted; reason for compare/contrast; features to be focused on

1. Feature 1: similarities and/or differences
   - Item A
   - Item B
2. Feature 2: similarities and/or differences
   - Item A
   - Item B
3. Feature 3: similarities and/or differences
   - Item A
   - Item B
4. Feature 4: similarities and/or differences
   - Item A
   - Item B

Conclusion: summary of similarities and/or differences; significance and value

Problem/Solution

Introduction: the problem's context

1. Description of the problem: importance, causes, effects, history, scope
2. Criteria for an effective solution: the benchmarks for judging possible solutions
3. Possible solutions worth considering
4. Presentation and support of best solution
5. Implementation plan

Conclusion: problem and solution restated and summarized

Argumentation/Persuasion

Introduction: question, concern, or claim

1. Strong argument for claim: discussion and support
2. Other arguments for claim: discussion and support for each
3. Objections, concerns, and counterarguments: discussion, concessions, answers, rebuttals
4. Strongest argument supporting claim: discussion and support

Conclusion: argument consolidated—claim modified, amplified

F.Y.I.

While in many writing situations the best organizational pattern may seem obvious, consider all options that help you achieve your goals. For example, if you are writing to a young driver to explain why her insurance premium is rising, you would likely explain how the *effects* (higher premiums) result from the *causes* (three traffic violations). However, you could cushion the bad news by also *comparing* the new rates with a competitor's policy that is even more expensive.

Lists

A list separates and orders information, making points memorable and clarifying their relationships. Follow these guidelines for developing and formatting lists.

Make all items in any list parallel in structure. (See page 263.)

Choose the appropriate format for your list.

- With a few short items, build the list into the paragraph, using transition words such as *first, second* or numbering/lettering such as *(1), (2)*.
- If you have many items, each item is long, or you want to stress each item, break out the items into a displayed list.

For a displayed list, arrange items sensibly and format the list correctly.

- Introduce the list with a lead-in. Generally, use a complete sentence that ends in a colon. If you must use an incomplete sentence, drop the colon.
- Select appropriate symbols for list items—such as bullets, numbers, or hyphens—and use these symbols consistently. Numbers stress ranking, sequence, priority, counting, or totals. Bullets or other symbols stress equality.
- Align all items within the list.
- From the symbol used, indent list items a moderate number of spaces (five maximum).
- Using hanging indentation, align each item's second and subsequent lines under the first line.
- Capitalize and punctuate items consistently. If list items are complete sentences, for each item capitalize the first word and use end punctuation. If list items are fragments (words, phrases, dependent clauses), generally keep first words lowercase and avoid end punctuation.

A Bulleted, Displayed List with Fragment Lead-in

On Wednesday, June 4, we would like you to be our guest at

- an open house from 8:30 a.m. to 4:00 p.m. with hourly tours of the new facilities
- a ribbon-cutting ceremony at 4:00 p.m. on the west lawn, with refreshments at 4:30 p.m.

A Numbered, Displayed List with Complete-Sentence Lead-in

To make this plan work, follow the procedure below for all your travel arrangements:

1. Book your flight through The Travel Center and charge it to the Rankin account.
2. When Sherri Pomerenki forwards the invoice to you, specify the account to be charged, sign the invoice, and return it to her.

F.Y.I.

Avoid putting too many lists on a page or splitting a list over two pages. Note that some of the formatting techniques recommended above may not work in e-mail programs.

STRUCTURING DOCUMENTS THROUGH PARAGRAPHING

The basic unit for organizing ideas in your documents is the paragraph. By isolating, developing, and supporting specific ideas, paragraphs structure your thinking into "bites" that readers can digest. A strong paragraph is a group of related sentences that does the following:

- It adds to your line of thinking, one step at a time.
- It stands by itself, but also connects with the paragraphs that come immediately before and after it.
- It is neither too long nor too short—for support paragraphs, generally 6–10 lines long or 4–5 sentences. (Long paragraphs tax the reader's concentration, whereas short ones suggest poor idea development.)
- It is set off by white space. (In most business documents, paragraphs are single-spaced, with double-spacing before and after them. The first line is not indented.)
- It performs a specific role in the message—introduction, support, transition, or conclusion.

To learn more about structuring your documents with paragraphs, review the following pages on types of paragraphs and on paragraph focus, completeness, and coherence.

Types of Paragraphs

Your documents should contain three or four types of paragraphs, each doing its job to push forward your message.

Introduction Paragraphs

Introduction paragraphs typically appear at the openings of documents. In longer documents, they may also be used at the beginnings of sections. These paragraphs announce topics and set direction:

- They greet and connect with the reader.
- They state the document's purpose and focus (and perhaps even the main point).
- They provide helpful background and context.
- They preview the content and organization of the message or section.

From an Invitation Letter
Welcome to the Municipal Sales Seminar. I hope that you have a stimulating, productive week. While you are here, please help us celebrate Rankin's birthday.

From a Periodic Report
The following report reviews 2004–2005 activities at Stewart Plastics' Joliet facility. Topics covered include manufacturing, safety, and quality. Based on this review, the report anticipates sales and production needs for 2005–2006.

Body or Supporting Paragraphs

These paragraphs structure the middle of a message or section by developing specific points related to the main point. (See the next page for a detailed discussion and models of these paragraphs.)

Transition Paragraphs

Transition paragraphs link sections in longer documents by first summarizing one discussion and then starting a new one. In a sense, such paragraphs help readers by looking backward, pointing forward, and establishing a connection between the two sections of the document.

From a Major Report

Earlier sections of this report show that Rankin's product portfolio (Chase, Ucopian, and Victor) compares favorably with the competition, that we have clear strategies developed for all four market segments (including niches), and that the 2004–2005 conditions for the small grains market look positive. However, Rankin's success also depends on its addressing key business issues related to marketing. This section reports on both progress and work-to-be-done in manufacturing, distribution, third-party support, product registrations, and public issues.

Concluding Paragraphs

Concluding paragraphs create closure at the ends of sections and documents. Typically, conclusions summarize key points, state recommendations, request action, and anticipate the future—politely helping readers exit the message or section with everything they need.

From a Letter Rejecting a Credit Application

I would be happy to reconsider your request in six months, after Nova has had time to address its current obligations. Until then, I encourage you to continue using MCE's 10 percent discount for cash purchases. Please call to inform me how you would like to proceed with your current order.

From a Periodic Report

The Joliet Plant is capable of $6 million in sales. Road blocks include the lack of a 5-Axis Router for custom-trimmed products and training for the use of this router. Therefore, I recommend the following:

1. Purchase a 5-Axis Router.
2. Fill the Process Engineering position with a person familiar with the 5-Axis Router and capable of conducting training.

From a Company Profile

For more information, contact our Customer Service Department at Abex Technologies, Inc., 121 Eldridge Dr., Flint, Michigan 48404–3497. Phone: (542) 555-2985.

Body Paragraphs' Structure

Body paragraphs carry the weight of information in most documents. To structure body paragraphs that readers will understand and accept, follow the pattern and tips below.

> As you know, the National Campaign for Literacy has spent 14 years helping millions of Americans to read. We work with more than 300 schools, neighborhood groups, and government agencies to combat illiteracy. Yet, illiteracy remains an enormous problem. To meet this need, we plan to enlarge our campaign this year, adding 29 new programs and expanding existing ones.

Topic sentence

Supporting sentences

Concluding sentence

The **topic sentence** creates direction and focus. By stating a single idea clearly, it forecasts the paragraph's discussion and answers the reader's question, "What's your point?"

FORMULA: A topic sentence = a limited topic + a specific point about it
EXAMPLES: Professionalism is more than technical skill.
To make the meeting productive, prepare in two ways.
Hope Services continues to improve its services to minority clients and minority communities in Reading.

Supporting sentences answer your reader's next question, "Can you explain or support your point?" Using facts, examples, and so on, these sentences logically unfold from the topic sentence. (For more on thinking patterns, see pages 195–213.)

Concluding sentences answer the reader's question, "Why is this point important?" Come to a logical stopping point by summarizing the discussion, clarifying its relevance, or pointing forward.

F.Y.I.

Generally, place the topic sentence first for your reader's sake. Sometimes, however, you may need to put it elsewhere:

- In the middle: To lead up to an idea, make your point, and then explore the idea's consequences (e.g., a bad-news paragraph)
- At the end: To build persuasively to the point (e.g., a sales letter)

Testing Your Supporting Paragraphs

Use these questions to evaluate your supporting paragraphs:

- What specific function does the paragraph fulfill? How does it add to your line of reasoning and supply readers with needed information?
- Would the paragraph work better if it broke earlier, was combined with another paragraph, was set up in list format, or was accompanied by a graphic?
- Does the paragraph flow smoothly from the paragraph that comes before it, and does it effectively lead into the paragraph that follows it?

Paragraph Focus and Unity

A well-structured body paragraph is focused and unified. The topic sentence creates a clear, workable idea, and all the supporting sentences relate well to that idea. Examine the unfocused and focused paragraphs below. Then review the tips for fixing focus problems in your paragraphs.

A Fuzzy, Fractured Paragraph

Hope Services is doing important work. The Cultural Diversity Program is reaching low-income minority households. Many headed by women. Our new Cultural Diversity Specialist has been an important addition to the staff and improved our presence in the community. The community is ideal for Hope Services' various programs, and our different staff members in the various programs continue to be excited about their jobs.

Notice the following problems with this paragraph:

- The topic sentence (italicized) is too broad to provide a center for the paragraph.
- The supporting sentences ramble, switching topics halfway through.

A Focused, Unified Paragraph

Hope Services continues to improve its services to minority clients. First, numbers indicate that HS is getting help to its target clientele for the Cultural Diversity Program low-income minority households headed by women. Second, hiring a Cultural Diversity Specialist has given HS's program a strong presence in the community. Finally, in the coming year, HS will focus on strengthening its outreach to minority communities, and training staff and volunteers in cultural diversity issues.

Notice the following fixes:

- The topic sentence now presents a workable center for the paragraph.
- All the supporting sentences develop that one idea without straying.

Focus and Unity Tips

Let your purpose rule your focus, and let your focus rule your organization.

To fix a fuzzy, fractured paragraph, do some or all of the following:

- *Rephrase* the topic sentence until it sets a clearer, more precise direction.
- *Cut* off-on-a-tangent phrases and sentences.
- *Reorder* sentences until they follow logically from the topic sentence.
- *Divide* the paragraph if it mixes or shifts focus.
- *Examine* the organizational pattern in the paragraph, making sure that one pattern (e.g., cause/effect) is not distorted by a competing pattern (e.g., problem/solution).

Paragraph Completeness

A well-structured body paragraph is complete when it fully develops the topic sentence by providing exactly what readers need to know. Such a paragraph avoids fragmentary content (too little support), overweight content (overkill), and redundant content (repetition). Study the fragmentary and complete paragraphs below; then review the fixes that will help you achieve completeness in your own paragraphs.

> **A Fragmentary Paragraph**
>
> This year, we have a lot to celebrate. Many good things have happened, so celebrate with us!
>
> **A Full Paragraph**
>
> This year, we have a lot to celebrate. Rankin Technologies turns 20, our head office expansion is finished, sales grew by 16 percent, and we received the Albany Chamber of Commerce Outstanding Business Achievement Award. On Thursday, June 6, we would like you to be our guest at an Open House. Join us between 8:30 a.m. and 5:00 p.m., tour our new facilities, and enjoy refreshments.

Notice in the fragmentary paragraph the broad, vague phrases "a lot" and "many good things," generalities that are never fleshed out. The revised paragraph provides complete details to support the general statement and satisfy the reader's need for more information.

Completeness Tips

Let your reader's need and understanding guide how many and what kind of details you include. Here are some choices, but see "Supporting Ideas Effectively," page 190, for more help.

Facts	Anecdotes	Analyses	Paraphrases
Statistics	Quotations	Explanations	Comparisons
Examples	Definitions	Summaries	Analogies

Order details based on your paragraph's purpose. Here's some advice, but see also "Using Thinking Patterns," pages 195–213.

- To *narrate* a process or anecdote, be sure to include all steps or events.
- To *describe* something, help readers see, smell, taste, touch, or hear it.
- To *explain* something (a concept, a process, a term), answer *what, how,* and *why* questions.
- To *persuade* (sell, recommend), construct a reasonable, logical, and correct argument.

If a paragraph gets too long and dense, break it up. Create a brief introduction (topic sentence + forecasting sentence). Then divide the paragraph into smaller paragraphs, perhaps listing points with bullets or numbers.

Paragraph Coherence

Well-structured paragraphs need to be coherent. Why? Because coherent writing hangs together. Instead of zigzagging, each sentence and each paragraph chart a clear course that readers can follow. Study the incoherent and coherent paragraphs below. Then use the tips that follow and the linking words list on the next page to achieve coherence in your paragraphs.

A Paragraph Lacking Coherence

I am honored that you asked Rankin Technologies to participate. Our resources are limited. We are choosing to give to environmental, urban renewal, and nonprofit organizations that focus on Third-World development. We must decline at this time. Our employees will continue their support through the United Way campaign.

Paragraph Improved for Coherence

I am honored that you asked Rankin Technologies to participate in *your project* to build this shelter. Because our resources are limited, *however*, we have chosen to commit ourselves to partnerships with nonprofit organizations that mesh with our own interest in the environment, urban renewal, and developing countries. *For this reason*, we must decline *your thoughtful request*. Rankin employees will, *nevertheless*, continue showing support for *your work* in the annual United Way campaign.

In the original paragraph, all of the sentences begin with "I," "our," and "we," and no connections are offered between the separate ideas. The revised paragraph uses linking words and repetition of key words to tie the ideas together.

Coherence Tips

1. Use verb tenses consistently to keep clear the connections between past, present, and future.
2. Use repetition. Key nouns, pronouns, verbs, and phrases, when repeated carefully or replaced with synonyms (different words with the same meaning), help readers track your points. Carry forward the discussion by carrying forward key terms.
3. Use parallelism. Parallel structure states similar or contrasting points using a consistent pattern (see page 263). As a coherence technique, parallelism shows relationships between ideas and pieces of information through balanced words, phrases, and clauses.
4. Use transitional words and phrases. Transitions provide "hooks" connecting a sentence with what came before—sentence with sentence, paragraph with paragraph. These "links" signal logical relationships (addition, contrast, and so on).
5. Use words (*first*, *second*, *third*) or numbers (*1, 2, 3*) to link steps in a process or actions in an event.

Transition and Linking Words

The words and phrases below can help you tie together sentences and paragraphs into a smooth, easy-to-read document.

Words that show location

Above	Behind	Down	On top of
Across	Below	In back of	Onto
Against	Beneath	In front of	Outside
Along	Beside	Inside	Over
Among	Between	Into	Throughout
Around	Beyond	Near	To the right
Away from	By	Off	Under

Words that show time

About	During	Next	Today
After	Finally	Next week	Tomorrow
Afterward	First	Second	Until
As soon as	Immediately	Soon	When
At	Later	Then	Yesterday
Before	Meanwhile	Third	

Words that compare things (show similarities)

Also	In the same way	Like *or* as	Likewise

Words that contrast things (show differences)

Although	However	On the other hand	Otherwise
But	In contrast	Still	
Even though	Nevertheless	Yet	

Words that emphasize a point

Again	In fact	The main point	To repeat
For this reason	Most importantly	To emphasize	Truly

Words that conclude, summarize, recommend, analyze, explain

All in all	Because	In conclusion	Therefore
As a result	Finally	In summary	To sum up

Words that add information

Additionally	And	Equally important	In addition
Again	Another	Finally	Likewise
Along with	As well	For example	Moreover
Also	Besides	For instance	Next

Words that clarify

For example	For instance	In other words	Put another way

CHECKLIST for Organization

Use this checklist to evaluate and revise the organization in your writing.

- ☐ The writing has a strong opening that introduces and focuses the piece, a middle that develops the main idea, and a closing that helps readers embrace or support the message.
- ☐ The organization (either direct or indirect) is appropriate for the topic, situation, and audience.
- ☐ The outline includes items stated in clear, parallel forms correctly formatted.
- ☐ Any summary effectively states the document's or the section's basic message.
- ☐ Subject lines, titles, and headings are clear, informative, parallel, and correctly formatted.
- ☐ Lists include items reasonably grouped, logically ordered, correctly bulleted or numbered, and properly formatted.
- ☐ Each paragraph is appropriately linked to the passages that precede and follow it.
- ☐ Each paragraph is unified, complete, and coherent.

CRITICAL THINKING Activities

1. Find two pieces of workplace writing: one fairly short and simple (e.g., a piece of correspondence) and the other fairly long and complex (e.g., a report, a manual). For each piece, identify and evaluate the organizational strategies used. Compare and contrast the pieces: How are the strategies similar and different, and why?
2. Find a journal article that includes most or all of the following: outline, summary, subject line, title, headings, and lists (check journals using APA documentation). Note the type of outline (alphanumeric or decimal numbering) used, and evaluate how effectively the elements above help deliver the message.

WRITING Activities

1. Review "Control the Flow of Information," page 217, and "Foolproof Organization Strategies," pages 218–224. Use their ideas to evaluate a piece of your writing for (a) an effective flow of information and (b) a clear opening, middle, and closing. Revise the writing as needed.
2. Choose a piece of your writing in which you use direct organization. Rewrite the piece using indirect organization. Share both models with the class, noting how the revision affects the writing's ideas and voice.
3. Review "Structuring Documents Through Paragraphing," pages 225–231. Use that information to revise a piece of your writing.

In this chapter

Weak Voice **234**
Strong Voice **235**
Making Your Writing Natural **236**
Making Your Writing Positive **238**
Developing "You Attitude" **240**
Checklist for Voice **242**
Critical Thinking Activities **242**
Writing Activities **242**

The seven traits

1. Strong Ideas
2. Logical Organization
3. **Conversational Voice**
4. Clear Words
5. Smooth Sentences
6. Correct Copy
7. Reader-Friendly Design

CHAPTER

12

Trait 3: Conversational Voice

When readers go through your writing, they do more than just see and make sense of words on the page. They hear a voice—your voice—even if they've never talked with you before. Voice is your personal presence on the page, the tone and attitude that come through. What's the big deal about voice? The *way* that you express yourself is your document's *implied* content. Voice is the between-the-lines message that your readers get whether you want them to get it or not. The result may be good or bad.

Imagine that you write a persuasive message with a voice that sounds ill informed or even manipulative. If the reader hears an ill-informed voice, he or she won't trust your message. If the reader hears a manipulative voice, he or she won't trust you. In either case, the voice that a reader hears in your writing colors your message and implies something about you—and even about your company.

What tone does *your* writing take? What attitude toward the topic and the reader does it convey? Can you tell? Do you know how to listen for voice in writing, or are you tone deaf? Moreover, how do you make the shift from a formal, academic voice appropriate for college writing to a voice more fitting for workplace writing? This chapter will help you to answer these questions, to train your writing ear, and to tune your computer keyboard. Learning to control your writing voice is tough, but this chapter will help you become more skilled at discerning voice in writing and producing workplace writing that conveys an appropriate voice—one that's positive, natural, polite, and professional.

> "Voice is the imprint of ourselves in our writing. Take that away . . . and there's no writing, just words following words."
>
> —Donald Graves

WEAK VOICE

Make your voice positively natural, real, and reader-focused by avoiding these tonal weaknesses and developing the tonal strengths described on the next page.

Insincere: Sounds unsympathetic, flattering, or exaggerated.

> Belles Lettres Books is exactly what this town needs. Your unique and daring specialty in New England authors will solve all of the cultural problems here! Success is yours to grasp!

Impersonal: Sounds disconnected from the subject and reader.

> Due to the fact that this monthly meeting requires commitment from the CEO and VP, it is important that attendance be prompt and regular.

Impolite: Sounds rude, blunt, irritated, frustrated, angry, impatient, demanding, or complaining.

> I put all that effort into discussing your project's financing. You might have told me earlier about your loan application with the competition.

Uncertain: Displays too much doubt; is wishy-washy.

> I think that maybe the fire hose does a somewhat poor clean-up job. I'm pretty sure it uses a lot of water.

Arrogant: Sounds overconfident, puffed up, or preachy.

> I'm sure you agree that our statements and references show that we deserve a $100,000 credit line. If you wish to receive an order for spring bedding plants, confirm this credit ASAP.

Flashy: Focuses on glitz, not substance.

> Congratulations! You did it! Your first order with In-Bloom!! Once again, congratulations!

Bored: Sounds listless and flat.

> The problem of illiteracy is a significant problem in this country. Reading is important, but many people can't read well.

Overly emotional: Tone is too flowery, melodramatic, or boisterous.

> Finally, and at long last! Our picture-perfect New Employee Orientation Checklist is finished! (I think I'm going to cry.)

Inappropriately funny: Joking or sarcastic tone is out of place.

> Thanks for giving us a chance to bid on Millwood Pharmaceuticals' waste removal. (Any chance of us finding some good drugs in that waste and having an impromptu party on the premises?)

Biased: Attitude demeans groups or individuals.

> Though you might end up getting some lazy college student, consider hiring a student intern.

Self-absorbed: Attitude is self-focused.

> Here's our catalog, showing our wonderful plants. We think that they are the best in the South.

STRONG VOICE

Sincere: Sounds honest and caring.

> Belles Lettres Books is a welcome addition to the city's economy, especially with your store's emphasis on New England authors. I wish you success.

Personal: Sounds conversational, person to person.

> Because this monthly meeting is a big commitment from Robert and Miriam, we owe them our regular attendance. If you can't make the meeting, please notify Richard by the Thursday before.

Polite: Shows respect.

> Thank you for your inquiry yesterday about financing your remodeling project. I appreciated your frankness about your loan application with Boulder National Bank.

Knowledgeable: Shows insight into topic, reader, policies, procedures, and so on.

> The fire hose works well to wash material from the floor into the drain, but it lacks the high pressure needed to remove dried material, and it uses 600 gallons per clean-up.

Confident: Sounds sure, capable, and helpful.

> By March 8, please check our statements and references, send information about your credit terms, and confirm our $100,000 credit line. In-Bloom will then submit its order for spring bedding plants.

Transparent (like glass): Tone is "see through," not glitzy.

> Because this was your first order with In-Bloom, we're sending you the enclosed gift—a 2003 Occasions Diary.

Energetic: Sounds enthusiastic.

> More than 20 percent of adults in this country cannot read at a third-grade level. Each year, more than 1 million young people leave high school functionally illiterate—many of them with diplomas.

Objective: Sounds clear and logical.

> To help new employees become familiar with our day-to-day procedures, Human Resources has developed the attached New Employee Orientation Checklist.

Appropriately humorous: Jokes fit the situation.

> Thank you for the opportunity to bid on Millwood Pharmaceuticals' waste removal. We would be very pleased to be at your "disposal."

Tolerant: Attitude is sensitive to different groups and individuals.

> Please consider having a student intern work as your assistant starting in the fall semester.

Reader- and topic-focused: Keeps attention where it should be.

> Because you indicated that you plan to expand your sales of bedding plants and silk flowers, I have also enclosed a current catalog with these sections flagged. Mr. Bostwick, we look forward to filling your orders . . .

MAKING YOUR WRITING NATURAL

Being natural with other people makes a good impression. So does natural writing. When writing is natural, it's graceful and real—not forced, stiff, or stuffy. To be natural in writing, choose the right tone and control that tone throughout the message.

Choose the Right Tone

Tone of voice refers to the emotion, attitude, and formality of your writing voice. To hit the right note, review the tips and Table 12.1.

Consider your role and your goal. Are you representing yourself, your department, your company? Are you guiding, managing, advising, reporting? For what purpose?

Assess your readers. How close are you? Are these readers inside or outside your organization, above or below you? What will they think of your message?

Judge your topic. How light or serious is your subject?

Clarify the form. Are you writing a complex, formal document such as a client proposal, or a simple, informal e-mail?

Table 12.1 Tone of Voice

Tone	Characteristics	Example
Formal *Use for* • Major documents • Messages to superiors • Messages to some people outside your company • Bad-news messages with legal implications	• No contractions • Few personal pronouns • Serious tone • Objective style • Somewhat complex sentence structure • Specific (sometimes technical) terms	The goal of Boniface Sanitation is serving the Tallahassee community. As indicated in the attached client list, Boniface contracts with 300 companies for waste removal.
Moderate *Use for* • Common documents • Messages to co-workers, equals, familiar people outside organization	• Occasional contractions • Liberal use of personal pronouns • Friendly, but professional tone • Varied sentence structure	Our goal is to serve you well. As Tallahassee's leading waste collector, Boniface serves more than 300 organizations like yours.
Informal *Use for* • Quick, "unpublished" documents (lists, questions) • E-mails and memos to very familiar co-workers	• Frequent contractions • Free use of personal pronouns • Appropriate humor • Loose sentences • Jargon, slang, in-house word choice	We'll be happy to deal with your garbage. As Tallahassee's top dogs in garbage, we haul waste from 300+ companies.

Control the Tone

A natural voice will help you connect with readers, whether you choose a formal, moderate, or informal tone. Once you've established a tone, however, you need to maintain it. Here are some practical tips that will help you strengthen the connection.

Be yourself. Make your writing reflect who you are at your working best—a professional representing a good department, company, or profession. Be confident in being yourself on paper. Don't try to sound smarter, more authoritative, less concerned, or wittier than you really are.

Say what you mean, and mean what you say. Be straightforward and sincere, letting your words carry your convictions and reflect your feelings, experiences, and intentions. Be up front, without being blunt or hurtful.

"Talk," don't "write." Picture your reader sitting across the table, and just say your piece. Rehearse it in your head, and then put it on paper.

Watch sentence length and complexity. Generally, keep your sentences short, but do vary the lengths and types (see page 268). Use active verb forms more often than passive forms to keep energy in your voice (see page 266).

Watch your contractions. One main difference between a formal and an informal voice is contractions (e.g., *it's* versus *it is*). Based on the situation's formality, either use or avoid contractions.

Show respect. Be friendly, but don't presume to be chummy. Instead, always maintain a professional relationship with readers—colleague to colleague, supplier to customer, subordinate to supervisor, and so on. Be careful especially with the reader's name and courtesy title (see "Forms of Address," page 391).

Watch impersonal pronouns.

- Avoid referring to yourself in the third person, which sounds mechanical: *The writer of this report first contacted the Department of Veteran Affairs.*
- Avoid using the indefinite pronoun *one*, which sounds pompous: *One must first soak one's feet in mineral salts before decalcifying one's skin.*

Begin and close with goodwill. Start by connecting with the reader through a polite greeting or compliment. Similarly, conclude with a polite comment. And don't hesitate to use *please* and *thank you*. This effort—when genuine—defeats a common tone problem: being so concise that you sound abrupt, cold, or even angry. While politeness may take a few more words, it's worth it.

F.Y.I.

Be carefully conversational in tone. Avoid the pauses, twists, over-informality, slang, grammar slips, and self-references that often go with a real conversation. In addition, learn your company's voice as shown in its public documents; and consider using this voice when speaking publicly for the company.

MAKING YOUR WRITING POSITIVE

You've heard of the power of positive thinking. This psychological principle applies to your workplace writing as well. By stressing successes over failures, solutions over problems, challenges and opportunities over hurdles, benefits over drawbacks, and advances over retreats, a positive tone communicates a productive attitude. How does positive thinking translate into a positive tone? Use the strategies below and on the next page.

Use Positive (Not Negative) Phrasing

Whenever possible, use wording that avoids negativity. Even if you can't be entirely positive, use a neutral tone.

- In explanations, state both pro and con reasons objectively.
- With problems and solutions, stress what *can* be done, not what *can't* be done and who's to blame.
- Generally, avoid building sentences on negatives: no, not, never, no one, nobody.
- When possible, choose words with positive associations (see page 255).

Negative Phrasing
Given these hazards currently being ignored by management, we cannot continue with the same ineffective methods. No undercoater should have to wear these ratty oxygen suits, which do not meet OSHA standards.

Positive Phrasing
To protect undercoaters from these hazards, I recommend that we purchase OSHA-approved oxygen suits.

Use Positive (Not Negative) Humor

Appropriate workplace humor has a light touch. It's insightful, gentle, perhaps self-deprecating—not hurtful, mocking, or sarcastic. However, make sure that your humor supports the message's clarity and consistency. Jokes shouldn't detract or distract.

Negative Humor
We apologize that your confirmed room was unavailable last night. Because several guests didn't hit the road when expected to (damn Yankees, no doubt!), we were forced to provide you accommodations elsewhere. For your tribulations, we were happy (well, almost) to pay for last night's lodging.

Positive Humor
Our goal now is to provide you with outstanding service and Southern hospitality as warm as the weather. To that end, we have upgraded your room at no expense to you. Please accept this box of chocolates as another gift to "sweeten" your stay with us.

Use Tact (Not Temper)

Shouting, accusations, and sermons show that you are out of control, make readers defensive, and destroy long-term goodwill. Conversely, tact helps you focus on issues (not personalities) and on solutions (not blame). Be especially careful to show tact in e-mail messages, where such temper-related violations are called flaming and shouting (see page 373).

> **Angry Tone**
>
> Quite frankly, we're sick of getting the run-around on this Snorkel Lift deal. We're sitting here six months after the due date, and you just haven't delivered on your promises. Tell me, are you going to get your act together?
>
> **Tactful Tone**
>
> We have been extremely disappointed with the lift's condition and performance. You originally promised a fully operational Snorkel Lift in "like new" condition ready in July 2000. That promise has not been met.

Offer Constructive (Not Destructive) Criticism

When poorly handled, criticism breaks down work and people. When handled well, it can build up and improve work, people, processes, and companies. When evaluating or offering criticism in writing, do the following:

- Balance discussion of weaknesses with praise of strengths.
- Focus on improvements; don't emphasize failure and disappointment.
- Direct criticism to the topic, not personalities.

> **Destructive Criticism**
>
> Your bid was rejected because it was weak on overall value. Better luck next time.
>
> **Constructive Criticism**
>
> Your bid was very competitive on cost estimates, experience, and references. However, what tipped the bid in favor of Earth-Scape Landscape Design was the overall value of that company's plan.

Be Confident (Not Doubtful)

State your ideas with appropriate concessions and qualifiers (see page 189), but without doubt, unnecessary hedging, or counterproductive apology.

> **Doubtful Tone**
>
> What do you think? I'm pretty sure that a power washer would make clean-up better, but maybe I'm missing something.
>
> **Confident Tone**
>
> The Douser Power Washer will keep the Cattle Processing Room cleaner at a lower cost than our present method.

DEVELOPING "YOU ATTITUDE"

"You attitude" is a writing approach that pays attention to the *reader's* perspective and builds that into the message. This strategy helps you as the writer connect with your readers and build long-term goodwill.

"You Attitude" Guidelines

Show empathy and concern. Understanding and caring about the reader's interests, values, worries, and needs is where "you attitude" starts. Always ask, "What does this message mean for my readers?" Then show genuine concern.

> Thank you for your patience and understanding as we investigated your concerns about the ATV16 drives. I apologize for the inconvenience caused to both you and American Linc. Here is the solution that we are putting into effect.

Think "team." When possible, stress cooperation, shared goals, compromise, and agreement so that you and your reader stand together by the message's end. Avoid the quicksand of misunderstanding, disagreement, and spiraling conflict.

> I have enclosed the information about Rankin Electrical Engineering that you asked for—a corporate brochure, current job list, past job list, some letters of recommendation, and more. This material shows, I believe, that we would be a solid match for the projects in western Illinois.

Handle names with care. Be especially sensitive with company names and the reader's name.

- Spell the reader's name correctly, use the proper courtesy and professional titles, and get the job title right.
- If appropriate, get your reader's name into the message at a key spot—a positive point, a question, the closing.
- Use the reader's first name only if your relationship warrants doing so.

> So please join the Purchase Awards Program! Just complete the enclosed form and return it in the postage-free envelope by April 12. A few weeks later, Ms. Wadsworth, you and your customers will be enjoying beautiful art!

Use personal pronouns to connect and protect. These pronouns, especially *you*, place people, particularly readers, front and center.

- Use *you* and *your* in positive and neutral situations.
- Avoid *you* and *your* if they accuse the reader, put the reader down, or imply that the reader should feel a certain way.
- Use *I*, *we*, *us*, and *our* when they create a sense of teamwork. Limit them when they focus solely on you, your company, and your interests.
- Use moderation. Too much *you* sounds like an overdone sales pitch. Too little *you* makes the reader disappear.

> In case you do not proceed with the Boulder National loan, our offer will be good for 60 days from today (until March 27). In addition, if lower rates are available at closing, you will receive that reduction.

"You Attitude" in Action

While the message below is whimsical, it illustrates the elements of "you attitude": empathy, teamwork, care for the reader's name, and careful use of personal pronouns.

What Not to Do

Hey Mr. Coyote:

We at ACME Corporation are pleased to announce the invention of the latest and best in laser technology: our ACME Roadrunner Beam, guaranteed to nail the fastest roadrunners in the world. Without a doubt, our laser leaves our competition in the dust. Our laser is lighter, more compact, and more accurate than theirs. In fact, it's probably twice as effective as the outdated piece of hardware you own now. We are absolutely confident that you will be more than pleased with our Roadrunner Beam because ACME's name and reputation stand alone. Wiley, what are you waiting for!!! Herein, find enclosed a special offer implementation form. Complete said form and return ASAP.

Eagerly yours,

Gary Gush

What to Do

Dear Mr. Coyote:

Is your present laser beam catching all the roadrunners you need? Or is it leaving you flattened and in a bad mood? ACME Corporation can help you.

Our new ACME Roadrunner Beam is light and compact. You can move it quickly and easily to avoid having it crush you at the bottom of a cliff. Moreover, the ACME Beam is accurate up to 10 miles so that you can reach your meal from a safe distance. Finally, the ACME Beam comes equipped with dual parachutes—one for the machine, and one for the operator in case you are standing on an overhanging cliff that you accidentally cut off.

If the ACME Roadrunner Beam sounds like a step forward in your search for a meal, Mr. Coyote, call one of our friendly operators at 1-800-ROADKIL. We'll send you, without obligation, the information that will put safer dining pleasure in your paws.

Sincerely yours,

Samantha Smooth

F.Y.I.

While empathy is important, avoid presuming to know the reader's feelings. For example, avoid saying, "I know you'll be happy to learn. . . ."

CHECKLIST for Voice

Use this checklist to strengthen the voice in your writing.

- ☐ The voice is sincere, personal, polite, knowledgeable, confident, and transparent.
- ☐ The voice is energetic but controlled, objective but engaged, and tolerant but honest.
- ☐ The level of formality fits the topic, audience, and occasion.
- ☐ The use of contractions, names, titles, and personal pronouns is appropriate for my purpose, message, and relationship with the reader.
- ☐ The word choice and phrasing create a positive tone.
- ☐ Any criticism sounds objective, warranted, measured, and constructive.
- ☐ The tone sounds informed and confident, not apologetic or arrogant.
- ☐ The writing shows an appropriate respect, empathy, and concern for the reader.
- ☐ The writing uses the reader's name carefully, as well as all personal pronouns.

CRITICAL THINKING Activities

1. Review "Making Your Writing Natural," pages 236–237, and "Making Your Writing Positive," pages 238–239. Then analyze a letter or e-mail that has poor tone. Read the document to the class and point out weaknesses in tone.
2. Analyze a business document for an appropriate "you attitude." Note strengths or weaknesses and discuss how the writer could improve the writing.
3. Collect and review samples of workplace writing from a specific business, nonprofit organization, or government agency. Characterize and evaluate the corporate voice presented. Is that voice clear, consistent, and fitting?

WRITING Activities

1. Working with two classmates, review the lists for "Weak Voice," page 234, and "Strong Voice," page 235. Exchange pieces of your writing, read them aloud, and listen for examples of weak or strong voice. Note the strengths and discuss how to correct the weaknesses. Revise your writing as needed.
2. Choose a piece of your writing that you want to polish. Read the piece aloud and note any weaknesses in voice. Revise the piece as needed; then develop a note that introduces the document, cites the changes that you made, and explains why the changes strengthen your writing.

WEB Link

For more help with developing a natural voice, a positive voice, and "you attitude," check the activities at http://college.hmco.com/english.

In this chapter

Cutting Unnecessary Words **244**
Selecting Exact and Fresh Words **248**
Avoiding Negative Words **255**
Checklist for Word Choice **258**
Critical Thinking Activities **258**
Writing Activities **258**

The seven traits

1. Strong Ideas
2. Logical Organization
3. Conversational Voice
4. **Clear Words**
5. Smooth Sentences
6. Correct Copy
7. Reader-Friendly Design

CHAPTER
13

Trait 4: Clear Words

"What do you read, my Lord?" asks Polonius.

"Words, words, words," responds Hamlet.

Hamlet answers as he does to foil a conniving character. However, you have probably said—or thought—the same phrase while reading a wordy report, a flowery advertisement, or a cliché-ridden business letter. Looking down at the words-without-meaning document, you think: Words that don't deliver substantive ideas in fresh, clear language are merely ink on the page or marks on a screen.

Which words should you choose when writing? Among other things, you want precise nouns, vivid verbs, and correct descriptors. You want fresh words rather than technical terms, business English, euphemisms, slang, or clichés. You want respectful language rather than negative or unfair terms. You want words that create verbal clarity rather than verbal fog, words that connect with readers rather than alienate or confuse them. You want words that communicate your message with precision and effect.

How do you learn to choose the best words? Practice, plain and simple. Through experience, you can learn to choose the best words for each writing task, to minimize the verbiage, and to select strong but tactful terms. For each message, you learn to put all the right words in all the right places.

CUTTING UNNECESSARY WORDS

Choosing the best words often means cutting unnecessary words—words that serve no real purpose in stating and advancing your ideas. The result of such cutting is concise writing, writing that's fat-free. But what exactly do you cut? For your readers' sake, you want to use as few words as possible, yet you don't want your writing to be so spare that it sounds abrupt and unclear. Study the problems below, identify them in your own writing, and then trim the real fat while leaving the meat alone.

Cut Deadwood

Deadwood is filler material—verbal "lumber" that can be removed without harming the sentence. Look for irrelevant information, obvious statements, and awkward phrases.

> **DEADWOOD:** The plant *must go through a thorough retooling process necessary* for producing *the most recent additions to the new line* of competitive products.
>
> **CONCISE:** The plant must retool to produce competitive products.

Cut Modifiers

When overused, adjectives and adverbs make your writing as thick as fudge. Trim back modifiers by practicing these principles:

- Don't use two modifiers when one will do.
- Use precise nouns and verbs to avoid using modifiers.
- Avoid intensifying adverbs (e.g., *extremely, intensely, awfully, especially*).
- Delete meaningless modifiers (e.g., *kind of, sort of*).

> **WORDY:** Our *licensed, professional, experienced* tree surgeon is *impatiently* waiting to *tenderly* look after your *deciduous, coniferous, flowering, and nonflowering* trees as if they were *almost* like her *very* own.
>
> **CONCISE:** Our tree surgeon is waiting to care for your trees as if they were her own.

Replace Phrases and Clauses

Phrases and clauses are groupings of words that frequently pad out sentences when fewer words will do. (For detailed definitions and explanations concerning phrases and clauses, see pages 718–720.) To get the padding out of your writing, follow these guidelines:

- Locate prepositional phrases (e.g., *at the beginning of the project*) and relative clauses (*who, which, that* clauses). Then replace these phrases and clauses, when possible, with simpler words.

> **WORDY:** *At the present time,* we are *proceeding as efficiently as is humanly possible* to modify the boxes, *which are the serial link boxes that you construct for American Linc in Raleigh, North Carolina,* by removing and replacing the *regulators that are faulty.*
>
> **CONCISE:** We are currently modifying American Linc's serial link boxes in Raleigh, North Carolina, to remove and replace the faulty regulators.

- Avoid wordy phrases that include filler nouns such as *case, fact, factor, manner,* and *nature.*

Replace	With
In many cases	Often
Aware of the fact that	Aware that
A crucial factor	Important
In a rapid manner	Rapidly
Of a dangerous nature	Dangerous

Cut Redundant Wording

Redundancy is unnecessary repetition. (Not all repetition is bad. For effective ways to use repetition, see "Use Parallel Structure," page 263.) Follow these practices to cut redundant wording from your writing:

- Check your sentences for words and phrases that say the same thing.

  REDUNDANT: The sound system that you installed helps *each and every* staff member in the family room by letting them focus on *any and all* conversations in the room as if *these staff members* were in the *family room themselves in the flesh.*

  CONCISE: The sound system that you installed helps staff in the family room focus on different conversations in the room as if they were in the room themselves.

- Check your writing for common redundancies:
  - Using *together* with verbs such as *assemble, combine, cooperate, gather, join, merge,* and *mix.*
  - Using *new* before *beginner, discovery, fad, innovation,* and *progress.*
  - Using *up* after *connect, divide, eat, lift, mix,* and *rest.*
  - Using *in color* with a color (e.g., *green in color*).
  - Using *in shape* with a shape (e.g., *round in shape*).
  - Using these common redundant phrases:

Added bonus	Completely eliminate	Plan ahead
Any and all	Cooperate together	Resume again
A week's time	Descend down	Small in size
Basic necessities	Each and every	Still remain
Brief in duration	End result	Very unique
Close proximity	On a temporary basis	Visible to the eye

WRITE Connection

Concise writing doesn't happen in one shot. While you should try to get to the point quickly while drafting, focus on conciseness during the refining stage of the writing process (see pages 32–36).

Wordy Versus Concise Wording: Sample Passages

Example 1

Concise writing is, in part, writing filled with action, writing that moves forward fast. The following example illustrates how wordy, repetitive writing can bog down a paragraph. Read the first passage and identify specific problems discussed on the previous pages. Then read the second passage and identify key changes.

> **WORDY:** With regard to benefit packages for employees of the company, next Thursday noon at 12:00 p.m., all members of your department must follow a requirement of being in attendance at a short, brief teleconference during which will be detailed the new health plan which will soon be available to all employees of Rankin Manufacturing. This event is a must. As was established in our prior agreement, Michael Farquahr and Susan Baldacci from Sunrise Health Systems will engage in interaction with us in "real time" over the newly installed telecommunications network. You and your employees will not be required to leave your workstations for this event. Because this is a trial run for the new network, I would greatly appreciate your feedback and response on the success of the experiment. I will circulate to you, so that you can provide this feedback, a survey-type questionnaire. Please kindly return it to me on or before January 18.
>
> **CONCISE:** On Thursday, January 10, at 12:00 p.m., your department needs to participate in a brief teleconference about the new health plan for Rankin employees. Michael Farquahr and Susan Baldacci of Sunrise Health Systems will present the plan and field questions over our new telecommunications network. Department members can simply stay at their workstations for this 30-minute discussion. Because this event is our first teleconference, I would appreciate your feedback. By next Monday, I will send a questionnaire; please complete and return it by January 18.

Example 2

Wordiness can mar any workplace document, including, for example, a job summary. Contrast the wordy and concise versions below.

> **WORDY:** While the Sexual Assault Specialist has a lot of really important duties that are part of helping individuals, or groups, or even families with problems related to sexual abuse or sexual assault, the job also includes responsibilities that are part of educating people in the community or staff at various businesses about how to deal with sexual abuse, or sexual assault, or even rape.
>
> **CONCISE:** The sexual assault specialist provides direct service to families or individuals affected by domestic abuse or sexual abuse, and the specialist does community education and staff training regarding sexual assault and rape prevention.

Wordy Versus Concise Phrasing

Table 13.1 offers a hit list of wordy phrases and effective, concise substitutes. Learn to recognize these problem phrases in your own writing.

Table 13.1 Wordy Phrases and Concise Substitutes

Wordy	Concise	Wordy	Concise	Wordy	Concise
Add an additional	Add	Few in number	Few	On a weekly basis	Weekly
Advance forward	Advance	Filled to capacity	Filled	On the grounds that	Because
Advance planning	Planning	Final conclusion	Conclusion	Over again	Over
A majority of	Most	Final outcome	Outcome	Past experience	Experience
A number of	Some/many	First and foremost	First/foremost	Past history	History
Any and all	Any/all	Foreign imports	Imports	Period of time	Period
Are of the opinion that	Believe	For the purpose of	For	Personal opinion	Opinion
Ask the question	Ask	For the reason that	Because	Personal in nature	Personal
Assembled together	Assembled	Free gift	Gift	Pertaining to	About
At an early date	Soon	Free of charge	Free	Plan ahead	Plan
At the conclusion of	After/following	Having the capacity to	Can	Postponed until later	Postponed
At the present time	Now	In connection with	About	Present status	Status
At this point in time	Now	In light of the fact that	Since	Prior to	Before
Attach together	Attach	In order to	To	Protrude out	Protrude
Based on the fact that	Because	In spite of the fact that	Although	Recur again	Recur
Basic essentials	Essentials	In the amount of	For	Reduce down	Reduce
Basic fundamentals	Fundamentals	In the event that	If	Refer back to	Refer to
Basic necessities	Necessities	In the vast majority	Most	Regress back	Regress
Both together	Together	In view of the fact that	Because	Reiterate again	Reiterate
Brief in duration	Brief	It is often the case that	Often	Repeat again	Repeat
Check up on	Check	It is our opinion that	We believe that	Resulting effect	Effect/result
Close proximity	Close	It is our recommendation	We recommend	Return back	Return
Collect together	Collect	It is our understanding	We understand	Round in shape	Round
Combine together	Combine			Rules and regulations	Rules/regulations
Completely unanimous	Unanimous			Serious crisis	Crisis
Connect together	Connect			Single unit	Unit
Consensus of opinion	Consensus			Small in size	Small
Depreciates in value	Depreciates	Joint cooperation	Cooperation	So as to	To
Descend down	Descend	Join together	Join	Specific example	Example
Despite the fact that	Although	Joint partnership	Partnership	Square in shape	Square
Disregard altogether	Disregard	Main essentials	Essentials	Subsequent to	After
Due to the fact that	Because	Make reference to	Refer to	Sum total	Total
During the course of	During	Meet together	Meet	Take into consideration	Consider
During the time that	During/while	Merge together	Merge	Throughout the entire	Throughout
		Mix together	Mix	True fact	Fact
Each and every	Each/every	More preferable	Preferable	Unresolved problem	Problem
End product	Product	Mutual cooperation	Cooperation	Until such time as	Until
End result	Result	New initiatives	Initiatives	Visible to the eye	Visible
Engaged in a study of	Studying	Of the opinion that	Think that	Vitally essential	Essential
Exact replica	Replica	On a daily basis	Daily	When and if	When/if
				With regard to	About

SELECTING EXACT AND FRESH WORDS

For most business writing, the best words are the simplest ones capable of communicating your meaning. These are quality words—precise, clear, and energetic. To select exact and fresh words, practice the advice that follows.

Replace or Clarify General Words

Alone, general words can be vague and imprecise. While they may effectively frame your topic, you need to use concrete, specific terms to flesh out more precisely what you mean. Check out the differences between the passages below. Then study the advice on using specific nouns, vivid verbs, and strong modifiers.

> **TIRED:** The conditions were poor, the pipe was damaged, and it was getting dusky.
>
> **FRESH:** Approximately 6–12 inches of mud covered the work area (including the ramps into the ditch), the ditch contained water up to the pipe's bottom, a steady rain was falling, and we had 3 hours of daylight left (with a Light Plant on site). The gas leak came from two quarter-inch cracks on the seam at 10 o'clock looking south.

- **Choose specific nouns.** General nouns create a fuzzy picture; specific nouns create a precise one. In the examples below, note how general nouns are replaced with specific terms.

    > **VAGUE:** The equipment cleans up some of the mess, but it doesn't have the strength needed to remove all the dirt from the room, and it uses a lot of water.
    >
    > **PRECISE:** The fire hose works well to wash material from the floor into the drain, but (1) it lacks the high pressure needed to remove dried material from walls and equipment, and (2) it uses a high volume of water—600 gallons per clean-up.
    >
    > **VAGUE:** While the power washer isn't cheap, eventually it would pay for itself.
    >
    > **PRECISE:** Savings realized by purchasing a power washer would pay for the washer in less than two months. While the current method costs $53.15 per day, a power washer would reduce that figure to $13.05 per day.

Table 13.2 shows examples of the general–specific difference.

Table 13.2 General Versus Specific Words

	Person	Place	Thing	Idea, Concept
General				
	Employee	Workplace	Heater	Problem
More specific				
	Assistant	Office	Furnace	Breakdown
Precise				
	HR assistant, Susan St. James	HR Department	Electric heat pump	Communication lag time

- **Choose vivid verbs.** Action-packed verbs make your writing clear and energetic. For example, *brush, scrub, sanitize, dissolve*, and *sweep* are all more precise and vivid than *clean*.

   BLAND: Clean-up involves doing a lot of hard work that is time-consuming.

   VIVID: Clean-up involves spraying an area, letting it soak, scrubbing with Rocal D, and rinsing.

F.Y.I.
Avoid using the "be" verbs (*is, are, was, were*) too often. A better verb can often be made from another word in the sentence.

- **Choose strong modifiers.** First, use fresh, specific adjectives to modify nouns: One strong adjective is better than several weak ones. Second, use adverbs only when needed to clarify the action of the verb. Don't use a verb–adverb combination when one vivid verb would be better.

   WEAK: The status-quo clean-up procedure in cattle processing just isn't state-of-the-art. It falls incredibly short of what's desperately needed by being a joke.

   VIVID: The current clean-up procedure in Cattle Processing falls short of what's needed by being labor-intensive, costly, wasteful, and unsanitary.

   Avoid especially these weak adjectives and adverbs:

   TIRED ADJECTIVES: bad, big, certain, cute, fun, funny, good, great, improved, long, neat, new, nice, old, several, short, small, various

   TIRED ADVERBS: absolutely, basically, certainly, essentially, exceptionally, extremely, greatly, highly, hopefully, incredibly, kind of, perfectly, sort of, very

F.Y.I.
The manager writing the paragraph below intended to deliver a strong recommendation of his colleague; however, notice how the weak nouns, verbs, and modifiers create only a vague, lifeless picture of her skills and work.

   WEAK: Mary came to Empire Estates in the spring of 2003 and joined my team some time later. As an accountant, she worked on some very important accounts and did really good work. For example, when we took on a new project last winter, Mary really helped us out. In fact, she even went to night school to get additional skills that we needed!

   STRONG: Mary joined Empire Estates in April 2003 and was assigned to my team in June 2004 as a Mortgage Specialist II. Her excellent skills at reconciling A/A and S/S accounts, as well as FHLMC P&I accounts, enabled Mary to make an immediate, strong contribution. Moreover, last January when our team needed help with CPI System balancing, Mary took a night class on the topic and voluntarily added CPI System balancing to her previous assignment. That training further helped our team.

Use Correct Words

Incorrect words confuse readers and make a bad impression. Be especially careful with the following terms:

- Your message's key words—making sure that both terms and synonyms are used accurately
- Names, numbers, percentages, and factual statements—making sure that you convey accurate data.
- Commonly confused pairs like *its* versus *it's* and *their* versus *there*.

  INCORRECT: A 20000 *pis pressure* treater *wood* use 75–08 *pints* per clean-up versus *60*.

  CORRECT: A 2,000 *psi* power washer *would* use 75–80 *gallons* per clean-up versus *600 gallons*.

  INCORRECT: With *it's* long power *chord* and hoses, the Douser is a *virtual* machine.

  CORRECT: With *its* long power *cord* and hoses, the Douser is a *versatile* machine.

 WRITE Connection

Check Chapter 49, "Using the Right Word," for a list of commonly confused words.

Restrict Technical Terminology

Technical jargon can be the specialized vocabulary of a subject, a profession, or even a company. Because technical jargon can make your writing difficult to read, use simpler terms or define those technical terms that must be used. Generally, follow these guidelines:

- Use technical terminology to communicate with insiders—people within the profession or company—as a kind of shorthand.
- Avoid "shop talk" with readers outside the profession or company.

  TECHNICAL: Bindford's Douser power washer delivers 2,200 psi p.r., runs off standard a.c. lines, comes with 100 ft. h.d. synthetic-rubber tubing, and features variable pulsation options through three adjustable s.s. tips.

  SIMPLE: Bindford's Douser power washer has a pressure rating of 2,200 psi (pounds per square inch), runs off a common 220-volt electrical circuit, comes with 100 feet of hose, and includes three nozzles.

WRITE Connection

For help with defining terms, see pages 202–203. For help with communicating technical information, see Chapter 6, "Communicating Technical Information."

Cut Flowery Phrases, Euphemisms, and Clichés

These tired terms, generally showing a lack of imagination, weaken your writing's clarity and energy. Use more forceful, direct, and fresh language.

Flowery Phrases

Flowery phrases are unnecessarily fancy and often sentimental.

> **FLOWERY:** The president's office is garnished in sundry shades of soothing azure and emerald, lifting the dark veil from the mahogany walls with desiccated floral sprays and photos of lush, blooming spring meadows.
>
> **FRESH:** The president's office is decorated in blues and greens, brightening the mahogany walls with dried floral sprays and photos of spring meadows.

Euphemisms

A euphemism is an overly polite expression that avoids stating or even hides an uncomfortable truth. Choose neutral, tactful phrasing, but avoid euphemisms. In particular, be careful to avoid them in bad-news situations (see page 425).

> **EUPHEMISTIC:** Her speaking style was "interesting."
>
> **CLEAR:** Her way of shrugging as she spoke distracted from her presentation.

Clichés

A cliché is a phrase that has become tired from overuse. Often, the cliché was once a fresh metaphor or simile that has now ceased to communicate anything surprising or insightful through the comparison. Writing dominated by clichés conveys a lack of originality and insight; readers quickly tune it out.

> **CLICHÉD:** We need all hands on deck because we have a tough road ahead.
>
> **PLAIN:** Everyone will need to work hard on the next project.

F.Y.I.

Playing creatively with a cliché may revive its energy. For example, check out this closing sentence in a sanitation company's waste-removal bid: "Until then, I'm at your 'disposal' to answer questions." Does the play with "at your disposal" work? You be the judge.

Some common clichés

Like a house on fire	Piece of cake	Rock the boat
Easy as pie	Stick your neck out	Bitter end
As good as dead	Burn bridges	Can of worms
Throw your weight around	Work like a dog	Go for broke
Wet blanket	Plant the seed	Pay through the nose
Rear its ugly head	Last but not least	Back to square one
Beat around the bush	Water under the bridge	Movers and shakers

Avoid Slang Terms

Slang is out of place in most business writing. Because slang terms are too familiar, time-bound, and peculiar to particular places and groups of people, they lack power and clarity.

> **SLANG:** The bucks we'd fork over for a primo power washer would pay off big time lickity split.
>
> **FITTING:** The savings realized by purchasing a power washer would pay for the washer in less than two months.
>
> **SLANG:** Our goal is to cut you in on some real cool deals and hot service. As Tallahassee's main man in garbage pushing, we serve up awesome service to over 300 businesses just like yours.
>
> **FITTING:** Our goal is serving you well. As Tallahassee's leading waste collector, Bonnitice serves over 300 organizations like yours. I've attached references, crew lists, and equipment brochures for your review.

Avoid "Business English"

"Business English" is language that sounds stuffy, dated, or trendy. From pompous vocabulary that seeks to sound weighty and official to buzz words that fly through the air out of the most recent trend, what's the solution to such language? Plain English. The simpler and more basic the vocabulary, the better.

> **BUSINESS ENGLISH:** In accordance with managerial policy, attached hereto note an indispensable enclosure containing highly informative delineations of policies and procedures pertaining to credit.
>
> **PLAIN ENGLISH:** I have enclosed a brochure that describes our credit policies and procedures in more detail.
>
> **BUSINESS ENGLISH:** The benchmark set is to maximize client pleasure by meeting demand with a supply of Southern hospitality. To facilitate maximization, management has upscaled your unit with no capital outlay required on your part.
>
> **PLAIN ENGLISH:** Our goal is to provide you with outstanding service and warm, Southern hospitality. To that end, we have upgraded your room at no expense to you.
>
> **BUSINESS ENGLISH:** Please be apprised as per request, that the disbursement in question was misdirected with the unfortunate result of tardy receipt. Our firm's humblest apology for this error. Appropriate corrective action has been implemented, including dehirements. Let's strategize some total quality solutions, especially if workforce conditions continue to precipitate problems.
>
> **PLAIN ENGLISH:** As you asked, I have looked into the late payment. Unfortunately, it was mailed to the wrong address. We apologize for this error. To correct the problem, we have talked to the employees responsible, corrected our files, and given the account a top priority. Please call me if you believe that other matters need to be addressed or if a similar problem happens.

Business English Versus Plain English

Table 13.3 offers contrasting examples of business English and plain English. Study the examples so that you can recognize weaknesses and strengths in your own phrasing.

Table 13.3 **Business English Versus Plain English**

Business English	Plain English	Business English	Plain English
Accordingly	So	Indispensable	Vital
According to our records	Our records show	In accordance with	As
Acquaint	Tell	In lieu of	Instead of
Activate	Start	In the amount of	For
Adhere	Stick	Interface	Discuss
Afford an opportunity	Allow/permit	Interminable	Endless
Along the lines of	Like	It has come to my attention	I have learned
Applicable to	Apply to	Kindly	Please
Apprise	Tell	Manifest	Show
Are in receipt of	Have received	Manipulate	Operate
Are of the opinion that	Think that	Modification	Change
Ascertain	Learn/find out	Necessitate	Require
As of this date	As of today	Necessity	Need
As per	According to	On a daily basis	Daily
As regards	Regarding	Operationalize	Start
Attached hereto	I have attached	Paradigm shift	Major change
At the present writing	Right now	Parameter	Limit
At your earliest convenience	By September 12	Paramount	Main
		Per se	As such
Call your attention to	Please note	Personnel surplus reduction	Layoffs
Case in point	Example	Pursuant to	Following up
Ceased functioning	Quit working	Per your request	As requested
Cognizant	Aware	Precipitate	Cause
Commence	Begin	Preliminary to	Before
Concur	Agree	Prioritize	Rank
Configuration	Shape	Procure	Buy/get
Delineate	Describe	Quantify	Measure/count
Disbursements	Payments	Ramification	Result
Do not hesitate to	Please	Recapitulate	Review
Due consideration	Careful thought	Remuneration	Pay
Enclosed please find	I have enclosed	Reproduction	Copy
Endeavor	Try	Salient	Important
Enumerate	List	Strategize	Plan/solve
Evacuate	Leave	Subsequent	Later/after
Expedite	Speed up	Terminate	End
Fabricate	Make	Thanking you in advance	Thanks for considering
Facilitate	Make easier		
Finalize	Settle/finish	Under separate cover	Separately
Fluctuate	Vary	Utilize	Use
Germane	Relevant	Vacillate	Waver
Herein	In this	Visualize	Picture
Heretofore	Until now	We beg to call your attention	Please note
Inasmuch as	As/because	Wherewithal	Means
Increment	Amount/step	You are hereby advised	Please know

Part 2 Benchmarking Writing with the Seven Traits

Fresh Word Choice in Action

The two examples below illustrate a poorly worded letter and a fresher version of the same letter. Read the first letter, and note the many examples of vague nouns, weak verbs, flowery phrases, jargon, and "business English." Then read the revised letter to see how the poor word choices have been fixed with fresh phrasing and plain English.

Poor-Quality Word Choices

Please be advised concerning an amendment to the manner in which disbursements of property taxes from escrow accounts shall be handled by this office. The particulars of the aforementioned modifications are delineated below.

As of the present writing, officials of the county courthouse mail statements of tax accrued directly to residents holding legal title on property. These property owners subsequently transport said statements to this institution, which then expeditiously forwards the designated funds to the appropriate county office. As of September of this year, a change will be necessitated. Customer tax statements shall now be posted directly to the bank, on or about the fifteenth day of the third and ninth months of the calendar year. Taxes in the amount indicated shall then be paid out accordingly in a direct fashion to the county taxation office 15 days after receipt of the tax statements under discussion. Copies of statements and receipts indicating signed-and-sealed payment shall then be forwarded under separate cover to applicable customers.

These procedural modifications, we trust, meet with your approval. They are, please be assured, in line with this bank's paramount concern for customer satisfaction and convenience. Kindly inform us of concerns.

Top-Quality Word Choices

I'm writing to let you know about a change in the way your property tax payments will be handled.

Right now, the county courthouse mails your property tax statement to you, and then you bring it to us. Beginning in September, this process will change. The county will mail your statement directly to the bank (on March 15 and September 15 each year). We will then make the payment from your escrow account on April 1 and October 1 and send you a copy of your statement and a receipt.

We believe that this procedure will be more convenient for you. If you have any questions or concerns about this change, please call me at 527-6328 between 9:00 a.m. and 5:00 p.m.

WRITE Connection

While you should try to choose exact and fresh words while drafting your document, focus on strengthening word choice at the refining stage (see pages 32–33).

AVOIDING NEGATIVE WORDS

Because effective business writing is ethical writing, choosing words carefully is a serious responsibility. Actions may speak louder than words, but some words can scream. In fact, negative words may destroy your reader's trust and even lead to legal action.

What is negative word choice? It can be any of the following:

- Insensitive phrasing that sounds insulting, abrupt, or cold
- Poor phrasing that leads to accidental misunderstanding
- Overly emotional phrasing that makes readers suspicious or turns them off
- Constructions that phrase points negatively
- Potentially libelous words that recklessly attack someone's character or behavior

> **WRITE Connection**
>
> For an extended discussion of using fair, respectful language—especially in connection with race, ethnicity, gender, disabilities, and age—see Chapter 4, "Writing for Diversity." Also check out the discussion of e-mail shouting and flaming on page 373.

Use Words Positively

To avoid negative words, do not write while you are angry and be aware of the reader's potential response to specific words. Make sure that each word in your message passes the politeness test.

Take Care of Names

Names of people, companies, products, and so on are words that people prize. If you abuse such key words, you put up a hurdle to your reader hearing and accepting your message.

> **CARELESS:** I read with some curiosity, Mrs. Daquincy, your letter concerning the Family Center's project to build a shelter that our city can enjoy.
>
> **CAREFUL:** I read with interest, Ms. DeQuincey, your letter concerning Family First Center's project. I commend your efforts to build a shelter for women and children victimized by domestic violence.

Replace Words Having Negative Overtones

Words have neutral "dictionary" definitions. However, they can also have positive or negative associations for readers. These associations, called *connotations*, can differ from person to person, from group to group, and from region to region.

> **POOR:** Although your bid was not *insufficient* with respect to cost and experience, it *failed* to offer the overall value of the bid from Earth-Scape Landscape Design. In particular, you *neglected* to consider stress on the Nooksack watershed.
>
> **BETTER:** Your bid was competitive with respect to cost and experience, but Earth-Scape Landscape Design's plan provides a sustainable landscape enhancing the Nooksack watershed.

Part 2 Benchmarking Writing with the Seven Traits

Seek Advice About Word Choice in Touchy Situations

When your message addresses a serious conflict or even a potential danger, restrain yourself from sending off a quickly drafted, poorly worded document. Instead, take the draft through a review—from a superior and/or your company's legal department. (*Note:* While such legal review is crucial for complaints, product instructions, and so on, make sure that your document doesn't become filled with legal jargon—stand firm for plain English.)

> **POOR:** While your past service may have been adequate, this Snorkel Lift fiasco just goes to show that Industrial Aggregate Equipment and project supervisor Nick Luther, in particular, are incompetent, unreliable, and, yes, even dishonest.
>
> **BETTER:** In the past, we have appreciated your service and assistance. However, we can only conclude from the problems with the Snorkel Lift rebuild that your level of service has seriously declined, and that you cannot provide what Rankin Manufacturing needs.

Accusatory words to avoid

Alcoholic	Drunk	Inferior	Pressure tactics
Bankrupt	Exorbitant	Insane	Psychotic
Bribery	Extortion	Insolvent	Senile
Bum	Fiscally irresponsible	Irresponsible	Stupid
Cheat/corrupt	Fool	Kickback	Suspicious
Crazy	Fraud	Lazy	Threaten to
Crook	Gullible	Liar/lies/lying	Ultimatum
Discrimination	Highway robbery	Misappropriate	Unbelievable
Dishonest	Incapable	Misconduct	Unheard of
Drug addict	Incompetent	Neurotic	Worthless

Documents That Require Special Sensitivity

Because of their sensitive topic or public nature, the documents below require special attention to negative word connotations. Watch your words!

Adjustment letters (Ch. 25)
Advice messages (Ch. 25)
Apologies (Ch. 25)
Claims or complaints (Ch. 26)
Collection letters (Ch. 27)
Communicating errors, changes, delays (Ch. 26)
Credit application rejection (Ch. 26)
Employee evaluations (Ch. 40)
Employee recommendations (Ch. 40)
Incident reports (Ch. 30)
Investigative reports (Ch. 30)
Job advertisements (Ch. 40)
Job descriptions (Ch. 40)
Job-offer rejection letters (Ch. 19)
News releases (Ch. 34)
Policy statements (Ch. 39)
Refusing favors or requests (Ch. 26)
Rejecting ideas, proposals, bids (Ch. 26)
Resignations (Ch. 26)
Troubleshooting proposals (Ch. 32)
Web pages (Ch. 36)

Chapter 13 Trait 4: Clear Words

Neutral and Positive Word Choice in Action

The two examples below show a letter that includes slanted and negative word choice as well as a more sensitive version of the same letter. Read the first letter, and note the negativity created through word choice. Then read the second letter to see how the word choices have been revised to create a neutral, sensitive message.

Slanted, Negative Word Choice

Ruby,

Unfortunately, I am writing to bring you up to speed on some major problems (which I'm sure come as no surprise to you) concerning your ability to get automobile insurance coverage.

To keep their premiums low, almost all insurance companies simply will not cover high-risk drivers like you (and many other teenagers, unfortunately). In just the past two years, you have run up a pretty dire driving record by being responsible for an accident and collecting four moving violations. Not surprisingly, your current auto insurance company, Greenwood Fire & Auto, has naturally decided to drop you as of February 15. I have managed to find two high-risk insurance companies that will cover you in spite of your very poor record.

Contact me soon so that we can deal with your problem.

Best of luck,

Larry

Neutral, Sensitive Word Choice

Dear Ms. Chen:

As your personal insurance agent, I want to update you on the status of your automobile insurance.

To keep their premiums low, some insurance companies will not cover high-risk drivers. In the past two years, you have had an at-fault accident and four moving violations. As a result, your current auto insurance company, Greenwood Fire & Auto, has chosen not to renew your policy (#46759) effective February 15, 2005. However, I do have two nonstandard auto insurance companies that are willing to provide you coverage.

Please call me at 555-1228 so that we can determine how best to meet your auto insurance needs.

Yours sincerely,

Larry Brown

CHECKLIST for Word Choice

Use this checklist to evaluate and strengthen the word choice in your writing.

- ☐ The sentences contain no irrelevant information or redundant wording.
- ☐ The words are concise rather than wordy, and precise rather than general.
- ☐ Bland nouns and verbs have been replaced with vivid words, and unnecessary modifiers have been cut.
- ☐ Where modifiers are needed, precise adjectives and adverbs clarify the meaning and enhance the message.
- ☐ Technical terms are used correctly, sparingly, and consistently.
- ☐ The writing uses common, precise words rather than slang or shop talk.
- ☐ Flowery language, business English, euphemisms, and clichés have been replaced with words that are plain English.
- ☐ The writing contains no unfair or disrespectful language.
- ☐ The word choice is confident and positive versus apologetic and negative.

CRITICAL THINKING Activities

1. Review two pieces of workplace writing—one short and one long. Analyze each for passages that are wordy, concise, or abrupt.
2. Analyze and evaluate three business documents for issues related to exact, fresh language: vague nouns, weak verbs, poor modifiers, business English, flowery language, euphemisms, jargon, and clichés. Highlight examples and explore how the word choice weakens the message.
3. Review your business-related e-mail messages to find examples of negative wording and disrespectful language that are or are not appropriate in e-mail. To guide your decisions, read Chapter 23, "Writing E-Mail Messages and Sending Faxes."

WRITING Activities

1. Select a sample piece of poorly worded workplace writing that you used in one of the Critical Thinking activities. Edit that piece to make it concise, fresh, and positive. Submit both the original and your edited version.
2. Working with two classmates, exchange pieces of writing and critique them for wordiness, tired wording, and negative phrasing. Where you find general or poor wording, suggest replacements. Where you find wordy passages, suggest cuts. After the discussion, revise your writing carefully and submit both the original draft and the revised copy to your instructor.

In this chapter

Smooth Sentences: Questions and Answers **260**
Rough Problems and Smooth Solutions **261**
Combining Choppy Sentences **262**
Energizing Tired Sentences **266**
Dividing Rambling Sentences **270**
Sentence Smoothness in Action **271**
Checklist for Sentences **272**
Critical Thinking Activities **272**
Writing Activities **272**

The seven traits

1. Strong Ideas
2. Logical Organization
3. Conversational Voice
4. Clear Words
5. Smooth Sentences
6. Correct Copy
7. Reader-Friendly Design

CHAPTER
14

Trait 5: Smooth Sentences

You write to conduct business—not to construct admirable sentences. Your work involves writing results-focused memos, reports, and project proposals—not novels. So why should you worry about something like sentence smoothness? Why, for example, must the sentences in your *daily e-mails* mimic Thoreau's furrow? The answer is simple: Like a deep, straight furrow that's easy to follow and pleasing to see, a smooth sentence is easy to understand and a pleasure to read.

Your readers need smooth sentences because they want to grasp your meaning quickly and correctly. Clumsy, incomplete, or muddled sentences slow the reading process, make the reader reread passages, and cause him or her to *guess* at your meaning rather than *get* it. By contrast, smooth sentences are energetic but energy efficient, complete but lean, carefully crafted but not flashy. Running straight ahead like that deep furrow, their meaning is clear at the first reading.

To write smooth sentences, start by believing in the quality of your thoughts and getting those thoughts on paper or on the computer screen. Then refine your ideas by refining your sentences. This chapter will help you in two ways. First, it will help you distinguish straight-furrow sentences from muddled, meandering, or incomplete ones. Second, it will help you sharpen your sentence-writing skills.

> "A sentence should read as if its author, had he held a plough instead of a pen, could have drawn a furrow deep and straight to the end."
> —Henry David Thoreau

SMOOTH SENTENCES: QUESTIONS AND ANSWERS

It's impossible to summarize in a few pages everything you need to know about sentences. We can start, however, by answering three questions.

What is a smooth, efficient sentence?
- It's direct—getting straight to the point.
- It's clear—easily understood in one read through.
- It's simple—emphasizing actors and actions, who is doing what.
- It's rhythmical—reading well both silently and aloud.

How do you write sentences that mean business?
- Generally, keep them short, aiming for twenty or fewer words.
- Aim for variety. While short is good, variety is important—not just for variety's sake, but to create balance and emphasis in your ideas.
- Write sentences that sound right when you speak them. Develop and trust your ear.
- Practice sentence-structuring techniques so that they become second nature. (Good writers don't say, "Oh, it's time for a complex sentence." They just do it.)
- Practice sentence modeling. Look closely at sentence patterns in letters, memos, and other documents. Learn what works and what doesn't. Then use the model sentences when you write, imitating them part by part.

   ORIGINAL SENTENCE: However, because all projects are formed by local people, I encourage you to share the enclosed literature with others in your community.

   IMITATION: However, because the problem affects both Human Resources and PR, I recommend that you share the proposal with the Communications Department.

Where can you get more help?
- "Stating Ideas Clearly" (pages 188–189) for help with making claim statements
- "Structuring Documents Through Paragraphing" (pages 225–231) for advice on joining sentences into coherent paragraphs
- "Editing for Sentence Smoothness" (page 34) for tips on strengthening sentences as you refine your writing
- "Correcting Faulty Sentences" (pages 278–283) for help fixing grammatical errors
- "Constructing Sentences" (Chapter 46) for explanations about sentence parts and structure
- "Using Punctuation" (Chapter 47) for guidance on properly punctuating sentences
- "Addressing ESL Issues" (Chapter 50) for instruction on sentence challenges facing English language learners (those learning English as a second or third language)
- The book's Web site at http://college.hmco.com/english

ROUGH PROBLEMS AND SMOOTH SOLUTIONS

Three problems that may plague your sentences are described below. The rest of this chapter addresses these problems more fully.

Problem 1: choppy sentences
- A series of short sentences with poor flow
- Simplistic structures
- Poor sentence connections

> I am responding to your job advertisement. It appeared in the Seattle Times. The date was June 12, 2005. I am applying for the Software-Training Specialist position.

Solution 1: sentence combining
- Coordinate related ideas.
- Subordinate secondary ideas.
- Add transitions.

> In response to your advertisement in The Seattle Times on June 12, 2005, I am applying for the position of Software-Training Specialist.

Problem 2: tired sentences
- Unenergetic flow
- Insecure or negative constructions
- Repetitive patterns

> I recently moved to Seattle and haven't neglected my career. There is a software-training specialist position that has opened up at Evergreen Medical Center. My application has been submitted.

Solution 2: sentence energizing
- Use active and passive voice properly.
- Change negative constructions to positive.
- Eliminate nominalized constructions and expletives.
- Vary sentence openings, structures, and types.

> Since my move to Seattle, I've been looking for opportunities in computer services. Recently, I applied for a position with Evergreen Medical Center as a software-training specialist.

Problem 3: rambling sentences
- Strung-out sentences
- Meandering, lazy flow
- Confusing, overpacked with ideas

> I enjoyed touring your facilities and meeting your office managers, and I would enjoy contributing to the work that you and other staff members do at Evergreen Medical Center, so I believe that my hospital training would be an asset.

Solution 3: sentence dividing
- Control ideas with lists and parallelism.
- Divide sentences or cut sentence parts.

> After touring your facilities and meeting your office managers, I am certain that I would enjoy contributing to your work at Evergreen Medical Center. I believe that my hospital training would be an asset.

COMBINING CHOPPY SENTENCES

When sentences are choppy, they have several flaws:

- They sound abrupt, static, or monotonous—not friendly and natural.
- They contain little information and repeat words needlessly.
- They give all ideas equal treatment when key points need more emphasis.

To fix choppy sentences, learn to combine them through coordination and subordination.

Combine Through Coordination

Coordination joins sentences and sentence elements to show equal relationships between ideas and details.

Create Compound Subjects

If you find different subjects doing a similar action in a series of sentences, then combine sentences by creating a compound subject.

> **CHOPPY:** The insurance adjustor examined hail damage to the roof. A contractor also checked the damage.
>
> **SMOOTH:** The insurance adjustor and a contractor examined hail damage to the roof.

Create Compound Predicates

If you find the same subject doing different actions in a series of sentences, combine sentences by creating a compound predicate.

> **CHOPPY:** The adjustor examined the roof. She checked the valleys closely. Finally, she inspected the carport.
>
> **SMOOTH:** The adjustor examined the roof, checked the valleys, and inspected the carport.

Create Compound Sentences

In a compound sentence, a coordinating conjunction (*and, but, or, nor, for, yet, so*) or a conjunctive adverb (*also, besides, therefore, however*) joins independent clauses into an equal relationship. The conjunction signals the relationship between the ideas. Remember to include a comma or semicolon between the two independent clauses.

> **CHOPPY:** The adjustor concluded that only the house roof was damaged. The contractor argued that the carport roof should also be replaced.
>
> **SMOOTH:** The adjustor concluded that only the house roof was damaged, but the contractor argued that the carport roof should also be replaced.
>
> **SMOOTH:** The adjustor concluded that only the house roof was damaged; however, the contractor argued that the carport roof should also be replaced.

Use Parallel Structure

Coordinated sentence elements should be parallel, following the same grammatical pattern. Parallel structure saves words, clarifies relationships, and sequences information.

Consistent Elements

For words, phrases, or clauses in a series, keep elements consistent.

> **UNPARALLEL:** I have instructed clients in Microsoft Windows XP, Corel WordPerfect 12, not to mention my familiarity with Adobe InDesign.
>
> **PARALLEL:** I have instructed clients in Microsoft Windows XP, Corel WordPerfect 12, and Adobe InDesign.
>
> **UNPARALLEL:** I have experience working in a hospital developing job descriptions, have even been recruiting technical employees, and in the training of human resources personnel.
>
> **PARALLEL:** I have experience working in a hospital developing job descriptions, recruiting technical employees, and training human resources personnel.

Correlative Conjunctions

Use correlative conjunctions (*either, or; neither, nor; not only, but also; as, so; whether, so; both, and*) so that both parts of the pattern are balanced.

> **UNPARALLEL:** Rankin not only turned 20 this year. It experienced 16 percent growth in sales in addition.
>
> **PARALLEL:** Rankin not only turned 20 this year but also experienced 16 percent growth in sales.

Consistent Modifiers

Make sure that modifiers apply consistently in a series.

> **CONFUSING:** MADD campaigns to severely punish and eliminate drunk driving because this offense leads to a great number of deaths and sorrow.
>
> **PARALLEL:** MADD campaigns to eliminate and severely punish drunk driving because this offense leads to many deaths and untold sorrow.

Arrangement of Sentence Elements

To stress a contrast, arrange sentence elements around differences.

> **WEAK CONTRAST:** The average child watches 24 hours of television per week and reads 36 minutes.
>
> **STRONG CONTRAST:** Each week, the average child watches television for 24 hours but reads for only 36 minutes.

WRITE Connection

Parallelism in headings, subheadings, lists, and outlines can effectively organize whole documents. See pages 220–224.

Combine Through Subordination

Because all ideas aren't equal, show the reader which ones are more or less important than others. Subordination helps you do so by placing key points in main clauses and playing down secondary points in words, phrases, or dependent clauses.

Shrink Ideas into Single Words

Pull a key word out of one sentence, integrate that word into another sentence, and cut what's left over.

CHOPPY: My previous position included administrative work. I administered some policies, as well as wages and benefits.

SMOOTH: My previous position included some policy and wage-and-benefit administration.

CHOPPY: I enjoyed touring your facilities. The meetings with office managers were especially helpful. They impressed me with their eagerness. Your staff's desire to improve its skills with technology really made an impact.

SMOOTH: After touring your facilities and meeting extensively with your office managers, I am impressed by your staff's eagerness to grow technologically.

CHOPPY: I found your computers up to date. Most computers were also functioning well. I also checked the printers and found most of them in good repair.

SMOOTH: In addition, I found most of your office equipment current and in good repair.

Shrink Ideas into Phrases

Look for phrases in one sentence that can be pulled easily into another sentence.

CHOPPY: I have done a lot of training in the health care system. That includes over 200 clients. My training experience spans seven years. I can work in both group and one-on-one situations.

SMOOTH: Within the health care system, I have trained approximately 200 clients over seven years in both group settings and one-on-one formats.

CHOPPY: I have extensive experience with software programs and systems. In addition, I have worked in a hospital. In that work, I developed job descriptions. Also, the recruitment of technical employees was part of that job.

SMOOTH: Having worked extensively with software programs and systems, I have also worked in a hospital setting developing job descriptions and recruiting technical employees.

CHOPPY: I have enclosed my current résumé. It details my education and qualifications in software programs and systems. The résumé also shows my experience working with software programs and systems.

SMOOTH: Enclosed is my résumé, which further details my education and experience working with software programs and systems.

Chapter 14 Trait 5: Smooth Sentences

Shrink Ideas into Subordinate Clauses

Sometimes, sentences can be combined by making one clause subordinate to (dependent on) a main clause. (Subordinate clauses are connected to the main clause with either a subordinate conjunction or a relative pronoun.)

SUBORDINATE CONJUNCTIONS: after, although, because, if, when, while, until, unless

CHOPPY: I would be happy to supply additional information. I would also appreciate an interview. Please call me at (554) 653-0087.

SMOOTH: If you want additional information or wish to set up an interview, please call me at (554) 653-0087.

RELATIVE PRONOUNS: who, whose, which, that

CHOPPY: Your evaluations made me a better worker. Those evaluations were always encouraging.

SMOOTH: Your evaluations, which were always encouraging, made me a better worker.

Vary Your Combining Techniques

Together, coordination and subordination are powerful methods of turning rough writing into smooth-flowing sentences. Check the differences in the passages below.

Choppy Passage

We discussed a starting date of August 24. That date works well for me. I'll now begin to search for housing. Thank you for the material you sent about moving costs. I also appreciated the information about real estate agencies. Could you please also send information about area schools?

Smooth Passage — *subordination* —

As we discussed earlier, the starting date of August 24 works well for me. Before then, I'll begin to search for housing. In addition to the material you've already provided about moving costs and real estate agencies, could you also please send information about area schools?   *coordination*   *subordination*

WRITE Connection

Writing a series of smooth sentences—a paragraph—requires that you create links or transitions from one sentence to the next. For help, check "Paragraph Coherence" and "Transition and Linking Words" (pages 230–231).

ENERGIZING TIRED SENTENCES

Tired sentences lack energy, variety, and rhythm. They limp along until your reader is hypnotized. If your sentences suffer from chronic fatigue syndrome, implement the cures below and on the following pages.

Rely on the Active Voice

Most verbs can be in either the active or the passive voice. When a verb is active, the sentence's subject performs the action. When the verb is passive, the subject is acted upon.

> **ACTIVE:** If you can't attend a meeting, notify Richard by the Thursday before.
>
> **PASSIVE:** If a meeting can't be attended by you, Richard must be notified by the Thursday before.

Positively Active

The passive voice tends to be wordy, unnatural, and sluggish because the verb's action is directed backward, not straight ahead like a plowed furrow. In addition, passive constructions tend to be impersonal—people can disappear.

> **PASSIVE:** The sound system can now be used to listen in on sessions in the therapy room. Parents can be helped by having constructive one-on-one communication methods modeled.
>
> **ACTIVE:** Parents now use the sound system to listen in on sessions in the therapy room. Therapists can help parents by modeling constructive one-on-one communication methods with children.

Positively Passive

The passive voice isn't an error. In fact, the passive voice has some important uses: (1) when you need to be tactful (e.g., in a bad-news letter), (2) if you wish to stress the object or person acted upon, and (3) if the actual actor is understood, unknown, or unimportant.

> **ACTIVE:** Our engineers determined that you bent the bar at the midpoint.
>
> **PASSIVE:** Our engineers determined that the bar had been bent at the midpoint. [tactful]
>
> **ACTIVE:** Congratulations! We have approved your loan.
>
> **PASSIVE:** Congratulations! Your loan has been approved. [emphasis on receiver; actor understood]

F.Y.I.

Here are two more tips about using active and passive voice:
- Avoid using the passive voice unethically to hide responsibility.
- Avoid passive constructions when phrasing actions, conclusions, and recommendations. State who should do *what*, or *what* won't get done.

Use Strong Verbs, Not Nominalized Constructions

Energetic sentences contain strong verbs. However, when strong verbs become nouns and are then replaced with weak verbs (e.g., *be, appear, seem, have, give, do, get*), the result is a nominalized construction—a sluggish, wordy, difficult sentence.

Nominalization	Strong verb
Give a description	Describe
Provide instructions	Instruct
Offer a recommendation	Recommend

> **SLUGGISH:** I had a discussion of the incident with the Undercoating Crew. They gave me their confirmation that similar developments had occurred before, but they had not proceeded to submissions of reports.
>
> **ENERGETIC:** I discussed the incident with the Undercoating Crew. They confirmed that similar problems had developed before, but they hadn't submitted reports.

Avoid Expletives

Expletives such as *it is* and *there is* are fillers that serve no purpose in most sentences, except to make them wordy and unnatural.

> **SLUGGISH:** *It is* certain that I am committed to Yellow Brick's standards of excellence in publishing, and *there is* also my appreciation of the company's supportive employee policies.
>
> **ENERGETIC:** I am certainly committed to Yellow Brick's standards of excellence in publishing, and I appreciate the company's supportive employee policies.
>
> **SLUGGISH:** There are far too many novels featured on the store's sale shelf; there must be room on this shelf for other sale items as well.
>
> **ENERGETIC:** The sale shelf presently features only novels, whereas it should display other sale items as well.

Avoid Negative Constructions

Sentences based on the negatives *no, not,* or *neither/nor* can be wordy and difficult to understand. It's simpler to state what is the case, as opposed to what is not; what can be done, as opposed to what cannot.

> **NEGATIVE:** During my seven years here, I have not been behind in making significant contributions to our company. My editorial skills have certainly not deteriorated as I have never failed to tackle not-so-simple assignments and responsibilities.
>
> **POSITIVE:** During my seven years here, I have made significant contributions to our company. My editorial skills have steadily developed as I have taken on more difficult assignments and responsibilities.

Use Sentence Variety Techniques

To energize your sentences, you may need to work on sentence variety. But what do you vary?

Vary Sentence Openings

Moving a modifying word, phrase, or clause to the front of the sentence stresses that modifier. However, avoid creating dangling or misplaced modifiers (see page 276).

NORM: Your confirmed room was unfortunately unavailable last night.

VARIATION: Unfortunately, your confirmed room was unavailable last night.

NORM: We are sorry for the inconvenience this may have caused you.

VARIATION: For the inconvenience this may have caused you, we are sorry.

Vary Sentence Lengths

Short sentences (ten or fewer words) are ideal for making points crisply. Medium sentences (ten to twenty words) carry the bulk of your discussion. When well crafted, occasional long sentences (more than twenty words) can deepen and expand your ideas.

SHORT: Welcome back to Magnolia Suites!

MEDIUM: We are pleased to host you during your stay in Memphis. Unfortunately, your confirmed room was unavailable last night when you arrived. For the inconvenience this may have caused you, we are sorry.

LONG: Because several guests did not depart as scheduled, we were forced to provide you accommodations elsewhere, but, for your trouble, we were happy to cover the cost of last night's lodging.

Vary Sentence Types

The most common sentence is declarative: It states a point. For variety, try questions, commands, conditional statements, exclamations, and even sentence fragments.

EXCLAMATION: Our goal now is providing you with outstanding service!

FRAGMENT: With warm, Southern hospitality.

DECLARATIVE: To that end, we have upgraded your room at no expense to you.

COMMAND: Please accept this box of chocolates as a gift to sweeten your stay.

QUESTION: Do you need further assistance?

CONDITIONAL: If you do, we are ready to fulfill your requests.

Vary Sentence Arrangements

Where do you want to place the main point of your sentence? You make that choice by arranging sentence parts into loose, periodic, or cumulative patterns. Each pattern has a specific effect.

> **LOOSE:** We have found an attractive reservation and payment plan to replace the old one, a plan that allows both you and the company to collect bonus miles and also gives you $150,000 life insurance per flight.
>
> **ANALYSIS:** This pattern is direct. It states the main point immediately, and then tacks on extra material.
>
> **PERIODIC:** Although this procedural change for making travel arrangements, which takes effect October 1, restricts your choices, both you and the company benefit.
>
> **ANALYSIS:** This pattern postpones the main point until the end. It builds to the point, creating an indirect, dramatic effect.
>
> **BALANCED:** Hiring a Cultural Diversity Specialist this year has given Hope Services a strong presence in the community; therefore, HS will focus in the coming year on strengthening its outreach to minority communities and increasing its cultural-diversity training.
>
> **ANALYSIS:** This pattern gives equal weight to complementary or contrasting points; the balance is often signaled by a conjunction (*and, but*), a conjunctive adverb (*however, nevertheless*), or a semicolon.
>
> **CUMULATIVE:** Because the plane ticket will be in your name, you will continue, as in the past, to accumulate miles on all flights you book through the Travel Center—that is, if you are a member of the airline's Bonus Miles Club.
>
> **ANALYSIS:** This pattern places the main idea in the middle, surrounding it with modifying words, phrases, and clauses. The result is a sentence that cushions, qualifies, or elaborates a point.

Use Positive Repetition

As you vary sentence styles and structures, strive especially to avoid repetitiveness. Instead, use emphatic repetition—repetition of a key word or grammatical construction to stress a point.

> **REPETITIVE:** Each year, more than a million poor-reading young people annually leave high school unable to read, functionally illiterate.
>
> **EMPHATIC** (repeats a word): Each year, more than a million young people leave high school functionally illiterate, so illiterate that they can't read daily newspapers, job ads, and safety instructions.
>
> **REPETITIVE:** As a result, these unskilled young people lack the skills needed to succeed in high-skilled jobs and high-skilled careers.
>
> **EMPHATIC** (repeats a construction—infinitive phrase): As a result, these young people lack the skills needed to find jobs, to keep jobs, and to advance in their careers.

DIVIDING RAMBLING SENTENCES

While choppy sentences dole out information in tidbits, rambling sentences offer readers platefuls to swallow whole. In other words, rambling sentences are more than long: They pile up too much information.

How Do You Recognize Rambling Sentences?

Use the following tests to identify ramblers:

- Scan your writing for sentences two or more lines long—longer than twenty words.
- Read sentences aloud, listening for ones that leave you breathless or lost.
- Check sentences for an overpopulation of coordinating conjunctions (*and, but, or*), subordinating conjunctions (*although, because, since, when*), and relative pronouns (*who, whom, which, that*).

How Do You Fix Rambling Sentences?

Once you've identified a rambler, you have a choice of repairs. Consider these strategies, and use the one that does the best job of addressing the issues of your specific sentence.

1. Ask, What's my basic point here? Then state that point directly and clearly, without the twists and turns of the rambling original.
2. Ask, What can my reader comfortably chew in one bite? Then do the following:
    - Divide your material into bite-sized pieces—one sentence per bite.
    - Order information by shaping it into lists. (See page 224 for help.)
    - Locate and cut say-nothing material from the sentence.
3. After circling coordinating conjunctions, subordinating conjunctions, and relative pronouns, use them to decide how to pull apart rambling sentences.

Rambling Sentence

I certainly agree that this arrangement, where you will be able to offer your clients "one-stop shopping" where they can meet all their computer needs with little difficulty, for hardware, software, training, and support, will benefit all parties so that I will have access to your established customer base without the difficulties that are so obvious and difficult to deal with of generating my own.

Rambling Sentence Fixed

I agree that this arrangement will benefit all parties involved. You will be able to offer your clients "one-stop shopping" where they can meet all their computer needs for hardware, software, training, and support. And I will have access to your established customer base without the difficulties of generating my own site for more information.

SENTENCE SMOOTHNESS IN ACTION

The key to sentence smoothness is getting it to happen in all your sentences, not just one. Contrast the passages below to see a variety of sentence smoothness techniques in action.

Original Rough Passage — *Short, choppy sentences*

Let me be the first to offer congratulations. We have made a decision to approve your loan. Next we need to receive an appraisal of the property, and then the closing can occur. I have attached, with these various factors in mind, a commitment letter, in which are listed the conditions that are necessary to be met to achieve successfully closure of this loan. Please review these conditions, calling me at 555-6900, in the event that questions you have may arise. — *Awkward syntax*

— *Long, overly complex sentence*

Awkward, passive construction

Preapproval for you to receive a Southeast credit card has also occurred. Taking advantage of this opportunity will result in a $20 reduction on closing costs associated with the completion of the paperwork regarding your mortgage loan. Taking advantage of this offer would require the completion of the enclosed application. Completion must occur five days prior to the date of the loan closing. This is why an application form and postage-free envelope have been enclosed. — *Impersonal passive voice*

Piled-up phrases and nominalized constructions

Edited Smooth Passage — *Varied sentence lengths and openings*

Congratulations! Your loan has been approved. Once we receive an acceptable appraisal of the property, we can proceed with the closing. With that in mind, I have attached a commitment letter listing the conditions that need to be met to close your loan successfully. Please review them and call me at 555-6900 if you have any questions.

— *Helpful transitions*

Natural syntax

Active voice

Southeast is also sending you a preapproved credit card. If you take advantage of this opportunity, you will receive a $20 reduction on the closing costs for your mortgage loan. To enjoy this reduction, simply return your completed credit card application to me five days before closing the loan. I have enclosed both an application and a postage-free envelope for your convenience.

— *Energetic verbs*

Varied sentence types and arrangements

CHECKLIST for Sentences

Use this checklist to evaluate and strengthen the sentences in your writing.

- ☐ All my sentences pass the read-aloud test.
- ☐ The sentences are clear, true, and straightforward—like a straight furrow.
- ☐ The writing includes no choppy, tired, or rambling sentences.
- ☐ Active and passive voice are used properly.
- ☐ Strong verbs replace nominalized constructions.
- ☐ Precise, energetic words and phrases replace expletives.
- ☐ Sentences are positive versus negative, focusing on *what is* versus *what is not*.
- ☐ Series of words or phrases within sentences are parallel.
- ☐ Sentences vary in structure, length, and type.

CRITICAL THINKING Activities

1. Working with two classmates, examine two business documents for tired sentences, nominalizations, expletives, negative constructions, and rambling sentences. Discuss why these items weaken the message.
2. Review a business-related e-mail message that you received. Examine the piece for the errors discussed in this chapter, and consider how the weaknesses affect your assessment of the writer and the message. Share your ideas with the class.

WRITING Activities

1. Working with a classmate, review the information regarding choppy sentences and discuss these corrective strategies: compound subjects, compound predicates, compound sentences, and parallel structure. Find a business document that contains choppy sentences, note how the structures weaken the writing, and revise the piece using subordination or coordination strategies. Share your work with the class.
2. Choose a piece of your own writing, look for the sentence problems discussed in this chapter, and revise the piece as needed.
3. Write a one-page memo on a topic related to your present job or a future job. Include in the memo at least six sentence problems cited in this chapter. Exchange papers with a classmate, identify the problems, and discuss how they could be fixed.

WEB Link

For further help on sentence smoothness issues, check http://college.hmco.com/english.

In this chapter

Basic Terms: A Primer for Correctness **274**
Correcting Unclear Wording **275**
Correcting Faulty Sentences **278**
Correcting Punctuation Marks **284**
Correcting Mechanical Difficulties **286**
Checklist for Errors **288**
Critical Thinking Activities **288**
Writing Activities **288**

The seven traits

1. Strong Ideas
2. Logical Organization
3. Conversational Voice
4. Clear Words
5. Smooth Sentences
6. **Correct Copy**
7. Reader-Friendly Design

CHAPTER
15

Trait 6:
Correct Copy

In accounting, a misplaced decimal point can be a disaster. Correctness counts, literally!

In writing, errors can be just as costly. Mistakes in grammar, punctuation, usage, mechanics, and spelling—even simple typos—can change your meaning and confuse readers. Often, the result is unnecessary and awkward follow-up. At the very least, such errors distract readers and make a bad impression.

How do you avoid these distracting errors, when should you even worry about them, and what if you just aren't that good at grammar? This chapter will answer these questions while showing you how to identify and fix the most common writing errors.

For a complete discussion of grammar, punctuation, usage, and mechanics, go to Part 9, "Proofreader's Guide." This chapter covers some of the most common writing errors.

BASIC TERMS: A PRIMER FOR CORRECTNESS

To improve the correctness of your workplace writing, start with the overview below before studying the errors on the pages that follow. The pages listed with the definitions alert you to the more complete explanations offered in the "Proofreader's Guide."

The **parts of speech** are the eight different ways words are used:

- A **noun** names something—a person, a place, a thing, or an idea (page 697).
- A **pronoun** is a word used in place of a noun (page 700).
- A **verb** expresses action, links words, or acts as a helper for a main verb (page 704).
- An **adjective** describes or modifies a noun or pronoun (page 710).
- An **adverb** describes or modifies a verb, an adjective, or another adverb (page 711).
- A **preposition** shows the relationship between its object (noun or pronoun following the preposition) and another word in the sentence (page 712).
- A **conjunction** connects words or groups of words (page 713).
- An **interjection** is a word that communicates strong emotion (page 713).

Sentences are groups of words that express a complete thought.

- The **subject** names the person or thing either doing the action in a sentence or being talked about (page 715). Example: *The ignition key*.
- The **predicate** is the sentence part that either tells what the subject is doing or says something about the subject (page 717). Example: *Got stuck because of a faulty tumbler*.
- A **phrase** is a group of related words functioning as a single part of speech and lacking a subject, predicate, or both (page 718). Example: *Of a faulty tumbler*.
- A **clause** is a group of related words having both a subject and a predicate; a clause can be independent (a complete, stand-alone thought) or dependent (an incomplete, reliant thought) (page 720). Independent: *The tumbler was faulty*. Dependent: *Because the tumbler was faulty*.

Punctuation refers to the marks or signals indicating relationships between sentences and sentence parts (page 723), such as periods, commas, and semicolons.

Mechanics refers to a range of basic issues—spelling, capitalization, plurals, numbers, abbreviations, acronyms, and initialisms (page 743).

Usage refers to problems associated with commonly confused and misused pairs and groups of words (e.g., *among* and *between*, *bring* and *take*) (page 755).

 F.Y.I.

Improve correctness by fixing these basic errors:

- Misspelling the reader's or another person's name.
- Missing typos not caught by the spell checker (e.g., "pubic/public meeting").

CORRECTING UNCLEAR WORDING

Each word should contribute to your message's clarity. When the errors on the next three pages creep into your writing, the result is unclear wording.

Unclear Pronoun Reference

An error in pronoun reference happens when what a pronoun refers to (its antecedent) is ambiguous, broad, or missing.

Who, which, and that clauses. Relative pronouns connect dependent clauses to the main clause. Like other pronouns, *who, which,* and *that* must refer to a specific noun, not to a whole sentence, clause, or phrase. In addition, that referent should come immediately before the relative pronoun.

> **UNCLEAR:** Turn in your surveys on Monday, which will complete the review process. [The pronoun *which* is vague because it doesn't refer to a specific noun.]
> **CORRECT:** To complete the review process, turn in your surveys on Monday.

> **UNCLEAR:** The review will be presented to the Administrative Council, which will be included in our annual report, in March 2006. [The phrase *which will be included in our annual report* is misplaced.]
> **CORRECT:** The review, which will be included in our annual report, will be presented to the Administrative Council in March 2006.

Vague *it, this, that,* and *they*. These pronouns should refer to a specific preceding noun. Avoid using them to refer generally to a clause or sentence.

> **UNCLEAR:** The building includes both a residence and a small coffee shop. Does *this* conform with zoning bylaws? [*This* refers generally to the preceding clause.]
> **CORRECT:** The building includes both a residence and a small coffee shop. Does *this use* conform with zoning bylaws? [*This* refers specifically to use.]

> **UNCLEAR:** Our attorney said that she had checked whether the plan was legal, and that they had approved the plan. [The pronoun *they* refers to no specific noun or pronoun.]
> **CORRECT:** Our attorney said that she had checked with the Building Inspector and City Manager, and that they had approved the plan.

Indefinite pronoun reference. If a pronoun can refer to two or more nearby nouns, readers may be confused.

> **UNCLEAR:** Rankin Technologies has developed an international reputation in ultraviolet technology for water purification. *It* will become even more significant in the next decade. [Does *it* refer to purification, technology, or reputation?]
> **CORRECT:** Rankin Technologies has developed an international reputation in ultraviolet technology for water purification. *This technology* will become even more significant in the next decade.

> **UNCLEAR:** Rankin will invest 6.2 million in development and marketing, and it will usurp most of those funds. [The pronoun *it* could refer to *development* or *marketing*.]
> **CORRECT:** Rankin will invest 6.2 million in development and marketing, and development will usurp most of those funds.

Unclear Modifiers

Unclear modifiers are adjectives and adverbs that create confusion because they are misplaced, dangling, or squinting.

Misplaced modifiers are unclear because they are placed incorrectly. To fix the problem, relocate modifiers, placing them as close as possible to what's being modified.

> **MISPLACED:** Please review the pamphlet describing my services and equipment *enclosed*.
>
> **CLEAR:** Please review the *enclosed* pamphlet describing my services and equipment.

F.Y.I.

Be especially careful with adverbs like *only, almost, just, even, hardly, barely,* and *regularly*. Generally, place these directly before the word modified.

> **MISPLACED:** These meetings will be held on the second Monday of each month from 8:30 to 9:30 a.m. *regularly* in the Human Resources Conference Room.
>
> **CLEAR:** These meetings will *regularly* be held on the second Monday of each month from 8:30 to 9:30 a.m. in the Human Resources Conference Room.

Dangling modifiers are unclear because the word being modified is missing or far removed. The fix? Insert the necessary noun or pronoun, or change the dangling phrase into a clause.

> **DANGLING:** Having committed to meeting with us, our regular attendance would be appreciated.
>
> **CLEAR:** Having committed to meeting with us, Robert and Miriam deserve our regular attendance.

F.Y.I.

Dangling modifiers are often verbal phrases placed at the beginning of a sentence. The phrase modifies the subject, but if the modified noun is missing or far removed, the verbal phrase "dangles" there.

Squinting modifiers are unclear because they are located between two sentence parts that could be referents. The reader is left unsure which part is being modified.

> **SQUINTING:** Any periodic reports that you have generated *promptly* submit to Richard. [Does *promptly* modify *generated* or *submit*?]
>
> **CLEAR:** Promptly *submit* to Richard any periodic reports that you have generated.
>
> **SQUINTING:** Richard will review all reports *carefully* assessing their content, grammatical correctness, and page layout. [The adverb *carefully* could modify *review* or *assessing*.]
>
> **CLEAR:** Richard will review all reports while *carefully* assessing their content, grammatical correctness, and page layout. [The adverb modifies *assessing*.]

Unclear Comparisons

When words that are needed to show exactly what is being compared to what are left out, the result is an incomplete, unclear comparison.

UNCLEAR: The Interconnect sound system is much more flexible and powerful than others.

CLEAR: The Interconnect sound system is much more flexible and powerful than our old Belclear system.

Adjective–Adverb Errors

An adjective–adverb error happens when an adjective is used where an adverb is necessary, or vice versa. Adjectives modify nouns; adverbs modify verbs, adjectives, or other adverbs.

INCORRECT: The new system *adequate* meets all our needs. [As an adjective, *adequate* is the wrong form to modify the verb *meets*.]

CORRECT: The new system *adequately* meets all our needs. [As an adverb, *adequately* correctly modifies *meets*.]

CORRECT: Both staff and clients has expressed their *enthusiastic* approval [adjective].

ALSO CORRECT: Both staff and clients have *enthusiastically* expressed their approval [adverb].

WRITE Connection

Be especially careful with *good* and *well*. See page 760.

General Vagueness

Vagueness results from wording that simply isn't accurate, complete, and specific.

VAGUE: You're invited to a social event on the first floor Thursday afternoon.

CLEAR: Health Services invites all employees to a "Sinful Sundaes" social in the main lounge next Thursday, March 18, between 3:00 and 5:00 p.m.

Factual Errors

Basic factual errors in your documents can cost your company time and money, and they can damage your reputation for reliability. Be especially alert to avoiding these errors:

- Misquoting someone
- Misplacing a decimal point
- Misspelling a proper name
- Getting dates, places, times, and amounts wrong

   WRONG: We are pleased to extend you *$300.00* in credit based on In-Bloom's *triple XXX* rating.

   CLEAR: We are pleased to extend you *$30,000* in credit based on In-Bloom's *triple A* rating.

CORRECTING FAULTY SENTENCES

Sentence errors create confusion. To help readers understand your writing, learn to spot and fix the agreement, completeness, and consistency errors presented on the next six pages. (For background information on constructing sentences, see Chapter 46, "Constructing Sentences.")

Subject–Verb Agreement Errors

Generally, agreement errors involve disagreements—in this case, mismatches between subjects and verbs. Singular subjects take singular verbs, and plural subjects take plural verbs. Simple enough, but mismatches can happen in six situations.

First, disagreement can happen when words, phrases, or clauses separate subject and verb.

> **INCORRECT:** Each Homes Unlimited affiliate, as well as HU's worldwide partners, *depend* on volunteers.
>
> **CORRECT:** Each Homes Unlimited *affiliate*, as well as HU's worldwide partners, *depends* on volunteers. [*Affiliate*, not *partners*, is the subject.]

F.Y.I.

Be careful when a modifying phrase with a noun comes between the subject and verb. (Watch subjects such as *series, type, part, portion,* or *one.*)

> **INCORRECT:** One of the pamphlets explain how local HU affiliates handle their own work groups.
>
> **CORRECT:** *One* of the pamphlets *explains* how local HU affiliates handle their own work groups. [*One* is the true subject.]

Second, disagreement can happen with compound subjects. Compound subjects joined by *and* need a plural verb. However, with compound subjects joined by *or* or *nor*, the verb agrees with the nearer subject.

> **INCORRECT:** Local HU affiliates or the regional director initiate the partnership paperwork.
>
> **CORRECT:** Local HU affiliates or the regional *director initiates* the partnership paperwork. [Use a singular verb because *director* is closer to the verb.]

Third, disagreement can happen with nouns that are plural in form but singular in meaning: *measles, news, mathematics, economics.* Exceptions are *scissors, trousers,* and *tidings*, which always take plural verbs.

> **INCORRECT:** Basic mathematics are needed for assembling this furniture. Scissors is optional.
>
> **CORRECT:** Basic *mathematics is* needed for assembling this furniture. *Scissors are* optional.

Fourth, disagreement can happen with indefinite pronouns. As subjects, singular indefinite pronouns take singular verbs, even when words come between them.

> **SINGULAR INDEFINITE PRONOUNS:** each, either, neither, one, everyone, everybody, everything, someone, somebody, anybody, anything, nobody, another
>
> **INCORRECT:** Everybody who buys HU homes contribute 500 hours of labor.
>
> **CORRECT:** *Everybody* who buys an HU home *contributes* 500 hours of labor.
>
> **INCORRECT:** For example, someone who can work well with tools *invest* hours doing carpentry work.
>
> **CORRECT:** For example, someone who can work well with tools *invests* hours doing carpentry work.

Some indefinite pronouns can be either singular or plural. The verb agrees with the noun in the phrase that follows the indefinite pronoun. Indefinite pronouns that can be singular or plural include *all, any, half, most, none,* and *some.*

> **INCORRECT:** Most of the homes is built from scratch. Some of the labor require expertise in plumbing and electricity. [*Is* does not agree with *homes; require* does not agree with *labor.*]
>
> **CORRECT:** *Most* of the homes *are* built from scratch. *Some* of the labor *requires* expertise in plumbing and electricity.
>
> **INCORRECT:** While any of the volunteers who *wants* to assist on a repair job may do so, professionals must supervise high-skill tasks. [*Wants* does not agree with *volunteers.*]
>
> **CORRECT:** While any of the volunteers who *want* to assist on a repair job may do so, professionals must supervise high-skill tasks.

Fifth, disagreement can happen with collective nouns. A collective noun, which names a group (e.g., committee, department, company), is singular when referring to the unit as one group, and plural when referring to group members as individuals.

> **INCORRECT:** The local committee set up the work schedule.
>
> **CORRECT:** The local *committee sets* up the work schedule.
>
> **INCORRECT:** While one crew will *wear* their hard hats on all jobs, another crew may do so only when the work requires climbing or lifting. [Because each person must wear a hat individually, both *crew* and *wear* are understood to be singular.]
>
> **CORRECT:** While one crew *wears* their hard hats on all jobs, another crew may do so only when the work requires climbing or lifting.

Sixth, disagreement can happen with measurements. Generally, treat measurements as singular. However, when the subject refers to individual parts of the whole measurement, use a plural verb.

> **INCORRECT:** In an HU home, 2,200 square feet are the average floor space. Seventy-five percent of these homes includes a garage.
>
> **CORRECT:** In an HU home, *2,200 square feet is* the average floor space. *Seventy-five percent* of these homes *include* a garage.

Pronoun–Antecedent Agreement Errors

As the name suggests, a pronoun–antecedent agreement error involves a disagreement—a mismatch—between a pronoun and its antecedent (the noun it replaces). Essentially, pronouns and antecedents must agree in number: Singular nouns take singular pronouns, plural nouns take plural pronouns. Sounds simple, but some problems may arise.

Norm. A singular pronoun (*its*) is used with a plural antecedent (*affiliates*).

INCORRECT: To complete its projects, HU affiliates follow procedures set up by the national organization.

CORRECT: To complete *their* projects, HU *affiliates* follow procedures set up by the national organization.

INCORRECT: A local organization promotes its work through a number of strategies including their public-service announcements.

CORRECT: A local *organization* promotes its work through a number of strategies including *its* public-service announcements.

Special case A. Use a singular pronoun to refer to such antecedents as *each, either, neither, one, anyone, everyone, everybody, somebody, another, nobody,* and *a person*.

INCORRECT: Everyone who volunteers is gladly accepted, regardless of their skill level.

CORRECT: *Everyone* who volunteers is gladly accepted, regardless of *his or her* skill level.

INCORRECT: Because the work assignments are so numerous and diverse, nobody is denied their opportunity to contribute to HU projects.

CORRECT: Because the work assignments are so numerous and diverse, *nobody* is denied *his* or *her* opportunity to contribute to HU projects.

Special case B. Two or more antecedents joined by *and* are considered plural.

INCORRECT: However, each project requires a carpenter, an electrician, and a plumber for his or her expertise.

CORRECT: However, each project requires *a carpenter, an electrician, and a plumber* for *their* expertise.

Special case C. Two or more antecedents joined by *or* or *nor* are considered singular. If one antecedent is masculine and the other is feminine, the pronouns should be masculine and feminine (*his or her, he or she*). If one antecedent is singular and the other plural, the pronoun agrees with the nearer antecedent.

INCORRECT: The regional director or the chair of the local affiliate applies for the permits required for the projects they supervise. Either the local chair or individual project supervisors make sure that construction meets all codes in his or her city or county.

CORRECT: The regional *director* or the *chair* of the local affiliate applies for the permits required for the projects *he or she* supervises. Either the local *chair* or individual project *supervisors* make sure that construction meets all codes in *their* city or county.

Sentence Completeness Errors

Sentence completeness errors include comma splices, run-ons, and sentence fragments. These errors prevent sentences from being single, complete thoughts.

Comma Splice

Connecting two independent clauses with just a comma isn't enough. To fix a comma splice, use one of the methods below.

SPLICED: I have enclosed material describing our financial services, I have also included a current schedule of interest rates.

CORRECT: I have enclosed material describing our financial *services. I* have also included a current schedule of interest rates. [Change the comma to a period.]

CORRECT: I have enclosed material describing our financial *services; I* have also included a current schedule of interest rates. [Change the comma to a semicolon.]

CORRECT: I have enclosed material describing our financial *services; however,* I have also included a current schedule of interest rates. [Change the comma to a semicolon and add a coordinating conjunction or a conjunctive adverb.]

CORRECT: I have enclosed material describing our financial *services, and I* have also included a current schedule of interest rates. [Add a coordinating conjunction.]

CORRECT: *In addition to* enclosing material describing our financial *services,* I have included a current schedule of interest rates. [Make one independent clause dependent.]

Run-On Sentence

When you join independent clauses without punctuation or a connecting word, the result is a run-on sentence.

RUN-ON: I have enjoyed serving you in a few days I will call to discuss your options.

CORRECT: I have enjoyed serving *you. In* a few days, I will call to discuss your options.

Sentence Fragment

A fragment is an incomplete thought—a phrase, a sentence missing a subject or verb, or a dependent clause. Don't let the period fool you!

FRAGMENT: Enclosing material on other valuable services. [This phrase lacks both a subject and a verb.]

CORRECT: *I am enclosing* material on other valuable services.

FRAGMENT: If you have questions regarding any of these services. [The subordinating conjunction *if* makes this clause dependent.]

CORRECT: If you have questions regarding any of those services, please call me at 515-555-6800. [While the sentence does not include a stated subject, *you* is the understood subject of the verb *call*. For more information on understood subjects, see page 716.]

Inconsistencies and Shifts in Construction

Sentence inconsistencies and shifts come in many forms: verb tense errors, active–passive shifts, faulty parallelism, pronoun–person shifts, and mismatched subjects and predicates. Essentially, these errors break various patterns established early in the sentence.

Shifts in Verb Tense

When you start a sentence or paragraph in one tense—for example, past tense—stick with it. Change tense only to show actual changes in time from one action to another.

> **SHIFTY:** We are currently replacing and testing the faulty regulators in the serial link boxes. These units operated satisfactorily. [*Are replacing and testing* is present tense, but *operated* is past tense.]
>
> **CORRECT:** We are currently *replacing* and *testing* the faulty regulators in the serial link boxes. These units *are operating* satisfactorily.

Shifts in Verb Voice (Active/Passive)

Generally, if you start a sentence in one voice, whether active or passive, all other verbs in the sentence should be in that voice. (See page 266 for more on active and passive voice.)

> **SHIFTY:** We will modify these serial link boxes and the proper components will be supplied to American Linc Company. [*Modify* is active voice, but *will be supplied* is passive voice.]
>
> **CORRECT:** We will *modify* these serial link boxes and *supply* American Linc Company with the proper components.
>
> **INCORRECT:** All repair costs will be paid by AC Drives, and we will send no invoice to American Linc. [*Will be paid* is passive voice, but *will send* is active voice.]
>
> **CORRECT:** AC Drives *will pay* all repair costs, and we *will send* no invoice to American Linc.

Faulty Parallelism

Words, phrases, and clauses arranged in a series need to be in parallel form: Match phrases with phrases, nouns with nouns, and so on. (See page 263 for more on parallel structure.)

> **NONPARALLEL:** With your support, AC Drives will continue to engineer, manufacture, and servicing of the best drives available. [The third item is not parallel because it is a phrase, not a verb.]
>
> **CORRECT:** With your support, AC Drives will continue to *engineer, manufacture,* and *service* the best drives available.
>
> **INCORRECT:** Our goals are to produce quality products, to charge competitive prices, and offering excellent service. [*To produce quality products* and *to charge competitive prices* are infinitive phrases; *offering excellent service* is a gerund phrase.]
>
> **CORRECT:** Our goals are *to produce* quality products, *to charge* competitive prices, and *to offer* excellent service.

Shifts in Pronoun Person

A pronoun is either first person (*I* or *we*—the person speaking), second person (*you*—the person spoken to), or third (*he, she, it,* or *they*—the person or thing spoken about). Avoid shifting incorrectly from one person to another.

> **SHIFTY:** Thank you for your patience and understanding as we investigated their concerns about the ATV16 drives.
>
> **CORRECT:** Thank *you* for *your* patience and understanding as we investigated *your* concerns about the ATV16 drives.

F.Y.I.

Be careful not to use *one* with a personal pronoun. In fact, avoid using *one* because it sounds awkward and impersonal in business writing.

> **SHIFTY:** When one seeks to provide the best drives, you must listen to customer concerns and suggestions about design and performance.
>
> **CORRECT:** When *you* seek to provide the best drives, *you* must listen to customer concerns and suggestions about design and performance.
>
> **SHIFTY:** While we always seek to provide excellent products and services, one does make errors on occasion.
>
> **CORRECT:** While we always seek to provide excellent products and services, *we* do make errors on occasion.

Mismatched Subjects and Predicates

Sentences follow patterns—specific arrangements of subjects and predicates. Avoid breaking those patterns. Make sure that subjects and predicates work together.

> **INCORRECT:** Serial link failure is when the voltage regulator malfunctions under the load. [The linking verb *is* must be followed by a predicate noun or a predicate adjective, not an adverb clause beginning with *when*.]
>
> **ALSO INCORRECT:** Serial link failure is where the voltage regulator malfunctions under the load. [The linking verb *is* must be followed by a predicate noun or a predicate adjective, not an adverb clause beginning with *where*.]
>
> **CORRECT:** Serial link failure is *a malfunction of the voltage regulator due to the load*.
>
> **CORRECT:** Serial link failure *occurs when the voltage regulator malfunctions under the load*.
>
> **INCORRECT:** Last March our factory switched component suppliers with the negative result due to the new voltage regulator was not strong enough. [The prepositional phrase *with the negative result* and the subordinating conjunction *due to* poorly link the ideas in the sentence.]
>
> **CORRECT:** Last March our factory switched component suppliers, *but* the new voltage regulator it provided has turned out to be less rugged than the old one we used.

CORRECTING PUNCTUATION MARKS

Punctuation marks are like traffic signs: They signal to readers where sentences end and how sentence parts relate. Incorrect punctuation sends the wrong signal and readers get confused. For more, see Chapter 47, "Using Punctuation." To fix the most common problems, check the next two pages.

Comma Problems

The most useful and therefore the most easily misused punctuation mark is the comma. Locate and fix these problems in your writing.

Omission after introductory word, phrase, or clause. When a sentence begins with a word, a long modifying phrase, or a clause before the main part, set off that material with a comma.

> For the past nine months˄In-Bloom has been ordering fresh, dried, and silk flowers from Cottonwood Hills.

> While we were initially reluctant to change wholesalers˄we have found that your services and products have warranted the change.

Omission between independent clauses in a compound sentence. Unless clauses joined by a coordinating conjunction are quite short, separate them with a comma.

> Our orders have always been accurately and quickly processed˄and we have been impressed with the quality of your products.

> In addition, we have appreciated your informative newsletter˄and we have gleaned many good ideas from your new-product seminars.

Inclusion in a compound subject or predicate. A compound subject or predicate with two elements should *not* have a comma separating the elements.

> Now, In-Bloom and its delivery service In-Bloom Express wish to increase their orders from Cottonwood Hills˄and boost sales to a growing clientele.

> However, we cannot achieve these goals˄and submit our orders until Cottonwood Hills increases our line of credit to $100,000.

Use of a single comma when none or a pair is needed. A single comma should not separate a subject and a verb or a verb and an object.

> The attached references and financial statements˄illustrate˄our strong market share.

> If you should need either additional credit information˄or marketing plans, please contact me.

Omission in a series. Sentence parts arranged in a series (more than two items) should be separated by commas. Although a comma is not necessary between the last two items, use it for clarity.

> We plan to expand our product offerings, particularly of silk flowers˄bedding plants˄and houseplants.

> By continuing to provide good products˄deliver superior service˄and offer an adequate line of credit, you will strengthen our already good working relationship.

Omission around nonessential elements, appositives, or other set-off items. If words, phrases, and clauses are simply explanatory or nonrestrictive, set them off with a pair of commas. (For a detailed explanation of restrictive versus nonrestrictive modifiers, see page 730.)

> In-Bloom‸Springfield's fastest growing florist shop‸ plans to open a second shop in Greenburg‸which is a growing bedroom community outside Birmingham.
>
> **CORRECT:** The products *that we will need you to provide for this store* include fresh flowers, silk flowers, and a variety of garden planters. [The italicized clause is not set off with commas because it is essential—the reader needs this information to understand the sentence.]

Omission of commas between equally modifying adjectives. Insert a comma between two or more adjectives that are interchangeable. (See page 726 for more on this rule.)

> During the last three holiday occasions, In-Bloom's clientele has frequently commented on the fresh‸wonderfully scented flowers from Cottonwood Hills.
>
> They have also appreciated the well-chosen‸vibrant colors of your silk-flower arrangements, particularly those introduced during early December.

Apostrophe Problems

Apostrophes have many uses (see pages 731–732). However, avoid these common abuses.

Omission in a contraction. Insert an apostrophe to show that one or more letters have been left out of a word to form a contraction.

> We̓re now planning to make larger orders. [*We're* is the contracted form of *we are*.]
>
> Our design staff have streamlined their ordering and storage procedures; as a result, they̓ve increased productivity by 20 percent.

***It's* or *its*.** Perhaps the most common error in writing is the "it's/its" confusion. With an apostrophe, *it's* functions as a contraction of *it is*. Without the apostrophe, *its* is a possessive form of *it*.

> In-Bloom is now almost a year old, and it̓s on solid financial footing. It̓s goal for the coming year is a 25 percent growth in sales.

Possession problems. To indicate ownership, add an apostrophe and an *s* after a singular noun, and add an apostrophe after the *s* of a plural noun. The word that comes immediately before the apostrophe is the owner. (See page 731 for more details.)

> Because a florist̓s success depends partly on suppliers̓ ability to quickly provide a large volume and selection of quality plants, In-Bloom requests a $100,000 line of credit.
>
> While Dale's Garden Center̓s product line is outstanding, its line of credit to In-Bloom must be increased.

CORRECTING MECHANICAL DIFFICULTIES

Mechanics issues range from capitalization to abbreviations. For a thorough discussion, see Chapter 48, "Checking Mechanics." The most common issues are discussed briefly here.

Basic Grammar Errors

Standard English requires that writers avoid some basic errors. In business documents, such errors (rightly or wrongly) imply that the writer is uneducated.

> **DOUBLE SUBJECT:** The design staff, they won't be happy with this decision. [The writer states the subject twice.]
>
> **CORRECT:** The design staff won't be happy with this decision.
>
> **DOUBLE NEGATIVE:** There wasn't no hint of trouble at the previous meeting. [The writer uses two negatives.]
>
> **CORRECT:** There was no hint of trouble at the previous meeting.
>
> **DOUBLE PREPOSITION:** The engineering staff went off to the convention on Thursday. [The writer uses two prepositions.]
>
> **CORRECT:** The engineering staff went to the convention on Thursday.
>
> **NONSTANDARD SUBSTITUTION:** Jessica must try and get here on time. Except for her tardiness, she would of had a spotless work review. [Avoid substituting *of* for *have* or *and* for *to*.]
>
> **CORRECT:** Jessica must try to get here on time. Except for her tardiness, she would have had a spotless work review.

Capitalization

Capitalizing the first word in a sentence is an obvious rule—one that should be followed even in e-mail messages. Here are two slightly more difficult rules:

- Capitalize proper names of people, places, and things. Be particularly careful with organizations, historical events, cultural and national groups, and job titles. However, don't capitalize general job titles, the seasons, and geographical directions.

    | Rankin Manufacturing Co. | The Tokyo Summit | African Americans |
    | VP of Manufacturing | Stephanie H. du Bois | Internal Revenue Service |
    | supervisor | autumn | southeast |

- In titles, headings, and subheadings, capitalize the first word, the last word, and all words in between except articles (*a, an, the*), prepositions (*to, by, with*), and conjunctions (*and, but, or, nor, for, so*).

    The North American Small Grains Market: 2003–2004 Situation Analysis for Herbicide Sales
    General Market Conditions for Small Grains
    Market Segments, Size, and Priorities

Numbers and Numerals

Generally, if you are using numbers extensively in a document, use numerals throughout. Otherwise, follow these rules:

- Write out whole numbers nine or lower. Indicate numbers 10 or higher with numerals.

  In 1996, Rankin selected wheat and barley as the top-priority markets. Each year, these two crops are planted in 570 million acres.

- If you use a number as the first word of a sentence, spell it out. If doing so makes the sentence awkward, rewrite the sentence.

  One hundred and fifteen million acres of the total global acreage is planted in North America.
  Of the total global acreage, 115 million acres is planted in North America.

- Use a numeral when a number doesn't stand alone as a whole number (e.g., fractions, percentages, decimals, time references, dates, page references, statistics, money amounts, addresses).

  In the Western Canada market segment, a 1.6 million acre potential exists for Chase Herbicide, a 0.5 million acre opportunity for Ucopian, and approximately 3–4 million acres for Victor.

F.Y.I.

APA documentation style states that, as a general rule, you should use words to express numbers nine or lower and numerals to express numbers 10 and above: three, nine, 55, 2,000, 110, 1,050 (see pages 749–750 for additional details).

Abbreviations, Acronyms, and Initialisms

Abbreviations, acronyms, and initialisms offer writers and readers helpful shorthand. Present these short forms correctly by following these rules:

- With the names of agencies and organizations, abbreviate only those words that are abbreviated in their official name (*Co., Corp., Ltd., Inc., &*).

  Rankin Manufacturing Inc.   Strunk & Whittle, D.D.S.

- Use acronyms or initialisms for well-known companies and organizations. For an explanation of the difference between acronyms and initialisms, see page 753. Generally, capitalize all letters and do not use periods. As a rule, write out the full name on first reference and place the acronym in parentheses.

  EXAMPLES: MADD, HUD, NAACP, IBM
  FIRST REFERENCE: Mothers Against Drunk Driving (MADD)

- In addresses, avoid abbreviating places, dates, and names. Abbreviate states only.

  Rankin Chemical Company, 2311 Industrial Road, Fargo, ND 58103-0682

CHECKLIST for Errors

Use this checklist to identify and correct errors in your writing.

- ☐ The antecedent or reference for each pronoun is clear.
- ☐ The word or words that each adjective or adverb modifies are clear.
- ☐ Sentences comparing or contrasting items clearly state the items addressed.
- ☐ Details such as quotations, names, dates, titles, places, times, measurements, and technical terms are correct.
- ☐ Subjects and verbs agree in number (both singular or both plural).
- ☐ Each pronoun agrees with its antecedent in number (both singular or both plural), in case (both nominative or both objective), and in sex (both masculine, both feminine, or both neutral).
- ☐ The writing includes no comma splices, run-ons, or incorrect sentence fragments.
- ☐ The sentences include no unclear or incorrect shifts in verb tense (present, past, future) or voice (active/passive).
- ☐ Words, phrases, or clauses within a series are parallel.
- ☐ The writing includes no substandard constructions such as double subjects, double negatives, double prepositions, or nonstandard substitutions.
- ☐ Sentences include no errors in spelling, capitalization, punctuation, or usage.

CRITICAL THINKING Activities

1. Working with two classmates, review this chapter's instructions and discuss any points that are new or unclear. To find additional information on a topic, use the book's index. Then exchange papers, proofread them, and discuss the errors.
2. Study Part 9, "Proofreader's Guide." Note any instructions that are new or unclear. Report what you find to the class, including any questions that you have regarding correctness issues.

WRITING Activities

1. Working with two classmates, exchange papers and proofread them for the correctness issues addressed in the "Proofreader's Guide." Discuss the errors and how to fix them. Revise your paper as needed.
2. Review five of your graded writing assignments, noting the correctness issues cited by your instructors. Then using this book's index and "Proofreader's Guide," check how you can fix the errors in your assignments.

WEB Link

Go to Houghton Mifflin's Web site at http://college.hmco.com./english, find the self-assessment test, take it, and note your errors. Then based on your test results, work through the related remedial activities on the Web site.

In this chapter

Weak Versus Strong Design **290**
Understanding Basic Design Principles **292**
Planning Your Document's Design **293**
Developing a Document Format **294**
Laying Out Pages **297**
Making Typographical Choices **302**
Checklist for Document Design **304**
Critical Thinking Activities **304**
Writing Activity **304**

The seven traits

1. Strong Ideas
2. Logical Organization
3. Conversational Voice
4. Clear Words
5. Smooth Sentences
6. Correct Copy
7. **Reader-Friendly Design**

CHAPTER

16

Trait 7:
Reader-Friendly Design

Appearances count. Dress for success. Look smart and be smart. Have you ever received tips like these? The fact that many workplaces have dress codes shows that many people in business believe that appearances count. The codes (or standards) remind employees to dress and groom professionally.

For similar reasons, many organizations have style guides to help employees groom their writing so that it has a professional, reader-friendly appearance. In fact, businesses have many reasons to care about document design. First, effective design meets broad professional standards, such as those leading to the creation of polished, attractive company correspondence. Second, such design meets industry standards, like those used for newspaper design. Third, professional design meets government standards, like those regulating the use of signal words such as WARNING or **DANGER** in instructions. Perhaps most importantly, professional design meets readers' standards: Effective design pulls readers into the document, helps them access information, and promotes understanding.

This chapter will help you distinguish between weak and strong design, learn key design principles, plan your own documents, and choose the best design elements, including format, page layout, and typography.

WEAK VERSUS STRONG DESIGN

A well-designed document invites reading, helping readers find what they need, understand it, and respond to it. How? Compare the poorly designed and effectively designed instructions below and on the next page.

Weak Design

Confidentiality Procedure: The Family and Children Confidentiality Policy states that *all clients have the right to confidentiality.* Conduct your work in keeping with the policy. During the client intake, the Coalition counselor will explain that information will not be shared *unless the client has given permission via the Consent to Release form.* The counselor will also explain that confidentiality may be limited or canceled *if Coalition staff have serious concerns about child abuse or neglect, or if the client may be of danger to self or others.* Coalition counselors will carefully handle outside requests for information on clients. *No client information will be shared without the client's permission.* Clients will be notified of any outside requests for information. If the client gives permission, he/she needs to sign the Consent to Release form and specifically indicate what information may be released and to whom. *A Coalition counselor also needs to sign the release form.* The Family and Children Coalition aims to treat fairly and without bias all persons seeking services. If a client believes that Coalition staff have not observed the confidentiality policy and procedure, the client should follow the Client-Grievance Process. If the Client-Grievance Process determines that a Coalition employee breached the client's confidentiality, *then the employee's work will be terminated.*

This document illustrates several design problems:

- Ineffective formatting for a procedure
- No paragraph divisions
- No listing or numbering
- No white space (with margins, between paragraphs, around headings)
- Lines too long and fully justified
- Poor highlighting techniques (e.g., no uppercase, underlining, bold, boxes)
- Poor typographical choices (sans serif for main text, overuse of italics, type small and compressed)

F.Y.I.

Both the content and design of a workplace document must help the reader understand and use the message. Reread the procedure above, noting specific design choices that impede the reader's ability to implement the procedure.

Strong Design

**Family and Children Coalition
Confidentiality Procedures**

The Family and Children Coalition Confidentiality Policy states that all clients have the right to confidentiality. Conduct your work in keeping with this policy by following the procedures below.

Client Intakes

During the client intake, the Coalition counselor will discuss the conditions of confidentiality with the client. The counselor will explain the following:

1. Information will not be shared unless the client has given permission via the Consent to Release form.
2. Confidentiality may be limited or canceled if Coalition staff have serious concerns about child abuse or neglect, or if the client may be of danger to self or others.

Outside Requests for Information

Coalition counselors will carefully handle outside requests for information on clients.

1. No client information will be shared without the client's permission.
2. Clients will be notified of any outside requests for information. If the client gives permission, he or she needs to sign the Consent to Release form and specifically indicate what information may be released and to whom. A Coalition counselor also needs to sign the release form.

Breaches of Confidentiality

The Family and Children Coalition aims to treat fairly and without bias all persons seeking services. If a client believes that Coalition staff have not observed the confidentiality policy and procedure, the following should occur:

1. The client should follow the Client-Grievance Process.
2. If the Client-Grievance Process determines that a Coalition employee breached the client's confidentiality, then the employee's work will be terminated.

The document shows several strengths:

Procedure parts are clear.

Paragraphing, lists, numbering, headings, and white space order information.

Lines are a moderate length, left justified, and right ragged.

Boldface the title and headings.

Typographical choices help reading: Main text is serif, and headings are sans serif; title is larger than headings, which are larger than text.

UNDERSTANDING BASIC DESIGN PRINCIPLES

To effectively design documents, start with clear goals. Aim to (1) make reading easy, (2) develop an attractive document, and (3) help readers use the information. With these goals in mind, practice the following design principles.

See Readers Reading

Understand the following:

- Readers are very busy and are surrounded by distractions.
- They often skim a document before reading closely.
- They may perform a task while reading (e.g., instructions).
- They read pages from left to right and from top to bottom.
- They make sense of information both verbally and visually.
- They digest information in small chunks, not whole documents.

Choose Fitting Design Features

Use a variety of design elements, but choose them logically so that they fit the document type and the audience.

- **Aim for simplicity.** Design elements should be both functional and attractive. Flash is usually distracting and out of place in most business documents.
- **Design the document so that all readers can access it.** With techniques ranging from headings to color, help readers enter and navigate your writing.

Make Pages Open and Balanced

Avoid lengthy passages of prose by effectively combining print and white space. Readers find big blocks of text daunting, so break up your message into manageable chunks.

Go with the Flow

Design pages so that information flows naturally from the upper left to the lower right. Keep this flow in mind when integrating text and graphics.

Develop Consistent Visual Cues

Design elements should distinguish between main and secondary points, as well as show how the points are connected. Create clear design patterns, and stick with them. For example, the type size, placement, and type style of headings should clearly indicate levels of information within the document.

Stress Connections and Contrasts

Information and ideas that are related should be clumped together, and their relatedness should be shown visually through similarities in size, shape, color, and alignment of material on the page. Content that stresses differences should be signaled through visual oppositions such as differences in size, shape, location, and color of elements on the page.

PLANNING YOUR DOCUMENT'S DESIGN

Often, designing a document is as simple as following a template. When you need to plan the design from scratch, however, consider these issues and make smart choices.

Readers
Who are they?
How will they read, access, and use the document?

Medium
Will the document be paper, electronic, or both?

Format
What type of document are you writing (e.g., letter, memo, report)?
What does your company expect by way of format?
Which access strategies make sense?

Numbering	Headers/footers	Table of contents
Index	Dividers, flaps, tabs	Boxes
Rules	Icons	Color
Links	Cross-references	Appendixes

Binding
How will pages be held together (e.g., paper clip, staple, binder)?
How will the document be copied, distributed, and/or posted?

Layout
What paper weight, size, and colors should you use for standard pages and special pages?
What size margins should you use? What line lengths should you set?
How many columns, what size, and what type?
What heading system and lists make sense?
Will you need to integrate text and graphics? How will you do so?

Typography
What typeface (e.g., Roman, Garamond) should you use for main and special text?
What type sizes should you use for main and special text?
What type styles should you use for emphasis—uppercase, boldface, italics, underlining, shadow, highlighting?
How should lines of text be justified—left only, fully, centered?

> **WRITE Connection**
> Web site and Web page design are special cases of document design. See Chapter 36, "Writing for the Web," for more help.

DEVELOPING A DOCUMENT FORMAT

Document format refers to the shape of the whole piece that you are writing. As you plan your document, develop a sound format.

Choose and Plan an Effective Format

In some situations, the choice of document format will be obvious. Standard formats for business documents are available throughout this book, as well as through your word-processing or desktop publishing program. In other situations, you may have to develop the format from scratch. Whatever the situation, you need to select the best format based on a variety of factors.

Document type. Are you writing a letter, a proposal, or a news release? What are that form's specifications?

Company practices. What format specifications does your company follow?

Reader expectations. Who are your readers, why will they read your document, and how will they use it?

Type of content. Is the information simple or complex?

Resources available. How much time and money do you have? What hardware and software? What expertise?

Production, reproduction, and transmission methods. How will the document be created—by hand, typewriter, computer, or print shop? Will paper have print on one side or both sides? Will multiple copies be printed or photocopied? In black and white or in color? Will the document become part of an electronic database or the Web?

While simple documents need little format planning, longer or more complex documents do. With an outline in mind, sketch the document on paper by hand or use a page-design program to create a document "map."

Choose Folds, Bindings, and Delivery Methods

How should you hold together and present your document? Consider the document's immediate, multiple, and long-term uses. Then select from these choices:

Panel folds (halves, thirds, quarters, and so on, as in pamphlets, brochures, and booklets).

Paper clips (for pages that the reader may need to separate).

Staples (whether one or more, to keep together short to medium-sized documents).

Ring binders (allow pages to be removed, replaced, added, and reordered).

Saddle binding (a booklet made from folded sheets stapled in two or more places along the fold).

Spiral binding (for medium to long documents, the spiral firmly keeps pages together; pages lie flat, but changing pages is inconvenient).

Perfect binding (pages glued to a cover spine, as with a paperback book; professional, published look for major documents).

Choose Paper

To select paper for your document, consider size, color, and type.

Size

Choose a paper size that makes your document easy to send, use, and store.

- Print most business documents (letters, memos, and reports) on standard $8^1/_2$-inch by 11-inch paper.
- Print legal documents on $8^1/_2$-inch by 14-inch paper.
- Print other documents ranging from instruction cards to promotional mailings on smaller stock, perhaps $5^1/_2$-inch by $8^1/_2$-inch paper.

Color

Choose a paper color that fits your document's personality and your reader's expectations.

- For most documents, use white, off-white, or cream.
- For special effects, section dividers, covers, and special sections (such as appendices), use colored paper. But choose colors wisely: Subtle, traditional colors (such as tan, blue, or green) imply strength and stability; pastels suggest lightness; flashy, neon colors connote "advertising."

Type

Paper is designated by its weight, called bond, and by its glossiness. Choose a weight and glossiness that match the document's importance, purpose, and use.

- For most documents, use 20-pound bond. It takes ink well and works well in printers, copiers, and fax machines.
- Consider heavier bond for covers, dividers, formal documents, and documents that will be heavily used (e.g., instruction cards).
- Use lighter paper for drafts and short-lived documents.
- Be careful with glossy paper. While gloss adds strength, it does create glare.

Create Access for Readers

"Access" elements help readers enter the document, find what they need, and exit.

Heading System

An informative system of headings and subheadings consistently presented throughout a document shows readers the structure of your thinking and allows them quick access to relevant sections. (See page 221 for more on headings.)

Headers and Footers

A definitive header or footer quickly identifies the document, chapter, or section for the reader. A header supplies a title, phrase, or other identifying word above the text on each page. A footer runs that information at the bottom of each page. Here are a few examples:

    Trait 7: Document Design   2005 Marketing Report   Ms. Alice Price 10/21/05

Page Numbering System

A document longer than one page should use numbering for easy reference. Different document types require different numbering systems.

- **Most documents.** Place arabic numerals in the upper-right corner of the second and subsequent pages. (Consider putting *continued* at the bottom center of each page except the last.)
- **Long documents.** Use lowercase roman numerals for front matter (i, ii, iii), arabic numerals for the body (1, 2, 3), and a letter–number combination with appendices (A-1, A-2).
- **Documents with individually numbered chapters.** Use two numbers. The first number indicates the chapter, and the second number indicates the page in the chapter (1-1, 1-2, 1-3).
- **Faxed or photocopied documents.** Indicate the page out of the total (1 of 6, 2 of 6, 3 of 6).

Table of Contents and Index

For longer documents, the table of contents and index are indispensable tools for locating topics and information quickly.

Dividers and Tabs

In longer documents, especially those that function as reference works, dividers and tabs help readers quickly find what they need. Dividers are typically pages of different paper weight and color from regular text, and they may have a tab for identification.

Icons

Clear symbols can sometimes communicate a concept or action better than words. In addition, icons that reappear in your document can guide readers and alert them to key information. Here are a few examples:

Color

Color changes (whether in paper or print) draw attention to material and establish patterns. If appropriate, use color with special tips, cautions, recommendations, headings, tabs, appendices, and so on.

LAYING OUT PAGES

Page layout—how you arrange material on a page—determines whether your document is attractive and readable. Avoid dense prose, cluttered pages, or largely empty pages.

Layout Options

Which page arrangement will best communicate your ideas? Consider orientation, columns, and the text/graphics mixture.

Orientation

Standard orientation for an $8^1/_2$-inch by 11-inch sheet is vertical. However, some documents, such as pamphlets, require a horizontal orientation. The same choices are available with other paper sizes. Select the orientation that best helps you present the information and that best meets your readers' expectations and use of the document.

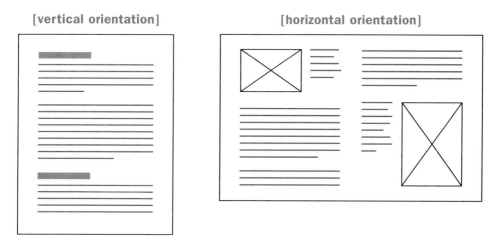

Columns

Think of a page as divisible into columns, and ask these questions:

What kind of columns should I use?

- With *newspaper columns,* text runs from the bottom of one column to the top of the next. This format is useful for booklets and newsletters.
- With *parallel columns,* the text in each column remains separate. This format is useful for instructions, résumés, itineraries, and so on.

How many columns should I use?

- *One column:* Standard for most documents, like memos, letters, and reports
- *Two columns, even:* Balances page nicely, allows comparisons and contrasts
- *Two columns, uneven:* Larger column useful for main text, smaller column for headings, side notes, icons, and so forth
- *Three or more columns:* Breaks up text into blocks or panels that give readers "chunks" of information; useful for pamphlets, brochures, booklets, newsletters, charts, and tables

Part 2 Benchmarking Writing with the Seven Traits

Text/Graphics Combination
If your document includes graphics such as line graphs, you need to integrate them effectively. See Chapter 5, "Using Graphics in Business Documents," for help.

Basic Page Layouts

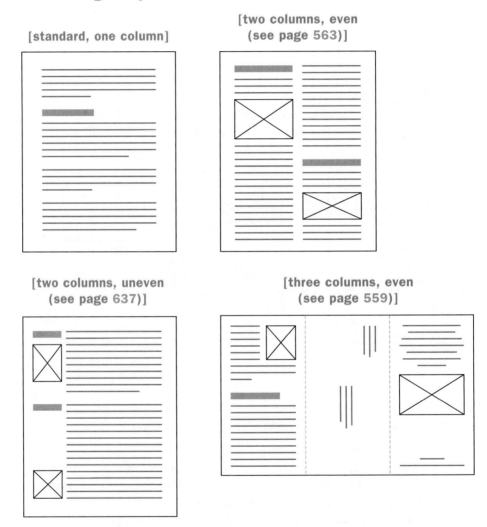

White Space

To balance dark print, build white space into your page layout. When used effectively, white space rests readers' eyes, "chunks" related information, and highlights key ideas and details. Follow these guidelines.

Use Healthy Margins

Frame your page with generous margins, generally 1 or $1^{1}/_{2}$-inches on all sides of a standard $8^{1}/_{2}$-inch by 11-inch sheet. Adjust margins to account for the type of page (e.g., instructions, brochure, memo) and the binding method (e.g., spiral, three-ring binder).

Break Up Text

Use techniques like these to keep your prose from getting heavy and dense on the page:

- Double space between paragraphs and keep paragraphs short.
- Surround headings with white space—generally, more space above than below the heading.
- Turn some material into lists.
- Increase line spacing to 1.5 or 2, if necessary.

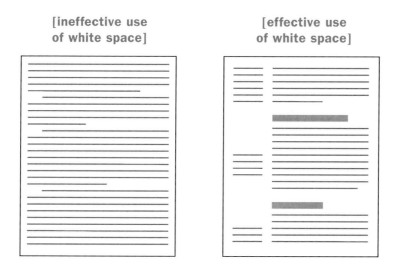

F.Y.I.
Avoid using too much white space. In some documents, an imbalance of white space over black print can suggest thin content.

Highlighting Options

As you lay out pages, decide *which* material to emphasize and *how*. Consider these techniques, remembering to use any technique moderately.

[boxes]
Framing a section of text or a graphic in a box isolates that material for special attention.

[shading]
A word, sentence, paragraph, or box that is shaded jumps out from surrounding material.

[color]
When used wisely, color shows patterns, highlights connections, and points readers to important headings, words, sentences, boxes, and graphics.

[rules]
Horizontal or vertical lines separate sections or columns of text.

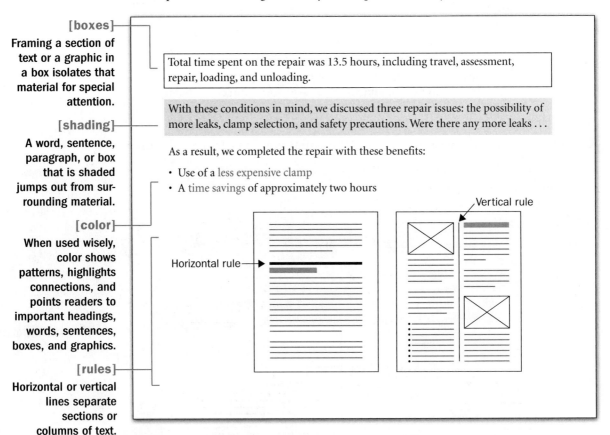

 WRITE Connection
For typographical methods of highlighting, see page 303.

Line Formatting

To effectively arrange your lines on the page, consider both length and alignment.

Line Length

In a standard, one-column business document, your lines should average 50–80 characters (10–15 words). Overly long lines make reading difficult, whereas overly short lines suggest lack of substance.

Alignment

Most text should be left justified (aligned) but right ragged. Fully justified lines (even at both left and right margins) make reading more difficult by introducing unnatural spaces between letters and words. Important lines and passages, such as titles and text in pamphlets, can be emphasized through centering.

Problems to Avoid

Before printing your document, check for common problems that detract from its appearance. A "print preview" or "make it fit" function in your word-processing program might help.

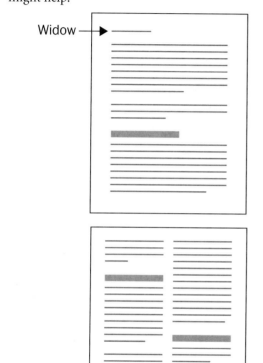

[widows]
Widows are single lines of text (e.g., the last line of a paragraph) that sit alone at the top of a page.

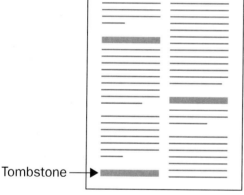

[tombstones]
Tombstones are headings or subheadings that sit alone at the bottom of a column or page.

[orphans]
Orphans are first lines of paragraphs left alone at the bottom of a page.

[split lists]
Split lists are lists divided between two pages. Especially if the items need to be tabulated, compared, or contrasted by the reader, keep the list on one page.

MAKING TYPOGRAPHICAL CHOICES

Typography refers to the actual print on the page or screen—those marks that ultimately communicate your message. With print, your aim is to create a positive impression, make reading easy, and clarify content. To achieve these goals, pay attention to typeface, type size, and type style.

Select Typefaces

Typeface refers to the special look shared by the letters and other symbols of a specific type. These typefaces cluster in families—Arial, Courier, and Times Roman, for example. In addition, typefaces are designated as *serif* or *sans serif*.

Serif typeface, like this, has finishes on the letters that make the type easier to read.

Sans serif typeface, like this, has no finishes. The print is clean and modern, but harder to read in extended passages.

A serifs

A no serifs

Here are some rules of thumb for selecting typefaces:

- Use a *serif* typeface for main text; use a *sans serif* typeface for headings and other special effects.
- Select a typeface that is readable given the specific conditions for your document.
- Select an attractive typeface that fits the document type. Avoid fancy or unusual typefaces, as well as frequent typeface changes.

Sample Passage

Assessment of Problem

The Eugene crew had exposed the leak area on both sides and the top of the pipe. As a first step, we assessed damage and conditions:

- Approximately 6–12 inches of mud covered the work area, including the ramps into the ditch.
- The ditch contained water up to the pipe's bottom . . .

Vary Type Size

Type size is measured in units called *points*, with 72 points equal to one inch. When selecting type size, consider the reading conditions.

Normal Reading Conditions

When readers will sit with the document squarely before them, make main text 10 or 12 points. For titles, headings, and subheadings, increase the size by 2–4 points; 12, 14, 16, or 18 points are good choices. Make the jump in size noticeable but not drastic.

Abnormal Conditions

For documents that will be read under difficult conditions and for document parts that are critical (e.g., warnings in instructions), consider using larger type.

Presentations

For slides, overhead transparencies, and PowerPoint presentations, think big: 18–36 points. Remember the people sitting in the back row.

Pacific Pipeline Repair Report	Title (18 points)
Assessment of Problem	Heading (16 points)
Site Conditions	Subheading (14 points)

Use Different Type Styles

Type style refers to special effects like those shown below. When using type styles, follow these rules:

- Use each stylistic effect for a reason, not for dazzle.
- Be moderate. Avoid combining several techniques.
- Use effects to set off special text from normal text.

UPPERCASE IS MORE PROMINENT THAN LOWERCASE BUT HARDER TO READ IN EXTENDED PASSAGES.

Boldface causes print to jump out. It's especially useful for headings, subheadings, warnings, and other key information.

Underlining is useful for indicating subheadings, key words, and key sentences, but avoid using it for extended passages.

Italics indicates book titles, words designated as words, and key statements. Avoid it, however, for extended passages because it is difficult to read.

Highlighting and similar techniques such as **shadowing** and outlining draw attention to key words and statements.

Color helps highlight headings, warnings, tips, and other key material. However, make sure that color combinations work well. (See page 86.)

CHECKLIST for Document Design

Use this checklist to evaluate and polish your document's design.

- ☐ The overall design is appropriate for the writing situation, type of document, message, and audience.
- ☐ The design follows guidelines in the organization's style guide.
- ☐ Design elements distinguish the document's topic, main point, and purpose.
- ☐ The layout effectively uses a heading system, lists, margins, columns, and white space to make the message clear and easy to read.
- ☐ Accent features such as highlighting, underlining, and color help bring out the message and are appropriate for the document's purpose and message.
- ☐ Design elements (such as headings, page numbers, and outlines) clearly identify and organize parts of the message.
- ☐ The document contains all appropriate elements, such as a title page, table of contents, introduction, outline, graphics, appendices, and index.
- ☐ The color, size, texture, and quality of paper are appropriate for the document's purpose, message, and audience.
- ☐ The type and size of fonts are attractive, emphasize information as needed, and fit the document's purpose and audience.
- ☐ The graphics (e.g., pictures, drawings, graphs) are well placed to support the message.

CRITICAL THINKING Activities

1. Review the first five pages of this chapter, particularly "Weak Versus Strong Design," pages 290–291. Look carefully at the two designs, along with the design notes accompanying each piece. Explain why you do or do not agree with the notes. Then choose a model document in this book, analyze the design, and list the design's strengths and weaknesses.
2. Working with a classmate, find three business documents of the same type (e.g., letters, memos, flyers, brochures) and compare and contrast the documents' (a) binding or folding, (b) use of color, (c) type and size of paper, (d) type and size of fonts, (e) headings, (f) headers and footers, (g) page numbering system, (h) table of contents, (i) index, (j) dividers and tabs, (k) icons, and (l) method of distribution. Decide which document's design most fully supports the document's purpose and message. Report your findings to the class.

WRITING Activity

Find a piece of writing that you recently completed. Using information from this chapter, review the format, page layout, and typography of this piece. Change the design of the piece, and bring both versions to class for comparison.

PART 3

The Application Process and Application Writing

17 Understanding the Job-Search Process

18 Developing Your Résumé

19 Writing Application Correspondence

20 Participating in Interviews

In this chapter

Overview of the Job-Search Process **308**
Assessing the Job Market **309**
Guidelines for Career Plans **312**
Conducting a Job Search **315**
Researching Organizations **316**
Using Web Resources **317**
Checklist for Career Planning **318**
Critical Thinking Activities **318**
Writing Activities **318**

CHAPTER
17

Understanding the Job-Search Process

During his or her lifetime, the average person changes jobs every 3.6 years, has ten different employers, and pursues three different careers. For these reasons, from now until you retire, you must think about job searching as a necessary ongoing process of planning, training, writing application documents, updating those documents, and looking for job opportunities. But don't let the scope of the task intimidate you! Just as climbing a stairway one step at a time is doable, so, too, is searching for a job one step at a time.

This chapter first presents an overview of the job-search process and explains how to assess the job market. That information will help you evaluate your career-related choices, including your selection of courses, internships, or part-time jobs. Then the chapter explains step-by-step how to write a career plan and how to conduct a job search. You can use that information to polish your career planning and pursue an internship, assistantship, or job. Finally, the chapter explains how to research organizations and how to use Web resources to find career-building information.

Part 3 The Application Process and Application Writing

OVERVIEW OF THE JOB-SEARCH PROCESS

Think of the job-search process as a series of doable steps. Then study and follow each step as explained in this chapter and the next three chapters.

1. **Develop and regularly update a career plan** that addresses these issues:

Skills	Career goals	Jobs and responsibilities	Certifications
Values, needs	Educational path		Preferred locations
Degrees	Internships	Special training	Interests, beliefs

2. **Prepare a job-search file** that contains the following items:

Résumé	Letters of recommendation	People to advise/ recommend you	Potential jobs and employers
Transcripts			

3. **Identify job vacancies and potential jobs** using sources like these:

College alumni listing	Employers (part-time, full-time, or internships)	Network of relatives, friends, and advisers
College employment office	Employer's campus visits	Newspaper ads/articles
Conferences	Government Web sites	Organizations' Web sites
Conventions	Job fairs	
Department's assistantships, job postings	Letters of inquiry	Trade journals
		Trade shows

4. **Research the employer** (before and after step 5):

Interview contacts within the organization.	Read the organization's Web site, newsletters, mission statement, financial report, brochures, and catalogues.	Study the organization's history.
Interview people outside the organization.		Learn about its worksites, products, markets, and competition.
Do library research.		

5. **Contact the employer:**

Send an application letter and résumé.	Send a follow-up letter or e-mail.	Call or leave a voice message.

6. **Interview and follow up:**

Review research.	Participate in interview.	Write follow-up letters and messages.
Plan clothing and travel.	Thank interviewers.	

7. **Respond to the employer's decision:**

If you accept an offer, inform the employer by phone and an acceptance letter.	If you reject an offer, inform the employer by phone and a job-rejection letter.	If you receive no offer, analyze feedback and revise your application efforts as needed.

ASSESSING THE JOB MARKET

Planning your career requires research regarding future job opportunities. The U.S. Labor Department identifies those opportunities by studying relationships between (1) the population, (2) the labor force, and (3) the demand for goods and services (see Figures 17.1 to 17.4). For example, Figure 17.1 shows that while the civilian population is currently growing, the labor force is growing even faster.

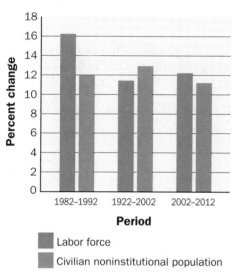

Figure 17.1 Percent Change in the Population and Labor Force

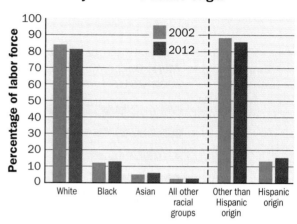

Figure 17.2 Percentage of Labor Force by Race and Ethnic Origin

Note: The four race groups add to the total labor force. The two Hispanic-origin groups also add to the labor force. Hispanics may be of any race.

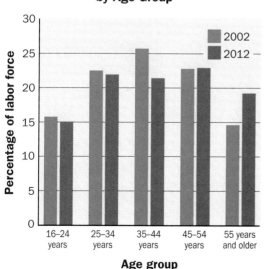

Figure 17.3 Percentage of Labor Force by Age Group

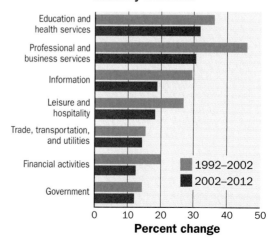

Figure 17.4 Percent Change in Wage and Salary Employment, Service-Providing Industry Divisions

Percent Change in Employment

Changes in one of the three components (population, labor force, or demand for goods or services) affect changes in the other component. For example, the Labor Department projects that between now and 2012, the size of the U.S. population age 60 or older will increase in relation to other age groups. Because that segment of the population generally requires more health care services than others, the department also projects an increase in the number of jobs related to health care.

Using this type of analysis, the Labor Department projects a variety of changes. For example, Figure 17.5 projects the percent change in total employment in each of the major occupational groups cited. Figure 17.6 projects the percent change in occupations expected to grow fastest.

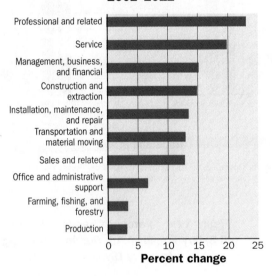

Figure 17.5 Percent Change in Total Employment by Major Occupational Group, 2002–2012

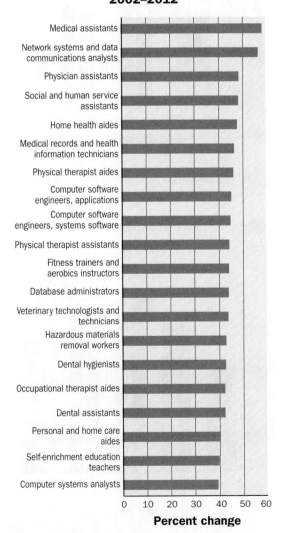

Figure 17.6 Percent Change in Employment in Occupations Projected to Grow Fastest, 2002–2012

Numerical Change in Employment

When reading charts like those shown in this chapter, do not confuse *percent* change with *numerical* change. For example, Figures 17.5 and 17.6 showed the *percent* change of employment in specific occupations. The next two charts show *numerical* change. Figure 17.7 shows occupations that should gain the largest number of jobs during 2000–2012. Figure 17.8 shows occupations that should lose the largest number of jobs over the same period.

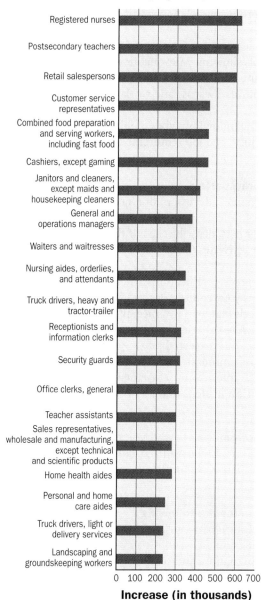

Figure 17.7 Occupations with the Largest Numerical Increases in Employment, 2002–2012

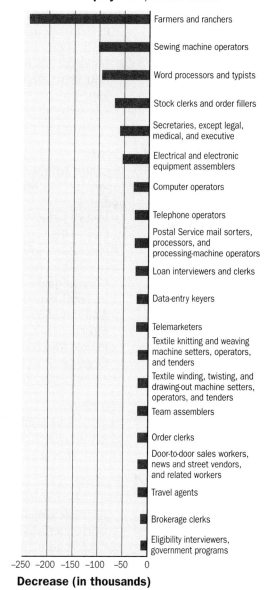

Figure 17.8 Job Decreases in Occupations with the Largest Numerical Decreases in Employment, 2002–2012

Part 3 The Application Process and Application Writing

GUIDELINES FOR CAREER PLANS

When writing a career plan, your goals are (1) to clarify your career objectives and (2) to spell out activities that will help you achieve those objectives.

1 Prewrite
Rhetorical analysis: Consider your purpose, your audience, and the context.
- ☐ What details will help you (and possibly a career counselor) identify your skills, qualifications, work-related experiences, and career objectives?
- ☐ Who could give you advice, a recommendation, or other career-related help?

Prepare to draft.
- ☐ Review your grade transcripts, notes from training programs, and job descriptions to identify details showing your qualifications.
- ☐ List details that describe your education (schools, training programs, instructors, degrees, certifications, awards) and work history (employers, jobs, projects).
- ☐ Research and take notes on the following: types of jobs in your career field, organizations offering jobs, and job requirements.

2 Draft
Organize your writing into three parts.

[opening] List personal data (name, address, phone/fax numbers, e-mail address).

[middle] In brief paragraphs or lists, describe each of the following:
- ☐ **Career field** and **job-related activities** that interest you
- ☐ **Potential employers** within your career field and the **jobs** they provide
- ☐ **Job requirements** such as degrees, certifications, internships, or skills
- ☐ **Current qualifications** such as your courses, degrees, certifications, training, work experiences, or awards that help you meet the job requirements
- ☐ **Additional qualifications** such as experiences, skills, or training that you must gain to fully prepare for your preferred jobs
- ☐ **People** who could advise or recommend you
- ☐ **Organizations** that could offer internships, assistantships, training, or jobs

[closing] State your plan of action in steps that accomplish your objectives.

3 Revise
Review your draft for ideas, organization, and voice.
- ☐ Is your overall plan clear, inviting, well organized, complete, and realistic?
- ☐ Is the tone positive, objective, and thoughtful?

4 Refine
Check your writing line-by-line for the following:
- ☐ Readable sentences and helpful transitions
- ☐ Correct terminology and clear, concise wording
- ☐ Correct grammar, usage, punctuation, and spelling

Model Career Plan

Rhetorical situation

Purpose: To spell out the writer's career objectives and a plan for achieving them
Reader: The writer, trusted colleagues, and possibly a career counselor
Format: Document that includes headings and divisions like those shown below

---

Personal Data

Rose Castillo
975 Elm Street
Bozeman, Montana 59718

Phone: 406-555-0243
Fax: 406-555-0244
E-mail: rosecastillo@earthlink.net

— List your contact information.

Career Field and Job-Related Activities

I want to work in graphic design (and possibly in animation) for these reasons. First, I enjoy drawing, designing posters, and looking at Web sites and magazines, so I think that I'll also enjoy the artistic aspects of graphic design. Second, I communicate quite well both orally and in writing, so I believe that I have the communication skills that a designer needs to negotiate with customers and interact with co-workers. Finally, because I like working with computers, scanners, printers, and cameras, I'm sure that I can handle the tools of the trade.

— Describe your career field and explain why you chose it.

Prospective Employers and Jobs

Many for-profit and not-for-profit organizations hire graphic designers to work on projects related to the following:

- Marketing (signage, logo designs, newspaper and magazine ads, brochures, fliers, posters, PowerPoint presentations, public Web sites, Flash animations, printing and Web-service negotiations)
- In-house communication (newsletters, internal Web sites, instructions, major reports, and employee handbooks)
- Product development (package design, concept sketches, idea boards, product labeling, design proposals, and printing negotiations)
- Product-support information (owner's manuals, instructional materials, Flash animated or Web-based courseware, product tags, printing and Web-service negotiations, and product-design proposals)

— Describe potential employers and the tasks they could ask you to do.

Job Requirements

While different organizations have different standards or requirements, many organizations require the following for full-time graphic artists:

- Portfolio that demonstrates skills in certain areas of design
- B.A. or B.F.A. in art with emphasis on graphic design
- Work experience such as internships, assistantships, or part-time jobs in graphic design
- For animation, coursework plus an internship or part-time job at a Web service or a studio that does interface design or Flash animation

— List common job requirements.

p. 1

p. 2

Current Qualifications

I've completed these goals:

- 60 semester credits toward a B.F.A. in graphic arts (3.5 GPA)
- Summer internship as a design assistant at Ezra's Advertising in Bozeman
- Summer job as a design assistant at Turpin Design Works in Seattle

Describe your job-related qualifications.

Goals

I want to do the following as part of my B.F.A. program:

- Develop my portfolio
- Take a marketing course from the Business Dept., a business-writing course from the English Dept., and a digital-photography course from the Media and Theatre Arts Dept.
- Design two or three Web sites for businesses in Bozeman or Ft. Collins
- Complete a one-semester internship at Turpin Design Works in Seattle
- Finish my B.F.A. in Graphic Design at Montana State University

Explain how you are increasing your qualifications.

People and Organizations

The people below could give me advice or write a recommendation for me; the organizations could offer an internship or possibly a job:

- Zachary Ulferts, Associate Professor at Montana State University
- Gidge Van Gorp, Art Director at Penguin Printing in Ft. Collins
- Jack Meyer, Project Director at Elijah Creative in Seattle
- Gabriel Kloosterman, Kreyko Design and Animation in Seattle
- KLMO TV in Ft. Collins

List people and organizations who can help you advance in your career.

Plan of Action

During the next three years, I plan to do the following:

- Finish my senior-thesis project and portfolio
- Finish my advanced-drawing course
- Finish courses in marketing, business writing, and digital photography
- Complete two or three Web sites for businesses
- Complete an internship at a Web-design studio in Seattle or elsewhere
- Write and format my résumé
- Request letters of recommendation from two or three key people
- Take "Interview Skills" from the Student Services Department

Describe how you will achieve your career goals.

CONDUCTING A JOB SEARCH

Conducting a job search is an important step in building your career. When you search, use all the appropriate resources described below.

College placement offices. Most college placement offices provide the following job-search services. Use them fully—they're free!

- **Advertise** job-training seminars and skill-development workshops
- **Counsel** students regarding careers
- **Help** with application documents
- **Offer** information about graduate-school and professional programs
- **Polish** your interview skills
- **Post** job openings
- **Present** information about internships
- **Refer** you to tutoring services
- **Share** lists of alumni who offer jobs, job referrals, and job-search tips
- **Sponsor** job fairs and campus visits for employers with job opportunities

City Web sites. Most cities have Web sites that describe area organizations and have links to their Web sites, along with links to local employment services.

Conferences and conventions. Professional organizations sponsor conferences and conventions at which employers advertise job openings and interview job seekers.

Department job postings. Most college departments have bulletin boards advertising jobs, internships, graduate programs, and conferences related to their disciplines.

Employment agencies. Public and private agencies offer help with application documents, interviewing, and job referrals, including part-time jobs.

Networking. People you know (employers, relatives, and friends) can give job-search tips, recommendations, and referrals.

Organizations' Web sites. Most organizations have Web sites featuring information on their products, services, and employment opportunities, along with contact information.

Federal government. Your regional office of the U.S. Office of Personnel Management offers information on employment. Call 912-757-3000 for details. (For more state and federal government sources, see the Web sites on page 317.)

Labor unions. Labor unions offer a variety of job-search support, including job ads.

Letters of inquiry. Letters are job-search tools. Write them well and often.

Public libraries. Libraries have newspapers and trade journals that include job ads. Libraries also have computers and research librarians to help you find information.

RESEARCHING ORGANIZATIONS

According to a survey cited in *Job Outlook 2003,* most college graduates looking for work neglect one essential step in the job-search process: researching the employer. As a result, the person presents himself or herself as one or more of the following:

- **Uninformed** about the organization's products, services, and competition
- **Uninterested** in the organization's history, mission, and goals
- **Unmotivated** to learn what the organization is or wants to become

The point is obvious: Why would an employer hire anyone who is uninformed, uninterested, and unmotivated?

Become informed by taking the time to research both the employer and the position. To guide your research, ask yourself questions like these:

- What are the organization's mission, values, goals, products, and services?
- When did the organization begin, how did it start, and what has it accomplished?
- Who are the organization's competitors, and how well are those organizations doing in comparison to them? Why?
- How is the organization structured: number of offices, geographic locations, layers of authority, and so on?
- What are the organization's benchmarks for measuring its own success or the success of its employees?
- How does the organization treat its employees in terms of working conditions, assignments, salary, benefits, training, advancement, profit sharing, and technical support?
- What is the organization's retention record and why does it have that pattern?
- How do your own interests, skills, and experiences match up with those of others with whom you would be working?

To get answers, use research strategies like these:

- Check out the organization's Web site and study its mission statement; annual report; catalogue of products or services; descriptions of its facilities, employees, and organizational structure; newsletters; and news releases.
- Do a Web search for stories about the organization that were published in area newspapers or broadcast on area radio or TV stations.
- Visit or write the organization, asking for a sampling of its catalogues, brochures, newsletters, and annual reports.
- Inquire at the Chamber of Commerce about the organization's role in and impact on the community, and ask for the organization's literature.
- Interview people both inside and outside the organization.

Chapter 17 Understanding the Job-Search Process

USING WEB RESOURCES

Using the World Wide Web to search for jobs has both advantages and disadvantages. The Web offers several benefits:

- It provides free access to excellent (particularly public) job-search resources.
- It can help you develop application documents or a personal Web site.
- It can help you contact organizations and learn about their job openings.
- You can submit your application promptly (often electronically).

The Web also has some disadvantages:

- Many jobs are not listed on the Web.
- Job-service agencies (especially for-profit organizations) may be shams.
- Because many jobseekers use the Web, your application may be overlooked.
- Contact only by e-mail may seem impersonal.

The following resources are fine job-search tools:

- Bureau of Labor Statistics (www.bls.gov), the federal government's primary fact-finding agency regarding labor, economics, and statistics, offers well-researched data on these topics.
- Career Guide to Industries (http://stats.bls.gov/oco/cg/), sponsored by the Department of Labor, offers information on careers in specific industries.
- CareerOneStop (www.careeronestop.org), a publicly funded site linked to the next three resources, offers many services for jobseekers and employers.
  - America's Career InfoNet (www.acinet.org/acinet/default.asp) offers information on careers, training, financial aid, skill development, and more.
  - America's Job Bank (www.ajb.org/) is a job-search engine linking state employment offices and showing job openings throughout the country.
  - America's Service Locator (www.servicelocator.org/) is a search engine that helps locate services and service organizations nationwide.
- Human Resources Development Canada (www.hrdc-drhc.gc.ca/common/home.shtml) is a Canadian government site that includes information on jobs and job training throughout Canada.
- Occupational Outlook Handbook (http://stats.bls.gov/oco/home.htm), produced by the U.S. Department of Labor, is both an online and paper publication with information on topics such as jobs, salaries, and industries.
- Occupational Outlook Quarterly (http://library.louisville.edu/government/federal/agencies/labor/outlook.html) is the U.S. Department of Labor's print and online quarterly publication with current information on jobs and careers.
- Monster.com (www.monster.com/) has both U.S. and international job postings and career advice.
- USAJOBS (www.usajobs.opm.gov/) is the official job site of the U.S. government and offers information on government jobs.

CHECKLIST for Career Planning

Use the seven traits to check your career plan and then revise as needed.

- ☐ **Ideas.** The career plan shows a thorough understanding of my qualifications, job opportunities, requirements for those jobs, and what I must do to meet the requirements.
- ☐ **Organization.** The document moves logically from an analysis of the career field and its related job requirements to a plan showing how I can achieve those requirements. Headings clearly indicate the contents of sections.
- ☐ **Voice.** The tone is objective, thoughtful, committed, and optimistic.
- ☐ **Words.** The words are clear and appropriate; technical terms or references to degrees or certifications are correct.
- ☐ **Sentences.** The plan has varied sentence structures linked with smooth transitions. Lists are introduced clearly and are stated in parallel form.
- ☐ **Copy.** The document includes no errors in grammar, usage, spelling, punctuation, and mechanics.
- ☐ **Design.** Information is presented in a clear, readable layout that includes appropriate sections, descriptive titles, short paragraphs, and an effective use of bulleted and numbered lists.

CRITICAL THINKING Activities

1. Read three periodicals in your career field looking for information on career opportunities, job responsibilities, and required skills and training. Write a few pages in your journal reflecting on how your career interests, skills, and training match those described in your reading.
2. Visit three Web sites listed on page 317 to find information on your career field, including job prospects, level of pay, and required education and certification. Does this information change your interest in the career? How and why?

WRITING Activities

1. Visit an area business related to your career field. Interview two people who do the type of work that you're training to do. Ask each person to trace his or her career path, to identify pivotal career choices, and to advise you on how to build a career. Take notes on your interviews and report your findings to the class.
2. Do a Web search on three organizations related to your career field but located outside your state, region, or country. From the Web information, learn all that you can about each organization's career opportunities, job responsibilities, and job qualifications. Then get additional information by e-mailing your unanswered questions to the personnel manager or head of human resources. Use your research to write a report or a career plan.

In this chapter
Guidelines for Résumés **320**
Checklist for Résumés **326**
Critical Thinking Activity **326**
Writing Activities **326**

CHAPTER
18

Developing Your Résumé

You write a résumé to persuade an employer that your training, experiences, and skills are so strong that he or she should ask you to come in for an interview. In other words, a résumé is a carefully crafted argument.

To win that argument, your résumé must show quickly and clearly that your *qualifications* match or exceed the employer's *expectations*. But how do you develop such a document? First, you find out all that you can about the employer and the job. Then, you use the guidelines and models in this chapter to spell out your qualifications in the language that the reader or scanner wants to see. (Many organizations use scanners to review résumés for key qualifications that match the job requirements.) Finally, you format the résumé as a paper document for readers or as an electronic document for scanners.

While the quality of your résumé will depend primarily on its content (your training, experiences, and skills), this chapter will help you present that content in a form that's both professional and effective. In addition, the chapter will help you design the document in multiple formats (chronological, functional, or electronic) to match each employer's preference.

Study the chapter now so you can develop a strong, persuasive résumé. Then after you learn about a job opening, review the chapter, using its advice to revise and update your résumé.

GUIDELINES FOR RÉSUMÉS

When writing a résumé, your goal is to show that your skills, knowledge, and experience match the requirements for a specific job. Each style of résumé presents that information differently.

1 Prewrite
Rhetorical analysis: Consider your purpose, your audience, and the context.
- What skills and experiences does the job require, and how do they match yours?
- What style of résumé best highlights your qualifications (chronological features experience; functional features skills)?
- What format (paper, electronic, or both) does the employer prefer?

Prepare to draft.
Gather the following details:
- Your career objective, worded to match the job description
- Your educational experiences (schools, degrees, certification)
- Your work experiences (employers and dates; responsibilities, skills, and titles; special projects, leadership roles, and awards)
- Activities and interests directly or indirectly related to the job
- Responsible people who are willing to recommend you, along with their addresses and phone and fax numbers

2 Draft
Organize the résumé into three parts.
[opening] List your contact information and job objective.
[middle] Write appropriate headings, and list educational and work experiences in parallel phrases or clauses. Refer to your training and skills with key words that match the job description (terms that could be identified by an employer's scanner).
[closing] List names, job titles, and contact information for your references; or state that references are available upon request.

3 Revise
Review your résumé for ideas, organization, and voice.
- Check for skills, training, and key words listed in the job description.
- Check for clear organization, correct details, and a professional tone.

4 Refine
Check your résumé line by line for the following:
- Strong verbs, precise nouns, and parallel phrases
- Correct names, dates, grammar, and punctuation
- Appropriate format (divisions, headings, lists, spacing)
- Correct formatting for a paper or e-document, and for a reader or scanner

> "A résumé is a balance sheet with no liabilities."
> —Robert Half

Chronological Résumé

Rhetorical situation

Purpose: To show that you have the skills and experiences required for the job
Audience: Director of human resources, department chair or foreman, scanner
Format: Paper document with details listed chronologically under key heads

---

LLOYD A. CLARK
1913 Linden Street
Charlotte, NC 28205-5611
(704) 555-2422
lloydac@erthlk.net

[opening]
Present contact information.

EMPLOYMENT OBJECTIVE
Law enforcement position that calls for technical skills, military experience, self-discipline, reliability, and people skills.

State your employment objective.

WORK EXPERIENCE
Positions held in the United States Marine Corps:

- Guard Supervisor—Sasebo Naval Base, Japan, 1999–2005
  Scheduled and supervised 24 guards.
- Marksmanship Instructor—Sasebo Naval Base, Japan, 1997–1999
  Trained personnel in small-arms marksmanship techniques.
- Company Clerk—Okinawa, Japan, 1996–1997
  Handled correspondence; prepared training schedules and assignments.

[middle]
List experiences, skills, and training (most recent first).

Use periods after clauses—including those with understood subjects.

Do not use periods after headings or phrases.

SKILLS AND QUALIFICATIONS

- In-depth knowledge of laws and regulations concerning apprehension, search and seizure, rules of evidence, and use of deadly force
- Knowledge of security-management principles and training methods
- Experience in physical-training management, marksmanship, and weaponry
- Computer word-processing and database skills on an IBM-compatible system
- Excellent one-on-one skills and communication abilities

EDUCATION

- Arrest, Apprehension, and Riot Control Course, Sasebo, Japan, 1998
- Marksmanship Instructor Course, Okinawa, Japan, 1997
- Sexual-Harassment Sensitivity Training, Camp Lejeune, NC, 1996
- School of Infantry, Camp Pendleton, CA, 1996

Keep all phrases and clauses parallel.

AWARDS AND HONORS

- Promoted meritoriously from Private (E-1) to Lance Corporal (E-3); promoted meritoriously to final rank of Corporal (E-4) in less than 2 years.
- Achieved "Expert" rating for pistol at annual marksmanship qualifications (3 years).
- Represented Marine Barracks, Japan, in shooting matches (placed in top half).

List awards and honors in order of importance.

References available upon request

[closing]
Offer references.

A résumé normally should not exceed one page; however, this candidate seeks a high-level management position requiring extensive experience.

Rhetorical situation

Purpose: To show the writer's high-level skills and broad experiences
Audience: Director of human resources and search committee
Format: Paper document; details listed chronologically under proper heads

---

PAGE 1 OF 2

JAMES A. KARSTEN
106 Lincoln Avenue SE • Waynesville, MO 65583 • (314) 555-2080 • jkn@busnet.com

[opening]
State the objective.

OBJECTIVE
Position as vice president or president of manufacturing in a metropolitan area.

EXPERIENCE

[middle]
List job history chronologically—most recent first.

1998–present   VICE PRESIDENT OF MANUFACTURING
                 UMI Manufacturing, Waynesville, MO

- Oversee construction, start-up, and operations at 50,000-sq.-ft. flagship manufacturing plant.
- Design and establish Total Quality Management (TQM) facility with progressive human-resource management policies and work teams.
- Hire and place all staff.
- Facilitate independent operation of all departments.
- Participate on UMI Resources Committee.

1992–1998   VICE PRESIDENT OF MANUFACTURING
              Allied Technology Incorporated, Lincoln, NE

- Directed manufacturing, quality, maintenance, engineering, and sales.
- Improved profit margins 25 percent through TQM and synchronous manufacturing.
- Rebuilt organization and implemented team leadership.
- Belonged to Allied's Product Planning and Manufacturing Board.
- Increased sales 65 percent.
- Earned Ford New Holland Q1 Preferred Supplier Award.

State tasks with strong verbs, clear data, and job-related terminology.

1990–1992   INTERNATIONAL PURCHASING MANAGER
              GFI Motors, Rochester, NY

- Negotiated all domestic and European contracts.
- Managed $150 million budget.
- Led cross-functional TQM task forces to implement synchronous techniques.
- Reduced annual material costs 20 percent and improved quality and delivery through redesign and process improvement.

Note range of tasks, including size of budget.

—continued—

JAMES A. KARSTEN

1988–1990   FUEL SYSTEMS PRODUCT MANAGER
GFI Motors, Rochester, NY

- Applied worldwide growth strategies.
- Organized five-year marketing plan to meet growth goals of $2.5 billion.
- Compiled a product portfolio that facilitated 20 percent growth in sales.

1986–1988   GENERAL SUPERVISOR OF OPERATIONS
GFI Motors, Rochester, NY

- Improved throughput by 60 percent; reduced overhead costs by 25 percent.
- Hired staff and incorporated the Human Resources Department.

1984–1986   MATERIALS MANAGER
Kaizen Inventory Management Incorporated, Jenkintown, PA

- Conducted supplier-operations analysis and synchronous implementation.
- Improved on-time delivery from 82 percent to 98 percent.
- Brought inventory turns to 300 percent; lowered transportation costs 25 percent.

1982–1984   SUPERVISOR OF QUALITY INSPECTION
Kaizen Inventory Management Incorporated, Jenkintown, PA

- Conducted machine debugging, qualification, and verification processes.
- Developed statistical process-control procedures for critical plant processes.
- Trained new workers in assembly techniques and statistical recording.

1979–1982   SUPERVISOR OF MANUFACTURING
Preston Company, Ft. Wayne, IN

- Participated in all phases of plant start-up and systems.
- Developed line balance, material layouts, and flowcharts for operating procedures.

EDUCATION
Rochester Institute of Technology—Rochester, NY

- Master's in Business Administration (MBA), 1992
- Bachelor of Science (BS) in Business Administration, 1979

ADDITIONAL TRAINING

- Continuous Improvement through Statistical Methods (Dr. Edwards Deming)
- World Class Manufacturing (Dr. Schonberger)
- Strategic Business Management (Wharton School of Business)

References available upon request

Functional Résumé

Rhetorical situation

Purpose: To feature the writer's skills, education, and experiences
Audience: Owner of a small company or HR director in a larger company
Format: Paper document featuring skills and listing experiences chronologically

---

[opening]
Present contact information and your employment objective.

Michelle Moore 3448 Skyway Drive   (406) 555-2166
Missoula, MT 59801-2883   E-mail: mimoore@erthlk.net

EMPLOYMENT OBJECTIVE   Electrical Engineer—designing or developing digital and/or microprocessor systems.

[middle]
Feature skills by referring to educational and work experiences.

QUALIFICATIONS AND SKILLS

Design
- Wrote two "C" programs to increase production-lab efficiency.
- Built, tested, and modified prototypes in digital and analog circuit design.
- Designed and worked with CMOS components.
- Wrote code for specific set of requirements.
- Helped implement circuitry and hardware for a "bed-of-nails" test.

Put the most important skills first.

Troubleshooting and Repair
- Repaired circuit boards of peripheral computer products.
- Helped maintain equipment using circuit-board testing.
- Improved product quality by correcting recurring problems.
- Debugged IBM-XT/Fox Kit Microprocessor Trainer (Z80).

Use periods after clauses—including those with understood subjects—but not after phrases.

Management
- Trained and supervised production technicians.
- Assisted in lab teaching for Microprocessors and Digital Circuits class.

EDUCATION
Montana State University, Bozeman, MT
- Bachelor of Science in Engineering, 2002
- Major: Electrical Engineering
- Independent Study: C programming, DOS and BIOS interrupts

Format for paper only:
- Boldface
- Underlining
- Bulleted lists
- Two columns

EXPERIENCE

October 2004 to present	Production Engineer (full-time)
	Big Sky Computer Products, Inc., Missoula, MT
June 2002 to September 2003	Engineer (part-time)
	Western Labs, Missoula, MT
September 2001 to May 2002	Engineering Assistant
	Montana State University, Bozeman, MT
May 2000 to September 2001	Engineering Intern
	Montana State University, Bozeman, MT

[closing]
Offer references.

References available upon request

Electronic Résumé

Rhetorical situation

Purpose: To present the writer's qualifications in terms sought by a scanner
Audience: Scanner programmed to find key terms related to or found in the job description
Format: Electronic document (formatting traits listed on the right)

Jonathan L. Greenlind
806 5th Avenue
Waterloo, IA 50701-9351
Phone: 319.555.6955
E-mail: grnlnd@aol.com

OBJECTIVE
Position as hydraulics supervisor that calls for hydraulics expertise, technical skills, mechanical knowledge, reliability, and enthusiasm.

SKILLS
Operation and repair specialist in main and auxiliary power systems, subsystems, landing gears, brakes and pneumatic systems, hydraulic motors, reservoirs, actuators, pumps, and cylinders from six types of hydraulic systems
Dependable, resourceful, strong leader, team worker

EXPERIENCE
Aviation Hydraulics Technician
United States Navy (1990–present)
* Repair, test, and maintain basic hydraulics, distribution systems, and aircraft structural hydraulics systems.
* Manufacture low-, medium-, and high-pressure rubber and Teflon hydraulic hoses.
* Perform preflight, postflight, and other periodic aircraft inspections.
* Supervise personnel.

Aircraft Mechanic
Sioux Falls International Airport (1988–1990)
Sioux Falls, South Dakota
* Performed fueling, engine overhauls, minor repairs, and tire and oil changes of various aircraft.

EDUCATION
United States Navy (1990–1994)
Certificate in Hydraulic Technical School; GPA 3.8/4.0
Certificate in Hydraulic, Pneumatic Test Stand School; GPA 3.9/4.0
Courses in Corrosion Control, Hydraulic Tube Bender, Aviation Structural Mechanics
Equivalent of 10 semester hours in Hydraulic Systems Maintenance and Repair

References available upon request

[opening]
Present contact information and employment objectives.

[middle]
List skills, experiences, and education using many key words.

Format for e-mail, scanner, and Web site:
- One column
- Asterisks as bullets
- Simple sans serif typeface
- Flush left margin
- No italics, boldface, or underlining
- ASCII or RTF text (readable by all computers)

[closing]
Offer references.

CHECKLIST for Résumés

Use the seven traits to check your résumé and then revise as needed.

- ☐ **Ideas.** My résumé shows that I understand the job and that I have the required training, skills, and experiences.
- ☐ **Organization.** I have organized my résumé either chronologically or functionally; headings clearly forecast the details that follow.
- ☐ **Voice.** The tone is confident and knowledgeable, but not arrogant.
- ☐ **Words.** My résumé includes strong verbs, key terms from the job description, and accurate terms for academic degrees, professional certifications, and job titles.
- ☐ **Sentences.** I use complete sentences only when needed to make a point; they provide information in parallel lists, phrases, and clauses.
- ☐ **Copy.** My résumé has periods after clauses (including those with understood subjects), and includes no errors in grammar, punctuation, capitalization, or spelling.
- ☐ **Design.** I use the proper design:
  - Paper only—bulleted lists, boldface, underlining, and business typeface.
  - Electronic—ASCII or RTF text (read by computers), sans serif typeface, single-column format, and wide margins, but no boldface or underlining.

CRITICAL THINKING Activity

Use your computer to list information that you will need to write a résumé. Organize details under the following headings: Contact Information, Employment Objectives (all reasonable possibilities), Qualifications, Skills, Work Experiences (including dates, duties, and special assignments or projects), Education (including key courses, training programs, dates, degrees, certifications, awards, and cumulative grade point average), and Possible References. Analyze the information that you gathered, noting your workplace strengths and weaknesses.

WRITING Activities

1. Find a job advertisement in your field; then research the employer, as well as the job's opportunities, responsibilities, and requirements. Using the information from the Critical Thinking activity, develop a paper résumé (either chronological or functional) that proves you are qualified for the job.
2. Find another job ad that is in your field but that has different responsibilities and requirements. Do the necessary research and write an electronic résumé appropriate for the job. Show both résumés to the class and explain how the content and form of each document match the employer's requirements.

In this chapter

Guidelines for Application Letters 328

Guidelines for Recommendation-Request Letters 330

Guidelines for Application Essays 332

Guidelines for Job-Acceptance Letters 334

Guidelines for Job-Rejection Letters 336

Guidelines for Thank-You and Update Messages 338

Checklist for Application Correspondence 340

Critical Thinking Activity 340

Writing Activities 340

CHAPTER
19

Writing Application Correspondence

Applying for a job usually requires communicating your interest in the job, requesting an interview, soliciting recommendations, sending a follow-up letter, and finally either accepting or rejecting the job offer. During the process, you must also update and revise your résumé so that it matches the specific job you are seeking.

This writing serves to represent you to your prospective employer. It communicates a great deal about who you are: your training, experiences, skills, knowledge, and even your personality. In addition, the documents show your ability to compose and present your thoughts in writing.

Employers commonly use application writing in three ways. First, they will evaluate your job-inquiry letter and résumé to determine whether to invite you for an interview. Second, they will assess all of your application documents to decide whether to offer you a job. Third, even after an employer hires you, he or she may use your application documents to assess whether you are ready for assignments that require writing or whether you must first complete remedial training in writing.

In summary, the quality of your application documents will affect whether you get the job and the assignments that you seek. To succeed, you must use this chapter's instructions well.

GUIDELINES FOR APPLICATION LETTERS

Your goal when writing an application (or cover) letter is to convince the reader to study your enclosed résumé and invite you for an interview.

1 Prewrite

Rhetorical analysis: Consider your purpose, your audience, and the context.
- What are the job's required skills, training, experiences, and tasks?
- What are the company's location, size, products or services, and mission?
- Who will read your letter, and what is his or her role in the selection process?
- How can you show that you're qualified and able to do the job?

Prepare to draft.
- Gather details about the company, its products or services, and the job.
- List your training and experience and describe how both match the job requirements.
- Find the name, title, and address of the person to whom you are writing.

2 Draft

Organize your letter into three parts.

[opening] Use a courteous but confident tone.
- Refer to the job and tell how you learned about it.
- State your main qualification.

[middle] Show that you are qualified.
- Tell how your education, experience, and skills fit the job. (Refer to your résumé.)
- Communicate your interest in and knowledge of the job and the company.

[closing] Close by encouraging contact.
- Explain when and where you might be reached.
- Request an interview.

3 Revise

Check your draft's ideas, organization, and voice.
- Have you explained why you can do the job well?
- Have you delivered your message in clear, well-organized paragraphs?
- Have you used a courteous, confident, business-like tone?

4 Refine

Check your writing line by line for the following:
- Correct names, titles, addresses, and application-related details
- Correct spelling, mechanics, usage, and grammar
- Appropriate fonts, proper letter design, and quality paper

> "Choose a job you love, and you will never have to work a day in your life."
> —Confucius

Chapter 19 Writing Application Correspondence

Application Letter

Rhetorical situation

Purpose: To show the writer's interest in and qualifications for the job
Audience: The director of human resources and the director of staff development
Format: Business letter formatted in full block and printed on quality paper

---

3041 45th Avenue
Lake City, WA 98125-3722
November 17, 2005

Ms. Marla Tamor
Human Resources Director
Evergreen Medical Center
812 University Street
Seattle, WA 98105-6152

Dear Ms. Tamor:

In response to your advertisement in the *Seattle Times* on November 11, I am writing to apply for the position of Software-Training Specialist. For the past seven years, I have worked as a trainer in the health care system at Pacific Way Hospital. ⎯ **[opening]** Name the job and the source of the ad. Introduce your qualifications.

I have instructed individuals and groups on how to use the following systems/software: Microsoft Office, WordPerfect Office, Lotus Millennium, and mainframe/business-specific programs. I am also trained to instruct clients in CorelDRAW and Adobe design products.

In addition to my work with software systems, I have developed job descriptions, recruited technical employees, and trained human resources personnel. I believe this experience would help me address the needs of a growing health care facility such as Evergreen Medical Center. ⎯ **[middle]** List specific training, experience, and skills.

Enclosed is my résumé, which further details my qualifications. I look forward to hearing from you and can be reached at (206) 555-0242 or at jmvrtz@aol.com. Thank you for your consideration. ⎯ **[closing]** Invite follow-up, provide contact information, and close politely.

Sincerely,

Jamie Vertz

Jamie Vertz

Enc.: Résumé

GUIDELINES FOR RECOMMENDATION-REQUEST LETTERS

Your goal when writing a recommendation-request letter is twofold: (1) to ask the reader to write a recommendation and (2) to help him or her write it. How do you help? By reminding the person about the strengths you showed during your work relationship and by explaining how those strengths match requirements for the job you are seeking.

1 Prewrite
Rhetorical analysis: Consider your purpose, your audience, and the context.
- ☐ Who could write a strong recommendation for you? Choose someone who knows your qualifications, is respected by the employer, and can write well.
- ☐ Which qualifications should the person cite in the recommendation?

Prepare to draft.
- ☐ Review details about the job's requirements, particularly training and experiences.
- ☐ List your skills, training, and accomplishments that match the job requirements.
- ☐ Note who must receive the recommendation and by when.

2 Draft
Organize the letter into three parts.

[opening] Establish a courteous, confident tone:
- ☐ Address the reader politely and correctly.
- ☐ State the job for which you are applying.
- ☐ Ask the person to write a recommendation.

[middle] Give the reader important details like the following:
- ☐ Skills, training, experiences, and certification required for the job
- ☐ A reminder about how your skills, training, and experiences match the job requirements
- ☐ Specifics about who must receive the recommendation and by when

[closing] Close by doing the following:
- ☐ Ask the person to inform you (give date) whether he or she will write the letter.
- ☐ Thank the reader for considering your request.

3 Revise
Check your draft for ideas, organization, and voice.
- ☐ Is the information complete, correct, and organized?
- ☐ Is the tone consistently polite, positive, and professional?

4 Refine
Check your writing and page layout for the following:
- ☐ Correct names, titles, dates, and job-related terminology
- ☐ Correct spelling, grammar, usage, punctuation, and formatting

Recommendation-Request Letter

Rhetorical situation

Purpose: To ask the reader to write a recommendation, and to provide details that will help the person write a strong letter
Audience: A respected person who can explain why you are qualified for the job
Format: A business letter formatted in full block and printed on quality paper

---

3041 45th Avenue
Lake City, WA 98125
November 21, 2005

Mr. Daniel Salazar
Director of Computer Services
Pacific Way Hospital
5483 Albany Avenue
Moses Lake, WA 98837

Dear Mr. Salazar:

Since my recent move to Seattle, I've been looking for opportunities to continue my career in computer services. Recently I applied for a position with Evergreen Medical Center as a software-training specialist, and I am writing to ask if you would write a letter recommending me for that position. Because you are familiar with my work at Pacific Way Hospital, your letter would be very helpful. — [opening] Identify the job and make the request.

As you may recall, my work at Pacific Way included training clients in software programs and systems, recruiting office and technical employees, and developing job descriptions. Your evaluations of my work were always positive; in fact, in my last evaluation, you wrote that I was "able to make even the most complex computer systems accessible to novices." Because the software-training position at Evergreen Medical Center involves many of the same responsibilities as my job in your department, your recommendation could weigh heavily in the application process. — [middle] Help the reader recall you and your work. Cite the job's requirements.

Please let me know whether you're willing to write this recommendation. I can be reached at (206) 555-0242 during daytime and evening hours or at jmvrtz@aol.com. The deadline for recommendations is December 16, 2005, and the person to whom you should address the recommendation is Ms. Marla Tamor, Human Resources Director. I've enclosed a preaddressed, stamped envelope for your convenience. Thank you for your time and consideration. — [closing] Request a reply; provide contact information.

Sincerely,

Jamie Vertz

Jamie Vertz

Enc.: Preaddressed stamped envelope

Part 3   The Application Process and Application Writing

GUIDELINES FOR APPLICATION ESSAYS

For some applications (e.g., graduate school or a professional program), you will be asked to submit an essay or personal statement. Whatever the situation, your goal is to address the assigned topic in a clear, well-organized discussion that communicates who you are, how you think and write, and why you are drawn to the position, program, or profession.

1 Prewrite
Rhetorical analysis: Consider your purpose, your audience, and the context.
- ☐ Who will read your essay? What skills or insights are they looking for?
- ☐ How does the topic relate to the organization and job?

Prepare to draft.
- ☐ Read the instructions carefully, noting the topic and what you are asked to do: analyze, explain, describe, or evaluate. Note the requested form and length.
- ☐ Use prewriting activities such as clustering, listing, freewriting, and outlining to generate and organize your ideas.
- ☐ Draft a sentence stating your main point or thesis regarding the topic.

2 Draft
Organize your writing into three parts.
[opening] Introduce the topic in an engaging way, provide a context, and state your thesis politely but with confidence.
[middle] Develop your thesis clearly and concisely, using details consistent with the instructions and fitting for the position and readers. Use each paragraph to advance the thesis, and avoid vague or unrealistic claims.
[closing] Stress an appropriate, positive point and look forward to participating in the program or organization.

3 Revise
Review your draft for ideas, organization, and voice.
- ☐ Is your main idea logical, clear, fully developed, and well supported?
- ☐ Are your ideas effectively shaped into unified, coherent, and complete paragraphs?
- ☐ Is the tone positive, but also objective, reasonable, and thoughtful?

4 Refine
Check your writing line by line for the following:
- ☐ Clear, concise, professional wording
- ☐ Readable sentences and helpful transitions
- ☐ Correct grammar, punctuation, and spelling—especially glaring typos
- ☐ Effective format (e.g., spacing, margins, fonts, type size)

> "Easy writing makes hard reading."
> —Ernest Hemingway

Application Essay

Rhetorical situation

Purpose: To explain how the writer became interested in social work, and why she wants to work for Regis Community Services in Tulsa, Oklahoma
Audience: The director of human resources, management, and senior counselors
Format: Essay with single-spaced lines and double-spacing between paragraphs

The Road from Bunde to Regis

During my early years, Bunde, Minnesota, was home. Now no more than a church and a dozen houses, then it was a cohesive community of about seventy people who looked after one another. For example, when Leo Folken lost his arm in a corn-picking accident, the community finished his harvest and cared for his livestock. When Jean O'Malley returned from Kansas City bruised by her husband's beatings and carrying a baby, the community adopted both mother and child. My family, along with the other residents of Bunde, taught me that helping those in need is not just one person's choice; it's everyone's responsibility.

When I left for Clark College in Kansas, my experiences in little Bunde prompted me to choose social work as my major. During the next four years, I took courses in which I learned about the history and practice of social work. But I also studied psychology and communication, and I volunteered for three semesters in Clark's Social-Outreach Program. The courses helped me understand the needs of those living without adequate housing, health care, or career skills; and the volunteer work helped me practice addressing those needs.

The following three years at the University of Nebraska provided further study of key topics related to social work, such as psychology, economics, and state and federal social policies and programs. In addition, my yearlong internship at Hope Services enabled me to work with a professional team in an urban community. In other words, earning my M.S.W. provided both the certification for a career in social work and the knowledge and skills to do that work.

While I am certainly prepared to join the staff at Regis Community Services, I am also eager to do so for a number of reasons. First, Regis's fine facilities and location enable it to serve clients throughout downtown Tulsa. Second, Regis's strong financial base supports the organization's excellent services. Most importantly, Regis's staff have both the professional skills and the passion to deliver those services.

In summary, little Bunde gave me the heart to become a social worker. Clark College and the University of Nebraska provided the knowledge and skills. But it is Regis Community Services that can offer the opportunity.

[opening]
Get the reader's attention and introduce the topic.

State the main idea.

[middle]
Develop that idea by citing details fitting the topic and position.

Present information in unified paragraphs linked with clear transitions.

Use a positive and thoughtful tone.

[closing]
Summarize and look ahead.

GUIDELINES FOR JOB-ACCEPTANCE LETTERS

As soon as you decide to accept a job offer, promptly write a job-acceptance letter. It allows you to express your appreciation for the offer, accept the job, and confirm your plans to start work.

1 Prewrite
Rhetorical analysis: Consider your purpose, your audience, and the context.
- ☐ Why did you choose to accept this job (list details, including who treated you well and how)?
- ☐ What are the job's title, duties, compensation, and starting date?
- ☐ What are your questions or concerns regarding moving services, temporary housing, the company's orientation program, schools for your children, and so on?
- ☐ When would you like to start the new job, and what are the impediments?

2 Draft
Organize your letter into three parts.

[opening] Establish a positive tone.
- ☐ Express appreciation for the offer (use the job title).
- ☐ Confirm the basic salary and accept the job offer.

[middle] Focus on the job transition.
- ☐ State the date, time, and place that you will report for work, or inquire about these details.
- ☐ Explain issues (housing, family schedule) that may require special consideration or assistance.

[closing] Repeat appreciation for the job offer and communicate enthusiasm for the job and the company.

3 Revise
Review the draft.
- ☐ Have you used a sincere and professional tone?
- ☐ Have you clearly expressed your thoughts and included accurate details?
- ☐ Have you clearly cited any details that the reader must address?

4 Refine
Check your writing line by line for the following:
- ☐ Well-chosen words, smooth sentences, and clear transitions between paragraphs
- ☐ Correct spelling, especially of all names and titles
- ☐ Correct grammar, punctuation, and usage
- ☐ Attractive layout formatted in block or semiblock and printed on quality paper

Job-Acceptance Letter

Rhetorical situation

Purpose: To thank the reader for the job, accept it, and make plans to start work
Audience: The director of human resources
Format: Business letter formatted in full block and printed on quality paper

---

3041 45th Avenue
Lake City, WA 98125
January 7, 2006

Ms. Marla Tamor
Human Resources Director
Evergreen Medical Center
812 University Street
Seattle, WA 98105

Dear Ms. Tamor:

Thank you for offering me the position of Software-Training Specialist at Evergreen Medical Center. I am happy to accept the position at the annual salary of $63,500. — **[opening]** Accept the job, noting key details.

As we discussed on the phone, my first day of work will be January 24. I will check in with the head of the computer center at 9:00 a.m. In the meantime, I'll search for housing closer to the medical center. Thank you for the material that you sent about real-estate agencies and moving-expense reimbursement. Could you please send information about the area schools as well? I have two children in elementary school this year. — **[middle]** Discuss starting date and transition information.

I look forward to joining the team at Evergreen and working with staff members in all areas involving technology training. — **[closing]** Look ahead.

Sincerely,

Jamie Vertz

Jamie Vertz

GUIDELINES FOR JOB-REJECTION LETTERS

If you decide to reject a job offer, promptly write a job-rejection letter that delivers the bad news forthrightly, but diplomatically.

1 Prewrite
Rhetorical analysis: Consider your purpose, your audience, and the context.
- ☐ What job were you offered, when, how, and by whom?
- ☐ How could you express your appreciation for the offer and the organization?
- ☐ Why did you reject the offer, what reasons are you willing to share, and how can you state these reasons diplomatically?
- ☐ How can you communicate your desire for good relations in the future?

Prepare to draft.
- ☐ Note the job title and reader's title.
- ☐ Focus your thinking by drafting a one-sentence statement explaining the rejection.

2 Draft
Organize the letter into three parts.

[opening] Establish a positive tone and buffer.
- ☐ Address the reader formally.
- ☐ Convey appreciation for the offer (include the job title).
- ☐ Relay thanks for how the company treated you.

[middle] Continue an indirect approach.
- ☐ Cite aspects of the job that you found attractive.
- ☐ Explain why you must decline. Be objective and brief.

[closing] Restate your appreciation for the offer.
- ☐ Convey your best wishes to the reader.
- ☐ Express willingness to be considered for future jobs.

3 Revise
Review your draft's ideas, organization, and voice.
- ☐ Have you used a courteous voice and an indirect approach?
- ☐ Have you clearly and briefly explained why you had to decline the offer?

4 Refine
Check your writing line by line for the following:
- ☐ Precise word choice and smooth sentences
- ☐ Correct spelling, punctuation, and grammar
- ☐ Proper elements of a business letter formatted in block or semiblock

> "It's a funny thing about life; if you refuse to accept anything but the best, you very often get it."
> —Somerset Maugham

Job-Rejection Letter

Rhetorical situation

Purpose: To thank the reader for the job offer and kindnesses shown, diplomatically reject the offer, and maintain good relations
Audience: Director of human resources and others involved in the interview process
Format: Business letter formatted in full block and printed on quality paper

---

3041 45th Avenue
Lake City, WA 98125
January 9, 2006

Ms. Marla Tamor
Human Resources Director
Evergreen Medical Center
812 University Street
Seattle, WA 98105

Dear Ms. Tamor:

Thank you for offering me the position of Software-Training Specialist at Evergreen Medical Center. I appreciate both your confidence in me and the professional treatment that I was shown throughout the application process. — **[opening]** Express thanks for the job offer.

In considering job opportunities, I was looking for a position in which I could grow professionally and acquire new skills. Because in many ways the job at Evergreen Medical Center is identical to the position I just vacated at Pacific Way Hospital, I have decided not to accept your generous offer. Instead, my career will take a new direction as I accept a position as a software consultant for a new marketing firm in Ballard. — **[middle]** Use an indirect approach to politely decline.

I enjoyed meeting you and becoming familiar with Evergreen Medical Center. Best wishes as you continue your search to fill the position. — **[closing]** Offer good wishes.

Yours sincerely,

Jamie Vertz

Jamie Vertz

GUIDELINES FOR THANK-YOU AND UPDATE MESSAGES

Your goal when writing a thank-you/update message is to thank the person who helped you in the job-search process, give an update on the process, and strengthen your relationship.

1 Prewrite
Rhetorical analysis: Consider your purpose, your audience, and the context.
- ☐ What did the reader do for you (write a recommendation, speak with the employer, offer advice or leads)?
- ☐ What information could you share that would show your appreciation and strengthen your professional and personal relationships?

Prepare to draft.
- ☐ Review previous correspondence with the reader regarding the application process.
- ☐ Review events in the job-search process, including correspondence from the employer.

2 Draft
Organize your message into three parts.

[opening] Use a friendly, conversational tone.
- ☐ Greet the reader using the title appropriate for your relationship.
- ☐ State your thanks, citing specifically what the reader did.

[middle] Describe how the application process is going.
- ☐ Briefly explain what has happened since your last correspondence.
- ☐ If the process is ongoing, explain when and how you think it might conclude.

[closing] Close politely by repeating your thanks and giving additional details.

3 Revise
Review your draft for ideas, organization, and voice.
- ☐ Are the reader's actions and the application process described clearly and accurately?
- ☐ Are paragraphs unified and complete?
- ☐ Is the voice courteous, professional, and positive?

4 Refine
Check your writing line by line for the following:
- ☐ Correct titles, names, and dates
- ☐ Smooth sentences and appropriate transitions
- ☐ Correct grammar, punctuation, usage, and spelling
- ☐ Effective format (e.g., spacing, margins, fonts, type size)

> "While a request letter conveys a student's need, a thank-you note conveys his or her maturity."
> —Professor Rhoda

Thank You and Update to Reference

Rhetorical situation

Purpose: To thank the reader for help given and describe the status of the job search
Audience: Person who helped you in the job-search process
Format: E-mail, memo, or letter

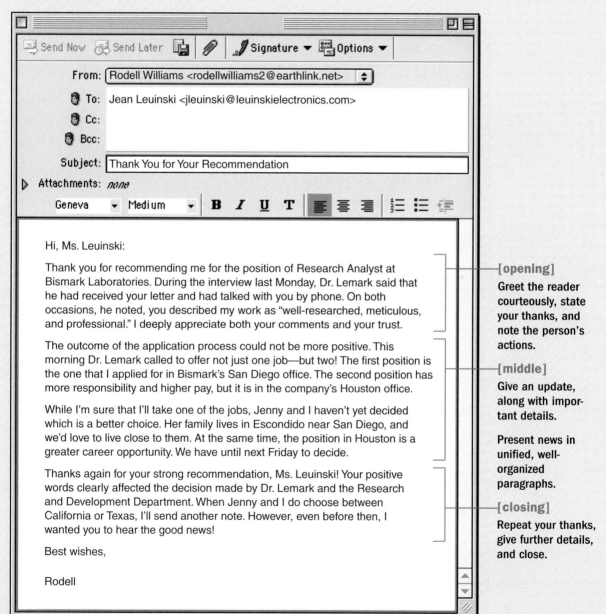

From: Rodell Williams <rodellwilliams2@earthlink.net>
To: Jean Leuinski <jleuinski@leuinskielectronics.com>
Subject: Thank You for Your Recommendation

Hi, Ms. Leuinski:

Thank you for recommending me for the position of Research Analyst at Bismark Laboratories. During the interview last Monday, Dr. Lemark said that he had received your letter and had talked with you by phone. On both occasions, he noted, you described my work as "well-researched, meticulous, and professional." I deeply appreciate both your comments and your trust.

The outcome of the application process could not be more positive. This morning Dr. Lemark called to offer not just one job—but two! The first position is the one that I applied for in Bismark's San Diego office. The second position has more responsibility and higher pay, but it is in the company's Houston office.

While I'm sure that I'll take one of the jobs, Jenny and I haven't yet decided which is a better choice. Her family lives in Escondido near San Diego, and we'd love to live close to them. At the same time, the position in Houston is a greater career opportunity. We have until next Friday to decide.

Thanks again for your strong recommendation, Ms. Leuinski! Your positive words clearly affected the decision made by Dr. Lemark and the Research and Development Department. When Jenny and I do choose between California or Texas, I'll send another note. However, even before then, I wanted you to hear the good news!

Best wishes,

Rodell

[opening] Greet the reader courteously, state your thanks, and note the person's actions.

[middle] Give an update, along with important details.

Present news in unified, well-organized paragraphs.

[closing] Repeat your thanks, give further details, and close.

CHECKLIST for Application Correspondence

Use the seven traits to check your document and then revise as needed.

- ☐ **Ideas.** The ideas in the letters show that I understand the job, job-search process, and writing purpose; the ideas in the essay show my qualifications and desire for the job.
- ☐ **Organization.** The job-rejection letter uses indirect organization, while other letters use direct organization. The essay's organization is set by the opening, topic sentences, and closing.
- ☐ **Voice.** The tone is informed, thoughtful, polite, and sincere.
- ☐ **Words.** The document includes precise nouns, strong verbs, and few modifiers (adjectives or adverbs); and job-related terminology is clear.
- ☐ **Sentences.** The sentences are complete, energetic, varied in structure, and clear.
- ☐ **Copy.** The document includes no errors in grammar, punctuation, capitalization, and spelling.
- ☐ **Design.** Letters are correctly formatted in full block or semiblock; the essay has single-spacing between lines, double-spacing between paragraphs, appropriate margins, and white space.

CRITICAL THINKING Activity

Choose one of the organizations that you researched for Critical Thinking activity 2 and Writing activity 1 on page 318, or find a job advertisement in your field and research that organization and job. List details about the company and job, highlighting those worth citing in a job-application letter.

WRITING Activities

1. Based on your assessment of the company and job addressed in the Critical Thinking activity, write an application letter for the job. Glean supporting details from your résumé and from the lists that you developed for the Critical Thinking activity on page 326.
2. Write a recommendation-request letter for the job addressed in Writing activity 1. Address the letter to an individual whom you identified in the Critical Thinking activity on page 326.
3. Write an application essay for the job that you addressed in the previous activities. Use details from your research on the organization, your résumé, and your lists for the Critical Thinking activity on page 326.
4. For the job addressed in the previous activities, write (a) a job-acceptance letter (using direct organization), (b) a job-rejection letter (using indirect organization), and (c) a thank-you/update message.

In this chapter
Interviewing for a Job or Program **342**
Inappropriate or Illegal Questions **344**
Common Interview Questions **346**
Guidelines for Interview Follow-Up Letters **348**
Interviewing a Job Applicant **350**
Checklist for Interview Follow-Up Letters **352**
Critical Thinking Activities **352**
Writing Activities **352**

CHAPTER
20

Participating in Interviews

During your career, you will be interviewed many times, and you will interview others even more often. Whenever you apply for a job, pursue a professional program, seek a promotion, or request a new assignment or project, you will likely be interviewed. Whenever your organization hires someone with whom you will work closely, you will likely help interview the applicants. How you perform when others interview you will affect whether you are chosen for the job, program, promotion, or project. How you perform while interviewing others will affect both the nature and the quality of your work situation.

This chapter will help you perform well on either side of the interview table. First, the chapter explains how to be interviewed, including what documents to study, materials to prepare, and questions to anticipate. Second, the chapter explains how to conduct an interview, including how to organize the event and how to design effective questions.

Study the chapter now to prepare for your class. Review the chapter later when you are preparing to be interviewed for an assistantship, internship, job, or program.

Part 3   The Application Process and Application Writing

INTERVIEWING FOR A JOB OR PROGRAM

While a good application letter (page 329), application essay (page 333), résumé (pages 321–325), and recommendation (page 331) can get you an interview, only a strong interview will actually win you the job or program. To interview well, follow the guidelines below.

Before the Interview

Research the Organization and the Job or Program

- Review information about the job and highlight references to required training, skills, and experiences.
- Using information published by the organization's public relations office or posted on its Web site, learn about the organization's products and services, special awards or accomplishments, number of employees, annual sales, network of offices, and mission statement.
- Check a public or college library for information such as objective analyses of the organization's products, services, management style, and markets.

Possible library sources

One Source (a CD-ROM database of corporate information)
Research Centers Directory
O'Dwyer's Directory of Public Relations Firms
CorpTech Directory
Gale Directory of Publications and Broadcast Media

Think About the Interview

- Review your application letter, application essay, résumé, and recommendations; then highlight items that match the job or program requirements. Consider how to emphasize these strengths.
- Underline requirements that you may not meet, and anticipate questions.
- Review "Common Interview Questions" (page 346), and prepare answers.
- Review "Inappropriate or Illegal Questions" (page 344), and prepare responses to them.

Think About Yourself

Review the following topics to identify your unique qualities and prepare for questions on these or related issues:

Career and educational goals	Tasks you enjoy, dislike, or fear
Talents or skills that you want to use	Interests and hobbies
Talents or skills you want to develop	People you admire
Disappointing workplace experiences	Places you like to visit
Challenging workplace experiences	Articles or books you have read

Plan for the Interview

- Select materials to take to the interview: work samples, audition tape, résumé, and so on.
- Prepare questions that will help you evaluate the employer and the job (see page 347).
- Make travel plans (including means of transportation, routes, and departure time) needed to arrive at the site about one hour early.
- Choose conservative, professional clothing, and do the needed cleaning and pressing (see "Checklist for Appropriate Dress," page 598).
- Snack lightly before the interview so that you feel comfortable and energetic.
- Before you step into the interview, recheck your appearance.

During the Interview

- Arrive at the interview *site* about an hour early so you have time to use the restroom, touch up your appearance, review key issues or questions, recheck your materials, and relax.
- Arrive at the interview *office* ten minutes early, introduce yourself to the secretary, explain why you are there, and complete forms neatly.
- Look the interviewer in the eye, shake hands firmly, and *smile*. (For tips on handling introductions, see "Practicing Workplace Etiquette," pages 594–595.)
- Listen to questions carefully, and answer them directly, positively, and clearly.
- If a question isn't clear, ask that it be clarified.
- Be yourself, but avoid negative behavior such as exaggeration, defensiveness, or vulgar or profane language.
- If your interview includes a meal (most do not), remember that you are being evaluated during the meal.
  - Avoid alcohol—even a little can dull your senses.
  - Order easy-to-eat food that enables you to keep your fingers and clothing clean (see "Eating and Drinking," page 597).
  - Choose an entree that is no more expensive than the item chosen by the host.
  - Ask relevant questions that help you learn about the host and give you time to eat.

After the Interview

- Tell when and where you can be reached.
- If appropriate, offer to send additional information.
- Thank the people involved in the interview.
- Retrieve personal items (e.g., coat, briefcase), say farewell, and leave.
- Write a follow-up letter or e-mail message (see pages 348–349).

INAPPROPRIATE OR ILLEGAL QUESTIONS

The U.S. government has strict guidelines for interviewers regarding *what* questions they may ask, as well as *when* and *how* they may ask them. For example, it is **illegal** for an interviewer to ask a direct or indirect question intended to assess your age.

> **ILLEGAL (DIRECT):** How old are you?
>
> **ILLEGAL (INDIRECT):** Which U.S. president was in office when you were in high school?

However, it is **legal** for the interviewer to "verify" that your age conforms to laws that specify age limits for a specific job.

> **LEGAL (VERIFICATION):** This job requires that you be at least 18 years old. Do you meet this requirement?

Table 20.1 is adapted from the U.S. Department of the Interior's "Interview Questions."

Table 20.1 Interview Questions

Subject	Illegal Questions	Legal If Job Related
Age	How old are you? What key historical events took place while you were in college?	Age may be verified, but only to confirm that an applicant's age conforms to the legal limits set for a specific job.
Citizenship	Are you a U.S. citizen?	Citizenship may be verified, but only to confirm that an applicant's citizenship status conforms to the legal requirements for a specific job.
Economic status	Do you have a good credit rating?	No legal question may be asked on the subject.
Education	What is the national, racial, or religious affiliation of your school?	Questions about an applicant's education are legal only if the information is used to assess his or her qualifications for a specific job.
Marital status/family	Are you married? Divorced? Widowed? With whom do you live? How old are your children? Is your spouse in the military?	No legal question may be asked on these subjects.
Military discharge	Were you honorably discharged?	No legal question may be asked on the subject. Interviewers may verify this information if legally required to do so.
Miscellaneous	Questions that are unrelated to the job or unnecessary for assessing the applicant's qualifications.	The applicant may be notified that statements, misstatements, or omissions of significant facts may lead to his or her not being selected.

Table 20.1 (continued)

Name	Did you ever legally change your name?	Interviewers may ask for the maiden name of a married woman.
National origin	What is your national origin, language spoken in the home, or native language?	No legal question may be asked on the subject.
Organizational affiliation	What organizations do you belong to?	Questions on the subject are legal only if they are related to the job being filled and to the applicant's qualifications.
Clubs or activities	Will you join the Lions' Club?	No legal question may be asked on the subject.
Personal plans	Do you plan to live in this area long?	No legal question may be asked on the subject.
Police record	Have you ever been arrested?	If the job has requirements such as bonding, interviewers may ask, "This job requires that you be bonded. Does this requirement present a problem?"
Race/color	What is your race or national origin?	No legal question may be asked about an applicant's race or color.
Religion	What church do you attend? What religious holidays do you observe?	No legal question may be asked about an applicant's religion.
Security clearance	Do you have a secret, top-secret, or other security clearance for the job?	If the job requires a security clearance, employers may ask, "This job requires a security clearance. Does this requirement present a problem?"
Sex	Do you plan on having children?	No legal question may be asked on the subject.
Work schedule/travel	Questions that relate to child care, ages of children, or other non-job-related issues.	If the job has requirements (e.g., travel, unusual hours), interviewers may ask, "This job requires out-of-town travel one week each month. Does this requirement present a problem?"

F.Y.I.

If you are asked an illegal question, consider these options:

- Ask why the question was asked, and then decide whether to answer it.
- Rephrase the question as you choose, and answer your revised form.
- Say that because the question is illegal, you prefer not to answer it.

COMMON INTERVIEW QUESTIONS

Questions from an Interviewer

Interviewers choose questions not only to learn your answers, but also to evaluate issues related to your answers. Below are ten common interview questions, along with notes indicating which related issues the person may be evaluating. On the following page are questions that you may want to ask the interviewer.

1. How did you learn about this job (or program), and why did you apply for it?
   - Evaluating how you conduct a search and what you expect from the job or program.
2. How do your experiences and training qualify you for this job (or program)?
   - Evaluating your assessment of the job's requirements and your ability to meet those requirements.
3. Describe one of the job's (or program's) more challenging tasks and how you would deal with it.
   - Evaluating what you consider challenging and how you respond to challenges.
4. Given the job description, how many hours would you use each day (or week) to complete the tasks?
   - Evaluating your ability to envision tasks and manage time.
5. What do you know about this organization's mission, products, and services; and how could you help the organization achieve its goals?
   - Evaluating whether you researched the organization and understand your role in it.
6. Describe one of your more rewarding (or disappointing) work-related or educational experiences, and explain why it was rewarding (or disappointing).
   - Evaluating whether you acknowledge your problems and how you deal with them.
7. Describe a good reading or learning experience, and explain why it was good.
   - Evaluating your commitment to research and learning.
8. Why did you leave your last position?
   - Evaluating your weaknesses or how you make choices.
9. What compensation do you expect for this position?
   - Evaluating whether your goals are realistic and whether you will resolve differences through negotiation.
10. Do you have any questions for us?
    - Evaluating your assessment of the job, organization, and personnel.

Questions to an Interviewer

At the conclusion of an interview, the interviewer will commonly ask whether you have any questions. Which questions you should ask depend on the job, program, organization, recruitment or interview process, and information that you have already received. Use the guidelines below to design appropriate questions.

Interviewers commonly give applicants information on the following topics: job description, hours, wages, benefits, opportunities for advancement, training programs, research opportunities, and special requirements like travel. You are welcome to ask for additional details about each topic.

- How often will I be asked to travel, and how many days will I be gone?
- What special training have people in my position requested and received?
- The company now offers full health benefits. How might that policy change and when?

Ask questions that help you understand the organization's strengths and structure.

- What is the organization's market share, and how might that situation change in the next three years?
- How does the company use the Internet to market its products? Will that change soon, and how will the change affect my assignment?
- Could you describe the organization's (or department's) management style, and how that style would affect my work?

Ask questions about your assignment.

- Could you describe one or two future projects to which I may be assigned?
- What are the chances that I'll be reassigned to a branch office? Which office could that be, and what support would I receive for the transition?
- Who will be my co-workers, and will those assignments change in the next year?
- What secretarial and technical support will I have?
- What office equipment and computer software do you provide?

Ask about personnel issues.

- How will my work be evaluated, how often, and by whom?
- What is your employee turnover rate?

Request a tour of the facility, and ask employees questions like these:

- What does your job entail, and how do you like it?
- What equipment do you use, and does it work well?
- How do you like the company's health plan, work schedule, or development programs?

GUIDELINES FOR INTERVIEW FOLLOW-UP LETTERS

After a job interview, it is both polite and prudent to write a follow-up letter or e-mail message. When drafting this document, your goal is to thank the employer, offer more information, and show that you're interested in the position.

1 Prewrite

Rhetorical analysis: Consider your purpose, your audience, and the context.
- What is your reader's role in the hiring process?
- How can you show your appreciation for the interview and your treatment?

Prepare to draft.
- Write a one-sentence summary of your qualifications.
- List interview highlights: people, setting, job responsibilities, and other topics discussed.
- List topics not covered in the interview that could help you get the job.

2 Draft

Organize the letter into three parts.

[opening] Establish a natural but respectful tone:
- Address the reader using his or her title.
- Thank the person for the interview and other specific courtesies.

[middle] Show your qualifications and continued interest:
- Note the job's key tasks, and briefly state how your training and experience enable you to do those tasks.
- Identify one or two key departmental or company goals, and state why you will be able to help achieve those goals.

[closing] Offer to provide additional information or (if appropriate) to pursue further training.

3 Revise

Review the draft for ideas, organization, and voice.
- Have you addressed relevant issues in a logical order?
- Have you supported main points with clear, correct details?
- Have you used a confident and polite tone?

4 Refine

Check your writing line by line for the following:
- Correct titles, names, and dates
- Correct spelling, punctuation, and grammar
- Attractive and appropriate layout, font size, and white space

Chapter 20 Participating in Interviews 349

Interview Follow-Up Letter

Rhetorical situation

Purpose: To thank the reader for the interview and show interest in the job
Audience: The director of human resources
Format: Business letter formatted in full block and printed on quality paper

In this case, the writer initiated the interview with her application letter shown on page 329.

3041 45th Avenue
Lake City, WA 98125
December 2, 2005

Ms. Marla Tamor
Human Resources Director
Evergreen Medical Center
812 University Street
Seattle, WA 98105

Dear Ms. Tamor:

Thank you for the opportunity to discuss the position of Software-Training Specialist yesterday afternoon. I enjoyed meeting you and the staff at Evergreen Medical Center and found our visit very informative. —[opening] **Thank the reader for the interview.**

After touring your facilities and talking with your office managers, I am certain that I would enjoy working at Evergreen Medical Center. I was impressed by your staff's eagerness to strengthen their technical skills, and I believe that my hospital-training experience would make me an asset to your team. In addition, I found most of your office equipment current and in good repair. —[middle] **Note why you could do the job well.**

Thank you again for the interview and for considering my qualifications. If you have other questions, I am available at (206) 555-0242 or at jmvrtz@aol.com. —[closing] **Thank the reader and give contact information.**

Sincerely,

Jamie Vertz

Jamie Vertz

INTERVIEWING A JOB APPLICANT

During your career, you will often be asked to help interview a job applicant. Use the process to measure his or her qualifications, skills, personality, and ability to work with others—including yourself.

Planning for the Interview

1. Review the job's requirements and compensation:
   - Schedule and tasks (particularly special duties such as lifting or traveling).
   - Skills, experience, and certifications
   - Wages, benefits, training programs, and career track
2. Read information about the candidate:
   - Application letter, résumé, and recommendation letters
   - Sample work or projects
3. Study the person's work record (jobs held, reasons for leaving), and write interview questions on issues that you want clarified (see tips on page 351).
4. If you are coordinating the interview, prepare information for the applicant:
   - A letter describing the interview schedule, people attending, required tests and work samples, travel and hotel arrangements, and so on
   - Handouts describing the job, compensation, and your organization
5. If you are leading the interview:
   - Prepare the schedule, reserve interview rooms, and make arrangements for refreshments or meals.
   - Inform the interview team about the schedule and where they can review application materials.

Communicating Clearly

1. Establish a positive atmosphere:
   - Welcome the applicant and introduce the interview team and schedule.
   - Invite the applicant to "tell us about yourself..."
2. Follow the introduction with questions from the team.
3. Listen carefully, noting how the applicant develops answers.
4. Invite the applicant to ask questions.

Following Up

1. Thank the applicant and explain how the selection process will proceed.
2. Record the interview team's responses.
3. Arrange for the evaluation of candidates.
4. Send follow-up communication (see guidelines and model on pages 348–349).

Designing Effective Questions

The job interview should be a friendly, issues-oriented conversation that helps you determine how well the applicant matches your hiring goals. To shape such a conversation, design questions that follow the guidelines below.

To learn what a candidate thinks, ask questions that are open-ended rather than "leading" or obvious.

> **POOR:** Is this job attractive to you?
>
> **GOOD:** Why is this job attractive, and how does it match your skills and experience?

To learn how a candidate will do the job, ask how he or she would complete a task.

> **POOR:** You have seen a copy of our newsletter. Can you produce this publication?
>
> **GOOD:** How would you produce our monthly newsletter in the time allotted?

To learn how well a person is qualified for the job, ask when and where he or she has used the required skills, experience, and training.

> **POOR:** Can you give presentations to large groups?
>
> **GOOD:** Tell us about a presentation that you've given to a large group. [Follow-up questions could ask about the content, preparation process, use of computer-driven visuals, or methods for dealing with a hostile audience.]

To assess an applicant's ability to work with others, ask about his or her work record.

> **POOR:** Tell us how you get along with colleagues at your present job.
>
> **GOOD:** Describe a recent project that you coordinated, and explain the steps you took to build your colleagues' commitment to the project.

To assess an applicant's long-term commitment to a project or your organization, ask about the person's record or goals.

> **POOR:** If we hire you, how long will you stay with us?
>
> **GOOD:** Describe your training and how it matches our department's development plan.

To avoid misleading or illegal questions, ask how the person would perform tasks that are part of the job. Do not pose hypothetical cases.

> **POOR:** If you had 15 assistants and 1,000 cartons of widgets to deliver, how would you delegate the work?
>
> **GOOD:** Describe one or two experiences that you've had in delegating work, and include details about the situation, workload, people involved, and results.

To assess an applicant's understanding of and commitment to ethical business practices, ask specific, objective questions.

> **POOR:** Do you think ethics are important in business?
>
> **GOOD:** Could you describe an ethically challenging business experience and how you addressed the problem?

CHECKLIST for Interview Follow-Up Letters

Use the seven traits to check your letter and revise as needed.

- ☐ **Ideas.** The letter shows that I understand and appreciate the employer, job, interview process, and my treatment during the process.
- ☐ **Organization.** The message is direct, thanking the reader for the interview, describing highlights in the process, and explaining how I can contribute to the department or company.
- ☐ **Voice.** The tone is respectful, lively, positive, and genuine—not flattering or unduly complimentary.
- ☐ **Words.** The message is delivered in common, but precise words. Job-related terminology is clear, and names and titles are correct.
- ☐ **Sentences.** The letter includes complete, energetic sentences that are varied in structure and read smoothly. Appropriate transitions link paragraphs.
- ☐ **Copy.** The letter includes no errors in grammar, punctuation, capitalization, and spelling.
- ☐ **Design.** The message is balanced on the page, includes adequate white space, and correctly follows a full-block or semiblock format.

CRITICAL THINKING Activities

1. Imagine that you have a job and are interviewing someone who is applying to be your co-worker. List five qualities that you want to evaluate in this person, and draft appropriate questions that will help you assess each quality.
2. Share with the class your answer for Critical Thinking activity 1. Explain why you think each quality is important and how your questions will help you assess these qualities.

WRITING Activities

1. Review the application letter that you wrote for Writing activity 1 on page 340, and imagine that you are preparing to interview for that job or another job. Draft ten questions that you may be asked, along with your answers.
2. Imagine that during the same interview, you were asked all of the inappropriate or illegal questions on pages 344–345. Write your answer to each question.
3. Review the application letter that you wrote for Writing activity 1 on page 340. Then imagine that your interview has taken place, and write an interview follow-up letter.

PART 4

Correspondence: Memos, E-Mails, and Letters

21 Correspondence Basics

22 Writing Memos

23 Writing E-Mail Messages and Sending Faxes

24 Writing Letters

25 Writing Good-News and Neutral Messages

26 Writing Bad-News Messages

27 Writing Persuasive Messages

28 Writing Form Messages

In this chapter

Writing Successful Correspondence **356**
E-Mail, Memo, or Letter: Which Should It Be? **357**
Three Types of Messages **358**
Correspondence Catalog **361**
Checklist for Correspondence Basics **362**
Critical Thinking Activities **362**
Writing Activities **362**

CHAPTER
21

Correspondence Basics

Doing business requires that you produce goods or services, and writing correspondence (e-mail messages, memos, and letters) is the medium through which you conduct much of that work. Whether you're developing a product, producing it, marketing it, or maintaining it, you have to communicate with people both inside and outside your organization. Writing helps you carry on that correspondence.

The key, of course, is corresponding in ways that help—rather than hurt—your organization, other people, and yourself. To do so, you must know how to analyze your writing situation; to assess whether your message is neutral, good news, bad news, or persuasive; to decide whether you should use direct or indirect organization; and to choose whether you should design the message as an e-mail, memo, or letter. In addition to making those decisions, you must write and send the message—activities that require many other choices and skills.

Part 4 of this book will help you understand those choices and develop those skills. For example, this chapter addresses core issues such as the principles of correspondence, the three basic types of correspondence, and ways to organize and find examples of each type. Later chapters give more detailed instructions, including guidelines and models for forty-seven forms of correspondence.

Work carefully through Part 4, asking regularly how the information could help an organization produce its goods or services.

Part 4 Correspondence: Memos, E-Mails, and Letters

WRITING SUCCESSFUL CORRESPONDENCE

To write successful correspondence analyze the situation and make smart choices.

Write for Good Reasons

Write—versus speak—in situations like these:

- Your message is complicated and detailed—readers need to see it on paper.
- You or your reader needs a written record—for example, when you make an official request, respond to a complaint, or accept a proposal.
- You have a large audience (in scattered locations)—writing offers efficiency.
- You do not need immediate feedback, voice contact, or body language.

Think About Your Goal and Focus on Your Readers

To gain your reader's positive response and long-term goodwill, state your goal clearly and be specific about what you want to happen. Consider each reader's knowledge, personality, needs, and attitudes.

Organize Your Message into Three Parts

The opening, middle, and closing should hold answers to the reader's key questions:

Why are you writing me? **Opening:** State the reason for writing and provide appropriate background.

What's it all about? **Middle:** Give supporting points and details.

What's next? **Closing:** End with who should do what and when, including information about making future contact.

Follow the Guidelines for Specific Forms

See memos (page 363), e-mail messages (page 369), and letters (page 377). For specific types of messages (e.g., routine responses, sales messages), rely on the guidelines and models supplied for good-news and neutral messages (page 397), bad-news messages (page 425), and persuasive messages (page 451).

Make Your Message Positive, Clear, and Prompt

Practice these tips:

- Address a person whenever possible, not a title or a department.
- Use an informative subject line or opening sentence.
- Avoid business jargon. Be conversational instead.
- Whenever possible, stress benefits for your reader (e.g., savings, help).
- Use *you* in positive situations, but avoid it in negative ones.
- Use a team approach (a *we* focus); avoid a *me versus you* feeling.
- Use lists, short paragraphs, headings, and white space for added accessibility.
- Send correspondence promptly—but never write in anger.

E-MAIL, MEMO, OR LETTER: WHICH SHOULD IT BE?

When deciding whether a message should be an e-mail, a memo, or a letter, remember that the medium you choose is part of your message. To choose your medium, consider the factors and advice given below.

The Medium: Issues to Consider

- **Cost.** Will e-mail or paper be more efficient and cost-effective?
- **Timing.** When should the message arrive? When do you need a response?
- **Seriousness.** How weighty is the topic? How important is this message?
- **Privacy.** How sensitive are the situation and the message?
- **Recordkeeping.** How important is a paper or an electronic record?
- **Readership.** What are the reader's needs? Which medium should be used for the reader's convenience? Is the reader inside or outside your company?
- **Impact of medium.** How will your readers respond to a given medium?
- **Company protocols.** What practices or procedures are expected in your company? When is e-mail acceptable over paper memos and letters?
- **Distribution.** How far or widely must the message travel?
- **Attachments.** Do attachments or enclosures go with the message? Are they paper or electronic, and what would be best?

Best Uses for Each Medium

E-Mails

Use e-mail primarily for routine, day-to-day business messages in these situations:

- Speed is a concern—both in sending and in receiving feedback.
- Wide distribution is needed.
- Electronic copies of the message and attachments are needed or valued.
- The reader is comfortable with the technology and software involved.
- Your company culture encourages the use of e-mail internally and externally.

Avoid e-mail primarily for highly serious, sensitive, or official messages as well as those requiring paper attachments.

Memos

Use a memo with internal documents that require some weight and authority.
Avoid a memo when the topic is routine business that can be handled more efficiently by e-mail, or when writing to someone outside your organization.

Letters

Use a letter primarily with external audiences (e.g., customers) when you need to send an official, signed company message.
Avoid a letter generally when it's an internal, routine message.

THREE TYPES OF MESSAGES

Written correspondence can take the form of good-news or neutral messages (e.g., a routine request or inquiry); bad-news messages (e.g., rejecting a proposal or application); or persuasive messages (e.g., special requests or sales letters). The category depends on the reader's likely response to your main point.

Good-News or Neutral Messages: Direct Approach

If your reader is likely to respond to your message as good news (or at least neutral news), be direct and get to your main point quickly.

Rhetorical situation

Purpose: To supply the information that the reader needs to act, get involved
Reader: College student motivated to volunteer time and energy
Format: Letter (official reply to an inquiry)

HOME BUILDERS

1650 Northwest Boulevard • St. Louis, MO 63124
314-555-9800 • FAX 314-555-9810 • www.homebuilders-stl.org

February 15, 2005

Philip Tranberg
1000 Ivy Street
St. Louis, MO 63450

Dear Philip:

[opening] *Open with goodwill and mention the context.*
Thank you for your interest in Home Builders. I appreciate hearing from a college student eager to help us provide decent, affordable housing.

[middle] *State the main point and follow up with details.*
As you requested, I have enclosed a list of Home Builders affiliates in Missouri. Each affiliate handles its own work groups and schedules. Because you are in college, I have also enclosed a brochure on Home Builders Campus Chapters. It shows you how to join or start a Campus Chapter and explains service learning for academic credit. Feel free to contact Ben Abramson, Campus Outreach Coordinator for your area, at the address on the enclosed material.

[closing] *Focus on future contact and action.*
Again, thank you for your interest in becoming a partner with Home Builders.

Sincerely,

Matthew Osgoode

Matthew Osgoode

Enclosures

Bad-News Messages: Indirect, "Soften the Blow" Approach

If your reader will likely be unhappy, angry, disappointed, or hurt to read your message, then be indirect and gradually build up to the main point—the bad news.

Rhetorical situation

Purpose: To inform the reader of dropped insurance coverage while offering alternatives and encouraging continued business
Reader: Teenage customer who has not been driving long
Format: Letter (creates official record and precedes conversation)

Wright Insurance Agency

3406 Capitol Boulevard, Suite 588
Washington, DC 20037-1124
Phone 612-555-0020
wright@insre.com

January 15, 2005

Ms. Virginia Beloit
32 Elias Street
Washington, DC 20018-8262

Dear Ms. Beloit:

Subject: Policy 46759

As your insurance agent, I need to update you on the status of your automobile insurance. —[opening] Start with a buffer, a neutral statement of the context.

To keep premiums low, some insurance companies refuse to cover high-risk drivers. In the past two years, you have had an at-fault accident and four moving violations. As a result, your current auto-insurance carrier, Greenwood Fire & Auto, has chosen not to renew your policy (#46759) effective February 15, 2005. However, I have found three companies that would be willing to offer you coverage. —[middle] Give reasons for the bad news. Then state it and follow up with options.

Please call me at (612) 555-0020 so that we can determine the best way to meet your auto-insurance needs. —[closing] Close by focusing on future contact and action.

Sincerely,

Eric Wright

Eric Wright

Persuasive Messages: Indirect-Argument Approach

When your reader may be indifferent or even resistant to your message, be indirect and build a convincing argument before making your main point.

Rhetorical situation

Purpose: To convince the reader to support arts in the community by joining the Purchase Awards Program
Reader: Local small-business person who recently opened a downtown store
Format: Letter (makes request official, formal, and weighty)

HANCOCK COUNTY ARTS COUNCIL
327 South Road • Bar Harbor, ME 04609-4216 • (207) 555-1456

March 15, 2005

Ms. April Wadsworth
Belles Lettres Books
The Harbor Mall
Bar Harbor, ME 04609-3427

Dear Ms. Wadsworth:

[opening] — *Create a context and gain the reader's attention.*

Does your store or office have a bare wall or corner that needs a painting, photograph, or sculpture?

[middle] — *Follow up with ideas and details to build a persuasive argument that focuses on benefits to the reader.*

For 14 years, your Hancock County Arts Council has sponsored ArtBurst—a fair where artists display and sell their work. Last year, ArtBurst attracted more than 90 artists and 15,000 visitors. This year, the fair will be held in Central Park on Saturday, May 7.

The Purchase Awards Program, supported by local businesses, is key to ArtBurst's success. Business people like you join the program by agreeing to purchase artwork. Your commitment helps us attract better artists and more visitors. And you personally win in two ways. First, you get a beautiful print, painting, drawing, photograph, or sculpture of your choice to decorate your business. Second, all ArtBurst publicity materials advertise your business.

[closing] — *State the main point, ask the reader to accept it, make action easy, and address any obstacles.*

So please join the Purchase Awards Program! Just complete the enclosed form and return it in the postage-paid envelope by April 12.

Yours sincerely,

Lawrence King

Lawrence King, Director

P.S. An Arts Council member can help you choose the right artwork for your business. Call for details.

CORRESPONDENCE CATALOG

Use Table 21.1 to (1) identify what type of message you're writing, (2) find helpful guidelines and models, and (3) select an approach.

Table 21.1   Correspondence Catalog

Type	Guidelines and Models	Approach
Good-News or Neutral Message The reader is receptive, positive, accepting.	Informative message (400) Routine inquiry or request (406) Positive response (410) Placing an order (414) Accepting a claim (416) Goodwill message (418) Strong employee recommendation (642) Applicant follow-up letters (338) Interview follow-up letter (348) Job-offer letter (635) Job-offer acceptance letter (335)	**Direct** Put the main point up front.
Bad-News Message The reader is disappointed, angry, hurt.	Refusing a favor or a request (428) or an inquiry (430) or a claim (433) or an application (434, 634) Rejecting an idea, a proposal, or a bid (436) Explaining errors, changes, or problems (440) Making claims or complaints (446) Resigning from a position (444) Employee recommendation with reservations (643) Job-offer rejection letter (336)	**Indirect** Start with a buffer and build up to the bad news.
Persuasive Message The reader is cautious, skeptical, resistant, indifferent.	Special request, inquiry, or promotional message (454) Sales message (460) Collection letter (466) Requesting a raise or a promotion (470) Application letter (328) Recommendation-request letter (330)	**Indirect** Construct an argument that leads to the point you want accepted.

F.Y.I.

All your correspondence writing must be persuasive in that it must convince the reader that your message is honest, worth reading, and worthy of the response requested.

CHECKLIST for Correspondence Basics

Use this checklist to evaluate and revise your correspondence.

- ☐ **Ideas.** The content is strong, clear, and accurate. It answers the reader's questions: Why are you writing? and What needs to be done?
- ☐ **Organization.** The structure is appropriately direct or indirect, based on the reader's likely response. It contains an informative subject line, and has an appropriate opening, middle, and closing.
- ☐ **Voice.** The tone is courteous throughout and focuses on the reader's needs.
- ☐ **Words.** Terms are precise, clear, and understood by the audience. Technical words are used carefully and defined where necessary.
- ☐ **Sentences.** Structures are short to medium in length; pass the "read aloud" test; and include transitions to link ideas.
- ☐ **Copy.** The writing contains no errors in grammar, spelling, mechanics, or usage.
- ☐ **Design.** The memo, e-mail, or letter follows all appropriate formatting rules such as spacing, margins, and alignment; contains short paragraphs with proper spacing; uses headings and bulleted or numbered lists wherever helpful; has a polished look that includes white space, clean typography, and good stationery; and includes initials, signatures, and attachments, if appropriate.

CRITICAL THINKING Activities

1. Review "E-Mail, Memo, or Letter: Which Should It Be?" (page 357). Based on this information, (a) review five pieces of business correspondence that you have written or received, and (b) explain why each of the five should or should not have been sent in the medium used.
2. Review "Correspondence Catalog," (page 361), noting the types of messages listed. Select one type from each category and study the designated guidelines and models. As you read, decide why the type is or is not correctly designated as (1) good news or neutral, (2) bad news, or (3) persuasive.
3. Review the seven guidelines listed in "Writing Successful Correspondence" (page 356). Is this advice relevant for the correspondence that you will have to write as part of your career? Why or why not?

WRITING Activities

1. Your college's recruitment office has asked you to share your experiences with potential students. Analyze what kind of message you will be writing. Then select an appropriate medium (memo, e-mail, letter) and draft the message.
2. Identify a business or other organization that interests you. Draft an appropriate piece of correspondence aimed at collecting information on this organization.

In this chapter

Guidelines for Memos 364
Basic Memo 365
Expanded Memo 366
Checklist for Memos 368
Critical Thinking Activity 368
Writing Activity 368

CHAPTER

22

Writing Memos

What is a memo (or memorandum)? When should you write one? Should all memos follow the same format? How is a memo different from an e-mail message or a letter? While these questions are addressed more fully later in this chapter and in Chapters 21–24, brief answers to the questions follow.

Essentially, a memo is a form of in-house correspondence in which the opening (listing the date, reader's name, writer's name, and subject) is similar to the opening of an e-mail; the body of the memo (developing the message) is similar to the body of a letter. But because a memo's opening looks like a list, and its body often uses lists rather than paragraphs, the format seems less personal than the format of a letter.

Whether they're distributed via interoffice mail, fax, or e-mail, memos are a useful medium for correspondence, particularly when you have the following goals:

- To remind others of deadlines, dates, and duties
- To create a record for yourself or for your reader
- To receive help or feedback
- To recommend or explain changes in policies and procedures
- To reinforce goals, opportunities, and achievements
- To report information on projects

Use this chapter to learn about the parts of a memo and to practice your memo-writing skills. Then work through Chapters 23 and 24 to learn similar information and skills regarding e-mails and letters.

GUIDELINES FOR MEMOS

When writing a memo, your goal is to make your point quickly, clearly, and effectively. If possible, keep your memo to a single page.

1 Prewrite

Rhetorical analysis: Consider your purpose, your audience, and the context.
- ☐ Why are you writing? What outcome (understanding, decision, action) do you want?
- ☐ Who are your readers? Why do they need this memo? What will their response and attitude be to the topic and your main point?
- ☐ When and how should the memo be sent?

Prepare to draft.
Gather necessary facts, figures, and attachments. Brainstorm details that need to be included, and make a list of your main points.

2 Draft

Organize your memo clearly.
[opening] Type *Memo* or *Memorandum* at the top of the page and complete the memo heading (name only one subject on the subject line).
[middle] Provide details that answer readers' questions: What is this memo about? What does it mean to me? Why is it important?
[closing] Clarify any action needed, especially who is responsible for what.
Note: For good or neutral news, put your key point in the subject line and introduction. For bad-news or persuasive memos, use a neutral subject line and then build to your main point.

3 Revise

Review your draft's ideas, organization, and voice.
- ☐ Have you given clear, accurate information?
- ☐ Have you arranged your points logically for the reader's understanding?
- ☐ Have you used a team attitude, focusing on company goals?
- ☐ Have you used a polite, professional voice?

4 Refine

Check your writing line by line for the following:
- ☐ Precise word choice, smooth sentences, and effective transitions
- ☐ A reader-friendly format and design that effectively uses lists, headings, tables, boxes, and white space
- ☐ Correct grammar, punctuation, and spelling

BASIC MEMO

Rhetorical situation

Purpose: To convince the reader that an employee's promotion is warranted and feasible
Audience: The director of human resources, with decision-making authority
Format: Memo (formalizes request and creates written record)

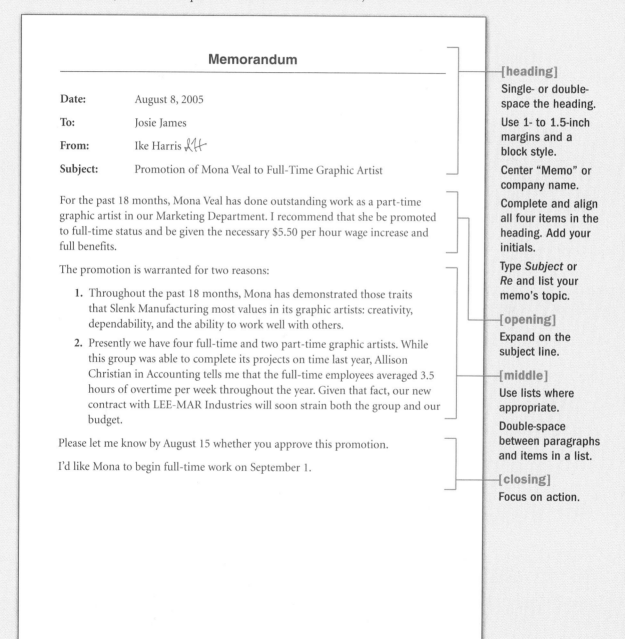

Memorandum

Date: August 8, 2005
To: Josie James
From: Ike Harris *IH*
Subject: Promotion of Mona Veal to Full-Time Graphic Artist

For the past 18 months, Mona Veal has done outstanding work as a part-time graphic artist in our Marketing Department. I recommend that she be promoted to full-time status and be given the necessary $5.50 per hour wage increase and full benefits.

The promotion is warranted for two reasons:

1. Throughout the past 18 months, Mona has demonstrated those traits that Slenk Manufacturing most values in its graphic artists: creativity, dependability, and the ability to work well with others.
2. Presently we have four full-time and two part-time graphic artists. While this group was able to complete its projects on time last year, Allison Christian in Accounting tells me that the full-time employees averaged 3.5 hours of overtime per week throughout the year. Given that fact, our new contract with LEE-MAR Industries will soon strain both the group and our budget.

Please let me know by August 15 whether you approve this promotion.

I'd like Mona to begin full-time work on September 1.

[heading]
Single- or double-space the heading.
Use 1- to 1.5-inch margins and a block style.
Center "Memo" or company name.
Complete and align all four items in the heading. Add your initials.
Type *Subject* or *Re* and list your memo's topic.

[opening]
Expand on the subject line.

[middle]
Use lists where appropriate.
Double-space between paragraphs and items in a list.

[closing]
Focus on action.

EXPANDED MEMO

While each memo includes the basic elements shown on page 365, sometimes you may need to add more elements. The guidelines below, which are numbered and modeled on the next page, show your options.

Heading

1. You can type the word *Memo* or the company's name at the top, but do not include the company's address or phone number.
2. For sensitive messages, label your memo as *confidential* and seal it in an envelope that is also marked *confidential*.
3. Complete your heading with job titles, phone numbers, e-mail addresses, or a checklist showing the memo's purpose. Initial the memo after your name in the heading or after your job title, if one is used. If you have more than one reader, use one of these options:
    - List the names after *To:* and highlight a different one on each memo copy.
    - Put *See distribution* after *To:* and list all of the readers at the end of the memo.
    - Type a department's name after *To:*.

Closing

4. Use quick-response options such as checklists, fill-in-the-blanks, or boxes.
5. Add an identification line showing the initials of the person who wrote the memo (in caps) and the person who typed it (in lowercase), separated by a slash.
6. If you're sending documents with the memo, type *Attachment(s)* or *Enclosure(s)*, followed by either (a) the number of documents or (b) a colon and the document titles listed vertically.
7. If you want to send copies to secondary readers, type *c* or *cc* and a colon; then list the names and job titles stacked vertically (when job titles are included). To send a copy to someone without the main reader knowing it, add *bc* (blind copy) only on the copy sent to the person listed after this notation.

F.Y.I.

If your memo is longer than one page, carry over at least two lines of the message onto a plain sheet of stationery. Use one of the heading formats shown below.

Ike Harris   Page 2   8 August 2005

Page 2
Ike Harris
8 August 2005

Model Expanded Memo

1 **Slenk Manufacturing**

Memorandum

2 **CONFIDENTIAL**

    Date: August 8, 2005

3   **To:**   Josie James, Director of Personnel
              Rebecca Tash, LAHW Representative

    From: Ike Harris, Graphic Arts Director *IH*

    Subject: Promotion of Mona Veal to Full-Time Graphic Artist

For the past 18 months, Mona Veal has done outstanding work as a part-time graphic artist in our Marketing Department. I recommend that she be promoted to full-time status. The promotion is warranted for two reasons:

1. Throughout the past 18 months, Mona has demonstrated those traits that Slenk Manufacturing most values in its graphic artists: creativity, dependability, and the ability to work well with others.

2. Presently we have four full-time and two part-time graphic artists. While this group was able to complete its projects on time last year, Allison Christian in Accounting tells me that the full-time employees averaged 3.5 hours of overtime per week throughout the year. Given that fact, our new contract with LEE-MAR Industries will soon strain both the group and our budget.

If you approve the promotion, please initial below and return this memo.

4   Yes, proceed with Mona Veal's promotion to full-time graphic artist. _____

5   IH/gm

6   Attachment: Evaluation report of Mona Veal

7   cc: Elizabeth Zoe
            Mark Moon

CHECKLIST for Memos

Use this checklist to revise and refine your memo before distributing it.

- ☐ **Ideas.** The memo creates a focused, cohesive message by stating the main point clearly and supplying complete and accurate information.
- ☐ **Organization.** The memo includes a complete heading; has a clear opening, middle, and closing; and is direct or indirect, based on the reader's likely response.
- ☐ **Voice.** The tone is courteous and natural; focuses on reader benefits; is appropriate for the writer's relationship with the reader; and is sensitive to secondary readers inside or outside the company.
- ☐ **Words.** The memo is clearly phrased; uses personal pronouns (especially *you*) positively; includes technical terminology familiar to readers; and defines unfamiliar terms.
- ☐ **Sentences.** The memo contains generally short but varied structures that read easily and contain clear transitions.
- ☐ **Copy.** The memo includes the following:
  - Correct punctuation (a colon follows each title in the memo heading)
  - Correct capitalization (including all major words in the memo heading and names of companies, people, job titles, and products)
  - Correct spelling (particularly names and key terms)
  - Accurate notations (reference numbers and attachments)
- ☐ **Design.** The memo has a properly formatted heading (date, to, from, subject) and effectively uses white space, margins, short paragraphs, headings, lists, bullets, numbers, alignment, and typographical features.

Last reminders

- ☐ Is the memo initialed or signed to make it more official?
- ☐ Are these details covered: attachments, copies, routing, and follow-up?

CRITICAL THINKING Activity

Find a memo written by a business person. Use the checklist for memos to evaluate the memo's content, and use this chapter's guidelines and models to evaluate the memo's form.

WRITING Activity

List organizations to which you have belonged: businesses, nonprofit groups, political parties, community groups, school clubs, and so on. Select one organization, and then (a) develop memo stationery that suits the organization's style and (b) draft a memo alerting readers to an upcoming significant event in the organization.

In this chapter

Guidelines for E-Mail Messages 370
E-Mail Model and Format Tips 371
Choosing and Using E-Mail 372
E-Mail Etiquette and Shorthand 373
Faxing Documents 374
Checklist for E-Mails and Faxes 376
Critical Thinking Activity 376
Writing Activities 376

CHAPTER
23

Writing E-Mail Messages and Sending Faxes

Today's communication technologies make possible different types of workplace correspondence (such as memos, e-mail messages, and letters), as well as many ways to create and send that correspondence. For example, you can send memos or letters through interoffice mail, the U.S. Postal Service, UPS, or FedEx. Alternatively, you can send memos and letters as faxes or as e-mail attachments. You can even post such messages on your company's intranet or on the Web. E-mail itself is perhaps the most powerful and popular medium for corresponding in the workplace today.

These technological options raise many questions: What is an e-mail message? How can you write a strong message that people will actually read, use, and respond to, rather than just delete? How is an e-mail message similar to or different from a memo or letter? When and how should you send e-mail attachments, including memos or letters? With respect to faxes, what protocols should you follow?

As a starting point, you can think of electronic communication—especially e-mail—as lying somewhere between words printed on paper and a conversation. In that sense, faxes and e-mail messages are not as weighty as signed paper and are not as dynamic as speaking. Conversely, both faxes and e-mail are effective media for disseminating information quickly and to many people simultaneously.

Chapter 22 explained what memos are and how to write them, and Chapter 24 gives parallel information about letters. This chapter focuses on e-mail and faxes, answering the questions listed above as well as others.

GUIDELINES FOR E-MAIL MESSAGES

In writing an e-mail message, your goal is to provide clear, concise information quickly and efficiently. Follow the steps on this page before clicking "Send."

1 Prewrite
Rhetorical analysis: Consider your purpose, your audience, and the context.
- Specify what you want your e-mail message to do—the measurable outcome.
- Think about your primary reader's position and possible reaction.
- Consider secondary readers: Who should be copied in on this message? (*Note:* Use the copy function with readers who need to be kept in the information loop.)
- Gather necessary information and arrange your points.

2 Draft
Write with the computer screen in mind.
- Complete the routing information required by your program.
- Give an informative subject line. Don't leave it blank.
- Limit the length of your message to one screen if possible.
- Send longer messages as attachments, pointing to them in the e-mail.

Organize the message into three parts.

[opening] If appropriate, use a greeting to personalize the message. Then state your reason for writing.

[middle] Provide details that answer readers' questions: What is this message about? What does it mean to me? Why is it important? (Try to restrict each e-mail message to a single topic.)

[closing] Indicate any follow-up needed. Who will be responsible for what? Then close politely.

3 Revise
Review your e-mail's ideas, organization, and voice.
- Have you included all of the necessary information?
- Is the information accurate and well organized?
- Have you used a friendly but professional voice or tone?

4 Refine
Check your e-mail for the following:
- Precise word choice, smooth sentences, and correct grammar.
- Use of short paragraphs, white or "screen" space, lists, and headings—keeping the format simple because the message may appear quite different on the reader's screen.
- Correctness—always reread even simple e-mails before hitting "Send."
- Attachments—have you attached the right files in a format that readers can open?

> "I try to leave out the parts that people skip."
> —Elmore Leonard

E-MAIL MODEL AND FORMAT TIPS

The model below shows a typical e-mail screen and message. As you write your own e-mails, follow these format tips as well:

- Use short paragraphs and lists, with double-spacing.
- Avoid italics, boldface, and special symbols that don't translate well.
- Use an appropriate greeting and a fitting closing.
- In the subject line, clarify the nature of your message and use labels when appropriate (FOR ACTION, FYI, URGENT). *Note:* Use "urgent" sparingly.
- Keep line lengths between 65 and 80 characters, using word-wrap.
- With attachments, consider sending them in plain text (ASCII) or rich text format (RTF). In addition, keep graphics files small enough to be downloaded and opened quickly and easily, or warn readers about the size.

Rhetorical situation
Purpose: To supply details about a meeting and to prompt preparation
Audience: Committee members already familiar with the situation and procedures
Format: E-mail (quickly outlines agenda and supplies attachments)

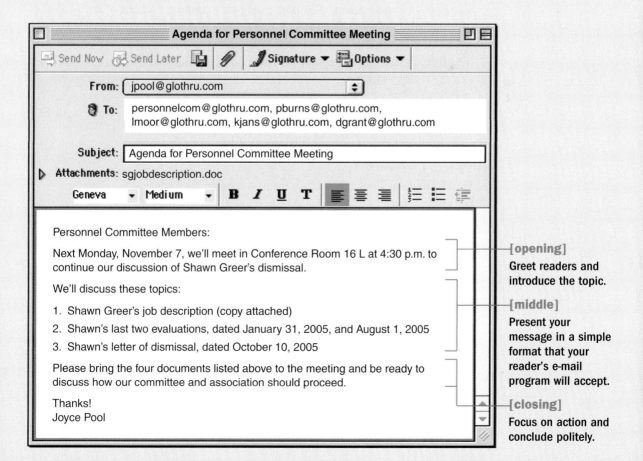

CHOOSING AND USING E-MAIL

E-mail has revolutionized business communication. But when should you choose e-mail over a phone call, letter, or memo? Check the tips below to help you decide.

Strong Points

Simplicity and speed. An e-mail message is composed on-screen through simple key strokes and travels instantly to the reader.

Flexibility and power. With e-mail, you can send, receive, and store messages easily, communicating with both on-site co-workers and far-off customers. In addition, you can embed Web address hyperlinks and attach files.

Clarity. E-mail replies can include the original correspondence, establishing a clear electronic trail.

Economy. Once an e-mail system is set up, maintenance costs are minimal.

Weak Points

Quality and reliability. Because e-mail is easy to use, writers may overload readers with poorly written messages. Also, e-mail systems occasionally malfunction.

Accessibility and respect. Reading e-mail on a computer screen is harder than reading print on paper. Moreover, some readers may look at e-mail as lightweight correspondence—easy to delete or ignore.

Confidentiality. E-mail can sometimes end up in the wrong place, so confidentiality cannot be guaranteed. In addition, your message may be forwarded, so think twice before sending sensitive information by e-mail.

Global issues. Because e-mail can travel around the globe to diverse readers, be sensitive to cultural differences. Avoid slang and colloquialisms.

Special Features

- **The address book** allows easy access to your e-mail addresses.
- **Mailing lists** allow you to distribute e-mail to groups of users.
- **Copy** allows you to send your message to more than your primary readers. **Blind copy** allows you to send your message to someone without the original reader's knowledge.
- **Reply** allows you to respond to an e-mail on-screen—with the option of including the original message. Use cut, copy, and paste with the original message: Don't automatically include the entire message or reply to "all."
- **Forward** allows you to distribute to others a message you have received.
- **A signature file** allows you to automatically attach contact information to your messages. Include your name, organization, department, and so on, stacked vertically.
- **Search** allows you to do a search of saved e-mail messages.
- **Folders** allow you to save, order, and group messages.
- **Attach documents** allows you to send digital files with your message.

E-MAIL ETIQUETTE AND SHORTHAND

How should you behave in the digital world of e-mail? Simply follow the rules of etiquette and use initialisms and emoticons carefully.

F.Y.I.
Emoticons and initialisms are e-mail shorthand symbols that are informal and can easily be misunderstood. Don't use either in business correspondence.

Etiquette

Appropriate use. Use e-mail for group projects, bulletins, routine messages, and immediate follow-up. Generally, avoid e-mail for sensitive issues, serious topics, or bad news. Always follow your company's policies about e-mail use.

Formality level. Use language that is appropriate for your reader. Distinguish between in-house e-mail and messages to people outside your organization.

Message checking. Check your e-mail several times a day. If you can't respond immediately, send a short message to indicate that you received the message and that you will reply by a specific time.

Distribution. Instead of distributing e-mail messages too widely, send them only to people who need them. Otherwise, readers may routinely delete your messages.

Flaming. Use of anger or sarcasm, called flaming (it is often signaled with uppercase), is never appropriate. Cool off and avoid sending an angry message.

Spamming. Avoid sending unsolicited ads by e-mail, a practice called spamming.

Forwarding. Think carefully before forwarding messages. When in doubt, get permission from the original sender.

Ethics. Because companies are legally responsible for their computer network activity, e-mail is company property. In addition, networks typically store messages for years. Only write messages you would not mind everyone seeing.

Emoticons

You make emoticons, often called smileys, with simple keyboard strokes. To get the picture, turn your head to the left.

:) or :-)	A smile
:D	A big smile
:-(	A frown, unhappy
:-O	Shocked or amazed
:-\	Undecided
:-/	Skeptical or puzzled
:-*	Oops!
ll*(	Handshake offered
ll*)	Handshake accepted

Initialisms

Initialisms are abbreviations written in all capital letters with no periods.

OTOH	On the other hand
F2F	Face to face
BTW	By the way
FYI	For your information
IMHO	In my humble opinion
TIA	Thanks in advance
WRT	With respect to
IOW	In other words
FWIW	For what it's worth

FAXING DOCUMENTS

A fax machine sends and receives black-and-white images of documents quickly, conveniently, and cheaply. To fax effectively, follow these tips.

Guidelines

- Fax when confidentiality isn't an issue. Most office fax machines are shared, so fax only routine messages or alert your reader that a fax is coming.
- Fax when your reader needs complex or graphic information (an order, drawing, table). Note that while line drawings transmit well, photos and shaded drawings don't.
- Remember color and size restrictions. Faxes transmit only black, white, and shades of gray. In addition, large documents do not fit through the standard machine. If possible, use a copier's shrink function to create a faxable document.
- Make sure your document is crisply printed in black ink with a type size of 10 to 12 points or larger. Because faxes come out fuzzier than the original, do not retransmit an already-fuzzy fax.
- To highlight material on a fax for your reader, use a circle, arrow, or star. Don't use a highlighter or colored pen.

Cover Sheets

For most in-house faxes, an informal note or cover sheet with your name and extension number may suffice. For formal faxes, use a full cover sheet. Design one with white space, room to write or type clearly, and headings in large, black print. Include some or all of these details, as numbered and modeled on the facing page:

1. Your company's letterhead.
2. The receiver's full name, title, company, and department.
3. The receiver's fax number and phone number.
4. Your name (clearly printed or typed), telephone number, and fax number. Including space for a signature allows the sender to "validate" the transmission as in a formal letter.
5. The date of transmission.
6. Message: a subject line, brief message, or special instructions (e.g., action requested, urgency, confidentiality).
7. The number of pages sent (specifying whether the number includes or excludes the cover sheet).

F.Y.I.

It's a good idea to include on the cover sheet a troubleshooting statement asking the receiver to contact a specific number if a transmission problem occurs, or a polite request to forward the fax or contact the sender if the fax has been misdirected.

Fax Cover Sheet

1

Langland, Hills, and Wooster
ATTORNEYS-AT-LAW

Facsimile Transmission

John H. Langland
Elizabeth L. Hills
Robert D. Wooster

450 Robin Run Road
Las Vegas, NV 89117-2374
(702) 555-8100

2 **To:** Mr. Vincent Malloy
Meridian Management Consultants

3 Phone: 702-555-3815
Fax: 702-555-3818

4 **From:** Elizabeth Hills
Phone: 702-555-8100, ext. 3
Fax: 702-555-8155

5 **Date:** February 12, 2005

6 **Message:** Please review the following documents concerning your real estate transaction and call me.

7 **Pages:** 11 (including cover sheet)

If you encounter any problems during transmission, please call 702-555-8100.

Important: This message is intended only for the use of the individual or entity to which it is addressed. The message is confidential and protected under applicable law, and any distribution or copying of this document is prohibited. If you receive this document by mistake, please notify us by telephone and return the document to the above address through the U.S. Postal Service. Thank you.

CHECKLIST for E-Mails and Faxes

For informal e-mails, do a quick check before clicking "Send." For an important e-mail or fax, examine the message more carefully using this list.

- ☐ **Ideas.** The e-mail or fax contains appropriate information (not confidential or sensitive); has a clear main point and supplies accurate data; refers to any attachments; and specifies any necessary action or follow-up.
- ☐ **Organization.** The e-mail contains a complete heading (or cover sheet for a fax), an informative subject line, and a clear three-part structure with an opening, middle, and closing.
- ☐ **Voice.** The tone is friendly and informal, fitting the relationship with the reader. The message avoids any hint of anger or sarcasm.
- ☐ **Words.** The message includes clear phrasing (personal pronouns used positively), vocabulary understood by readers, and needed definitions.
- ☐ **Sentences.** The message contains short and smooth constructions that read easily, include fitting transitions, and are grouped in unified paragraphs.
- ☐ **Copy.** The message is free of grammatical errors, typos, and factual mistakes.
- ☐ **Design**
  - Faxes use short paragraphs, double-spacing between paragraphs, bulleted or numbered lists, and crisp, large print for clear transmission.
  - E-mail messages use short paragraphs, double-spacing between paragraphs, bulleted or numbered lists, plain text (ASCII) or rich text format (RTF), line lengths of 65 characters or fewer, and the word-wrap feature (not "Enter").

Last reminders

- ☐ Have you double-checked that the reader's e-mail address is correct?
- ☐ Are all of the appropriate readers receiving this message?
- ☐ Have you considered details: copies, blind copies, follow-up, and so on?

CRITICAL THINKING Activity

Use the information in this chapter to assess the quality of three business-related e-mail messages that you have received.

WRITING Activities

1. Using information in this chapter, revise a business-related e-mail that you have written. Submit both the original draft and the revised draft to your instructor.
2. With a partner, use the information in this chapter to draft an "E-Mail and Fax Use Policy Statement" for the workplace. If possible, choose a specific organization with which you are familiar. (For more on policies, see page 614.)

In this chapter

Guidelines for Letters **378**
Professional Appearance of Letters **379**
Basic Letter **380**
Expanded Letter **382**
Letter Formats **384**
Letters and Envelopes **388**
Forms of Address **391**
Checklist for Letters **396**
Critical Thinking Activity **396**
Writing Activity **396**

CHAPTER

24

Writing Letters

A letter typically sends a message from you to someone outside your organization. While occasionally connecting you with colleagues within your company, more often letters link you with suppliers, customers, potential customers, and others affected by your work.

As information vehicles, letters vary greatly in form, content, and delivery methods. For example, a letter's form may be as simple as a handwritten note on personal stationery addressed to one person, or as complex as a semiblock message on company letterhead with multiple readers and several formal notations. Similarly, while the content of a note may be a basic thank-you message, the content of a letter may be a detailed report accompanied by multiple attachments. Finally, while the note may be delivered in an unstamped envelope through interoffice mail, the letter may be delivered in a secure package requiring same-day courier service and the signature of the recipient.

This chapter explains many options for designing a letter's form and completing details for its delivery. Among other things, you will learn the elements of a basic letter and an expanded letter, standard letter formats, U.S. and international postal abbreviations, and 150 courtesy titles and salutations.

Study this chapter now, but don't try to memorize its details. Rather, think through the details, understand them, and complete the end-of-chapter activities to practice using them. Later, when you're working on the content issues addressed in Chapters 25–27 (good-news, bad-news, and persuasive messages), come back to this one periodically to review its details. Together, Chapters 24–27 will help you produce excellent letters, all of which have an appropriate form, content, and delivery method.

GUIDELINES FOR LETTERS

When writing a business letter, your goal is to communicate your message clearly while creating a positive impression of yourself and your organization.

1 Prewrite

Rhetorical analysis: Consider your purpose, your audience, and the context.
- Ask yourself what you want the letter to accomplish. What understanding, decision, or action should result?
- Explore the reader's concerns about, knowledge of, and history with your organization, as well as his or her knowledge of the topic and likely reaction to your message.
- Consider whether you or someone else should "authorize" the letter through a signature; clarify when and how to send the letter.

Plan your message and the letter's format.
- Gather information from files and other appropriate resources.
- Jot down your main points in a logical order—a sequence sensible to the reader.
- Select a full-block, semiblock, or simplified format based on what your company prefers and what would appeal to your readers. (See pages 384–387.)

2 Draft

Organize the body of your letter into three parts.

[opening] State the situation (reason for writing, background).

[middle] Give the full explanation, supporting points, and details. If your message is good or neutral news, make your key point early. For a bad-news or persuasive message, build up to the main point.

[closing] End with a call for action (who should do what, when) and, if appropriate, mention future contact.

3 Revise

Review your draft's ideas, organization, and voice.
- Are all names, dates, and details accurate? Are explanations clear?
- Is information presented in a logical order?
- Do you use a friendly but professional tone tuned into the reader's perspective?

4 Refine

Check your letter line by line for the following:
- Precise wording and positive use of personal pronouns, especially *you*
- Smooth sentences that pass the "read aloud" test
- Correct spelling, especially of people's and companies' names
- Correct grammar, punctuation, usage, and mechanics
- Appropriate format and design—complete letter parts, fitting courtesy elements, balanced type and white space, crisp print on quality stationery

> "Be yourself when you write. You will stand out as a real person among robots."
> —William Zinsser

PROFESSIONAL APPEARANCE OF LETTERS

Before your readers catch a word of your message, they've already read your letter's overall appearance. Use the following guidelines to make a good first impression.

First Impressions
Choose Your Look

Do you want your letter to be traditional and conservative or friendly and contemporary? (See "Letter Formats," pages 384–387.)

Frame Your Letter in White Space

Make your margins 1 to 1.5 inches on the left and right, top and bottom. Create a balanced, open look by centering the message vertically and adjusting the space between parts.

Make Reading Easy

Direct your reader's attention to the message by using sensible type sizes and styles.

- Keep the type size at 10–12 points.
- Choose a user-friendly typeface. Serif type has fine lines finishing off the main strokes of the letters. (This is serif type.) Sans serif type has a block-letter look. (**This is sans serif type.**) Serif typefaces are easier to read; sans serif typefaces work well for headings.
- Avoid <u>flashy</u> and frequent type changes, *as well as overuse of italics and* **boldface.**

Print for Quality

Use a quality printer and avoid any handwritten editing changes. Always print a clean final copy.

Letter Perfect

The weight, size, and color of your paper should be chosen with care.

- **Use 20- to 24-pound bond paper.** Thinner, lighter paper feels weak; thicker, heavier paper feels formal and stiff. The 20- to 24-pound paper folds cleanly, takes ink crisply, and works well in most office machines.
- **Use 8.5- by 11-inch paper.** This size is the standard and files easily. Other sizes (monarch or baronial) may be used for personal correspondence, executive letters, or mass mailings.
- **Use white paper for most letters.** For a softer look, go to cream or ivory. Be careful with other colors. Light, subtle colors mean business (light gray, blue, tan). Bold colors scream, "I'm an ad!"

Letterhead pages, continuation sheets, and envelopes should match in paper weight, size, color, and design.

BASIC LETTER

All letters should include a clear message and information about the writer and the reader. Details for both are listed below and modeled on the facing page.

1. The **heading** provides the reader with a return address. Type the address (minus the writer's name) at the top of the letter. Spell out words such as *Road*, *Street*, and *West*. Omit the address if you are using letterhead stationery.

2. The **date** shows when the letter was drafted or dictated. Write the date as *month, day, year* for U.S. correspondence (August 5, 2004—with a single comma); write *day, month, year* for international or military correspondence (5 August 2004—no comma). Don't abbreviate the month or use a number rather than a word.

3. The **inside address** gives the reader's name and complete mailing address. Type it flush left and include as many details as necessary, in this order:
   - Reader's courtesy title, name, and job title (if the job title is one word)
   - Reader's job title (if two or more words)
   - Office or department
   - Organization name
   - Street address/P.O. box/suite/room (comma precedes *Northeast* or other directional)
   - City, state, ZIP code (or city, province, postal code)
   - Country (if not the United States)

4. The **salutation** personalizes the message. Capitalize all first letters and place a colon after the name. (See "Forms of Address," pages 391–395.)

5. The **body** contains the message, usually organized into three parts:
   - An opening that states why you are writing
   - A middle that gives readers the details they need
   - A closing that focuses on what should happen next

6. The **complimentary closing** provides a polite word or phrase to end the message. Use closings such as *Yours sincerely* or *Best wishes*. Capitalize the first word only and add a comma after the closing.

7. The **signature block** makes the letter official. Align the writer's name with the complimentary closing. Place a job title below the typed name.

8. Use an **enclosure note** when you enclose something with the letter. Type *Enclosure(s)* or *Enc(s)*. Then indicate the number of enclosures or list the enclosures by name, with the names stacked vertically. (See an alternative format for the vertical list on page 527.)

9. A **postscript** contains a personal or final note. Type *P.S.* (with periods but no colon) followed by the message.

Basic Letter Model and Format Tips

This model shows the letter parts described on page 380. Note these tips:

Paragraphing. Single-space within paragraphs; double-space between paragraphs. Do not indent the first line of the paragraph.

Justification. Use left justification and leave the right margin ragged (uneven).

Margins. Set margins at the left and right, top and bottom, at 1 to 1.5 inches. Adjust spacing between parts to center the letter vertically on the page.

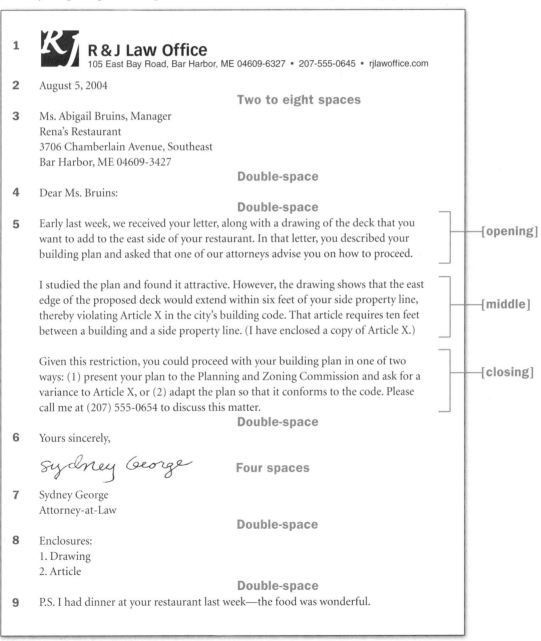

EXPANDED LETTER

When you, your reader, a typist, a filing clerk, or future readers need additional information (more than what's in the basic letter), include one or more of the following items, which are numbered and modeled on the facing page:

1. A **method of transmission note** indicates how a letter should be or has been sent: via facsimile, via registered mail, via overnight courier.
2. A **reference line** begins with a guide word and a colon (*Reference:, In reply to:*) followed by a file, an account, an invoice, or a database number.
3. When appropriate, use a **confidential notation** on both the letter and the envelope. Capitalize or underline the word *confidential* for emphasis.
4. In the **inside address,** if you have two or more readers (not a married couple), stack names by alphabetical order or position. For two readers at separate addresses, stack the addresses (including names) with a line between.
5. The **attention line** designates a reader or department but encourages others to read the letter. Place the notation two lines below the inside address, flush left or centered. Capitalize or underline it for emphasis. An attention line is not shown in the model on the facing page. See page 447 for an example.
6. The **subject line** announces the topic and is placed flush left two lines below the salutation. Capitalize or underline it for emphasis.
7. The **signature block** may include the writer's courtesy title (typed in parentheses before the name) to clarify his or her gender or a preferred form of address. If two people must sign the letter, type the second name beside the first starting at the center of the page—or type the second name four spaces below the first name.
8. In the **identification line,** type the writer's initials in capitals and the typist's initials in lowercase, separated by a slash (but no spaces).
9. Use the **copies notation** by typing *c* or *cc,* followed by a colon and a vertical list of people (with job titles in parentheses). To send a copy to someone else without the reader knowing it, type *bc* or *bcc* (blind copy), but only on the copy sent to the person listed.
10. **Continuation pages** follow a letter's first page. On blank stationery, carry over at least two lines and use a heading in one of the formats below.

10 Abigail Bruins   Page 2   August 5, 2005
    Paul Meyer

                                          10 Page 2
                                              Abigail Bruins
                                              Paul Meyer
                                              August 5, 2005

Expanded Letter Model

 **R & J Law Office**
105 East Bay Road, Bar Harbor, ME 04609-6327 • 207-555-0645 • rjlawoffice.com

August 5, 2005

1. Via facsimile

2. Reference: Article X

3. CONFIDENTIAL

4. Ms. Abigail Bruins
   Mr. Paul Meyer
   Rena's Restaurant
   3706 Chamberlain Avenue, Southeast
   Bar Harbor, ME 04609-3427

Dear Ms. Bruins and Mr. Meyer:

6. BUILDING PERMIT

Early last week, we received your letter, along with a drawing of the deck that you want to add to the east side of your restaurant. In that letter, you described your building plan and asked that one of our attorneys advise you on how to proceed.

I studied the plan and found it attractive. However, the drawing shows that the east edge of the proposed deck would extend within six feet of your side property line, thereby violating Article X in the city's building code. That article requires ten feet between a building and a side property line. (I have enclosed a copy of Article X.)

Given this restriction, you could proceed with your building plan in one of two ways: (1) present your plan to the Planning and Zoning Commission and ask for a variance to Article X, or (2) adapt the plan so that it conforms to the code. Please call me at (207) 555-0654 to discuss this matter.

Yours sincerely,

Sydney George

7. (Ms.) Sydney George
   Attorney-at-Law

8. SG/mbb
   Enclosures 2

9. cc: Leah Theodore (Senior Partner)

LETTER FORMATS

You can arrange a letter in a semiblock, full-block, or simplified format. The specific rules of each are listed in Table 24.1 and modeled on the pages that follow. Choose the letter format that best fits the formality of the situation, your organization's guidelines, and your sense of the reader's preference. Review the "Best Uses" section in Table 24.1 to decide when to use each format.

Table 24.1 Letter Formats

Semiblock Format	Full-Block Format	Simplified Format
Rules		
Date line, method of transmission line, reference line, complimentary close, and signature block align with a vertical line at the center of the page; all other parts of the letter are flush left	All parts flush left	All parts flush left. No salutation or complimentary closing. Subject line and writer's name in capital letters; dash between the writer's name and title
Character		
Professional, traditional	Professional, clean, contemporary	Bare bones, functional
Plus		
Balanced appearance on the page	Easy to set up and follow	Easy to set up
Minus		
More difficult to set up than full block or simplified	May appear unbalanced to the left of the page	Impersonal format due to lack of courtesy elements
Best Uses		
International and traditional letters, as well as executive and social letters	Routine letters where courtesy elements are still important (unlike simplified format), not social and executive letters	Basic, simple notices where courtesy isn't important—regular reminders, bulletins, orders, mass mailings; not appropriate for high-level or persuasive letters
Note		
You may indent the subject line and all paragraphs to further soften the form. In addition, you may drop the space between paragraphs.	More traditional and international readers may not prefer this format.	You may drop courtesy titles from the inside address.

Semiblock Format

R & J Law Office
105 East Bay Road, Bar Harbor, ME 04609-6327 • 207-555-0645 • rjlawoffice.com

August 5, 2005

Ref. A. Bruins #2

CONFIDENTIAL

Ms. Abigail Bruins
Mr. Paul Meyer
Rena's Restaurant
3706 Chamberlain Avenue, Southwest
Bar Harbor, ME 04609-3427

Dear Ms. Bruins and Mr. Meyer:

BUILDING PERMIT

Early last week, we received your letter, along with drawings of the deck that you want to add to the east side of your restaurant. In that letter, you described your building plan and asked that one of our attorneys advise you on how to proceed.

I studied the plan and found it attractive. However, the drawings show that the east edge of the proposed deck would extend within six feet of your side property line, thereby violating Article X in the city's building code. That article requires ten feet between a building and a side property line.

Given this restriction, you could proceed with your building plan in one of two ways: (1) present your plan to the Planning and Zoning Commission and ask for a variance to Article X, or (2) adapt the plan so that it conforms to the code. Please call me at (207) 555-0654 to discuss this matter.

Yours sincerely,

Sydney George

(Ms.) Sydney George
Attorney-at-Law

SG/mbb
Enc.: Article X
cc: Leah Theodore

Full-Block Format

R & J Law Office
105 East Bay Road, Bar Harbor, ME 04609-6327 • 207-555-0645 • rjlawoffice.com

August 5, 2005

Ref. A. Bruins #2

CONFIDENTIAL

Ms. Abigail Bruins
Mr. Paul Meyer
Rena's Restaurant
3706 Chamberlain Avenue, Southwest
Bar Harbor, ME 04609-3427

Dear Ms. Bruins and Mr. Meyer:

BUILDING PERMIT

Early last week, we received your letter, along with drawings of the deck that you want to add to the east side of your restaurant. In that letter, you described your building plan and asked that one of our attorneys advise you on how to proceed.

I studied the plan and found it attractive. However, the drawings show that the east edge of the proposed deck would extend within six feet of your side property line, thereby violating Article X in the city's building code. That article requires ten feet between a building and a side property line.

Given this restriction, you could proceed with your building plan in one of two ways: (1) present your plan to the Planning and Zoning Commission and ask for a variance to Article X, or (2) adapt the plan so that it conforms to the code. Please call me at (207) 555-0654 to discuss this matter.

Yours sincerely,

Sydney George

(Ms.) Sydney George
Attorney-at-Law

SG/mbb
Enc.: Article X
cc: Leah Theodore

Simplified Format

R & J Law Office
105 East Bay Road, Bar Harbor, ME 04609-6327 • 207-555-0645 • rjlawoffice.com

August 5, 2005

Ref. A. Bruins #2

CONFIDENTIAL

Abigail Bruins
Paul Meyer
Rena's Restaurant
3706 Chamberlain Avenue, Southwest
Bar Harbor, ME 04609-3427

BUILDING PERMIT

Early last week, we received your letter, along with drawings of the deck that you want to add to the east side of your restaurant. In that letter, you described your building plan and asked that one of our attorneys advise you on how to proceed.

I studied the plan and found it attractive. However, the drawings show that the east edge of the proposed deck would extend within six feet of your side property line, thereby violating Article X in the city's building code. That article requires ten feet between a building and a side property line.

Given this restriction, you could proceed with your building plan in one of two ways: (1) present your plan to the Planning and Zoning Commission and ask for a variance to Article X, or (2) adapt the plan so that it conforms to the code. Please call me at (207) 555-0654 to discuss this matter.

Sydney George

(MS.) SYDNEY GEORGE—ATTORNEY-AT-LAW

SG/mbb
Enc.: Article X
cc: Leah Theodore

LETTERS AND ENVELOPES

Sometimes, a letter can be faxed or sent as an e-mail attachment. Most often, however, letters are officially signed and delivered by envelope. To deliver your letter properly, follow the tips below on folding and the U.S. Postal Service guidelines that follow.

Folding Letters

A Standard Fold

To put a letter in its matching envelope, place the letter face up and follow these steps:

1. Fold the bottom edge up so that the paper is divided into thirds.
2. Fold the top third down over the bottom third, leaving one-fourth inch for easy unfolding, and crease firmly.
3. Insert the letter (with the open end at the top) into the envelope.

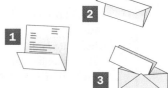

A Large Sheet in a Small Envelope

If you must place a letter in a small envelope, follow these steps:

1. Fold the bottom edge up so that the paper is divided in half.
2. Fold the right side to the left so that the sheet is divided into thirds; crease firmly.
3. Fold the left third over the right third.
4. Turn the letter sideways and insert it (with the open end at the top) into the envelope.

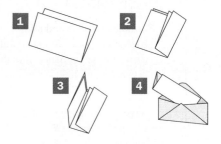

A Window Envelope

Position the inside address on the letter so that it will show through the window. Then place the letter face up and fold it as follows:

1. Fold the bottom edge up so that the paper is divided into thirds, and create a clean crease.
2. Turn the letter face down with the top edge toward you and fold the top third of the letter back.
3. Insert the letter in the envelope and make sure that the entire address shows through the window.

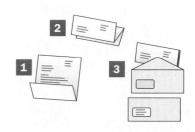

 F.Y.I.

Follow these tips when considering which envelope to use for your letter.

- **Use standard business envelopes.** Avoid odd shapes and sizes. If you must use a dark envelope, use a white label for the address.
- **Match envelopes with letters.** Match paper quality, weight (bond), color, and size.

U.S. Postal Service Envelope Guidelines

To be sure that your letters are delivered quickly and correctly, follow all U.S. Postal Service (USPS) guidelines as explained and shown below.

1. Type the receiver's name and address in black ink on a light-colored envelope. Use an all-capital-letters style for everything in the address. Make sure all lines are horizontal and lined up flush left. Leave out all punctuation except the hyphen in the ZIP code.
2. Type the receiver's address—including the type of street (ST, AVE), compass points (NE, SW), and full ZIP code—in the order shown. Place the suite, room, or apartment number on the address line, after the street address.
3. Use USPS abbreviations for states and other words in the address. Use numerals rather than words for numbered streets (9TH AVE). Add ZIP+4 codes. (Go to www.usps.com for all ZIP codes used in the U.S.)

Tips for International Mail

When sending international mail, print the country name in capital letters alone on the last line. As long as the country, city, and state or province are in English, the name and address may be in the language of the country listed.

Pattern	Example: United Kingdom	Example: Canada
Name of receiver	MR BRUCE WARNER	MS TAMARA BEALS
Street address or P.O. box	2431 EDEN WAY	56 METCALFE CRES
City, state/province, code	LONDON W1P 4HQ	MONTREAL QC J7V 8P2
Country	UNITED KINGDOM	CANADA

 WEB Links

- U.S. Postal Service: www.usps.com (English)
- Canada Post: www.canadapost.ca (English and French)

Standard Postal Abbreviations

Table 24.2 Abbreviations for States, Provinces, and Territories

U.S. States and Territories

Alabama	AL	Minnesota	MN	Virgin Islands	VI
Alaska	AK	Mississippi	MS	Washington	WA
Arizona	AZ	Missouri	MO	West Virginia	WV
Arkansas	AR	Montana	MT	Wisconsin	WI
California	CA	Nebraska	NE	Wyoming	WY
Colorado	CO	Nevada	NV		
Connecticut	CT	New Hampshire	NH		
Delaware	DE	New Jersey	NJ		
District of		New Mexico	NM		
Columbia	DC	New York	NY		
Florida	FL	North Carolina	NC		
Georgia	GA	North Dakota	ND		
Guam	GU	Ohio	OH		
Hawaii	HI	Oklahoma	OK		
Idaho	ID	Oregon	OR		
Illinois	IL	Pennsylvania	PA		
Indiana	IN	Puerto Rico	PR		
Iowa	IA	Rhode Island	RI		
Kansas	KS	South Carolina	SC		
Kentucky	KY	South Dakota	SD		
Louisiana	LA	Tennessee	TN		
Maine	ME	Texas	TX		
Maryland	MD	Utah	UT		
Massachusetts	MA	Vermont	VT		
Michigan	MI	Virginia	VA		

Canadian Provinces, Territories

Alberta	AB
British Columbia	BC
Manitoba	MB
New Brunswick	NB
Newfoundland and Labrador	NL
Northwest Territories	NT
Nova Scotia	NS
Nunavut	NU
Ontario	ON
Prince Edward Island	PE
Quebec	QC
Saskatchewan	SK
Yukon Territory	YT

Table 24.3 Abbreviations for Use on Envelopes

Annex	ANX	Lake	LK	Route	RTE
Apartment	APT	Lakes	LKS	Rural	R
Avenue	AVE	Lane	LN	Rural Route	RR
Boulevard	BLVD	Meadows	MDWS	Shore	SH
Building	BLDG	North	N	South	S
Causeway	CSWY	Northeast	NE	Southeast	SE
Circle	CIR	Northwest	NW	Southwest	SW
Court	CT	Office	OFC	Square	SQ
Drive	DR	Palms	PLMS	Station	STA
East	E	Park	PARK	Street	ST
Expressway	EXPY	Parkway	PKWY	Suite	STE
Floor	FL	Place	PL	Terrace	TER
Fort	FT	Plaza	PLZ	Throughway	TRWY
Freeway	FWY	Port	PRT	Turnpike	TPKE
Harbor	HBR	Post Office Box	PO BOX	Union	UN
Heights	HTS	Ridge	RDG	Viaduct	VIA
Highway	HWY	River	RIV	View	VW
Hospital	HOSP	Road	RD	Village	VLG
Junction	JCT	Room	RM	West	W

FORMS OF ADDRESS

Usually, you can rely on common sense to tell you how to address your reader with respect. When you're unsure, use the following guidelines and tables to find a fitting title, salutation, and complimentary close.

Guidelines for Addressing Readers

Choose elements that have the same level of formality, as shown in Table 24.4.

Table 24.4 Levels of Formality in Forms of Address

Level of Formality	Title	Salutations	Complimentary Closings
Formal	Sir/Madame The Honorable	Dear Sir/Madam: Your Excellency:	Very truly yours, Respectfully,
Standard	Mr./Ms./Mrs./Miss Dr./Reverend	Dear John/Jane Doe: Dear Mr./Ms. Doe:	Yours sincerely, Cordially,
Informal	John/Jane Doe	Dear John/Jane: Jane,	Best wishes, Regards,

Choose the title, salutation, and closing based on these factors:

- **The situation.** Consider your relationship with the reader as well as the seriousness of the message (and its legal consequences).
- **The reader's preference.** Check a letter that he or she has sent you.
- **The reader's profession.** Use a formal title (*Senator, General*) rather than a standard courtesy title (*Mr., Ms.*) with certain professionals.

Address individuals whenever possible, honoring names and titles.

- Avoid writing to positions, titles, or departments (*Dear Personnel Manager, Attention: Marketing Department*). To get a name, call the company or visit its Web site.
- Present names in the form that the reader prefers (*Elizabeth C. Moize*, not *Beth Moize*).
- As a general rule, abbreviate courtesy titles (*Mr., Ms., Mrs.*) and spell out professional titles (*Professor, Reverend*). The exceptions are *Miss* and *Dr.*
- Don't guess your reader's professional title (*Professor, Dr.*) and gender (*Ms. Robin Tate, Mr. Pat Randle*). Be sure by double-checking.
- Avoid outdated courtesy forms (*Mr. and Mrs. Al Smith, Ladies, Gentlemen, To Whom It May Concern*).

F.Y.I.

In e-mail messages to familiar colleagues, very informal courtesy elements are fitting. Use a salutation like *Hi, Verne!, Sherry, Greetings,* or *Hi, team!* Add a complimentary closing like *Thanks!, Cheers!,* or *All the best.*

Professional Titles

Using professional titles acknowledges your reader's achievements and expertise (Table 24.5). But don't overdo it: Avoid using two professional titles that mean the same thing (e.g., *Dr. Paula Felch, M.D.*).

Table 24.5 Professional Titles

Profession	Titles in Address	Salutations
Business		
CEO	Ms. Sarah Falwell Chief Executive Officer	Dear Ms. Falwell:
Vice president	Dr. David Levengood Vice President	Dear Dr. Levengood:
Company official	Ms. Susan Cook, Comptroller	Dear Ms. Cook:
Education		
President or chancellor of university (Ph.D.)	Dr. Joe Smith, President	Dear Dr. Smith: (or) Dear President Smith:
Dean of a school or college (Ph.D.)	Dr. Marjorie Stone, Dean School of Life Sciences	Dear Dr. Stone: (or) Dear Dean Stone:
Professor (Ph.D.)	Dr. Patricia Monk Professor of Psychology	Dear Dr. Monk: (or) Dear Professor Monk
Instructor (no Ph.D.)	Mr. Art Linkman Instructor of Physics	Dear Mr. Linkman:
Legal		
Lawyer	Mr. Daniel Walker Attorney-at-Law	Dear Mr. Walker:
	Daniel Walker, Esq.	Dear Daniel Walker, Esq.:
Medical		
Physician	Dr. Sarah McDonald Sarah McDonald, M.D.	Dear Dr. McDonald:
Registered nurse	Nurse John Seguin John Seguin, R.N.	Dear Nurse Seguin:
Dentist	Dr. Leslie Matheson Leslie Matheson, D.D.S.	Dear Dr. Matheson:
Veterinarian	Dr. Manuel Ortega Manuel Ortega, D.V.M.	Dear Dr. Ortega:

Male, Female, Multiple, and Unnamed Readers

Paying attention to the way you address men, women, and groups of people pays off. Courtesy titles show that you're professional—and courteous. See Table 24.6 for appropriate forms of address.

Table 24.6 Courtesy Titles for Male, Female, Multiple, and Unnamed Readers

Reader	Titles in Address	Salutations
One Woman (avoid showing marital status)		
Preferred	Ms. Barbara Jordan	Dear Ms. Jordan:
Married or widowed	Mrs. Lorene Frost	Dear Mrs. Frost:
Single	Miss Adriana Langille	Dear Miss Langille:
Two or More Women (alphabetical)		
Standard	Ms. Bethany Jergens Ms. Shavonn Mitchell	Dear Ms. Jergens and Ms. Mitchell:
Formal	Mmes. Bethany Jergens and Shavonn Mitchell	Dear Mmes. Jergens and Mitchell:
One Man		
Standard	Mr. Hugh Knight	Dear Mr. Knight:
With *Jr., Sr.,* or roman numeral	Mr. Brian Boswell, Jr. Mr. Brian Boswell III	Dear Mr. Boswell:
Two or More Men (alphabetical)		
Standard	Mr. Alex Fernandez Mr. Nate Shaw	Dear Mr. Fernandez and Mr. Shaw:
Formal	Messrs. Alex Fernandez and Nate Shaw	Dear Messrs. Fernandez and Shaw:
One Man and One Woman (alphabetical)		
	Ms. Paula Trunhope Mr. Joe Williams	Dear Ms. Trunhope and Mr. Williams:
Married Couple		
Same last name	Mr. William and Mrs. Susan Lui	Dear Mr. and Mrs. Lui:
Different last names	Mr. William Bentley Ms. Sinead Sweeney	Dear Mr. Bentley and Ms. Sweeney:
One Reader (gender unknown)		
	M. Robin Leeds Robin Leeds	Dear M. Leeds: Dear Robin Leeds:
Mixed Group Company, Department, Job Title, or Unknown Reader		
	Acme Corporation Human Resources Dept.	**Formal** Dear Sir or Madam: **Informal** Dear Manager:

Government Officials and Representatives

To properly address government officials follow this pattern:

INSIDE ADDRESS: The Honorable *(full name; full title on second line)*
FORMAL SALUTATION: Dear Sir/Madam: or Dear Mr./Madam *(position)*:
INFORMAL SALUTATION: Dear Mr./Ms. *(last name)*: or Dear *(position + last name)*:

Table 24.7 Courtesy Titles for Government Officials and Representatives

Official	Titles in Address	Salutations
National		
President	The President	Dear Mr./Madam President:
Vice President	The Vice President	Dear Mr. Vice President:
Speaker of the House	The Honorable Steven Kudo	Dear Mr. Speaker:
Cabinet members, undersecretaries	The Honorable Jane Doe	Dear Madam: Dear Attorney General Doe:
Senators (U.S. or state)	The Honorable Bill Johnson	Dear Senator Johnson:
Representatives (U.S. or state)	The Honorable Joan Walker	Dear Ms. Walker: Dear Representative Walker:
Heads of offices and agencies	The Honorable John Hillman Postmaster General	Dear Mr. Postmaster General: Dear Mr. Hillman:
Chief Justice (U.S. or state)	The Honorable Shelby Woo Chief Justice of California	Dear Madam Chief Justice:
State/Local		
Governor	The Honorable Mary Lee	Dear Governor Lee:
Mayor	The Honorable Mark Barne	Dear Mayor Barne:
Council member	The Honorable Corey Springs	Dear Mr. Springs:
Judge	The Honorable Grace Kim	Dear Judge Kim:

Military Professionals

With service personnel in the Army, Air Force, Marine Corps, Navy, and Coast Guard, you need to show special attention to rank and branch. Follow these guidelines:

- Address most military personnel by following this pattern:

    INSIDE ADDRESS: *full rank + full name + service branch abbreviation*
    Major General Karl P. Bastion, USAF
    FORMAL SALUTATION: Sir/Madam:
    INFORMAL SALUTATION: Dear *(short version of rank + last name)*:
    Dear General Bastion:

- If needed, add a comma and *Retired* after the service branch abbreviation.
- For Warrant Officer, Chief Warrant Officer, and Junior Officers (Lieutenant, Lieutenant Junior Grade, Ensign), use *Mr.* or *Ms.* in the salutation, not rank.

Service branch abbreviations

Army	USA	Navy	USN	Marine Corps	USMC
Air Force	USAF	Coast Guard	USCG		

Note: Add "R" for reserves.

Religious Titles

To address religious leaders from any faith with titles that fit their positions, note the following and consult Table 24.8.

- The use of *The* before *Reverend* differs from church to church. Follow the organization's preference.
- In some religious orders, the title in the salutation is followed by the reader's first name. Other orders prefer the last name.
- If the person has a Doctor of Divinity degree, add a comma and *D.D.* after his or her name in the address (but not in the salutation).

Table 24.8 Religious Titles

Clergy	Titles in Address	Salutations
Roman Catholic		
Cardinal	His Eminence, Edward Cardinal Romero	Your Eminence: Dear Cardinal Romero:
Archbishop and bishop	The Most Reverend Henri Crétien	Your Excellency: Dear Bishop/Archbishop Crétien:
Priest	The Reverend Morris Franklin	Reverend Sir: Dear Father Franklin:
Nun	Sister Mary Jennsen	Dear Sister Mary: Dear Sister Jennsen:
Monk	Brother Atticus Bartholemew	Dear Brother Atticus: Dear Brother Bartholemew:
Protestant		
Bishop (Anglican, Episcopal, Methodist)	The Right Reverend Samuel Wolfe	Right Reverend Sir: Reverend Sir: Dear Bishop Wolfe:
Dean (head of cathedral or seminary)	The Very Reverend Nicholas Cameron	Very Reverend Sir: Dear Dean Cameron:
Minister or priest	The Reverend Susan Edwards	Dear Reverend Edwards:
	Pastor Edwards	Dear Pastor Edwards:
Chaplain	Chaplain Adam Carp Captain, USMC	Dear Chaplain Carp:
Jewish		
Rabbi	Rabbi Joshua Gould	Dear Rabbi Gould:
Rabbi with Doctor of Divinity degree	Rabbi Joshua Gould, D.D.	Dear Dr. Gould:

CHECKLIST for Letters

Before you send a letter, revise it using the seven-traits benchmarks.

- ☐ **Ideas.** The discussion is complete, clear, and accurate, answering the reader's questions: *Why are you writing? What's it all about? What needs to happen?*
- ☐ **Organization.** The letter's message has a three-part structure. It is direct or indirect, based on the reader's likely response.
- ☐ **Voice.** The tone is natural and courteous. It displays a level of formality appropriate to the reader and situation.
- ☐ **Words.** Phrasing is precise and clear, uses personal pronouns (especially *you*) in a positive way, avoids in-house terms, and includes needed definitions.
- ☐ **Sentences.** Each sentence is complete, clear, and smooth.
- ☐ **Copy**
  - Punctuation is correct (e.g., colon after salutation, comma after closing).
  - Capitalization is correct (e.g., cities, months, names, titles, companies, "Dear" in salutations, and only the first word of closings).
  - Spelling is correct (e.g., names of people, cities, streets, and companies).
  - Abbreviations are used only if appropriate and familiar to the reader.
  - Forms of address are correct and appropriate to the recipient.
- ☐ **Design**
  - The formatting is correct and reflects company style (e.g., semiblock, full block).
  - It uses elements well (e.g., white space, margins, lists, numbering).
  - It is printed on quality stationery in a crisp, professional typeface.

Last reminders

- ☐ Enclosures, copies, routing procedures, and mailing method are appropriate.
- ☐ The final letter is signed in black ink.

CRITICAL THINKING Activity

Review the instructions regarding forms of address on pages 391–395. In three sample letters, evaluate the courtesy titles, names, salutations, and complimentary closings used. Did the writers make appropriate choices?

WRITING Activity

Choose a recently purchased product. Find an appropriate reader at the company, and write a letter complimenting the product. Develop three versions of the letter (semiblock, full-block, and simplified formats). Assess which format works best for this situation.

In this chapter

The Art of Being Direct **398**
Guidelines for Informative Messages **400**
Guidelines for Routine Inquiries and Requests **406**
Guidelines for Positive Responses **410**
Guidelines for Placing Orders **414**
Guidelines for Accepting Claims **416**
Guidelines for Goodwill Messages **418**
Checklist for Good-News and Neutral Messages **424**
Critical Thinking Activities **424**
Writing Activities **424**

CHAPTER
25

Writing Good-News and Neutral Messages

There's no news like good news (and neutral news isn't far behind). That's as true in the workplace as it is in the rest of life. A good-news message relays information that is clearly positive. A neutral-news message delivers information that is helpful and welcome, though it may not seem distinctly good or bad. Regardless of whether the message is good or neutral news, readers are usually happy to receive it. They commonly respond, "It's good to know this!" or "It's my job to deal with this, so I'll take care of it."

This chapter includes guidelines and models for seventeen good- or neutral-news messages, which are grouped into six categories. Some models are in letter format, others in e-mail format, and still others in memo format. In addition, the chapter includes "The Art of Being Direct" and the SEA formula. The first piece offers tips on how to get to the point, build goodwill, and be helpful. The second piece provides a template for organizing good-news messages. The chapter's checklist and activities will help you practice these tips while writing specific forms of good or neutral news.

Work through this chapter to get a good grasp on delivering good or neutral news. When you need detailed help with format-related issues, refer to Chapters 22–24 on memos, e-mail messages, and letters, respectively. Together, these four chapters will help you write effective good or neutral news.

THE ART OF BEING DIRECT

Good-news and neutral messages should simply "tell it like it is"—no rambling, no fluff. Because these routine messages are both common and key to business success, do them right.

Get to the Point

Put key points first to "frontload" your message. Specifically, position the main point early—in the first paragraph or early in the second. In addition, state the main point crisply in a clear, positive sentence.

> We are pleased to extend you $100,000 in credit based on Dale's Garden Center's strong financial condition.

Build Goodwill

Don't be so direct that you sound abrupt, impolite, unconcerned, or impersonal. Instead, build *goodwill*—the oil that lubricates business activity.

- Be confident but not pushy, businesslike but not stuffy.
- Use a person-to-person tone by including personal pronouns, especially *you*.
- Use *please* and *thank you*, as well as greetings and complimentary closings.
- Use the reader's name (spelled correctly) with respect.
- Use neutral or positive phrasing, not negative words. For example, refer to *the situation*, not *your complaint*; say *please do the following*, not *you must*; talk about *opportunities* and *challenges*, not *headaches*.

Offer Help

Provide truly useful information that prevents confusion and unnecessary follow-up:

- Supply all the details that your reader needs, not everything you know.
- By anticipating further questions or projecting additional needs, offer unexpected bonus material that benefits your reader.
- Organize the details for easy digestion by using short paragraphs, displayed lists, and white space.
- Supply specific, accurate facts—not vague, general statements.
- Focus on reader benefits, stressing why your main point is good news.
- Use attachments or enclosures for the reader's convenience (not your own).
- Provide easy-response options: postcards, phone numbers, fax numbers, e-mail addresses, forms, self-addressed stamped envelopes.

F.Y.I.

To write direct, polite, helpful messages, consider these additional issues:

- **Secondary Readers:** Who might see your message, with what reaction?
- **Timing:** When would it be productive for readers to get the message?
- **Medium:** Should it be e-mail, memo, or letter? See page 357.

Use the **SEA** Formula

Move smoothly from *situation* to *explanation* to *action*, as mapped out below.

Situation

Explain your reason for writing, following these tips:

- Supply helpful background and context.
- State your main point—a clear and simple statement, request, or question.
- Avoid abrupt or rambling openings.
- Avoid stating the obvious, as in this opening:

  I have received and read the letter that you wrote dated August 10, 2005, in which you asked about the terms of the warranty for the Triad 35 mm camera that you purchased on July 25, 2005. You asked (1) . . . , (2) . . . , and (3) . . . .

  Example Situation

  Thank you for your interest in Home Builders. I appreciate hearing from a college student eager to help us provide decent, affordable housing.

Explanation

Expand on the main point in one or more well-developed paragraphs:

- Present points that support and expand the main point.
- Order supporting points logically—for the reader's convenience. In general, put more important information early.
- Explain what the details mean and why they are important.
- Use displayed lists, if appropriate, to make information easy to scan and use.

  Example Explanation

  As you requested, I have enclosed a list of Home Builders affiliates in Missouri. Each affiliate handles its own work groups and schedules. Because you are in college, I have also enclosed a brochure on Home Builders Campus Chapters. It shows you how to join or start a Campus Chapter and explains service learning for academic credit. Feel free to contact Ben Abramson, Campus Outreach Coordinator for your area, at the address on the enclosed material.

Action

Focus positively on what's next:

- If helpful, restate the main point and its importance.
- Stress upcoming contact or offer more help.
- Explain precisely any steps that you, your reader, or others should take.

  Example Action

  Again, thank you for your interest in becoming a partner with Home Builders. If I can provide other information, please let me know.

> "Words set things in motion. I've seen them doing it. Words set up atmospheres, electrical fields, charges."
>
> —Toni Cade Bambara

GUIDELINES FOR INFORMATIVE MESSAGES

Your goal when writing an informative letter, memo, or e-mail is to provide and make sense of ideas and details your reader needs. You'll also want to make the information clear enough to avoid unnecessary follow-up. As you'll see from the models on the following pages, informative messages can offer advice, make announcements, confirm plans, cover attachments, and provide updates.

1 Prewrite

Rhetorical analysis: Consider your purpose, your audience, and the context.
- ☐ What is your "informing" objective—the practical result you want?
- ☐ Who are your readers, what do they presently know about the topic, and what do they need to know?
- ☐ What's the best delivery method for the information—e-mail, memo, letter, or attachment (e.g., brochure, report)?

Prepare to draft.
- ☐ Locate and/or generate the data needed for the message. Study the information to ensure that you understand it. In addition, check it for accuracy.
- ☐ Select the information needed to describe and explain the topic; then order points in a sequence that allows readers to digest the information.
- ☐ If you find an existing document (e.g., report, brochure) that contains the needed information, attach the document to a cover message, highlighting passages and sections of the document.

2 Draft

Organize your message into three parts.
[opening] Explain what information you are sending and why.
[middle] Present information in easy-to-read chunks (lists and short paragraphs); include only those details your reader truly needs.
[closing] If a response is needed, clarify the action. If fitting, invite feedback or offer more help.

3 Revise

Review your draft's ideas, organization, and voice.
- ☐ Have you included accurate details—in the best order?
- ☐ Is the information explained at a level appropriate to your reader?
- ☐ Have you anticipated your reader's information needs?
- ☐ Have you used a confident, knowledgeable tone?

4 Refine

Check your writing line by line for the following:
- ☐ Clear, objective wording and readable sentences with helpful transitions
- ☐ Consistency in lists (parallelism)
- ☐ Correct grammar, punctuation, and spelling
- ☐ Effective format (parts, spacing, typography)

Advice Message

In an advice message, you inform your reader typically by sharing your opinion, perspective, wisdom, or expertise so that the reader understands the situation and can decide how to proceed. A lawyer offers such advice in the model below.

Rhetorical situation

Purpose: To accurately convey results of research into property purchase
Audience: New homeowner in the community seeking to upgrade property
Format: Formal letter on firm letterhead (to give advice official weight)

Bui, Gappa & Longman

408 6th Street NE, St. Cloud, MN 56303 • Telephone 800-555-4264, 320-555-8430
Fax 320-555-8436 • ybui@bgw.com

June 10, 2005

Dr. Shonna Williams
1305 Bayshore Drive
Bear Lake, MN 56935

Dear Dr. Williams:

Thank you for stopping by my office to discuss your interest in purchasing the city-owned property adjacent to your east lot line at 1305 Bayshore Drive. I have now researched the matter and have the following advice on how to proceed.

Specifically, I reviewed your property's abstract, checked the plait drawings at the Kanabec County Office, and visited with Bear Lake City Manager, Jean Quamba. From the abstract and plait drawings, I learned that the city-owned property is 18 feet on the north and south sides and 196 feet on the east and west sides.

Ms. Quambe reported that in 1988, the city purchased this property as part of its plan to extend the city's bike trail on the south to Birdview Park on the north. However, in 1997, the city purchased a section of the abandoned railroad and built the bike trail there instead. She also said that the city council may be willing to sell the property adjacent to yours for a "reasonable price."

Purchasing the property at this time could be a wise decision. The property would extend your living space, give you greater privacy on the north, and enhance the value of your home.

Please call me at 800-555-4264 so we can discuss how to proceed.

Sincerely,

Yen Bui

Yen Bui

[opening]
Situation: Introduce the topic and its context.

[middle]
Explanation: Describe your research or other basis for your advice.

Present information objectively.

State your advice clearly, along with supporting details.

[closing]
Action: Offer further help.

Part 4 Correspondence: Memos, E-Mails, and Letters

Announcement or Notice

Typically, an announcement or notice is distributed to people to keep them in the loop on developments, changes, and events. For example, in the e-mail below, the writer shares with all employees details of a policy change concerning company travel.

Rhetorical situation

Purpose: To explain a new policy clearly so that it can and will be followed
Audience: All company employees, but especially pertinent to staff who travel regularly
Format: E-mail (convenient for sharing a basic change)

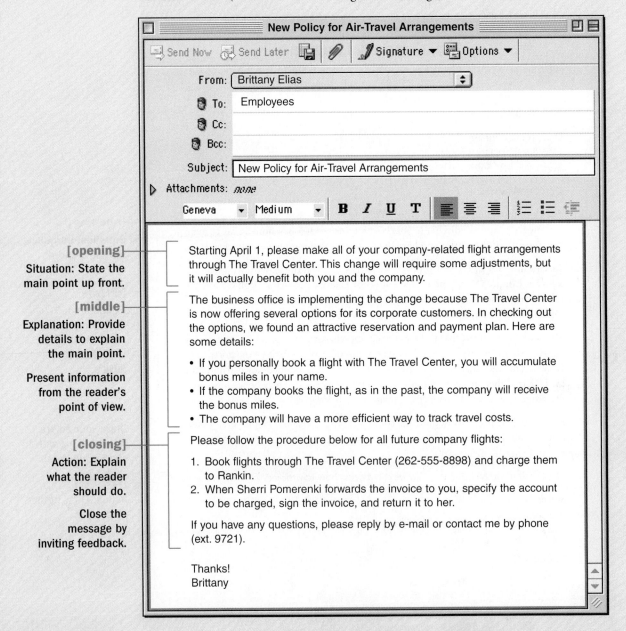

[opening]
Situation: State the main point up front.

[middle]
Explanation: Provide details to explain the main point.

Present information from the reader's point of view.

[closing]
Action: Explain what the reader should do.

Close the message by inviting feedback.

Subject: New Policy for Air-Travel Arrangements

Starting April 1, please make all of your company-related flight arrangements through The Travel Center. This change will require some adjustments, but it will actually benefit both you and the company.

The business office is implementing the change because The Travel Center is now offering several options for its corporate customers. In checking out the options, we found an attractive reservation and payment plan. Here are some details:

- If you personally book a flight with The Travel Center, you will accumulate bonus miles in your name.
- If the company books the flight, as in the past, the company will receive the bonus miles.
- The company will have a more efficient way to track travel costs.

Please follow the procedure below for all future company flights:

1. Book flights through The Travel Center (262-555-8898) and charge them to Rankin.
2. When Sherri Pomerenki forwards the invoice to you, specify the account to be charged, sign the invoice, and return it to her.

If you have any questions, please reply by e-mail or contact me by phone (ext. 9721).

Thanks!
Brittany

Confirmation Message

A confirmation message informs the reader that a statement is accurate or that an event has been scheduled. By putting it in writing, such a message develops mutual understanding.

Rhetorical situation

Purpose: To ensure that all workshop-related details are covered in advance
Audience: Workshop leader, expert on topic, but with less knowledge of the company and city
Format: E-mail (quickly shares key details and allows for timely feedback)

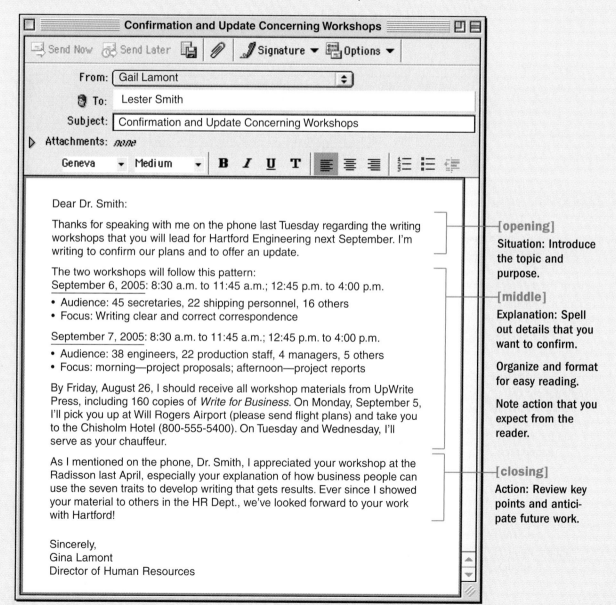

[opening]
Situation: Introduce the topic and purpose.

[middle]
Explanation: Spell out details that you want to confirm.

Organize and format for easy reading.

Note action that you expect from the reader.

[closing]
Action: Review key points and anticipate future work.

Part 4 Correspondence: Memos, E-Mails, and Letters

Cover Message

You provide "cover" for an attachment or an enclosure by pointing to and explaining that document or item. In addition, your cover message helps readers use that attachment.

Rhetorical situation

Purpose: To alert readers to a procedural and documenting change; to ensure proper use of the attachment

Audience: Department managers responsible in part for new employee training and evaluation; appreciate smooth and effective review process

Format: E-mail (offers convenient, timely distribution plus electronic copy of document for review and use)

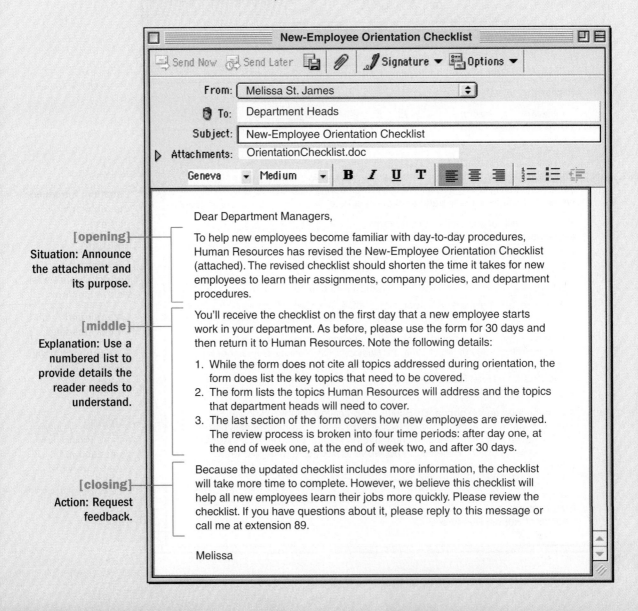

[opening]
Situation: Announce the attachment and its purpose.

[middle]
Explanation: Use a numbered list to provide details the reader needs to understand.

[closing]
Action: Request feedback.

Update

Providing readers an update on specific tasks, projects, problems, or issues keeps those readers in the loop on the topic's status, communicates your progress, and creates an opportunity for useful feedback.

Rhetorical situation

Purpose: To report progress made on revising a process; to solicit feedback on changes
Audience: Immediate supervisor; familiar with original process and called for changes
Format: E-mail (sufficient for sharing ideas and soliciting feedback)

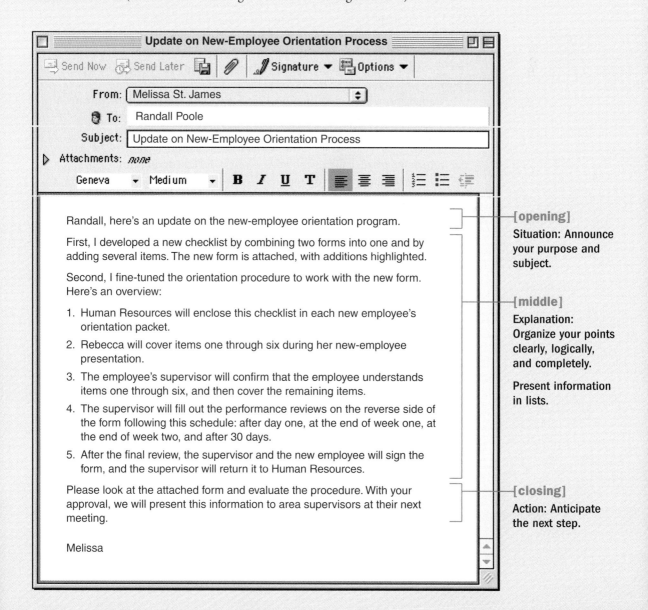

GUIDELINES FOR ROUTINE INQUIRIES AND REQUESTS

Your goal when you make a routine inquiry or a request is to state clearly and politely what you want and to make it easy (even enjoyable) for the reader to respond—whether you're asking for information, cooperation, or participation. (Note: If your request must be more persuasive, see page 454.)

1 Prewrite
Rhetorical analysis: Consider your purpose, your audience, and the context.
- ☐ What exactly do you need? Why? How will you use it?
- ☐ To whom are you writing? What do you know about this person? Why and how will he or she see your request or inquiry as routine?
- ☐ By when do you need a response? By what means or method?

Prepare to draft.
- ☐ Gather information you require to be clear and precise about your need. Jot down background information related to your request. List exact questions you want answered, or specific items you need.
- ☐ Decide which easy-response options you should offer your reader: a return postcard, a fax number, an e-mail address, or an SASE (self-addressed, stamped envelope).

2 Draft
Organize your thoughts with the reader in mind.

[opening] Provide needed background and make your inquiry or request.

[middle] List the main points (or questions) to explain what you need and why. Order them sensibly for your reader. If possible, link the request to mutual benefits or the reader's benefits.

[closing] Discuss how to respond. Supply any information the reader may need (a payment method, a telephone number, a deadline). Thank the reader for attending to your request.

3 Revise
Review your draft's ideas, organization, and voice.
- ☐ Have you clearly stated your need, with supporting points logically ordered?
- ☐ Did you explain the benefits to your reader and how to respond?
- ☐ Did you use a polite, person-to-person tone (including a simple *please* and *thank you*)? Is the tone confident but not pushy?

4 Refine
Check your writing line by line for the following:
- ☐ Precise and clear terms, especially concerning what you've requested
- ☐ Readable sentences with good transitions
- ☐ Correct grammar, punctuation, usage, and spelling
- ☐ Effective format (parts, spacing, typography)

Information Request

In a routine information request, your reader simply has the details that you need—and it's easy and appropriate for that reader to deliver the goods. While you can get much information in person or over the phone, sometimes you put your request in writing for the reader's convenience or to get a written record (as shown in the sample letter below from a lawyer). At other times, you put your request in writing because the questions are complex or extensive.

Rhetorical situation

Purpose: To gather accurate zoning information about a specific property to advance a sale
Audience: Civil servant with access to government records
Format: Formal letter (creates official, recorded request and leads to official reply)

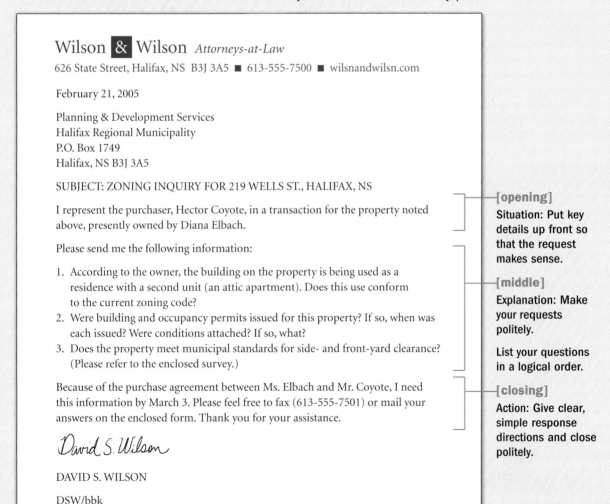

Wilson & Wilson *Attorneys-at-Law*
626 State Street, Halifax, NS B3J 3A5 ■ 613-555-7500 ■ wilsnandwilsn.com

February 21, 2005

Planning & Development Services
Halifax Regional Municipality
P.O. Box 1749
Halifax, NS B3J 3A5

SUBJECT: ZONING INQUIRY FOR 219 WELLS ST., HALIFAX, NS

I represent the purchaser, Hector Coyote, in a transaction for the property noted above, presently owned by Diana Elbach.

Please send me the following information:

1. According to the owner, the building on the property is being used as a residence with a second unit (an attic apartment). Does this use conform to the current zoning code?
2. Were building and occupancy permits issued for this property? If so, when was each issued? Were conditions attached? If so, what?
3. Does the property meet municipal standards for side- and front-yard clearance? (Please refer to the enclosed survey.)

Because of the purchase agreement between Ms. Elbach and Mr. Coyote, I need this information by March 3. Please feel free to fax (613-555-7501) or mail your answers on the enclosed form. Thank you for your assistance.

David S. Wilson

DAVID S. WILSON

DSW/bbk
Enclosures 2

[opening]
Situation: Put key details up front so that the request makes sense.

[middle]
Explanation: Make your requests politely.

List your questions in a logical order.

[closing]
Action: Give clear, simple response directions and close politely.

Routine Request for Cooperation

Cooperation involves working together toward a common goal. When you make a routine request for cooperation, you call your reader toward that shared task and shared goal. To do so effectively, be precise about what cooperation involves and be clear about the benefits of that united effort.

Rhetorical situation

Purpose: To encourage readers to prepare for and effectively participate in a regular meeting
Audience: Human Resources Department members, familiar co-workers
Format: E-mail (provides quick and convenient distribution)

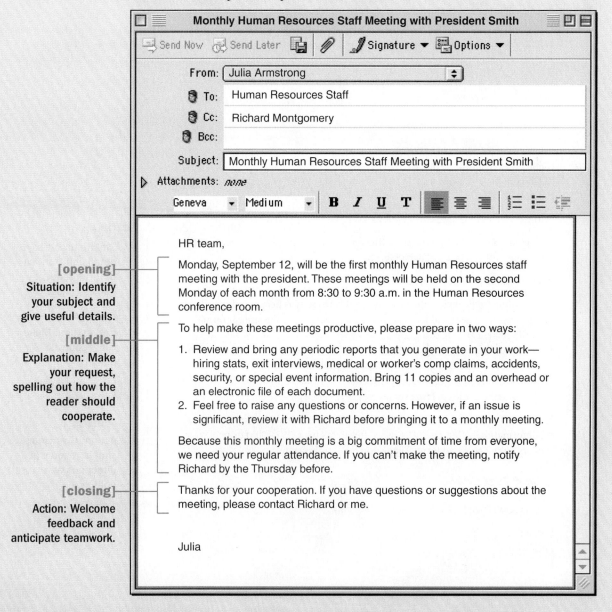

[opening]
Situation: Identify your subject and give useful details.

[middle]
Explanation: Make your request, spelling out how the reader should cooperate.

[closing]
Action: Welcome feedback and anticipate teamwork.

Routine Invitation

Generally welcome, invitations inform people about events, request their attendance, and encourage their participation. Good invitations are quite literally "inviting"—positive, personal, and clear.

Rhetorical situation

Purpose: To encourage the reader's attendance at a company anniversary celebration
Audience: Regional sales representative for company in town for a sales seminar
Format: Letter from management (gives invitation a friendly "weight")

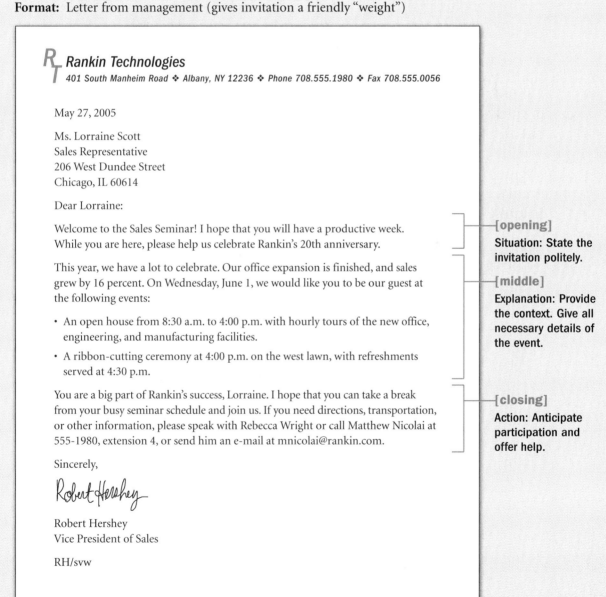

Rankin Technologies
401 South Manheim Road ❖ Albany, NY 12236 ❖ Phone 708.555.1980 ❖ Fax 708.555.0056

May 27, 2005

Ms. Lorraine Scott
Sales Representative
206 West Dundee Street
Chicago, IL 60614

Dear Lorraine:

Welcome to the Sales Seminar! I hope that you will have a productive week. While you are here, please help us celebrate Rankin's 20th anniversary.

This year, we have a lot to celebrate. Our office expansion is finished, and sales grew by 16 percent. On Wednesday, June 1, we would like you to be our guest at the following events:

- An open house from 8:30 a.m. to 4:00 p.m. with hourly tours of the new office, engineering, and manufacturing facilities.
- A ribbon-cutting ceremony at 4:00 p.m. on the west lawn, with refreshments served at 4:30 p.m.

You are a big part of Rankin's success, Lorraine. I hope that you can take a break from your busy seminar schedule and join us. If you need directions, transportation, or other information, please speak with Rebecca Wright or call Matthew Nicolai at 555-1980, extension 4, or send him an e-mail at mnicolai@rankin.com.

Sincerely,

Robert Hershey

Robert Hershey
Vice President of Sales

RH/svw

[opening]
Situation: State the invitation politely.

[middle]
Explanation: Provide the context. Give all necessary details of the event.

[closing]
Action: Anticipate participation and offer help.

GUIDELINES FOR POSITIVE RESPONSES

Your goal when writing a positive response is to give the reader (in a timely and helpful way) what he or she requested. Such help can involve responding positively to an inquiry, a request, a proposal, or a credit application.

1 Prewrite
Rhetorical analysis: Consider your purpose, your audience, and the context.
- ☐ What primary and secondary objectives should your response achieve?
- ☐ Who is your reader, and what has he or she requested? Why does the reader need this? How will he or she use it?
- ☐ By when does the reader need a response, and in what form?

Prepare to draft.
- ☐ Review the original request and highlight items you need to cover.
- ☐ Gather answers, facts, and useful documents (such as reports, pamphlets, and brochures).
- ☐ Choose the best form for your response. (For frequent requests, use a form letter, postcard, or cover letter.)

2 Draft
Organize the response according to the request.
[opening] Refer to the reader's request, and state your response.
[middle] Provide all the information needed in easy-to-read paragraphs or lists. Consider answering questions in the order they were given.
[closing] Express thanks, invite further contact, and, if appropriate, offer additional information that the reader may find useful.

3 Revise
Review your draft's ideas, organization, and voice.
- ☐ Is your information accurate, complete, and carefully organized?
- ☐ Have you anticipated further needs from your reader?
- ☐ Have you used a helpful tone that avoids clichés and jargon?

4 Refine
Check your writing line by line for the following:
- ☐ Exact, clear wording (matching the level of language to the request)
- ☐ Sentences that read well and have strong transitions
- ☐ Correct grammar, punctuation, usage, and spelling
- ☐ Effective format (parts, spacing, lists, typography)

> "Short paragraphs put air around what you write and make it look inviting."
> —William Zinsser

F.Y.I.
Ask yourself: Are you the right person to respond? Do you have what the reader needs and the authority to share it? If not, forward the request to someone who does, and inform the reader of your action.

Positive Reply to an Inquiry

When a colleague, client, or another person interested in your work makes an inquiry, he or she is asking you to check an issue or answer some questions. A positive reply supplies the response that the reader needs.

Rhetorical situation

Purpose: To make a clear loan offer and encourage continued contact
Audience: Potential clients who have expressed some initial interest in a loan
Format: Letter (provides official weight to offer)

ASPEN STATE BANK
4554 Ridgemount Boulevard, Aspen, CO 81225-0064, PHONE 459-555-0098, FAX 459-555-5886
contact@aspenstatebank.com

February 24, 2005

Christine and Dale Shepherd
1026 11th Avenue, NE
Aspen, CO 81212-3219

Dear Christine and Dale:

Thank you for your inquiry yesterday about financing your resort project. I enjoyed discussing your project and appreciated your frankness about your current loan with Boulder National Bank. ⎯ **[opening]** Situation: State the reason for your response and your appreciation for the inquiry.

Although you commented that you will seek an extension of your loan from Boulder National, I have enclosed Aspen State Bank's commitment letter, subject to the terms we discussed. Perhaps you will consider our package. Rates available are as follows:

5-year fixed rate	4.5%
10-year fixed rate	4.875%
20-year fixed rate	5.375%

⎯ **[middle]** Explanation: Provide the reader with the desired information and stress its value.

In case you do not proceed with the Boulder loan, this commitment will be good for 60 business days from today (February 24). If lower rates are available at closing, you will receive the benefit of that reduction.

Thank you for your interest. I hope that your project goes well. If we can't work together on this project, please keep us in mind for future credit needs. ⎯ **[closing]** Action: Anticipate and invite future contact.

Yours sincerely,

Cara Harrison

Cara Harrison
Loan Officer

Enclosure: Commitment Letter

Request or Proposal Acceptance

When you accept a request or a proposal from a reader, you do two things. First, you acknowledge that you agree with the reader. Second, you promise to deliver on your agreement by acting accordingly—fulfilling the request or following through on the proposal.

Rhetorical situation

Purpose: To finalize an agreement to create a business relationship, clarifying all pertinent details
Audience: Computer retailer seeking technical and training support
Format: Letter (makes response formal, official, and detailed)

Juanita Guiverra, Computer Consultant

368 Palm Palace Boulevard
Miami, FL 33166-0064
Telephone: 305.555.0010
FAX: 305.555.0500
E-Mail: jguiverra@cnsult.com

March 21, 2005

Mr. Gavin Farnsworth
Miami Computer Enterprises
1202 South Benton
Miami, FL 33166-1217

Dear Mr. Farnsworth:

[opening] *Situation: State your acceptance positively.*

I have reviewed your letter of March 14. In response to your proposal, I am happy to offer my consulting services to Miami Computer customers.

[middle] *Explanation: Stress the benefits of the decision and cover details that need to be clarified or recorded.*

This arrangement will benefit all parties involved. Together, we will be able to offer your clients "one-stop shopping" for all their computer needs—hardware, software, training, and support. And I will be able to work with your established customer base without having to generate my own.

Therefore, I accept your proposed rate of $45 per hour (minimum of 20 hours per week) as indicated in the amended agreement (outline enclosed). Please note that the bold items on the outline indicate additions to the original proposal. I simply added the items covered in your letter.

[closing] *Action: Explore the next step and anticipate a positive outcome.*

Please let me know of any specific information or documentation that you need to see on my invoices. I look forward to a productive partnership in which we will serve each other and your clients.

Yours sincerely,

Juanita Guiverra

Juanita Guiverra

Enc.: Agreement Outline

Credit Approval

A credit approval is written in response to a credit request (page 459). When an existing customer or a potential customer requests a credit line or wishes to open an account to purchase your products and services, you approve that request (typically after a credit check) and spell out the conditions of such credit.

Rhetorical situation

Purpose: To confirm approval of credit line, clarify conditions, and encourage orders
Audience: Gardening retailer with relatively brief but successful history and good credit
Format: Letter (records and formalizes offer)

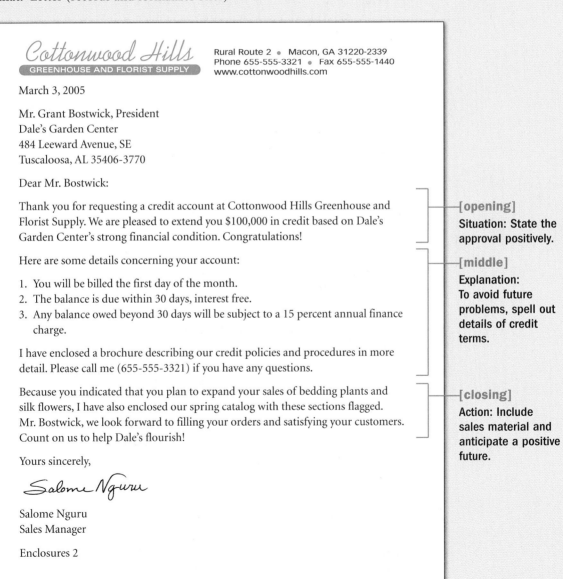

GUIDELINES FOR PLACING ORDERS

Your goal when placing an order by memo, letter, fax, or computer is to give complete, correct information so that you get the right product shipped by the right method and on time.

1 Prewrite
Rhetorical analysis: Consider your purpose, your audience, and the context.
- ☐ What precisely do you need and why? How many of each?
- ☐ From whom are you ordering? Are you familiar with this company?
- ☐ When do you need the order, and why is that deadline important? When can the order be received during the day?
- ☐ Where do you need the items, and is that site prepared for the order—both physically and in terms of staff?

Gather details about the product or service.
- ☐ List the title and catalog code.
- ☐ Give each product's, name, description (e.g., style, color, size), and technical specifications.
- ☐ State the number of items or units, price per item, total price for items, and total price of order (including tax, shipping, and handling).
- ☐ List special conditions, limits, or requirements.

Clarify shipping and receiving details.
- ☐ Include the payment method, account number, and purchase-order number.
- ☐ Note the delivery method, handling instructions, and address (including the receiver's name and phone number).
- ☐ List deadlines and substitution instructions if appropriate.

2 Draft
Organize the order as simply and logically as you can.
[opening] Get right to the point—that you are placing an order.
[middle] In a list or table, give all details needed to process the order.
[closing] Offer final instructions and authorize the order.

3 Revise
Review your draft's ideas, organization, and voice.
- ☐ Is the information complete? Are any key details missing?
- ☐ Did you double-check your numbers to avoid costly errors?
- ☐ Have you used a friendly, yet businesslike tone?

4 Refine
Check your writing line by line for the following:
- ☐ Precise terms, numbers, and abbreviations
- ☐ Short, clear sentences
- ☐ Correct grammar, punctuation, usage, and spelling
- ☐ Effective format (readable type size, spacing)

Faxed Purchase Order

Rhetorical situation

Purpose: To order and receive the correct number and type of equipment
Audience: Discount furniture retailer being used for the first time by writer
Format: Faxed order form (in keeping with the reader's ordering system)

NAMASTE HOTEL/CONVENTION CENTER
250 State Street • Madison, WI 53593 • 608/555-1256

PURCHASE ORDER #7963

Date:	October 17, 2005	Number of pages 1
To:	Order Department Drummond Discount Furniture FAX: 1-748-555-8273	
From:	Oscar Martinez, Convention Manager	
Account:	478-91-4554	

[opening]
Situation: Offer complete contact information, and make your request.

Please send us the items below from your 2005 catalog (#SSD-217).

Quantity	Description	Item Cost	Total
3	Conference Cabinet Model #CTT-338	$525.00	$1,575.00
4	Overhead Projector Model #BQ-220	395.00	1,580.00
3	Wall/Ceiling Screen Model #PSC-8484	169.00	507.00
			$3,662.00

[middle]
Explanation: List full product information in table form.

Ship to: Julie Postwright, Convention Coordinator, at the address above

Bill to: Bernie Ford, Accounts Payable, at the address above

Clarify order-processing details.

Instructions:
If you cannot ship these items without substitutions for delivery by November 11, please inform me immediately (phone: 1-608-555-1256; fax: 1-608-555-1257).

Oscar Martinez
Authorized signature

[closing]
Action: List contact number and specific instructions.

Part 4 Correspondence: Memos, E-Mails, and Letters

GUIDELINES FOR ACCEPTING CLAIMS

When someone is dissatisfied with a product, service, development, or change, he or she might file a claim or complaint (see page 446). When you accept a claim or complaint, your goal is to solve the problem promptly by offering a solution, credit, or adjustment. You'll also want to win back the other party's confidence and build a reputation for fairness.

1 Prewrite

Rhetorical analysis: Consider your purpose, your audience, and the context.
- ☐ What can you do to solve the problem, repair damage to the company, restore goodwill with the reader, and avoid similar problems in the future?
- ☐ To whom are you writing, and what do you know (or need to know) about this person? What is the source of his or her dissatisfaction? What inconveniences and other effects is the person experiencing?
- ☐ What procedures should you follow to investigate and reply?

Research the problem.
- ☐ Review the original complaint, plus pertinent policies and records.
- ☐ Investigate the issue, conduct tests, and explore possible solutions.

2 Draft

Organize your response.
[opening] Provide the context and state what you've done to fix the problem.
[middle] Concisely explain the problem's causes, following these guidelines:
- ☐ Apologize once, early in the message, and be businesslike. Don't grovel.
- ☐ Stress the unusual nature of the problem, and explain what's been done to make sure it doesn't happen again (but don't promise it never will).
- ☐ Avoid making excuses or attacking co-workers.

[closing] Thank your reader for helping you resolve the problem and, if fitting, anticipate future contact.

3 Revise

Review your draft's ideas, organization, and voice.
- ☐ Have you expressed agreement with the claim early in the message?
- ☐ Have you provided a clear explanation in a positive, ungrudging tone?

4 Refine

Check your writing line by line for the following:
- ☐ Precise, neutral wording and easy-to-read sentences with transitions
- ☐ Correct grammar, punctuation, usage, and spelling
- ☐ Effective format (parts, spacing, typography)

F.Y.I.
To create a positive tone, use phrases such as *the situation, you state that,* and *your patience and understanding*. Avoid phrases such as *your complaint, your problems, you claim that,* and *we will do what we can.*

Positive Adjustment

A positive adjustment resolves a reader's claim by making changes. Possible positive adjustments include repairs, replacements, discounts, credits, upgrades, procedural changes, and gifts.

Rhetorical situation

Purpose: To offer a practical solution to the client's claim; to apologize and maintain a good working relationship
Audience: Major customer with whom writer has had a good past history
Format: Formal letter on company letterhead (gives weight to message)

1400 NW Academy Drive
Phone 412/555-0900

Atlanta, GA 30425
Fax 412/555-0054

July 8, 2005

Mr. Jamaal Ellison
Southeast Electric
1976 Boulder Road, Suite 1214
Charlotte, NC 28216-1203

Dear Mr. Ellison:

Thank you for your patience and understanding as we investigated the malfunction of the ATV16 drives that you had installed for American Linc Company. I apologize for the inconvenience caused to both your company and American Linc. Below is a description of the problem, along with our solution. — **[opening]** Situation: Provide necessary background, apologize, and offer solutions.

Problem: Serial link failure. We determined the cause to be a voltage regulator failing in the serial link box. Last March, our factory switched component suppliers and the new voltage regulator was not as rugged as the old component.

Solution: We are currently modifying American Linc's serial link boxes in Raleigh to remove and replace the faulty regulators. These units have been tested and operate satisfactorily. We will continue to modify these boxes and supply American Linc with the proper components. — **[middle]** Explanation: Discuss causes and solutions clearly in neutral language.

Thank you, Mr. Ellison, for helping us find these solutions. With such support, AC Drives will continue to provide the best drives available to its customers. I look forward to continuing business with Southeast Electric. — **[closing]** Action: Express appreciation and focus on future business.

Yours sincerely,

Elaine Hoffman

Elaine Hoffman
Product Manager

GUIDELINES FOR GOODWILL MESSAGES

Your goal when writing a goodwill message is simply to show the reader that you appreciate him or her in the context of a specific event—an award, a promotion, a personal loss, and so on. Such appreciation can involve a range of goodwill expressions: apology, congratulations, sympathy, thanks, and welcome.

1 Prewrite
Rhetorical analysis: Consider your purpose, your audience, and the context.
- ☐ How can you recognize, encourage, and affirm your reader?
- ☐ Who is your reader, and what relationship do you have?
- ☐ What background matters? When and how should you send the message?

Prepare to write.
- ☐ Gather details about the situation—the award, achievement, arrival, mistake, or loss. What is the occasion, and why is it important?
- ☐ Order the details naturally, and jot down a few phrases that you want to include in the message.

2 Draft
Organize the goodwill message into three parts.
[opening] State your main point—thank you, congratulations, welcome, apology, or sympathy.
[middle] Follow with details that show your interest or concern.
[closing] Conclude with a forward-looking statement.

3 Revise
Review your draft's ideas, organization, and voice.
- ☐ Is your reason for writing clear? Are your details accurate?
- ☐ Is your tone sincere—no flattery, pity, exaggeration, or excuses?

4 Refine
Check your writing line by line for the following:
- ☐ Precise and sensitive wording, not clichés; effective use of personal pronouns such as *you, I,* and *we*
- ☐ Smooth-reading sentences with strong transitions
- ☐ Correct grammar, spelling, and punctuation
- ☐ Personal yet professional format (parts, spacing, typography)

F.Y.I.
Choose a form that fits the goodwill message and situation.
- Write a letter on company letterhead when you represent the company.
- Write a note on personal letterhead, plain paper, or a handwritten card when you address business associates about personal matters.
- Write a memo when addressing co-workers about business matters.
- Write an e-mail in routine situations when a record isn't critical.

Apology

Workplace writers apologize to acknowledge an error, problem, or inconvenience, and to make amends, if possible. A good apology is professional and sincere.

Rhetorical situation

Purpose: To express regret for inconvenience and offer compensatory services
Audience: Hotel customer forced to stay elsewhere until a room became available
Format: Friendly, signed letter on hotel's stationery (offers official concern)

 **MAGNOLIA GRAND**
2580 Peach Tree Court
Memphis, TN 64301
901-555-5400 maggrand@hytp.com

July 7, 2005

Ms. Joan Meyer
605 Appleton Avenue
Green Bay, WI 53401

Dear Ms. Meyer:

Welcome back to Magnolia Grand! We are pleased to host you during your stay in Memphis. — **[opening]** Situation: Use a positive tone.

We apologize that your confirmed room was unavailable last night, and we are sorry for the inconvenience this may have caused you. Because several guests did not depart as scheduled, we were forced to provide you accommodations elsewhere. For your trouble, we were happy to cover the cost of last night's lodging.

Our goal now is to provide you with outstanding service and warm, Southern hospitality. To that end, we have upgraded your room at no expense to you. Please accept this box of chocolates as a gift to sweeten your stay with us. — **[middle]** Explanation: Apologize and explain what you are doing to make amends.

Should you need any assistance, please contact our Manager on Duty by dialing the hotel operator. You can contact me directly at extension 408.

We hope that the remainder of your stay with us is enjoyable. Thank you for your patience, understanding, and patronage. — **[closing]** Action: Stress further assistance and continued satisfaction.

Yours sincerely,

Mary-Lee Preston

Mary-Lee Preston
Front Office Manager

MP/am

Congratulations Message

A good congratulations message celebrates the reader's achievement, whether individual or group. A strong message is genuine and specific; it avoids any hints of envy, sarcasm, or self-interest.

Rhetorical situation

Purpose: To offer congratulations and praise to an award-winning company
Audience: Seven-year business contact with a fast-growing high-tech firm
Format: Signed letter on company letterhead (makes message official)

NOBEL LABORATORY SUPPLY
3765 Briarwood Road, Albany, NY 12211-1663
Phone 518-555-8800 • Fax 518-555-0241

February 12, 2005

Dr. Gail Kim
Head of Engineering
Rankin Technologies
3020 Brantley Road
Albany, NY 12211-4361

Dear Gail:

[opening] — Situation: State your congratulations.

I read in yesterday's *Albany Free Press* that Rankin has won the 2004 Albany Chamber of Commerce Outstanding Business Achievement Award in the large company category. Congratulations!

[middle] — Explanation: Provide precise details without exaggeration.

For seven years, I have been aware of Rankin's excellent work using ultraviolet technology for water purification. I've enjoyed watching the company grow, develop international markets, and refine environmentally friendly technology.

[closing] — Action: Focus on future success and cooperation.

I expect Rankin will continue to grow and be innovative in the years to come. Congratulations again, Gail, to you and your staff.

Yours sincerely,

Robert Jordan

Robert Jordan
President

Sympathy Message

Although it may at first sound strange, a sympathy message is a goodwill message and, therefore, is good news. While the *situation* (the death of a loved one, a divorce, an accident) may be painful or grief-filled for the reader, the *sympathy message* is likely welcome to the reader because it shows concern and caring. A strong sympathy message is personal, specific, heartfelt, and sensitive; it avoids clichés, platitudes, and glib advice.

Rhetorical situation

Purpose: To show care and concern; to offer support in the face of grief
Audience: Grieving co-worker whose son has died of leukemia
Format: Handwritten note (offers personal quality to message)

February 4, 2005

Dear Barb,

Andrea and I were saddened to learn of Gary's death. We know how courageously he faced the past four years with leukemia. I know his struggle was yours and Robert's, too.

Gary was a fine young man. I remember all the stories you shared through the years. I also remember Gary's teenage antics at company picnics and bowling tournaments (especially his Groucho Marx imitation). He had such a sense of humor. We won't forget him.

If you would like to talk, please give me a call. Perhaps Andrea and I can take you and Robert out for dinner sometime next week.

Please know, Barb, that Andrea and I wish you, Robert, and the rest of your family peace even in the midst of your grief.

With sympathy,
Matt

[opening]
Situation: Express your sympathy, but don't presume to know the reader's feelings.

[middle]
Explanation: Share significant memories. Be specific and avoid clichés.

[closing]
Action: Offer what help you can, but don't give advice. Simply show that you care.

Thank-You Message

Like other goodwill messages, a thank-you note, e-mail, or letter is not directly related to conducting business. Indirectly, however, thanking someone for a job well done, for help offered and delivered, or for goals met and perhaps exceeded is good for business. Why? Because giving thanks validates the reader and strengthens working relationships.

Rhetorical situation

Purpose: To express appreciation for a newly installed sound system
Audience: Owner of an electronics retailer, with employees as secondary readers
Format: Letter on company letterhead (creates professional tone)

Hope Services *Child Development Center*
2141 South Fifth Place, Seattle, WA 90810 • Telephone 436-555-1400
www.hopeserv.org

May 16, 2005

Mr. Donald Keebler
Keebler Electronics
466 Hanover Boulevard
Penticton, BC V2A 5S1
CANADA

Dear Mr. Keebler:

[opening] — *Situation: State your thanks directly.*

On behalf of the entire staff at Hope Services, I want to thank you for helping us choose a sound system that fits both our needs and our budget. Thanks, too, for working around our schedule during installation.

[middle] — *Explanation: Provide clear, specific details. Be personal and professional in tone.*

We have found that the system meets all our needs. Being able to adjust sound input and output for different uses in different rooms has been wonderful. The system helps staff in the family room with play-based assessment, and team members are tuning in to different conversations as if they were in the room themselves. As a result, children who might feel overwhelmed with too many people in the room can relax and play naturally. In addition, parents use the sound system to listen in on sessions in the therapy room as therapists model constructive one-on-one communication methods with children.

[closing] — *Action: Use the reader's name and stress cooperation and future contact.*

Thanks again, Donald, for your cooperation and excellent work. I would be happy to recommend your services to anyone needing sound equipment.

Yours sincerely,

Barbara Talbot

Barbara Talbot
Executive Director

Welcome Message

A welcome message aims to make the reader feel at home—in your company, community, and so on. For anyone joining a group, a good welcome message creates a sincere, personal connection and helps orient him or her to the group.

Rhetorical situation

Purpose: To make an employee feel welcome and to prepare for orientation
Audience: New production manager with several years of experience in industry
Format: E-mail (quickly and conveniently conveys basic message)

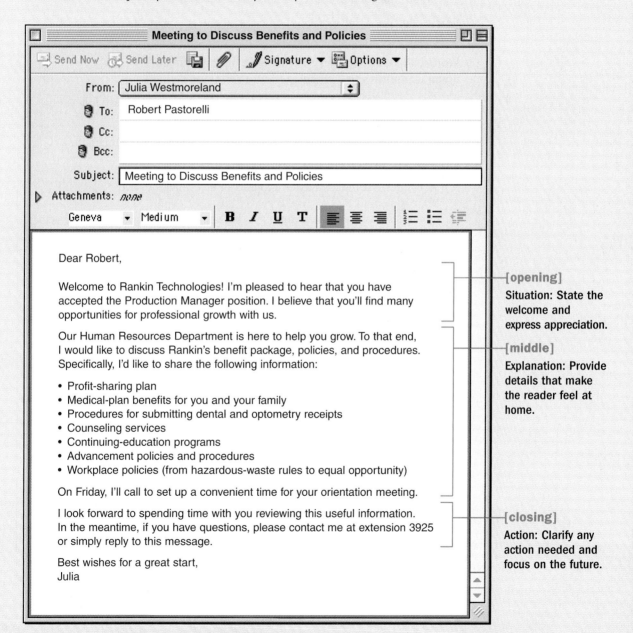

Dear Robert,

Welcome to Rankin Technologies! I'm pleased to hear that you have accepted the Production Manager position. I believe that you'll find many opportunities for professional growth with us.

Our Human Resources Department is here to help you grow. To that end, I would like to discuss Rankin's benefit package, policies, and procedures. Specifically, I'd like to share the following information:

- Profit-sharing plan
- Medical-plan benefits for you and your family
- Procedures for submitting dental and optometry receipts
- Counseling services
- Continuing-education programs
- Advancement policies and procedures
- Workplace policies (from hazardous-waste rules to equal opportunity)

On Friday, I'll call to set up a convenient time for your orientation meeting.

I look forward to spending time with you reviewing this useful information. In the meantime, if you have questions, please contact me at extension 3925 or simply reply to this message.

Best wishes for a great start,
Julia

[opening]
Situation: State the welcome and express appreciation.

[middle]
Explanation: Provide details that make the reader feel at home.

[closing]
Action: Clarify any action needed and focus on the future.

CHECKLIST for Good-News and Neutral Messages

Use this checklist to evaluate and revise your good-news and neutral messages.

- ☐ **Ideas.** The message states main and supporting points clearly, precisely, and accurately, supplying readers with all the information needed.
- ☐ **Organization.** The message follows the SEA formula (situation, explanation, action) outlined on page 399.
- ☐ **Voice.** The tone is businesslike, polite, confident, and open; it is not rushed, disrespectful, abrupt, or manipulative.
- ☐ **Words.** Phrasing includes everyday language understood by readers; unfamiliar phrases and obscure technical terms are defined.
- ☐ **Sentences.** Patterns are clear and concise. Sentences read smoothly, follow one another logically, and are linked by helpful transitions.
- ☐ **Copy.** The message is free of errors in grammar, spelling, and punctuation.
- ☐ **Design.** The message uses
    - Correct formatting for a letter, a memo, or an e-mail message.
    - White space and easy-to-read type.
    - Easy-access features such as numbers, bullets, or graphics.

CRITICAL THINKING Activities

1. Briefly explain why each type of message in this chapter (informative, etc.) is good or neutral news. Support your answer with details from the models.
2. Find three good-news or neutral pieces of correspondence. For each, identify the type of message, evaluate its effectiveness, and revise it as needed.
3. Choose any three models from this chapter. For each, highlight the main-point sentence and explore how the message does or does not illustrate the "art of being direct" (see page 398).

WRITING Activities

1. Identify a situation at your present or past job or internship when an employee should write a good-news or neutral message. Use the appropriate guidelines and model to write that message.
2. Identify a situation in your college work (e.g., communication with a professor, a staff member or administrator, fellow students, or someone in the community) that would require a good-news or neutral message. Use the appropriate guidelines and model to write the message.
3. Identify someone who has helped you in your education or career, and write the person a thank-you message.

In this chapter

The Art of Being Tactful **426**

Guidelines for Denying Requests **428**

Guidelines for Rejecting Suggestions, Proposals, or Bids **436**

Guidelines for Explaining Problems **440**

Guidelines for Resigning **444**

Guidelines for Making Claims or Complaints **446**

Checklist for Bad-News Messages **450**

Critical Thinking Activities **450**

Writing Activities **450**

CHAPTER
26

Writing Bad-News Messages

Being the bearer of bad news used to be dangerous. In ancient times, the legend goes, people might kill the messenger if they disliked the message! While today business may be more civilized, the fact remains that a bad-news message is one that your reader doesn't want to receive.

To deliver bad news, you have two major choices regarding how to do so. First, you can state the bad news right away. Second, you can soften it by leading up to it with a "buffer" and an explanation. If the bad news is minor, if your reader expects it, or if your relationship can take it—go ahead and be direct. In most cases, however, the second approach is preferable.

While no one likes to deliver bad news, this chapter will help you do it effectively and diplomatically. It opens with "The Art of Being Tactful" and the BEBE formula. The first piece offers tips on how to be indirect, thoughtful, honest, and conciliatory; the second piece offers tips on how to organize bad-news messages. Following these pages, a series of guidelines and models illustrate how you can write sixteen types of bad-news messages, which are classified into five categories.

Study the chapter carefully so that you understand how to write bad-news messages, and then complete the end-of-chapter activities to practice the task.

THE ART OF BEING TACTFUL

> "Tact is the art of making a point without making an enemy."
> —Howard W. Newton

Because bad-news messages share what your reader won't like hearing, they require tact. To write tactful messages, be indirect, thoughtful, and honest.

Be Indirect

Resist blurting out the bad news right away, an approach that alienates readers and may close their ears to whatever you say afterward. Instead, soften the blow by building up to the bad news with an explanation that readers can understand and accept.

Be Thoughtful

Give readers a coated bad-news pill, not a sharp needle. For example, offer some good news or affirm the reader in some way, if possible. Similarly, avoid stressing blame or engaging in destructive criticism. Be careful, in particular, with the pronoun *you*, avoiding it when it seems to attack the reader. Instead, establish common ground, a *we* focus rather than *us versus you*. Focus on agreement, cooperation, and choices. Finally, use the passive voice of verbs to guard the reader's ego. (See page 266.)

> **TOO DIRECT:** Your offer simply wasn't good enough to tempt me.
>
> **TACTFUL:** Although your proposal was attractive, I must turn down your offer.

Words and phrases to avoid

I am surprised	I question your	Measures must be taken
Company policy prohibits	You failed to	You must admit/accept
Company policy plainly states	You obviously did not	I regret to inform you
Are not able to	I cannot understand your	Unjustified
Must refuse/reject	Contrary to what you say	Misinformed
You claim that	I trust you will agree	Obviously your problem
Has never happened before	Negligent	Unreasonable expectations

Words and phrases to use

Thank you	Proceed	I encourage	Explore all the options
I agree	I appreciate	Look forward	I commend
Your concern	We can address	Consensus	Improvement
Prior commitments	Take positive steps	The policy aims to	I have reviewed

Be Honest

While being thoughtful, always state the truth. State the bad news clearly and firmly—no vague, weak, or deceptive phrasing. Moreover, be sincere, not syrupy. Avoid, for example, overdone apologies. Avoid, as well, hiding behind company policy. If you refer to a policy, explain its relevance. Finally, express regret for the bad news and its effects (e.g., inconvenience), but don't apologize for the decision or the bad news itself.

Use the BEBE Formula

Bad news is softened when you follow this pattern: *buffer, explanation, bad news, exit*.

Buffer

Open with a buffer—a neutral statement indicating the message's purpose. For example, offer agreement, understanding, or appreciation. Be sincere, but not so upbeat that the reader expects good news to follow.

> As your insurance agent, I need to update you on the status of your automobile insurance.

Explanation

Build toward the bad news with an explanation that follows naturally from the buffer. Offer legitimate, researched business reasons, not personal ones. Be objective and factual in tone, not aggressive or defensive. Moreover, anticipate and answer the reader's likely questions. Finally, avoid double-talk, jargon, vagueness, and insensitive word choice (see pages 248–257).

Bad News (plus alternative)

State the bad news honestly but tactfully, being especially careful with *you*. Position the bad news, if possible, in the middle of a paragraph, not at the beginning or end (positions of emphasis). If possible, offer the reader an alternative or a compromise. Focus on what *can* be done, not what *can't*.

> To keep premiums low, some insurance companies refuse to cover high-risk drivers. In the past two years, you have had an at-fault accident and four moving violations. As a result, your current auto-insurance carrier, Greenwood Fire & Auto, has chosen not to renew your policy (#46759) effective February 15, 2005. However, I have found three companies that would be willing to offer you coverage.

Exit

End as positively as possible, creating closure for the situation. For example, anticipate future work or contact. For any alternative offered, give details that the reader needs to act. In particular, avoid sounding apologetic, wishy-washy, or too upbeat. Be polite and encouraging, but don't assume how the reader will respond.

> Please call me at (612) 555-0020 so that we can determine the best way to meet your auto-insurance needs.

F.Y.I.

Here's some additional advice for writing tactful bad-news messages:

- Is writing the best choice, or should this issue be handled in person?
- Be timely. Hearing no news is worse than getting bad news.
- Consider possible secondary readers. How would they respond?

GUIDELINES FOR DENYING REQUESTS

Your goal when denying a request is to write a gracious refusal that phrases your "no" carefully. You may say no to a funding request, an inquiry, a donation request, a claim or complaint, or a credit application.

1 Prewrite

Rhetorical analysis: Consider your purpose, your audience, and the context.
- How can you convince the reader that your "no" is necessary and fair while still maintaining his or her goodwill?
- Who is your reader? What did he or she ask for, and why? How strong or personal is the issue for the reader?
- Is there important background or history that lies behind this request? What might be the consequences of saying no?

Prepare to write.
- Review the original request, looking for specific issues that you must address.
- Develop your rationale for turning down the request.
- If possible, explore other options for the reader (e.g., alternatives).

2 Draft

Organize your thoughts with the reader in mind.
[opening] Normally begin with a buffer, a neutral statement.
[middle] Build up to the request denial with sound reasoning:
- Be brief. One good reason is preferable to several weak ones.
- Offer reasons that are business-related, not personal.
- When referring to company policies, explain them; don't hide behind them.
- Follow up with an alternative, if one exists.

[closing] Express regret (without apologizing) and end politely.

3 Revise

Review your draft's ideas, organization, and voice.
- Have you supplied accurate information that helps the reader say, "I understand"?
- Have you followed the BEBE formula (page 427) to soften the request denial?
- Have you used a sincere, gracious tone that avoids a *we versus you* attitude?

4 Refine

Check your writing line by line for the following:
- Neutral, exact, and sensitive wording
- Easy-to-read sentences with smooth transitions
- Correct grammar, punctuation, usage, and spelling
- Effective format (parts, spacing, typography)

WRITE Connection

For more help with word choice, see "Showing Respect for Diversity" (page 78) and "Avoiding Negative Words" (page 255).

Funding-Request Denial

Turning down a request for funding—whether for a project, equipment, personnel, or other resources—involves balancing tact and encouragement with an explanation that the need isn't presently a top priority or that the funding isn't available.

Rhetorical situation

Purpose: To turn down a training request while exploring other options
Audience: Sales manager committed to sharpening sales staff skills
Format: E-mail (appropriate reply to initial e-mail request)

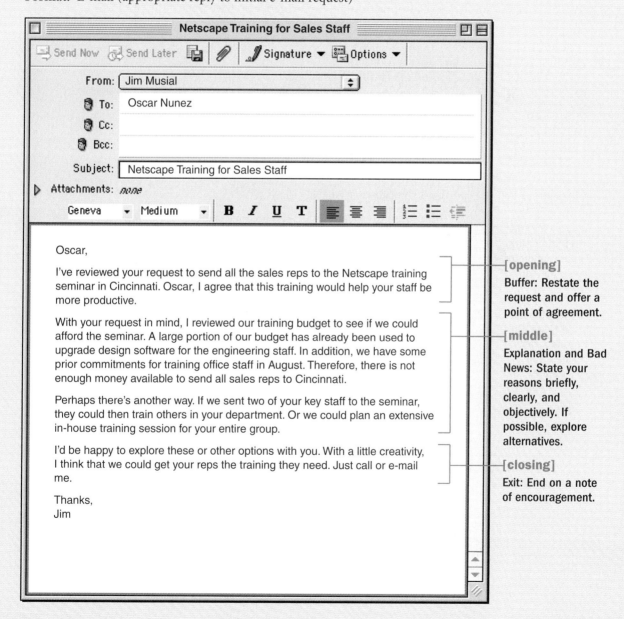

Email content:

Oscar,

I've reviewed your request to send all the sales reps to the Netscape training seminar in Cincinnati. Oscar, I agree that this training would help your staff be more productive.

With your request in mind, I reviewed our training budget to see if we could afford the seminar. A large portion of our budget has already been used to upgrade design software for the engineering staff. In addition, we have some prior commitments for training office staff in August. Therefore, there is not enough money available to send all sales reps to Cincinnati.

Perhaps there's another way. If we sent two of your key staff to the seminar, they could then train others in your department. Or we could plan an extensive in-house training session for your entire group.

I'd be happy to explore these or other options with you. With a little creativity, I think that we could get your reps the training they need. Just call or e-mail me.

Thanks,
Jim

Annotations:
- **[opening]** Buffer: Restate the request and offer a point of agreement.
- **[middle]** Explanation and Bad News: State your reasons briefly, clearly, and objectively. If possible, explore alternatives.
- **[closing]** Exit: End on a note of encouragement.

Negative Response to an Inquiry (Personal Letter)

As the opposite of a positive response to an inquiry (page 410), the negative response to an inquiry essentially argues that the writer or the company does not have or cannot share the information requested, or that the writer cannot give a positive answer to the reader's question.

Rhetorical situation

Purpose: To tactfully explain that an employee's treatment at a specific practice is not covered by the company's medical program
Audience: Chiropractic doctors, partners who have treated the employee
Format: Formal, signed letter by management (gives weight to response)

R T Rankin Technologies
401 South Manheim Road ❖ Albany, NY 12236 ❖ Phone 708.555.1980 ❖ Fax 708.555.0056

May 21, 2005

Brandon Chiropractic Center
119 Franklin Street
Pasadena, CA 91103-0417

Dear Dr. Brandon and Dr. Willoughby:

[opening] — *Buffer: Show appreciation for the inquiry.*

Thank you for your letter of May 14 concerning Brad Sneed's treatment. I appreciate your concern for our employees.

[middle] — *Explanation and Bad News: Give sound, factual business reasons and explain policies. State the bad news politely.*

Because Rankin Technologies has certain legal rights and obligations under worker's compensation law, we have developed a company policy for Rankin employees who require medical care due to a work-related accident. They must visit designated medical providers. If employees wish to discontinue care with designated providers, they become responsible for the subsequent cost of their care. The list of designated providers (on which your agency's name does not appear) has been put together to meet our employees' needs and our worker's comp obligations.

Our medical coordinator, John Anderson, enforces Rankin's Designated Medical Providers Policy consistently. His decision not to approve Brad's treatment at your clinic was a matter of company policy and in no way intended to discourage any of our employees from using your services.

[closing] — *Exit: End positively.*

Thank you, again, for your concern. Please accept my best wishes for the continued success of your practice.

Sincerely,

Miriam Helprin

Miriam Helprin
Director of Human Resources

Negative Response to an Inquiry (Form Letter)

When you get frequent inquiries that require a bad-news response, you can develop a polite, tactful form letter that reaches a diversity of readers.

Rhetorical situation

Purpose: To inform readers that a project to help them doesn't exist in their area
Audience: Generally low-income individuals and families in need of decent, affordable housing
Format: Standard form letter (seeks to speak to each reader's perspective)

HOME BUILDERS

1650 Northwest Boulevard • St. Louis, MO 63124
314-555-9800 • FAX 314-555-9810 • www.homebuilders-stl.org

Date

Name
Address
City, State ZIP Code

Dear :

Thank you for asking Home Builders to help you purchase a home. I appreciate your need for decent, affordable housing. —[opening] **Buffer: Connect with the reader's need.**

The literature that I have enclosed stresses partnership. To help people purchase adequate housing, Home Builders arranges a partnership between the homebuyer and local Home Builders volunteers who help build the home. The buyer must help to plan and build the house (usually about 500 hours) and must accept full responsibility for the cost of the house. Home Builders charges the buyer only the actual costs of materials and does not charge interest. —[middle] **Explanation and Bad News: Provide details to put the bad news in context.**

In North America, Home Builders assists people in areas where there are local projects. Because we do not have a project in your area, we cannot help you purchase a home at this time. However, because all projects are started by local people, I encourage you to share the enclosed literature with leaders in your community. If you find others interested in starting a project, please contact me, and I'll send information about setting up a Home Builders group in your area. **Present the bad news and follow up with an alternative.**

I hope the enclosed material answers your questions and gets you interested in starting a local project. —[closing] **Exit: End politely and positively.**

Yours sincerely,

Staff Name
Staff Title

Enclosures (3)

Negative Response to a Donation Request

Businesses and their employees are often asked to donate money, time, and other resources to causes. In a "no" response to a donation request, your aim is to politely turn down the request while showing respect for the cause. In your explanation, stress your present commitments—your commitment of available funds to other causes, or your commitment to causes closely related to your business, community, and beliefs.

Rhetorical situation

Purpose: To turn down a domestic-violence nonprofit group's request
Audience: Executive director of the organization who sent the original request; someone deeply concerned with women's and children's welfare and justice
Format: Management letter on letterhead (makes response official but sincere)

Rankin Technologies
401 South Manheim Road ❖ Albany, NY 12236 ❖ Phone 708.555.1980 ❖ Fax 708.555.0056

April 18, 2005

Ms. Marlis DeQuincey
Executive Director
Family First Center
468 Provis Way
Fairfield, NY 12377-2089

Dear Ms. DeQuincey:

[opening] — Buffer: Express interest in the reader's cause.

I read with interest your letter about Family First Center's project. I commend your efforts to build a shelter for women and children victimized by domestic violence.

[middle] — Explanation and Bad News: Provide clear reasons for not participating. State the refusal tactfully. If possible, offer an alternative.

I am honored that you have invited Rankin Technologies to participate in your project. Rankin seeks to be a good corporate citizen and a positive force in the community. To that end, we have already committed ourselves to partnerships with nonprofit organizations that mesh with Rankin's interests in the environment, in urban renewal, and in developing countries. For this reason, we cannot participate in your project at this time. Rankin employees will, however, be encouraged to continue to support your work in the community campaign. In fact, I will distribute materials about your project to our employees so that individuals may choose to get involved.

[closing] — Exit: Affirm the reader.

I wish you well, Ms. DeQuincey, in your important work of helping the victims of physical and emotional violence in this community.

Yours sincerely,

Barbara Reinholdt

Barbara Reinholdt
Office Manager

BR/dn

Negative Response to a Claim or Complaint

Saying no to a reader's claim or complaint amounts to disagreeing with his or her conclusions about fault or responsibility and communicating that you will not do what the reader wants. As the opposite of a positive response to a claim (page 417), such a negative response requires (a) great tact; (b) sound, compelling reasoning; and (c) sometimes a willingness to compromise.

Rhetorical situation

Purpose: To explain the source of a customer's problems and offer alternatives
Audience: Familiar client using product in electrical engineering application
Format: Formal, signed letter (creates record and makes response official)

1400 NW Academy Drive   **AC Drives**   Atlanta, GA 30425
Phone 412/555-0900   Fax 412/555-0054

June 16, 2005

Mr. Jamaal Ellison
Southeast Electric
1976 Boulder Road, Suite 1214
Charlotte, NC 28261-1203

Dear Mr. Ellison:

We have finished investigating your concerns about the ATV16 drives that you installed for American Linc Company. We do understand that the drive and serial-link failures have inconvenienced both you and American Linc.

— **[opening]**
Buffer: Restate the problem and show concern.

After testing the drives you returned, our line engineer determined that they failed because the temperatures in the cabinet exceeded the maximum operating temperature of the drives, leading to electronic-component failure. As noted in the ATV16 manual, the drive may malfunction under such conditions. For this reason, we cannot repair the drives without charge. We would be happy, however, to consider the following solutions:

1. We could remove the drive's plastic cover and install a stirring fan in the enclosure to moderate the temperature.
2. We could replace the ATV16 drives with the ATV18 model, a model more suitable for the machine you are using. (If you choose this option, we would give you a 15 percent discount on the ATV18s.)

— **[middle]**
Explanation and Bad News: Use sound evidence and state the rejection clearly.

Offer helpful alternatives.

Please let me know how you would like to proceed. I look forward to hearing from you and to continuing our partnership.

— **[closing]**
Exit: Focus on the next step and on future business.

Yours sincerely,

Elaine Hoffman

Elaine Hoffman
Product Manager

Negative Response to a Credit Application (Business to Business)

When you reject a credit application, you aim to tactfully communicate that credit cannot be extended while still keeping the customer as a cash-only client. Arguing that the decision is in the best interest of both businesses and that you have used sound, widely accepted financial benchmarks is an important strategy. Consider, for example, these reasons typically cited for denying credit: lack of credit history, lack of business experience, problems with financial statements, a history of problems, large debt obligation, and unstable economic conditions.

Rhetorical situation

Purpose: To turn down a credit request based on company's large debt obligations
Audience: Office manager of advertising company; a repeat computer and software client
Format: Signed letter (creates record of denial while encouraging continued business)

---

miami computer enterprises
South Benton Mall, Miami, FL 33107
Phone: 305.555.9001 Fax: 305.555.9845
E-mail: gfarnsworth@mce.com

July 11, 2005

Mr. Alexander Bennitez
Office Manager
Nova Advertising
664 Helene Boulevard, Suite 200
Miami, FL 33135-0493

Dear Mr. Bennitez:

[opening] — *Buffer: Thank the reader for his or her business.*

Thank you for your recent order of Mac G4s. These machines have the speed and power to meet the graphics needs you specified.

[middle] — *Explanation and Bad News: Provide brief, sensitive reasoning for the rejection.*

I have reviewed your request to make this purchase on credit. Our credit check showed that at present Nova's obligations are full. Although I would like to extend credit, creating more obligations would not be good for Nova or Miami Computer Enterprises.

[closing] — *Exit: Stress the reader's value as a customer, and keep the door open for future business.*

I would be happy to reconsider your request in six months or so, after Nova has had time to address its current obligations. Until then, I encourage you to continue using MCE's 10 percent discount for cash purchases. Please call to let me know how you would like to proceed with your current order.

Yours sincerely,

Gavin Farnsworth

Gavin Farnsworth
Owner

Negative Response to a Credit or Loan Application (Business to Consumer)

Turning down a consumer's credit or loan application requires special tact because of individuals' concern for their creditworthiness. Sound business reasons and accepted benchmarks like those listed on the previous page are a good start, but you also want to be helpful, when possible, by encouraging solutions and alternatives—for example, taking steps to improve a credit rating, reapplying after certain conditions are met, or applying for a different type of loan.

Rhetorical situation

Purpose: To turn down the customer's loan application based on lack of credit history
Audience: Young woman seeking to open a gift shop by borrowing capital
Format: Signed letter on bank letterhead (makes denial official, but invites future contact)

LONE STAR BANK
5550 North Adeline Road, Houston, TX 77022
Phone 547.555.0100, FAX 547.555.7024, E-Mail contact@loanestar.com

May 5, 2005

Ms. Mary-Lou Twain
780 East 41st Street, Apartment 712
Houston, TX 77022-1183

Dear Ms. Twain:

Thank you for meeting with loan officer Jean Olms last Friday and applying for a loan to open your gift shop. —[opening] **Buffer:** Express appreciation for the application.

When we review an application, one of the factors that we consider is the applicant's credit history. A good credit history shows a pattern of paying obligations. At this time, because you have not established a credit history, we cannot approve your request to borrow $95,000. However, you can establish a good credit history in one of two ways:

- Apply for, use, and make prompt payments on a credit card.
- Take out and repay a smaller loan at Lone Star Bank. Just a $5,000 loan successfully repaid would establish a positive financial record.

—[middle] **Explanation and Bad News:** Provide objective reasons for the rejection and offer suggestions.

We hope that these suggestions will help you begin to establish a good credit history. Then you may reapply for the loan that you requested. —[closing] **Exit:** If appropriate, encourage applying when conditions change.

Sincerely,

Rodney Thayer

Rodney Thayer
President

RT/bjh

GUIDELINES FOR REJECTING SUGGESTIONS, PROPOSALS, OR BIDS

Your goals when rejecting suggestions, proposals, or bids are to remember that someone's pride is on the line and to state the rejection with tact.

1 Prewrite
Rhetorical analysis: Consider your purpose, your audience, and the context.
- ☐ How can you reject the proposal, not the person? What understandable reasons can you provide? What other options (if desirable) can you suggest?
- ☐ Who is your reader? Why did he or she share the suggestion, proposal, or bid? How is he or she invested in it?
- ☐ Does an important background and history lie behind the proposal? Will there be important consequences to saying no? Is timing an issue for the rejection?

Gather information and develop an explanation.
- ☐ Develop your rationale based on facts and sound reasoning. Collect details from the proposal itself, policy statements, comparisons with alternatives, and other forms of research.
- ☐ Shape your argument, select supporting details, and order your thoughts into a tactful but persuasive pattern.

2 Draft
Organize the rejection into three parts.

[opening] Begin with a buffer—a neutral reference to the proposal. Affirm the proposal in some way or commend the initiative shown.

[middle] Build up to the rejection with an explanation. State business-related reasons concisely, without apology. Avoid dwelling on the proposal's failures. Accept parts of the proposal or suggest alternatives if possible.

[closing] Thank the reader. If appropriate, request refinements, encourage future proposals, or suggest a meeting.

3 Revise
Review your draft's ideas, organization, and voice.
- ☐ Have you provided a logical explanation that the reader will understand and accept?
- ☐ Have you followed the BEBE formula (page 427) to create an indirect rejection?
- ☐ Have you used a natural, conversational tone throughout? Have you been especially tactful in the rejection statement?

4 Refine
Check your writing line by line for the following:
- ☐ Precise and sensitive wording (especially use of the pronoun *you*)
- ☐ Easy-to-read sentences with strong transitions
- ☐ Correct grammar, punctuation, usage, and spelling
- ☐ Effective format (parts, spacing, typography)

Suggestion Rejection

When rejecting a suggestion, you need to show that you genuinely considered and studied the reader's idea but that circumstances—either internal shortcomings or external forces—make adopting the idea inadvisable or impossible.

Rhetorical situation

Purpose: To reject the idea of telecommuting based on its disadvantages and a past study
Audience: Employee who has worked in the business office for six years
Format: E-mail (offers fitting method of response to original e-mail suggestion)

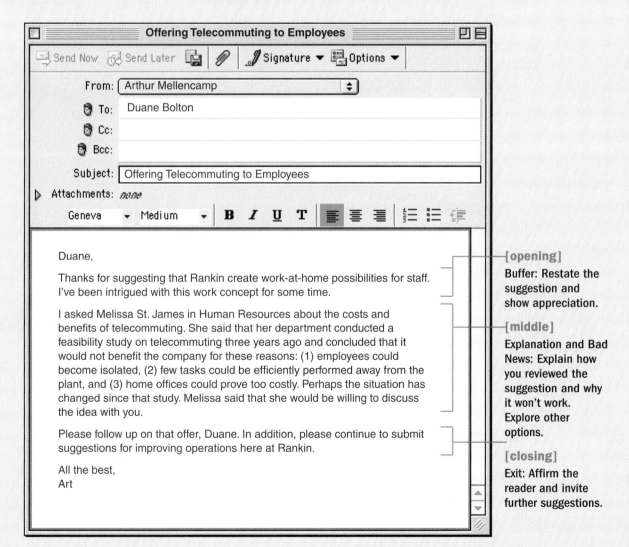

Proposal Rejection

Rejecting a proposal—especially when the reader put a lot of time, thought, and energy into it, hoping to bring about a positive change—requires that you tactfully argue that you cannot accept the proposal. Your reasons might include the following: The description of the problem or need isn't complete or accurate; the proposal needs to address alternative solutions; or the proposal isn't workable or desirable.

Rhetorical situation

Purpose: To politely decline a business offer while continuing a positive working arrangement
Audience: Owner of computer retail store; in need of consistent service and support
Format: Signed letter (makes response official)

Juanita Guiverra, Computer Consultant

368 Palm Palace Boulevard
Miami, FL 33166-0064
Telephone: 305.555.0010
FAX: 305.555.0500
E-Mail: jguiverra@cnsult.com

March 24, 2005

Mr. Gavin Farnsworth
Miami Computer Enterprises
South Benton Mall
Miami, FL 33107-1217

Dear Gavin:

[opening]
Buffer: Show appreciation.

Thank you for your proposal that I join your Customer Training Department. I appreciate your confidence in my ability to provide Miami Computer Enterprises' clients with instruction and technical support.

[middle]
Explanation and Bad News: Give your reasons objectively, stress positives in the proposal, state the rejection tactfully, and explore other options.

While considering your proposal, I reflected on the reasons that I started my own computer-consulting service two years ago. One of the reasons was flexibility. As an independent consultant, I could regulate my work activities around family demands. Although your proposal was financially attractive, I must turn down your offer, at least for now.

In 17 months (August 2006), my youngest child will enter grade school. If you are still interested in me at that time, I would be happy to reconsider your proposal. Until then, I hope you will want me to continue doing contract projects for MCE, especially with your Spanish-speaking clients.

[closing]
Exit: End positively.

Thanks again for your generous proposal. I wish MCE continued growth and success.

Yours sincerely,

Juanita Guiverra

Juanita Guiverra

Bid Rejection

A bid is a type of client proposal submitted to a potential customer in an effort to win a contract or business (see page 530). When you reject a bid, your aim is to tactfully convey that the reader has not won the business he or she desired. Do so by objectively describing which factors you considered, how they weighed in your decision, and why the winning bid was stronger.

Rhetorical situation

Purpose: To explain politely and objectively why another bid won the job
Audience: Owner/landscape architect of large regional landscaping firm
Format: Signed letter (fitting official response to formal bid)

EVERSON CITY PLANNING AND DEVELOPMENT COMMITTEE

Everson City Council · Everson, WA 98247-2311 · 306/555-2134 · www.eversonpdc.org

February 12, 2005

Mr. Felix Grove
Sea-to-Mountain Landscapers
8900 Coast Road
Seattle, WA 98134-6508

Dear Mr. Grove:

SUBJECT: Bid 4459 Everson City Park

Thank you for your bid to design and develop Everson's eight-acre city park adjacent to Kingston Elementary School and the Nooksack River.

Your bid was competitive for several of the criteria outlined in our original Request for Proposals (RFP). Your cost estimates, experience, and references were as strong as those from other bidders. However, Earth-Scape Design's overall plan tipped the bid in its favor. By including a variety of native plant species, Earth-Scape's natural, sustainable landscape will require less long-term care and create less stress on the Nooksack watershed. Because its plan contained a variety of plants, it also offered added educational value.

The Planning and Development Committee appreciates the work that you put into your proposal. We look forward to your interest in future Everson projects.

Yours sincerely,

Alice Potter

Alice Potter
Development Committee Chair

[opening]
Buffer: Specify the bid and thank the bidder.

[middle]
Explanation and Bad News: Highlight the reader's strengths objectively, but specify why another bid won.

[closing]
Exit: If appropriate, encourage bidding on future projects.

GUIDELINES FOR EXPLAINING PROBLEMS

Your goal when explaining a problem—a negative change, poor results, a crisis—is to share the bad news accurately and calmly. Delays, price increases, product recalls, losses, layoffs, injuries, closings, and strikes are bad news that no one wants to share. But if you have to do it, follow these steps.

1 Prewrite
Rhetorical analysis: Consider your purpose, your audience, and the context.
- ☐ Your aims are to accurately and honestly communicate the problem and its nature; to promptly short-circuit gossip and fear; and to seek cooperation to help overcome the difficulty. How can you accomplish these aims?
- ☐ To whom are you writing? Why is the news bad particularly for them? What do they already know and need to know about the situation?
- ☐ What aspects of the situation are sensitive or confidential? Should you consider important background, history, procedures, policies, and laws?

Research the problem.
- ☐ Gather accurate details about what went wrong, the causes, and the effects (especially those relevant to readers).
- ☐ Develop strategies for addressing the problem and moving forward.
- ☐ If appropriate, seek legal counsel before writing.

2 Draft
Organize the bad-news message tactfully.
[opening] Start with a buffer—a neutral statement clarifying the topic.
[middle] Explain the problem that has developed:
- ☐ Be honest and take ownership of the problem, while stressing the positive.
- ☐ Clarify what the news means for readers. If appropriate, apologize for any harm the news may cause.
- ☐ Focus on developing solutions, not assigning blame or making excuses.

[closing] Explain what to do with the bad news, and request feedback (if appropriate).

3 Revise
Revise your draft's ideas, organization, and voice.
- ☐ Have you provided a clear and accurate explanation?
- ☐ Have you followed the BEBE formula (page 427) to share the news indirectly?
- ☐ Have you used a natural tone that avoids not only gloom and doom but also false cheerfulness?

4 Refine
Check your writing line by line for the following:
- ☐ Careful, sensitive phrasing that avoids clichés
- ☐ Logical transitions that make the sentences flow smoothly
- ☐ Correct grammar, punctuation, usage, and spelling
- ☐ Effective format (lists, illustrations, spacing, typography)

Negative Change Announcement

A negative change is something that causes disruption, inconvenience, increased work, stress, hardship, or financial cost for your reader. In such situations, a tactful explanation of what has caused the change is a good starting point. In addition, you should consider ways to help the reader adjust to the change.

Rhetorical situation

Purpose: To explain an insurance-policy cancellation and offer alternatives
Audience: Young driver who is developing a spotty driving record
Format: Letter on company letterhead (offers official weight to message)

Wright Insurance Agency

3406 Capitol Boulevard, Suite 588
Washington, DC 20037-1124
Phone 612-555-0020
wright@insre.com

January 15, 2005

Policy 46759

Ms. Virginia Beloit
72 Elias Street
Washington, DC 20018-8262

Dear Ms. Beloit:

Periodically, insurance companies review their policies, assess the cost of offering the policies, and make changes where needed. When that happens, it's my responsibility as an insurance agent to inform my clients and help them make necessary adjustments. —[opening] **Buffer:** Introduce the topic and its context.

Last week Hawkeye Casualty, the company with which you have your auto insurance policy, discontinued all policies for drivers considered "high risk." Because you have had a traffic accident within the past 12 months and have received two speeding tickets during the same period, the company has relabeled your status as "high risk." As a result, Hawkeye Casualty has cancelled your auto-insurance policy effective January 31, 2005. However, I have found another company that will offer you auto insurance. While the cost of this new policy is somewhat higher than your present policy, the coverage is comparable, and the company is reliable. —[middle] **Explanation and Bad News:** Give rationale for the change.

State the change. If possible, offer help.

Please call me at 612-555-0020 within the next week so we can discuss the situation and decide how to proceed. —[closing] **Exit:** Explain what the reader should do.

Sincerely,

Eric Wright

Eric Wright
EW/rn

Poor Results Explanation

When sales are lower than projected, complaints are higher than normal, quality is less than excellent, or production is off target, these poor results need to be addressed. A good explanation shares the results objectively, analyzes causes, and promotes action to turn around the situation.

Rhetorical situation

Purpose: To press for improvements in customer service
Audience: Sales managers of each auto dealership within the company
Format: E-mail (shares message and attachments quickly)

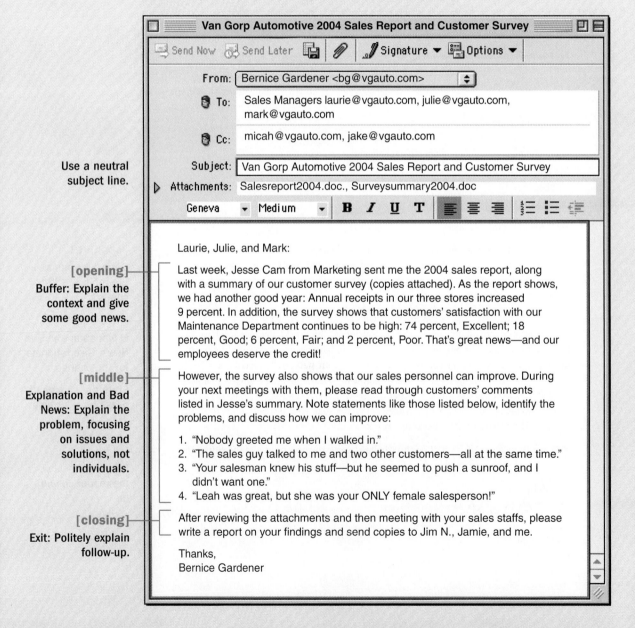

Crisis Management

In a message addressing a crisis, you want to avoid two extremes. On the one hand, you do not want to downplay the situation's seriousness. On the other hand, you do not want to sound so alarmed that you elicit panic. Instead, you must manage the crisis—a serious economic downturn, a breach of security or confidentiality, a lawsuit, wrongdoing by an employee, and so on—by facing the situation squarely, communicating information that is appropriate to share, and convincing your readers that you are acting forcefully to address the difficulty.

Rhetorical situation

Purpose: To share poor inspection results, along with the planned response
Audience: All employees, all concerned about errors and the company's future
Format: Paper memo (reaches all employees; lends managerial weight to message)

MEMO

Date: July 21, 2005
To: All Staff
From: Lawrence Durante, President
Subject: Recent FDA Plant Inspection

As you know, this past Monday, July 18, the FDA came to our plant for a spot inspection. I'm writing to share the inspection results and our response.

The good news is that the FDA inspectors did not find problems warranting a shutdown of Premium Meats. The bad news is that the inspectors cited us for three major violations, resulting in a fine of $90,000.

The FDA is sending us a clear message. We must take immediate steps to protect our customers, our jobs, and our company. To that end, I have taken the following steps:

1. The Executive Committee met with me to review the FDA report and determine the problem areas in our production process.
2. I have directed the Production Management Team to review quality-control procedures and conduct two retraining sessions immediately.
3. I have appointed a Quality Task Force of both management and production staff to study the production process and make further recommendations.
4. I have briefed Sales and Public Relations staff and directed them to contact customers and the media.

If you have any suggestions or questions, please speak to your immediate supervisor or a member of the Quality Task Force. (The Task Force membership, mandate, and time line are listed on the back.)

[opening]
Use a neutral subject line.

Buffer: State your reason for writing.

[middle]
Explanation and Bad News: State the case factually and calmly.

Focus on solutions: what has been done and what needs to be done.

[closing]
Exit: Stress a positive future, but be realistic. Ask for feedback, if appropriate.

GUIDELINES FOR RESIGNING

Your goal in writing a letter of resignation is to resign with tack and dignity. You'll also want to avoid writing anything that may come back to haunt you.

1 Prewrite

Rhetorical analysis: Consider your purpose, your audience, and the context.
- How can you tactfully communicate your decision to resign, while leaving the door open to future contact and cooperation?
- Which supervisor will read your resignation first? How good is your relationship with that reader? Will your resignation come as a surprise? Which secondary readers might see your resignation, and what would they think?
- How positive or negative has your experience with the organization been? Is there important history related to your resignation?

Brainstorm and research content for your resignation.
- Review your company's resignation policy, and make sure you follow all proper procedures, including setting a date for your last day.
- List your length of time in the position and nature of your experience.
- Note your reasons for leaving (considering your reader's response).
- Consider ways to make the transition a smooth one.

2 Draft

Organize the resignation notice carefully.

[opening] State your intention to resign politely, providing a date consistent with company policy. If you have already spoken to the reader about your resignation, remind the reader of that discussion.

[middle] Provide explanatory details.
- Express appreciation by focusing on the company's best qualities, your co-workers, or your growth.
- Give clear, concise reasons, resisting the temptation to vent or celebrate.
- Express regret, but don't be overly apologetic.
- Focus on the future—where you are going and why.

[closing] Offer to help train your replacement, and thank the reader.

3 Revise

Review your draft's ideas, organization, and voice.
- Have you provided a clear, concise explanation that doesn't reflect negatively on the reader?
- Have you given complete, accurate details consistent with company policy?
- Have you used a sincere but objective tone?

4 Refine

Edit the resignation message line by line for the following:
- Sensitive wording and easy-to-read sentences with strong transitions
- Correct grammar, punctuation, usage, and spelling
- Effective format (parts, spacing, typography)

Resignation Memo

Because of its official nature, a resignation should be formatted as an initialed or signed memo or as a signed letter. Consider, too, what protocol to follow. Generally, communicate your decision to resign in person. You can either submit your written resignation at that time or follow up the meeting by confirming in writing your resignation. If the situation with your supervisor is highly charged and you are afraid that you might say things you'll regret later, consider submitting a written resignation first—a resignation that is indirect, tactful, and brief.

Rhetorical situation

Purpose: To officially resign and clarify related details not covered in an initial meeting
Audience: Production manager who has been a good mentor
Format: Initialed memo (fitting for team member to manager situation)

> The hardest thing to learn in life is which bridge to cross and which to burn.
>
> —David Russell

MEMORANDUM

Date: June 10, 2005

To: George Grant
Production Manager

From: Michael Harrison *MH*
Production Engineer

Subject: Decision to Pursue Other Career Options

George, as we discussed earlier today, I have decided to resign my position as Production Engineer. In keeping with company policy, I will stay on for three more weeks. My last day of work at Micro Solutions will be July 1.

Working at Micro Solutions for the past three years has tested and stretched my college training, giving me some excellent practical work experience. I also enjoyed working with the people of MSI in a relaxed atmosphere.

I do believe, however, that I need a change. To that end, I have decided to form a partnership with a colleague from college. We plan to start our own software design firm in Dallas.

Please let me know how I can assist with transition issues during the next three weeks. Of course, I would be glad to participate in training my replacement. If you would like to schedule an exit interview, let me know.

Thank you, George, for your understanding and for the years we have worked together.

[opening]
Bad News: Communicate your decision in a neutral tone.

[middle]
Explanation: Compliment the reader and the company. Then, briefly outline your reasons without criticizing the company.

[closing]
Exit: Offer to help, and close with goodwill.

GUIDELINES FOR MAKING CLAIMS OR COMPLAINTS

Your goal when you write a claim or complaint, whether it's basic or serious, is to convince the reader to take responsibility for the issue and to resolve it. While doing so, you also want to maintain positive relations.

1 Prewrite

Rhetorical analysis: Consider your purpose, your audience, and the context.
- ☐ Aim to accurately and objectively describe the problem you experienced, and to prompt a concrete resolution.
- ☐ Identify the most appropriate reader (e.g., a sales representative, manager, department head): Who has the authority to address and resolve your complaint? Why should this person care about your claim?
- ☐ Clarify important background for the claim, and decide whether you should approach it as routine or serious.

Prepare to draft.
- ☐ List reasons supporting your claim, collect details about the situation, copy key documents, and outline steps you have taken so far.
- ☐ Especially for a serious claim, research the company and your rights by contacting the Better Business Bureau, the Chamber of Commerce, regulatory agencies, or a lawyer.

2 Draft

Organize your information with the reader in mind.

[opening] State the problem clearly and objectively.

[middle] Describe the problem, supplying relevant evidence:
- ☐ Dates of purchase, shipping, and conversations
- ☐ Copies of receipts, canceled checks, warranties, and correspondence
- ☐ Model, serial, catalog, purchase order, and invoice numbers
- ☐ Names of sales or customer-service representatives
- ☐ The problem's impact in terms of delays, inconveniences, or losses

[closing] Spell out what you want the reader to do—replace the product, refund your cost, credit your account, repair the product, cancel an order. Never threaten, but do note what will happen if the claim isn't settled.

3 Revise

Review your draft's ideas, organization, and voice.
- ☐ Is your message clear and complete?
- ☐ Have you used a firm but fair tone that asks for a reasonable solution?

4 Refine

Check your complaint line by line for the following:
- ☐ Precise wording, smooth sentences, and effective transitions
- ☐ Correct grammar, punctuation, usage, and spelling
- ☐ Effective format (spacing, typography, lists)

Complaint (Basic)

A basic claim or complaint outlines a fairly minor, routine problem—something is broken, missing, poorly serviced, late, and so on. In such situations, the business will likely respond positively to your claim in an effort to maintain your business. Businesses know, for example, that for every complaint received, eight or nine more exist but are never submitted. Such problems must be fixed and prevented from happening again.

Rhetorical situation
Purpose: To get an account credited properly due to some shipping errors
Audience: Shipping manager responsible for smooth processing of orders
Format: Signed letter in simplified format (communicates official weight)

---

brunewald systems design

5690 Brantley Boulevard, P. O. Box 6094
Trenton, NJ 08561-4221
451.555.0900

February 25, 2005

BHC Office Supply Company
39 Davis Street
Pittsburgh, PA 15209-1334

ATTENTION: Shipping Manager

I'm writing about a problem with the purchase order #07-1201. Copies of the original PO plus two invoices are enclosed.

[opening]
Buffer: Establish the claim's context.

Here is the sequence of events concerning PO 07-1201:

Dec. 16, 2004:	I faxed the original purchase order.
Jan. 7, 2005:	Because I hadn't heard from your office, I spoke with Kim in customer service. Then I re-sent the PO because she could not find the original in your system.
Jan. 17, 2005:	I received a partial shipment, with the remaining items back ordered (invoice 0151498).
Jan. 21, 2005:	I then received a second shipment that was complete (invoice 0151511). Noting the duplication, I contacted Kim, and she cancelled the back-ordered items.

[middle]
Bad News and Explanation: Tactfully spell out the facts.

Point out the results of the problem in a neutral tone.

I am returning the partial order (duplicate items) by UPS. Please credit our account for the following: (1) the duplicate items listed on invoice 0151498 ($563.85), (2) the shipping costs of the partial order ($69.20), and (3) the UPS costs to return the duplicate items ($58.10). The total credit comes to $691.15.

Specify the adjustment that you want.

I look forward to receiving an adjusted statement and to continued cooperation.

Gary Sheridan

GARY SHERIDAN—OFFICE MANAGER

GS/mc
Enclosures 3

[closing]
Exit: Anticipate future business.

Complaint (Serious)

Sometimes, making a claim isn't routine—that is, it is *not* simple and straightforward. If you believe that your reader must be strongly convinced to act, then write a persuasive claim that objectively builds a logical argument and firmly presses for a fair action.

Rhetorical situation

Purpose: To get repeatedly problematic mechanical repairs fixed as originally agreed; to seek compensation

Audience: Primary reader is director of operations for a heavy-equipment company; secondary readers may be managers and employees responsible for project, perhaps lawyers

Format: Signed, two-page letter on company letterhead, with copies sent to company's president (creates official weight and urgency)

 Rankin Technologies
401 South Manheim Road ❖ Albany, NY 12346 ❖ Phone 708.555.1980 ❖ Fax 708.555.0056

January 14, 2005

Mr. Steven Grinnel
Director of Operations
Industrial Aggregate Equipment Company
4018 Tower Road
Albuquerque, NM 87105-3443

Dear Mr. Grinnel:

[opening]
Bad News: Specify the problem and the reason for concern.

I am very concerned about the 40-foot Snorkel Lift that we contracted with you to rebuild when we traded in our old Marklift. Continued delays in the rebuilding schedule and subsequent problems with the lift itself leave me uncertain about Industrial Aggregate's ability to provide Rankin Technologies with continued service.

Here is an overview of the problem:

[middle]
Explanation: Provide a detailed outline of the problem and its history.

Keep your tone neutral.

1. We ordered the Snorkel Lift in April 2004, and you promised delivery in July. We did not receive the lift until September.

2. When the lift arrived, we noticed several key parts had not been replaced, and the boom did not operate correctly. Your project supervisor, Nick Luther, assured us that the parts would be fixed in a timely manner, and he provided a substitute lift free of charge.

3. Two months later, Mr. Luther called to say that everything was fixed. However, when we visited your facility on December 17, the gauges and tires on the lift had not been replaced, and the dual fuel unit had not been installed.

Page 2
Steven Grinnel
January 14, 2005

4. When we finally received the Snorkel Lift on December 22, several items that we noticed on December 17 still had not been fixed. In fact, the lift still had these deficiencies:

- Several oil leaks
- Missing "on/off" switch in the basket
- No dual fuel capabilities
- Boom vibration when retracted after full extension

We have been extremely disappointed with the lift's condition and overall performance. Your original promise of a fully operational Snorkel Lift in "like new" condition by July 2004 (agreement copy enclosed) has not been met.

In the past, we have appreciated your service and assistance. From our experience of the past nine months, however, we can only conclude that you are experiencing problems that make it difficult for you to provide the service that Rankin Technologies needs.

We want to resolve this issue. By February 14, please provide us with a lift that meets all the specifications agreed to and that has no operational deficiencies. If you are unable to provide the lift by that date, we will cancel our order and seek reimbursement for the used Marklift we traded in April 2004.

Sincerely,

Jane Ballentine

Jane Ballentine
Maintenance Project Engineer

JB/rd
Enc.: copy of agreement
cc: Andrew Longfellow
President, Industrial Aggregate Equipment Company

[middle]
Be specific and factual.

Give needed background and attach relevant support.

[closing]
Exit: State the proposed solution clearly and firmly.

Distribute copies to appropriate authority figures and concerned parties

> "Yesterday was the deadline for all complaints."
> —Jobo proverb

WRITE Connection

To avoid negative, potentially libelous phrasing, see "Avoiding Negative Words," pages 255–256.

CHECKLIST for Bad-News Messages

Use this checklist to evaluate and revise your bad-news messages.

- ☐ **Ideas.** The message presents all the facts accurately and focuses on solutions.
- ☐ **Organization.** The message follows the BEBE formula (buffer, explanation bad news, exit) as outlined on page 427.
- ☐ **Voice.** The tone is sensitive, honest, accurate, and objective (but firm when needed). It avoids defensiveness and anger.
- ☐ **Words.** Phrasing conveys the bad news clearly but tactfully; avoids the pronoun *you* if it sounds accusatory; and uses neutral or positive language.
- ☐ **Sentences.** The sentences read well aloud, are succinct, and use the passive voice to soften negative or difficult statements.
- ☐ **Copy.** The message is free of grammar, spelling, and punctuation errors.
- ☐ **Design.** The message follows correct memo, e-mail, or letter format; features an attractive layout with ample white space; and organizes points and details with numbers, bullets, or graphics.

CRITICAL THINKING Activities

1. Briefly explain why each form in this chapter is a bad-news message. Support your answer with details from the models.
2. Choose three models from this chapter. For each, identify the sentence containing the bad news and then explain why the message does or does not illustrate the "art of being tactful" (see page 426).
3. Find three bad-news pieces of correspondence. For each, identify the type of message, evaluate its effectiveness, and revise it as needed.

WRITING Activities

1. Recall (or create) a workplace situation that warranted a bad-news message. List details about the situation and then write the document.
2. Recall (or create) a college-related situation that involved bad news (e.g., an unsuccessful scholarship or financial-aid application, a tuition hike, a cancelled student-club event). Write the message, using an appropriate medium (e.g., memo, e-mail, letter). Then ask a classmate to use the checklist above to assess your message's quality. Revise as needed.
3. Recall (or create) a situation in which you, as a consumer, were dissatisfied with a product or service you purchased or experienced. Draft an appropriate claim or complaint letter for this situation.

In this chapter

The Art of Persuasion 452

Guidelines for Special Requests and Promotional Messages 454

Guidelines for Sales Messages 460

Guidelines for Collection Letters 466

Guidelines for Requesting Raises or Promotions 470

Checklist for Persuasive Messages 472

Critical Thinking Activities 472

Writing Activities 472

CHAPTER
27

Writing Persuasive Messages

All persuasive messages are sales pitches—whether you're selling an idea, a product, a service, or a special request. However, even though you may have a great idea, a dynamite product, or a noble cause, your readers won't necessarily see that. They may be indifferent to the issue or even resistant to your message.

How, exactly, do you persuade readers to accept your point of view? How do you sell the value of your idea, product, service, or request without resorting to a hard sell, scare tactics, or an overly aggressive manner? If you can speak to your readers' needs and focus on how they will benefit, then your letter, memo, or e-mail may well produce the results you want.

This chapter opens with "The Art of Persuasion" and the AIDA formula. The first section will help you focus on the reader's needs, find a fitting appeal, build credibility, prove your point, and say it well. The second section will help you get the reader's attention, pique his or her interest, and ask the person to take action. The remainder of the chapter offers guidelines, models, and a checklist for the range of persuasive messages that you might have to write at work.

Many persuasive messages (or sales pitches) are unconvincing. With the help of the advice in this chapter, your persuasive messages can be clear, genuine, and compelling.

THE ART OF PERSUASION

The heart of persuasive writing is offering a convincing argument while satisfying readers' needs. To do so, practice the strategies that follow.

Focus on Readers' Needs

How does what you are "selling" satisfy readers? Using Table 27.1, (1) locate needs that match your "product" and (2) use the corresponding appeals to develop a persuasive theme—a strong value or benefit—as the focus of your sales pitch. For example, your electric drill could help a customer become good at woodworking (the need). You could focus on the self-fulfillment (an appeal) of completing woodworking projects.

Table 27.1 Persuasive Appeals Focused on Readers' Needs

Reader Needs	Persuasive Appeals
To make the world better by helping others	To values and social obligations
To achieve by being good at something or getting recognition	To self-fulfillment or status
To belong by being part of a group	To group identity, acceptance, love
To survive by having things run smoothly, avoiding threats, or having necessities	To quality, reliability; security, savings, losses prevented; or value, sensations, satisfaction

Build Credibility

A persuasive message is credible—so trustworthy that readers can change their minds painlessly. To build your credibility, observe three rules. First, be thoroughly honest. Don't falsify data, spin information, or ignore facts. Second, make realistic claims and promises. Avoid emotionally charged statements, pie-in-the-sky forecasts, and undeliverable deals. Third, develop and maintain trust—in your attitude toward the topic, in your treatment of readers, and in your respect for the opposition.

Prove Your Point

Choose types of evidence from the following list that match your product and will convince your reader.

Facts	Descriptions	Visuals	Personal testimonies
Statistics	Examples	Comparisons	Expert evaluations
Test results	Samples	Experiences	Stories or anecdotes

Say It Well

Here are some examples of powerful persuasive words:

Reader's name	Endorsed	Money	Solution	Supported
You	Discover	Tested	Benefit	Offer
Please	Ease	New	Value	Reduce
Thank you	Guarantee	Proven	Increase	Privilege
Postage-paid	Health	Results	Growth	Credit
Complimentary	Love	Save	Approved	Discount

Use the AIDA Formula

Developed for sales letters, the AIDA pattern—*attention, interest, desire, action*—can work well, when modified, for any persuasive message.

Attention

Grab the indifferent or resistant reader by using creative openings (and avoiding weak openings).

Creative openings

Tell a brief story	Explore a comparison	Offer a challenge or puzzle
Use a powerful quote	Ask an interesting question	Play with a familiar saying
Speak of a current event	State a startling fact	Show knowledge of the reader

Weak openings

Obvious statement	Cliché	Writer-centered comment
Negative comment	Boring description	Gimmick
Bland question	Sterile statistic	

Interest and Desire

Create curiosity about your cause, product, or service; then move the reader toward "taking ownership" of it. Encourage him or her to care about what you're promoting, to want what you are selling. Try these strategies:

- Outline a need or problem. Then show your product, service, idea, or plan solving that problem or meeting that need.
- Picture the reader participating in the cause or using the product or service.
- Avoid clichés, exaggeration, manipulation, and attacks on competition.
- Balance logical and emotional appeals in a way fitting both your reader and your product, service, or cause.
- Counter obstacles that your reader might raise. For example, cite guarantees, free trials, discounts, service agreements, and public relations benefits.

Action

Ask the reader to take a realistic step. Select an action that is a good match for your cause, product, or service, and for your client. What can you realistically expect the reader to commit to?

Make a phone call	Complete a form	Request more information
Try the sample	Visit a Web site	Write a check
Bring in the coupon	Place an order or sign up	Review some literature

In addition to your request, it is best to

- Link acting to your key selling point—the reader's main benefit.
- Provide incentives to act quickly—a deadline or dated offer, for example.
- Put an incentive to act, a special offer, or a key selling detail in a postscript. (Most readers review the postscript after the introductory paragraph.)

GUIDELINES FOR SPECIAL REQUESTS AND PROMOTIONAL MESSAGES

Your goal when writing a special request or promotional message is to convince the reader that your cause is worthwhile so that saying yes seems both right and reasonable. Such special requests can press readers to cooperate, assist, participate, donate, or grant credit.

1 Prewrite
Rhetorical analysis: Consider your purpose, your audience, and the context.
- How can you educate your reader about your cause? Consider appropriate evidence—facts, statistics, examples, stories, quotations.
- To whom are you writing? What do they have that you need? What do they need to know about your cause and the situation?
- What background factors might influence the message? Is timing an issue? What ethical challenges will you face as you seek to persuade the reader?

Consider your reader's motivations, interests, and needs.
- What positive feelings can the reader gain by contributing?
- What social good is associated with your cause?
- What recognition, publicity, or contacts might the reader gain?
- What are the economic benefits (such as tax advantages)?

2 Draft
Organize the special request into three parts.
[opening] Get your reader's attention and create a theme (the need, the value).
[middle] Explain your need, make the request, and stress reader benefits through persuasive appeals. Follow these strategies:
- Use facts, quotations, examples, and stories to help the reader see, understand, and identify with your cause.
- Address obstacles and counter these objections if possible.

[closing] Confidently ask for a reasonable action. Supply forms, response cards, tear-off sections, postage-free envelopes, e-mail addresses, Web sites, toll-free numbers, or other appropriate choices.

3 Revise
Review your draft's ideas, organization, and voice.
- Have you provided a complete explanation that speaks to your reader's motivations and needs?
- Have you effectively followed but modified the AIDA formula?
- Have you used a sincere tone that avoids bullying, flattery, and exaggeration?

4 Refine
Check your writing line by line for the following:
- Exact, fresh phrasing rather than clichés
- Easy-to-read sentences that are generally short, in the active voice, and energetic
- Correct grammar, punctuation, spelling, and usage
- An attractive, professional format (e.g., margins, spacing, typography)

Persuasive Request for Cooperation

Sometimes, a request for cooperation is routine because the reader will quickly comply with it (see page 408). At other times, the request needs to be more persuasive because the cooperation may involve extra time, greater effort, or more personal cost. To prompt cooperation, stress mutual benefits and consider including incentives (e.g., bonuses, rewards, recognition, time off).

Rhetorical situation

Purpose: To encourage employees to complete a training program in a timely way
Audience: All employees of the airline; training is mandatory but has varying degrees of relevance to members of the group
Format: Printed company bulletin (gives weight to request and creates easy reference)

> *It is our attitude more than anything else that will effect a successful outcome.*
> —William James

EMPLOYEE BULLETIN

Date: November 19, 2005

From: Preston Emerson, President & CEO

Subject: WORKPLACE HAZARDOUS MATERIALS INFORMATION SYSTEM (WHMIS)

Our goal at Pacific Air is to create a safe workplace for all employees. One way we can achieve this goal is by following federal regulations for handling hazardous materials.

The Health and Safety Committee will therefore be training everyone on proper handling of hazardous materials. This general training has been approved by the Portland Health and Safety Board.

[opening]
Attention: Create a request theme that focuses on cooperation.

You should receive your training packet during the first week in December. Because effective training is the key to safety, please follow the instructions in the packet and complete the program and follow-up test by January 30. Once you complete the program, you will receive a WHMIS certificate.

As an incentive, if you complete your test by January 5 with all the answers correct, your name will be entered in a drawing for a getaway weekend:

- Hotel accommodations in Seattle and dinner for two
- Tour of Olympic National Park
- Confirmed travel on Pacific Air

[middle]
Interest and Desire: Provide background and instructions. Stress deadlines and benefits. Encourage cooperation.

So look for your packet early in December. Then get to work on upgrading your safety training to ensure our company's compliance and a safe workplace.

This program is a joint effort of the Health and Safety Committee and Pacific Air management. We appreciate your cooperation!

[closing]
Action: Specify the next step and anticipate compliance.

Persuasive Request for Assistance

When the assistance you need requires your reader's extra effort—a time or resource investment above his or her ordinary work assignment—then your request must be persuasive. Be clear and precise about the help needed, and when possible stress benefits for the reader and thanks for the benefits you yourself will receive.

Rhetorical situation

Purpose: To encourage helpful, interesting contributions to an open house
Audience: Engineers who have developed and tested the company's products
Format: Memo (creates some official weight and offers easy response section)

MEMO

Date: August 11, 2005
To: Felton Engineering Staff
From: Jilliane Seaforth *JS*
Subject: Ideas for Open House Displays

[opening]
Attention: Create a sense of shared purpose.

As you know, we will be moving to our new location on August 18, and we have scheduled an open house for September 1. To help visitors at that event learn what Felton Engineering does, I plan to set up displays showing samples of your unique heater designs and interesting product applications.

[middle]
Interest and Desire: Give any needed background information.

As you pack for the move, please help me by doing the following:

(1) identify products that would interest visitors, and (2) look for blueprints, sketches, small models, and prototypes that illustrate those products. (Remember that visitors may understand commercial applications more easily than technical military or aerospace designs.)

Ask for specific help.

Then, list below (1) your name, (2) the name of the product, (3) the product number, and (4) the type of display materials that you have.

[closing]
Action: Set a deadline, make cooperation simple, and be positive.

Please return this memo to me no later than August 22. I will pick up your materials, set up the displays, and return the materials to you after the open house. I look forward to turning your ideas and samples into great displays!

1. Your Name: _____

2. Product Name   3. Product Number   4. Display Materials
   _____   _____   _____
   _____   _____   _____
   _____   _____   _____
   _____   _____   _____

Request for Participation (Persuasive Invitation)

A persuasive request for participation aims to convince the reader to "sign up" for something or attend an event when he or she might at first be resistant or uninterested. Such an invitation needs to show how the benefits of participation outweigh the costs.

Rhetorical situation

Purpose: To encourage readers to "hire" university interns
Audience: Graphic designers—busy, artistic, project-oriented professionals who may feel uncertain about giving time and responsibility to an intern
Format: E-mail (allows for efficient review and response, especially with attachments)

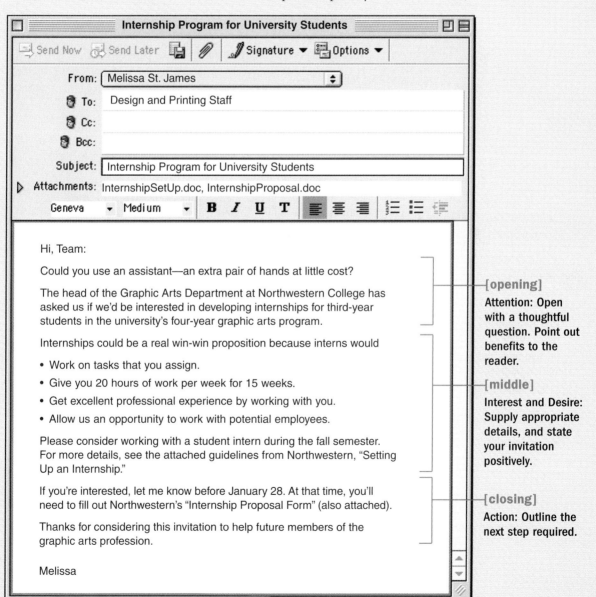

Fund-Raising Message

The aim of a fund-raising message is clearly to prompt a donation. To encourage such action, your message should get the reader to care enough about your cause and to see its benefits so that he or she becomes willing to part with a fitting level of funds.

Rhetorical situation

Purpose: To solicit a repeat (and increased) donation to an annual nonprofit campaign
Audience: CEO of successful publishing company; interested in literacy
Format: Signed, formal letter on organization's letterhead (gives request weight)

NATIONAL CAMPAIGN FOR LITERACY
1516 West Elizabeth Terrace, Wadsworth, IL 60421
Phone 431-555-9000, Fax 431-555-1066, Web Site ncl.org

November 11, 2005

Mr. Cecil Featherstone, CEO
Words, Words, Words, Inc.
541 West 34th Street
New York, NY 10001-7352

Dear Mr. Featherstone:

[opening — Attention: Use startling facts.]
More than 20 percent of adults in this country cannot read at a third-grade level. Each year, more than a million students leave high school functionally illiterate (some with diplomas).

[middle — Interest and Desire: Sell your cause with key details. Create a sense of urgency. Request a donation politely. List benefits for both giver and receiver.]
As you know, the National Campaign for Literacy has spent 14 years helping millions of citizens learn to read. We work with more than 300 schools, neighborhood groups, and government agencies to combat illiteracy. Yet, illiteracy remains an enormous problem. To address this need, we plan to fund 29 new programs this year, as well as to expand existing ones.

We appreciate your past generosity and hope we can count on your continued support. In addition, to enable us to help more people, we are asking that you please consider raising your donation level.

Of course, your gift will bring you recognition, including an acknowledgment in more than 100,000 promotional brochures. However, the greatest benefit comes as you help millions of people get better jobs and earn personal dignity.

[closing — Action: Make donating simple.]
Please continue supporting our effort to promote adult reading. You may make out your check to National Campaign for Literacy and return it to me in the enclosed envelope, or you may call me at 431-555-9000, ext. 0786.

Sincerely,

Gail Goldstein

Gail Goldstein
Associate Director

Credit Application

In a credit application, you seek to open a favorable line of credit with a business. To ensure success, you seek to prove that you are creditworthy and that the credit will facilitate a profitable business relationship.

Rhetorical situation

Purpose: To establish a healthy credit line with a supplier
Audience: Manager of supply company; relationship good but less than a year old; reader would be interested in increased business but is cautious about extending credit
Format: Signed letter on company letterhead (makes request formal and official)

Dale's Garden Center
484 Leeward Avenue, SE, Tuscaloosa, AL 35406-3770
Phone 205/555-8900 FAX 205/555-1600
E-mail grant@dalesgardencenter.com

January 3, 2005

Ms. Salome Nguru, Manager
Cottonwood Hills Greenhouse and Florist Supply
R.R. 2
Macon, GA 31220-2339

Dear Ms. Nguru:

For the past nine months, Dale's Garden Center has been ordering fresh, dried, and silk flowers from Cottonwood Hills. We have been impressed with the quality of your products, most recently with those we sold for the Christmas holidays.

We are now planning to expand our product offerings, particularly of silk flowers and bedding plants. For this reason, we expect to make larger orders more frequently. However, before we can submit the orders, we need Cottonwood Hills to set up an account for us with a $100,000 line of credit.

Dale's Garden Center has been in business for almost a year and is on solid financial footing. The attached references and financial statements show that we are strong and growing.

By January 24, we hope that you will be able to check our statements and references, send information about your credit terms, and confirm a credit line of $100,000. Dale's will then submit an order for spring plants.

Thank you for considering our request.

Yours sincerely,

Grant Bostwick

Grant Bostwick
President

Enclosures 4

[opening]
Attention: Stress positive aspects of the current relationship.

[middle]
Interest and Desire: Explain the need for credit.

Stress benefits for the reader.

Establish your credit record.

[closing]
Action: Ask for reasonable action and suggest further benefits.

GUIDELINES FOR SALES MESSAGES

Your goal when writing a sales message is to convince readers that they need something, that a particular product or service will satisfy their need, and that you can help them meet that need.

1 Prewrite
Rhetorical analysis: Consider your purpose, your audience, and the context.
- ☐ How can you convince readers to act, while promising only what you can deliver?
- ☐ To whom are you writing? How can you overcome their resistance or indifference to your product or service? What are their values?
- ☐ What background and circumstances relate to your message? What ethical issues relate to your sales pitch?

Know your product and the competition.
- ☐ List readers' needs that can be met by your product or service.
- ☐ List features and benefits of your product or service, comparing and contrasting these with your competitors' features and benefits.

2 Draft
Organize the message using the AIDA formula: attention, interest, desire, action (see page 453).

[opening] Get attention with a reader-centered statement that establishes your sales theme. Consider using a quotation, question, or startling fact.

[middle] Generate interest in and desire for your product.
- ☐ Outline a problem and show how your product or service solves it.
- ☐ Stress features and benefits, avoiding clichés, exaggerations, and attacks on the competition.
- ☐ Stress cost from the beginning if it's a key selling point; put it later if it's not, and stress value instead.
- ☐ Anticipate and counter objections to buying your product or service.

[closing] Call for a realistic action and provide incentives. (Use a postscript for an incentive, special offer, or a key selling point.)

3 Revise
Review your draft's ideas, organization, and voice.
- ☐ Have you included all key selling points in a persuasive order?
- ☐ Have you used a confident, sincere tone that fits what you're selling and who's buying?

4 Refine
Check your writing line by line for the following:
- ☐ Lively and fresh phrasing, including personal pronouns (especially *you*) and the reader's name, if possible
- ☐ Correct grammar, spelling, usage, and punctuation
- ☐ Attractive design (white space, boldface, bullets, graphics, color)

Sales Letter (First Contact)

In an initial sales letter, you aim to connect your product or service with a real need that the reader experiences, as well as to establish your credibility and to convince the reader that you can deliver the product or service efficiently.

Rhetorical situation

Purpose: To convince the reader to inquire about outsourcing projects
Audience: Chief creative executive of advertising agency
Format: Signed letter on quality letterhead (communicates writer's credibility)

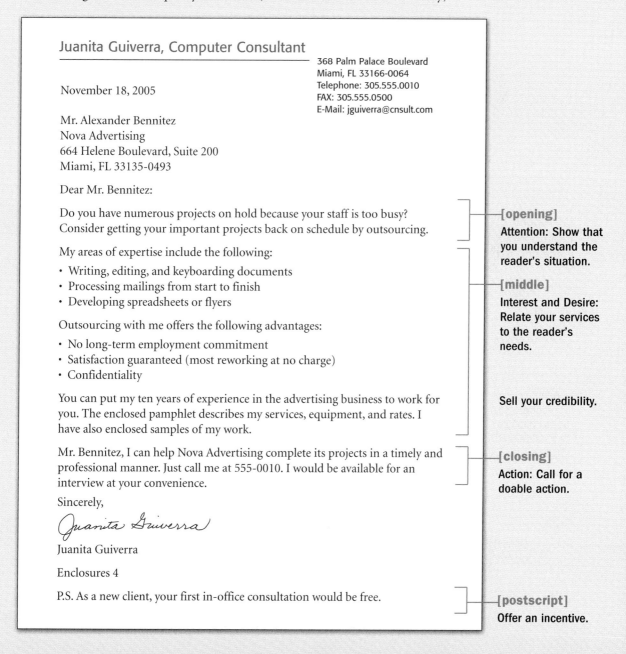

Juanita Guiverra, Computer Consultant

368 Palm Palace Boulevard
Miami, FL 33166-0064
Telephone: 305.555.0010
FAX: 305.555.0500
E-Mail: jguiverra@cnsult.com

November 18, 2005

Mr. Alexander Bennitez
Nova Advertising
664 Helene Boulevard, Suite 200
Miami, FL 33135-0493

Dear Mr. Bennitez:

Do you have numerous projects on hold because your staff is too busy? Consider getting your important projects back on schedule by outsourcing.

My areas of expertise include the following:

- Writing, editing, and keyboarding documents
- Processing mailings from start to finish
- Developing spreadsheets or flyers

Outsourcing with me offers the following advantages:

- No long-term employment commitment
- Satisfaction guaranteed (most reworking at no charge)
- Confidentiality

You can put my ten years of experience in the advertising business to work for you. The enclosed pamphlet describes my services, equipment, and rates. I have also enclosed samples of my work.

Mr. Bennitez, I can help Nova Advertising complete its projects in a timely and professional manner. Just call me at 555-0010. I would be available for an interview at your convenience.

Sincerely,

Juanita Guiverra

Juanita Guiverra

Enclosures 4

P.S. As a new client, your first in-office consultation would be free.

[opening]
Attention: Show that you understand the reader's situation.

[middle]
Interest and Desire: Relate your services to the reader's needs.

Sell your credibility.

[closing]
Action: Call for a doable action.

[postscript]
Offer an incentive.

Form Sales Letter

When you want to reach many potential clients through a mass mailing, you write a form sales letter. Even though it goes to many readers, this letter should still speak to each reader personally through personal pronouns, attention to shared needs, and mail merging (if possible).

Rhetorical situation

Purpose: To encourage flower orders for Valentine's Day
Audience: College students; of dating age, concerned about relationships, perhaps lacking disposable income
Format: Form letter on letterhead (offers efficiency; more effective than spam e-mail)

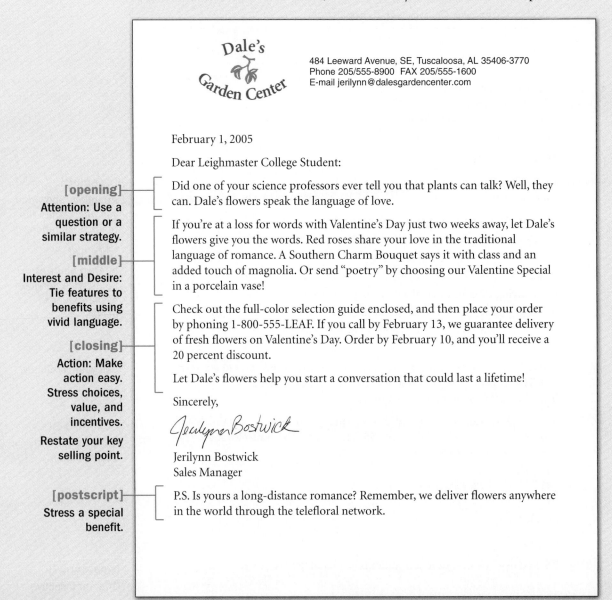

Dale's Garden Center
484 Leeward Avenue, SE, Tuscaloosa, AL 35406-3770
Phone 205/555-8900   FAX 205/555-1600
E-mail jerilynn@dalesgardencenter.com

February 1, 2005

Dear Leighmaster College Student:

[opening]
Attention: Use a question or a similar strategy.

Did one of your science professors ever tell you that plants can talk? Well, they can. Dale's flowers speak the language of love.

[middle]
Interest and Desire: Tie features to benefits using vivid language.

If you're at a loss for words with Valentine's Day just two weeks away, let Dale's flowers give you the words. Red roses share your love in the traditional language of romance. A Southern Charm Bouquet says it with class and an added touch of magnolia. Or send "poetry" by choosing our Valentine Special in a porcelain vase!

[closing]
Action: Make action easy. Stress choices, value, and incentives. Restate your key selling point.

Check out the full-color selection guide enclosed, and then place your order by phoning 1-800-555-LEAF. If you call by February 13, we guarantee delivery of fresh flowers on Valentine's Day. Order by February 10, and you'll receive a 20 percent discount.

Let Dale's flowers help you start a conversation that could last a lifetime!

Sincerely,

Jerilynn Bostwick

Jerilynn Bostwick
Sales Manager

[postscript]
Stress a special benefit.

P.S. Is yours a long-distance romance? Remember, we deliver flowers anywhere in the world through the telefloral network.

Chapter 27 Writing Persuasive Messages

Sales Letter (Following a Contact)

When you've made an initial contact with a prospective client—by phone, through a visit, at a convention—a follow-up letter seeks to capitalize on your initial discussion. In a sense, the letter picks up the discussion where it left off and offers a further persuasive argument concerning your product or service and the reader's needs.

Rhetorical situation

Purpose: To build on a convention contact by providing company information that builds credibility
Audience: Construction manager of a firm that does large industrial projects
Format: Signed letter with enclosures (creates strong, persuasive package)

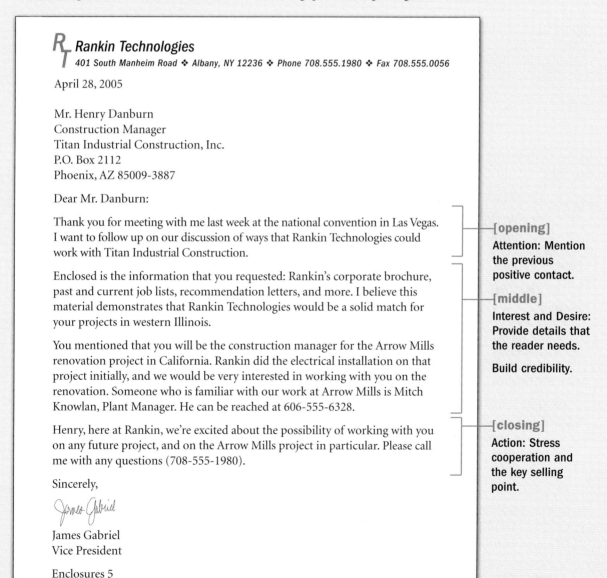

Sales Letter (Following a Sale)

When a new client has used your product or service, you have an opportunity to accomplish two goals in one letter: thank the client and generate more business. Do so with genuine goodwill and a gentle offer that shows continued attention to the reader's values and needs.

Rhetorical situation

Purpose: To follow up a flower order with material that encourages further orders
Audience: First-time female customer
Format: Friendly, signed letter (communicates service focus and points to enclosures)

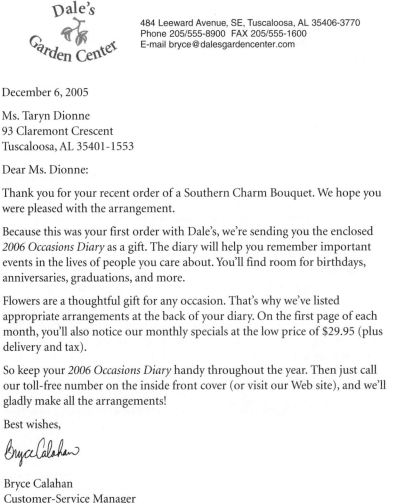

Dale's Garden Center
484 Leeward Avenue, SE, Tuscaloosa, AL 35406-3770
Phone 205/555-8900 FAX 205/555-1600
E-mail bryce@dalesgardencenter.com

December 6, 2005

Ms. Taryn Dionne
93 Claremont Crescent
Tuscaloosa, AL 35401-1553

Dear Ms. Dionne:

[opening] — Attention: Thank the reader for previous business.

Thank you for your recent order of a Southern Charm Bouquet. We hope you were pleased with the arrangement.

[middle] — Interest and Desire: Introduce other products without high-pressure tactics. Stress value and benefits.

Because this was your first order with Dale's, we're sending you the enclosed *2006 Occasions Diary* as a gift. The diary will help you remember important events in the lives of people you care about. You'll find room for birthdays, anniversaries, graduations, and more.

Flowers are a thoughtful gift for any occasion. That's why we've listed appropriate arrangements at the back of your diary. On the first page of each month, you'll also notice our monthly specials at the low price of $29.95 (plus delivery and tax).

[closing] — Action: Invite the reader to take a simple step.

So keep your *2006 Occasions Diary* handy throughout the year. Then just call our toll-free number on the inside front cover (or visit our Web site), and we'll gladly make all the arrangements!

Best wishes,

Bryce Calahan

Bryce Calahan
Customer-Service Manager

[postscript] — Emphasize a special offer.

P.S. I've also enclosed a *2005 Christmas Floral Selection Guide* filled with gift-giving ideas for friends and family.

Sales Letter (Inactive Customer)

When specific customers stop buying your products or using your services, a focused sales letter might accomplish two things. First, it might help you learn things about your own products or services or about the competition that will help you become more attractive and competitive. Second, the letter might help you open a dialogue that will bring the former customer back to your business.

Rhetorical situation

Purpose: To encourage a former customer to reconsider landscaping services
Audience: General manager of a furniture retailer
Format: Signed letter on company letterhead (fitting for customer contact)

∼ VERDANT LANDSCAPING ∼
1500 West Ridge Aveune
Tacoma, WA 98466
253-555-1725

January 6, 2005

Ms. Karen Bledsoe
Blixen Furniture
1430 North Bel Air Drive
Tacoma, WA 98466-6970

Dear Ms. Bledsoe:

We miss you! While reviewing contract renewals, I noticed that Verdant Landscaping has not been scheduled to care for your grounds since fall 2003. You were a valued customer. Did our service fall short in some way? Whatever prompted you to make a change, we would like to discuss ways we could serve you better.

[opening]
Attention: Express concern for the former customer.

During the past year, Verdant has expanded its services to help clients enhance the appearance of their businesses. A full-time landscape architect is available for consultations about improving your grounds with flower beds, hardy shrubs, and blooming trees. A tree surgeon is on call to care for trees that need special attention. Our lawn crews now offer regular cutting and cutting with mulching. All this at competitive rates!

[middle]
Interest and Desire: Explore possible reasons for the lost business and offer solutions.

I'd like to call you next week to discuss whatever concerns you may have, and to offer you a 10 percent discount on your first month of lawn care. At that time, I can answer any questions you may have about our new services as they are described in the enclosed brochure.

Sell your product or service by focusing on improvements.

[closing]
Action: Encourage a first step, and spell out your own follow-up.

Sincerely,

Stephen Bates

Stephen Bates
Customer Service

Enclosure

GUIDELINES FOR COLLECTION LETTERS

Your goal when writing collection letters is to persuade the customer to make a payment and to continue doing business with you. To do so effectively, you may need to write two or three letters, moving from a friendly reminder to a firm last request.

1 Prewrite

Rhetorical analysis: Consider your purpose, your audience, and the context.
- ☐ Aim to encourage open communication, create a written record, and resolve the debt.
- ☐ Think about the best way to approach this particular customer. How long has he or she been a customer? What is the relationship's history?
- ☐ What ethical considerations apply to this case? What short-term and long-term consequences may develop for this collection process?

Gather information and strategies necessary for collection.
- ☐ Collect complete details: account number, amount owed and for what, purchase dates, past-due date, interest and charges, payment history.
- ☐ Review policies for late payments, penalties, and payment plans.
- ☐ For early letters, consider positive incentives such as discounts, upcoming sales, and good credit standing. For later letters, consider stronger incentives such as lost credit privileges, harmed credit record, and collection action.

2 Draft

Organize the letter into three parts.

[opening] State the account's status in neutral language.

[middle] Review the account details. Then stress the positive results of payment (extended credit, peace of mind). In later letters, describe objectively the consequences of not paying.

[closing] Politely ask for payment, and make responding easy. Empower the reader to pay by offering choices for resolving the debt, if appropriate.

3 Revise

Review your draft's ideas, organization, and voice.
- ☐ Have you included accurate details and logical policy statements?
- ☐ Have you used a polite tone (no threats or lectures)?

4 Refine

Check your writing line by line for the following:
- ☐ Precise phrasing with words such as *help, cooperation,* and *fairness*
- ☐ Easy-to-read sentences with logical transitions
- ☐ Correct grammar, punctuation, usage, and spelling
- ☐ Effective format (e.g., lists, spacing, typography)

> "Don't find fault, find a remedy.
> —Henry Ford

F.Y.I.

When developing a sequence of collection letters, seriously consider having a lawyer review the content and phrasing.

Chapter 27 Writing Persuasive Messages 467

Collection Letter (First Notice)

The first letter in the collection sequence functions as a straightforward (and even friendly) reminder. It uses mild and positive appeals to encourage payment.

Rhetorical situation

Purpose: To prompt a client to pay an amount past due on a large order of flanges
Audience: Controller of a company with a record of prompt, full payment
Format: Formal, signed letter (creates official weight and record)

HANFORD BUILDING SUPPLY COMPANY, INC.
5821 North Fairheights Road, Milsap, CA 94218, Phone 567-555-1908

June 1, 2005

Account: 4879003

Mr. Robert Burnside, Controller
Circuit Electronics Company
4900 Gorham Road
Mountain View, CA 94040-1093

Dear Mr. Burnside:

This letter is a reminder that your account is past due (presently 60 days).

As of today, we have not yet received your payment of $1,806.00, originally due March 31. A copy of the March 1 invoice #QR483928 is enclosed. It refers to your January 10, 2005, order #S95832 for 3,000 mitered flanges that we shipped January 28.

Hanford appreciates your business, Mr. Burnside. Please give this matter your prompt attention so that Hanford Building Supply Company and Circuit Electronics can continue their good relationship. Your check for $1,828.58 (past due amount, plus 1.25 percent interest) will keep your account in good standing and avoid further interest charges and penalties. We have enclosed a postage-paid envelope for your convenience.

If there are any problems, please call (567-555-1908, ext. 227) or e-mail me (marta@hanford.com). As always, we look forward to serving you.

Sincerely,

Marta Ramones

Marta Ramones
Billing Department

Enclosures 2

[opening]
Attention: State the account's status.

[middle]
Interest and Desire: Review the account's history. Focus on keeping a positive relationship.

[closing]
Action: Urge the reader to contact you with any problems.

F.Y.I.
For many situations, a preprinted card or a sticker on the invoice or bill will serve well as the first reminder.

Part 4   Correspondence: Memos, E-Mails, and Letters

Collection Letter (Second Notice)

The second letter in the collection sequence, while maintaining a polite tone, indicates more concern than the first letter. Now you use more serious appeals to prompt payment and urge contact—to avoid a damaged credit rating, for example.

HANFORD BUILDING SUPPLY COMPANY, INC.
5821 North Fairheights Road, Milsap, CA 94218, Phone 567-555-1908

July 2, 2005

Account: 4879003

Mr. Robert Burnside, Controller
Circuit Electronics Company
4900 Gorham Road
Mountain View, CA 94040-1093

Dear Mr. Burnside:

[opening] — *Attention: Express concern about the account.*

Despite the reminder we sent on June 1, your account is now 90 days past due.

As of today, your payment of $1,828.58 has not arrived. A copy of your March 1 invoice #QR483928 is enclosed. It refers to your January 10, 2005, order #S95832 for 3,000 mitered flanges that we shipped January 28.

[middle] — *Interest and Desire: Review the account's history. Outline benefits of good credit and offer to help.*

Because of your excellent credit rating, you have enjoyed substantial discounts, convenient payment terms, and positive credit references from us. If you wish to maintain your good credit rating, we need your payment.

Circuit Electronics has been one of Hanford's most valued customers for more than five years. You have always paid your bills promptly. We are concerned about this uncharacteristic tardiness. Is there a problem we can help solve?

[closing] — *Action: Request payment or contact; stress cooperation.*

Please send your payment of $1,851.44 today (includes 1.25 percent interest) or contact me at 567-555-1908, ext. 227, so that we can resolve this matter.

Sincerely,

Marta Ramones

Marta Ramones
Billing Department

Enclosure 1

Collection Letter (Final Notice)

In the third and final letter in the collection sequence, you focus objectively on cause and effect—the customer's silence and nonpayment will cause collection proceedings or a similar action on a specific date. Even at this late stage, however, you empower the reader to pay by offering solutions.

HANFORD BUILDING SUPPLY COMPANY, INC.
5821 North Fairheights Road, Milsap, CA 94218, Phone 567-555-1908

August 1, 2005

Account: 4879003

Mr. Robert Burnside, Controller
Circuit Electronics Company
4900 Gorham Road
Mountain View, CA 94040-1093

Dear Mr. Burnside:

On January 28, 2005, we shipped you the 3,000 mitered flanges you ordered (#S95832). On March 1, we sent you the invoice for $1,806.00 (#QR483928). Copies of your purchase order and our invoice are enclosed. — **[opening]** Attention: Recap the facts.

Each month since then, Hanford has sent Circuit Electronics a reminder urging payment, and asking you to contact us. We have not heard from you, and your account is now 120 days past due with a balance of $1,874.58 (includes l.25 percent interest per month). Consequently, we must begin collection proceedings. — **[middle]** Interest and Desire: Outline the steps taken. State the next step in clear, neutral terms.

However, you can still resolve this matter, Mr. Burnside. Either call me now to discuss this problem (ext. 240 at the number above), or send a check by August 15 for the balance owed. By choosing either option, you can prevent this account from being turned over to a collection agency. — **[closing]** Action: Offer one final way to cooperate by a specific date.

Sincerely,

Floyd Kovic

Floyd Kovic
Vice President
Finance Division

Enclosures 2

WRITE Connection

To avoid negative and potentially libelous phrasing in your collection letters, see pages 255–256.

GUIDELINES FOR REQUESTING RAISES OR PROMOTIONS

Your goal when asking for a raise or promotion is to convince the reader that you deserve what you're requesting. Your message's immediate objective is to prompt a meeting with your supervisor to discuss the matter.

1 Prewrite

Rhetorical analysis: Consider your purpose, your audience, and the context.
- ☐ Are you aiming for a raise, a promotion, or both? Why?
- ☐ What does your supervisor know about you and your work? What does he or she value?
- ☐ What are current conditions in your department, your company, and beyond?

Do your homework.
- ☐ Review your job description, evaluations, recent accomplishments (projects, sales figures, client endorsements), and specifics of your last raise or promotion.
- ☐ Study company pay and promotion policies, considering negative factors (e.g., company finances) and researching comparable compensation.

2 Draft

Organize the request into three parts.

[opening] Provide background information. Then indicate politely and positively your wish to advance.

[middle] Focus on your contributions to the organization.
- ☐ Cite highlights of your performance.
- ☐ Stress your improvements and loyalty.
- ☐ Explain what you would bring to the new position (promotion).
- ☐ Point to industry standards as appropriate.

[closing] Request the raise or promotion, thank the reader for considering your request, and ask for a meeting.

3 Revise

Review your draft's ideas, organization, and voice.
- ☐ Have you included all the details necessary to convince your reader?
- ☐ Have you organized your details to build a strong case?
- ☐ Have you kept your tone positive? Have you avoided boasting, whining, and threatening—as well as apologizing for your request?

4 Refine

Check your writing line by line for the following:
- ☐ Energetic verbs and precise nouns
- ☐ Easy-to-read sentences with strong transitions
- ☐ Correct grammar, punctuation, usage, and spelling
- ☐ Attractive design (proper memo or letter format, white space, typography)

Request for Raise or Promotion

Rhetorical situation
Purpose: To press for a promotion to senior editor based on seven years of strong work
Audience: Director of the self-help division at a publisher; familiar with the writer's work
Format: Printed, initialed memo (gives weight to request)

MEMO

Date: November 15, 2005

To: Wendy Waite
              Director, Self-Help Division

From: Brittany Schwartz *BS*
              Associate Editor

Subject: Senior Editor Position

I have served as an Associate Editor with Yellow Brick Books for four years. Before that, I was an Assistant Editor for nearly three years. I believe that I am now prepared and qualified for a promotion to Senior Editor. —[opening] Attention: Put the request in context.

During my seven years here, I have made significant contributions to this company. My editorial skills have steadily developed as I have taken on more difficult assignments and responsibilities. —[middle] Interest and Desire: Stress length of service, performance, and loyalty.

For the last two years I have supervised development and production of the Yin-Yang Imprint. As you know, this line has enjoyed more than critical success; it has generated more than $1,000,000 in annual sales.

Cite specific achievements and evaluations.

I have appreciated and am committed to Yellow Brick's standards of excellence in publishing, and I value the company's supportive employee policies. As my evaluations show, I work diligently and efficiently. Furthermore, I have worked well with my colleagues and have earned their respect.

Explain benefits of the promotion to the company.

My promotion to Senior Editor would be more than a recognition of my contributions to Yellow Brick. It would put me in a position to develop a series of books on improving family life through time management, communication skills, financial planning, and design of living space.

Thank you for considering my request. I look forward to meeting with you to discuss this matter and my additional ideas for future books. Please call me at 6328 to arrange a convenient meeting time. —[closing] Action: Express thanks and ask for a meeting.

CHECKLIST for Persuasive Messages

Use this checklist to benchmark and revise your persuasive messages.

- ☐ **Ideas.** The message states the main point clearly and convincingly; uses accurate and compelling details; and connects with the reader's needs and concerns.
- ☐ **Organization.** The message follows the AIDA formula (attention, interest, desire, action) as outlined on page 453.
- ☐ **Voice.** The tone is polite and personal, demonstrating sensitivity to the reader's needs and concerns. It is not hesitant, apologetic, or aggressive.
- ☐ **Words.** Phrasing includes precise nouns, vivid modifiers, and energetic verbs; not clichés, jargon, flowery phrases, and "business English."
- ☐ **Sentences.** The constructions are generally short to medium length, read well aloud, and use transitions to tie ideas together.
- ☐ **Copy.** The message is free of grammar, spelling, punctuation, usage, and typing errors.
- ☐ **Design.** The message uses page layout, white space, and type style to make content accessible. It organizes points with headings, numbers, bullets, and graphics.

CRITICAL THINKING Activities

1. Working with a classmate, study the guidelines and models for special requests and promotional messages, sales messages, collection letters, and requests for raises or promotions. Briefly explain why each form is a persuasive message. Support your answers with details from the models.
2. Choose any three models from this chapter. For each, identify the sentence containing the call to action and then explain why the model does or does not illustrate the "art of persuasion" (see pages 452–453).
3. Collect a sales or promotional letter, a full-page advertisement from a magazine, an ad or some other persuasive piece from a Web site, a persuasive message sent from an office on your campus, and a persuasive e-mail you receive. (*Caution*: Avoid opening spam messages!) For each piece, analyze and evaluate the persuasive techniques.

WRITING Activities

1. Based on your workplace experience, identify a situation that required a persuasive message. Write the message using fitting persuasive strategies and the AIDA formula, as well as an appropriate guidelines page and model.
2. Identify a situation at your college (e.g., club event, request for faculty help or participation, help from a community member on a project) that requires a persuasive message. Using the "art of persuasion" strategies in this chapter, as well as an appropriate guidelines page and model, draft the message.

> **In this chapter**
>
> Guidelines for Form Messages 474
> Standard Form Message 475
> Menu Form Message 476
> Guide Form Message 477
> Checklist for Form Messages 478
> Critical Thinking Activities 478
> Writing Activities 478

CHAPTER 28

Writing Form Messages

Writing messages well takes time in any setting. In the workplace, where time is money, writing unnecessary messages is a serious waste. For example, if you commonly face the same situation and you compose a new document each time, you waste time. Rather than draft a new message for each occasion, you should write a form message—a "recyclable" piece of correspondence.

The words *form message* may make you groan. You find poorly written form letters and e-mails as easy to spot and as unpleasant to look at as cold, half-decayed fish. Form messages, in other words, have these disadvantages:

- **Impersonal quality.** When used too often or at the wrong time, form messages suggest that you don't care about the needs of your individual readers.
- **False economy.** A badly written form message fails to get results—wasting both the reader's time and your money.

Nevertheless, a form message can be distributed widely as a mass mailing through paper or electronic media, and used over and over as the need arises. A well-written form message has several advantages:

- **Savings and flexibility.** Because you prepare a form message only once and reuse it often, writing one saves you work and leads to efficient communication.
- **Accuracy and consistency.** A well-written form message delivers the right message every time. Like a reliable watch, you can count on it.

As you can see, a form message need not repulse you. In fact, the best form message looks and feels as friendly and precise as if it were written for you alone. This chapter presents the keys to writing such messages.

GUIDELINES FOR FORM MESSAGES

Your goal when creating a form message is to shape your message so that each reader feels that it speaks to him or her.

1 Prewrite
Rhetorical analysis: Consider your purpose, your audience, and the context.
- Aim to develop a message that meets the needs of many readers, sounds personal, keeps readers' goodwill, and can be used for a mass mailing or reused in the future.
- Study the size, diversity, and complexity of your audience. Who are your readers, and how can you successfully address all of them?
- Select a form (standard, menu, guide) and a medium (letter, memo, e-mail) that fit the situation.
- Determine which type of message you are writing. (See "Correspondence Catalog," page 361, for appropriate guidelines and models.)
- Decide which hardware, software, and other resources will help you prepare, duplicate, send, and reuse this message.

2 Draft
Organize the form letter logically.
[opening] State the message's purpose.
[middle] Provide a full explanation that reaches your entire audience.
[closing] Clarify any action needed by whom and when. Be direct if the message is good or neutral news (page 397), and indirect if it is a bad-news (page 425) or persuasive (page 451) message.

3 Revise
Review your draft's ideas, organization, and voice.
- Have you included all of the details that all of your readers will need? Will your explanations be clear to your least knowledgeable reader?
- Have you ordered information in a sequence that all readers can digest?
- Have you used a natural, conversational style?

4 Refine
Check your writing line by line for the following:
- Effective wording, especially personal pronouns like *you*
- Fresh nouns and strong verbs—plain English (whenever possible) for a diverse audience, not difficult jargon
- Easy-to-read sentences that all of your readers can understand
- Correct grammar, punctuation, usage, and spelling—standard English
- Attractive format (white space, highlighting techniques, quality stationery)

WRITE Connection
For help reaching a wide variety of readers, check Chapter 4, "Writing for Diversity."

STANDARD FORM MESSAGE

When all of your readers need to or can get the exact same message, a standard form message will suffice. A standard form message contains a fixed text.

Rhetorical situation

Purpose: To confirm approval of a mortgage loan and encourage application for a credit card
Audience: All customers who have successfully applied for a mortgage
Format: Standard form message on company letterhead (creates official record)

[Letterhead]

[Date]

[Address]

Dear [Name]:

Congratulations! Your loan has been approved. Once I receive an acceptable property appraisal, I will schedule your closing. With that in mind, I have attached a commitment letter listing all the conditions that must be met to successfully close your loan. Please review them and call me at (207) 555-6900 if you have any questions. — **[opening]** Be personal, state the main point, and supply key details.

You have also been preapproved for a Southeast credit card. If you take advantage of this opportunity, you will receive a $50 credit toward the closing costs of your mortgage loan. To receive this credit, simply return your completed credit-card application to me five days before your closing. I have enclosed both the application form and a postage-paid envelope for your convenience. — **[middle]** Add helpful secondary information.

Thank you for choosing Southeast Mortgage. If I can serve you in any other way, please let me know. — **[closing]** Be positive and clarify action.

Sincerely,

[Signature]

[Name]
Loan Processor

Enclosures 3

MENU FORM MESSAGE

A menu message gives you options. It lets you select which paragraph you want for the opening, the middle, and the closing. (Each paragraph is coded for easy reference.) The challenge here is to ensure that opening, middle, and closing options logically mesh with one another.

Rhetorical situation

Purpose: To follow up on various forms of customer inquiry concerning financing
Audience: The entire range of bank clientele
Format: Menu options (allows individual letters to be shaped according to type and degree of contact)

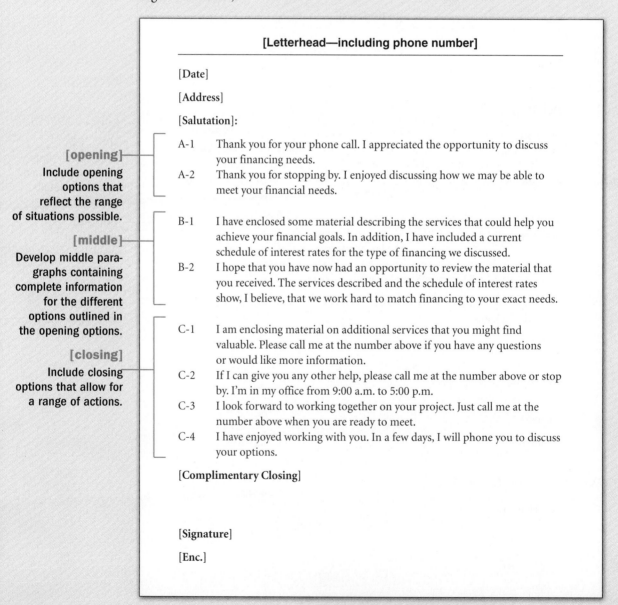

[opening] — Include opening options that reflect the range of situations possible.

[middle] — Develop middle paragraphs containing complete information for the different options outlined in the opening options.

[closing] — Include closing options that allow for a range of actions.

[Letterhead—including phone number]

[Date]

[Address]

[Salutation]:

A-1 Thank you for your phone call. I appreciated the opportunity to discuss your financing needs.

A-2 Thank you for stopping by. I enjoyed discussing how we may be able to meet your financial needs.

B-1 I have enclosed some material describing the services that could help you achieve your financial goals. In addition, I have included a current schedule of interest rates for the type of financing we discussed.

B-2 I hope that you have now had an opportunity to review the material that you received. The services described and the schedule of interest rates show, I believe, that we work hard to match financing to your exact needs.

C-1 I am enclosing material on additional services that you might find valuable. Please call me at the number above if you have any questions or would like more information.

C-2 If I can give you any other help, please call me at the number above or stop by. I'm in my office from 9:00 a.m. to 5:00 p.m.

C-3 I look forward to working together on your project. Just call me at the number above when you are ready to meet.

C-4 I have enjoyed working with you. In a few days, I will phone you to discuss your options.

[Complimentary Closing]

[Signature]

[Enc.]

GUIDE FORM MESSAGE

A guide message is a fill-in-the-blank form that lets you add (through keyboarding or merging) details that fit each specific reader. As such, the guide form offers greater flexibility than the standard form and more personal content than the menu form. One challenge of a guide message, however, is ensuring that the various options are compatible with the grammar and syntax of the sentences.

Rhetorical situation

Purpose: To sort out for buyers the legal details of property transactions
Audience: Any fellow lawyer representing the party selling the property
Format: Guide form message (allows insertion of details specific to given transaction)

[Letterhead]

[Date]

[Lawyer]
[Address]

Dear [Lawyer]:

Subject: [Transaction]

I am acting for the purchaser, [**Name(s)**], in the transaction noted above. I understand that you are acting on behalf of the vendor, [**Name(s)**].

I now have a copy of the Agreement of Purchase and Sale and have advised [**Name(s)**] of [**his/her/their**] rights and obligations, particularly with respect to precedent conditions. Once these conditions have been met by [**Final date**], we will be able to proceed with closing this transaction on [**Day and date**].

Therefore, please send me a copy of the property description with book and page references, as well as any back title information you can provide. I would also appreciate receiving a copy of any location certificate your client may possess.

Please note two more items:
1. I have enclosed draft adjustments. Please review them and provide me with your response.
2. My [**client/clients**] [**wishes/wish**] to take title as follows: [**Name(s), of place, as type of tenant(s)**].

Thank you for your attention to these details. I look forward to working together to conclude this transaction for the mutual benefit of our clients.

Yours sincerely,

David S. Wilson

David S. Wilson

Enc.: Draft adjustments

[opening]
Clarify your purpose. When possible, create places to integrate names.

[middle]
Present all necessary details.

Indicate fill-in-the-blank choices with brackets. Use bold or italics to feature these option fields.

Make sure that all choices work grammatically.

[closing]
End politely and positively.

CHECKLIST for Form Messages

Use this checklist to benchmark and revise your form messages.

- ☐ **Ideas.** The message supplies all the information needed by all readers; makes the main point and supporting points clear; and uses solid, relevant details to support each point.
- ☐ **Organization.** The message uses the pattern that fits the specific type of message (e.g., informative, sales, collection).
- ☐ **Voice.** The tone is friendly, informed, and positive—not distant or impersonal.
- ☐ **Words.** Phrasing includes plain English that all readers can understand; defines terms that may be unfamiliar; uses personal pronouns effectively (especially *you*); and integrates the reader's name into the message, if possible.
- ☐ **Sentences.** Constructions are short to medium in length and read smoothly.
- ☐ **Copy.** The message is free of grammar, punctuation, spelling, and mechanical errors; and merges information manually or electronically without creating grammar errors or awkward sentences.
- ☐ **Design**
  - The message is properly formatted as a letter, memo, or e-mail message (as fits the situation).
  - It is effectively designed as a standard, menu, or guide message.
  - It appears attractive and accessible thanks to the use of white space, typographical features, lists, quality stationery, and so on.

CRITICAL THINKING Activities

1. Collect three form messages, and use this chapter's instructions to analyze how these messages are put together and to evaluate their quality.
2. Think about the profession that you are currently in or training for. In what situations might you routinely use form messages, and why? If necessary, interview someone working in this field about the messages that he or she has to write regularly. Collect sample messages, if the person is willing to share them.

WRITING Activities

1. Working with a classmate, develop a workplace situation that could be addressed by a form message. Write that message in an appropriate format (standard, menu, or guide), and select an appropriate medium (memo, e-mail, or letter). Present your message to the class, along with an explanation of the following: (a) why a form message is appropriate for your workplace situation, (b) why you chose the given format, and (c) why you chose the specific medium.
2. Identify a nonprofit community organization or college club whose cause you could help by writing an appropriate form message. Research the situation and write the message.

PART 5

Reports and Proposals

29 Report and Proposal Basics

30 Writing Short Reports

31 Writing Major Reports

32 Writing Proposals

33 Designing Report Forms

> **In this chapter**
>
> Writing Successful Reports and Proposals **482**
> Types of Reports and Proposals **483**
> Checklist for Report and Proposal Basics **484**
> Critical Thinking Activities **484**
> Writing Activities **484**

CHAPTER

29

Report and Proposal Basics

In many ways, reports and proposals are the heavy-lifting documents in workplace writing. Often based on significant research and commonly packed with critical details and data, these documents deliver the substantive information and analysis that organizations need to plan, produce, evaluate, and advance their work, including all major projects.

Because the documents are significant, you may be worried about writing them. Relax! This chapter and the four that follow will help you tackle the task of writing reports and proposals. For example, this chapter offers eight tips for writing these documents as well as a table listing the seven major types of reports and proposals (including the goal of each). It also identifies the pages in this book where you will find instructions for writing these types of documents.

Use this chapter now to get an overview of what reports and proposals are and what they do in the workplace. Later, you can use the chapter as a quick reference for finding information on specific types of reports and proposals. Chapters 30–33 contain detailed help for writing each type.

WRITING SUCCESSFUL REPORTS AND PROPOSALS

While reports and proposals can range from two pages to more than one hundred pages, and can be either routine or special, they all work when you follow these guidelines:

1. **Know your assignment.** What type of report or proposal are you writing? What should it deliver? Here are three possible assignments:
   - Supply plain information—just the facts.
   - Share conclusions—your insights about the information.
   - Offer recommendations—steps to take based on your conclusions.

2. **Know your audience.** One reader may want all the fine details, another just the bottom line. Your audience may be multiple and complex, so consider these issues:
   - **Your readers' work and responsibilities.** How does your report or proposal relate to or impact this work?
   - **Your readers' values, needs, and priorities.** What matters to your reader and why? How do these issues relate to the topic and context of your report or proposal?
   - **Your readers' knowledge of the subject.** What do they already know, not know, and need to know?
   - **Your readers' use of the report or proposal.** Will the piece prompt understanding, decisions, and/or actions?

3. **Research the topic.** Using focused research questions, gather accurate data so that you can create an information-rich document. Objectively look at all sides of the subject, using sound thinking patterns like cause/effect, compare/contrast, and so on. For more on research, see pages 117–168.

4. **Organize for understanding.** Structure your report or proposal in three sections that answer your reader's key questions:

   Introduction: What's the report or proposal about?
   Findings: What are the details?
   Key points: What's the bottom line?

   To be direct, reverse the findings and key points or add a summary to the introduction. In fact, pay careful attention to developing a summary for your report or proposal, as the summary will likely be the document's most-read portion (see page 128).

5. **Be formal when necessary, but never stiff.** Generally, the less familiar your reader and the longer the report or proposal, the more formal your approach should be. However, formal doesn't mean impersonal. Be energetic, natural, and matter-of-fact.

6. **Design for clarity, access, and impact.** Help readers find information:
   - Use tables, graphs, charts, maps, photos, lists, and numbers.
   - Create visual snap with white space, boldface, italics, and underlining.
   - Develop coherence with headings, introductions, and transitions.

7. **Rely on models and templates.** Use the outlines and models in this book. Check your company's files, as well as computer software, for similar guides.
8. **Remember the big picture.** Why is the report or proposal important? Will it result in change? Will it be passed on to secondary readers? Consider its ripples—both within and beyond the company.

TYPES OF REPORTS AND PROPOSALS

The type of report or proposal you need to write is determined by (1) your purpose, (2) your readers' expectations, and (3) the information that your readers need. Table 29.1 identifies various types of reports and proposals, along with their goals. You will find appropriate guidelines, outlines, and models in Chapters 30–33.

Table 29.1 Types of Reports and Proposals

Type	Examples/Other Names	Writer's Goal
Short Reports (Chapter 30)		
Incident (486–489)	Accident, breakdown, error, occurrence, spoilage, stoppage, or trouble reports	Examine an accident, problem, or delay to determine causes, effects, and solutions.
Investigative (490–495)	Appraisal, audit, evaluation, experiment, feasibility, lab, impact, inspection, interview, research, or test reports	Provide the results of research and testing procedures.
Periodic (496–499)	Weekly, monthly, quarterly, annual reports; activity, compliance, control, department, evaluation, monitor, operations, or status reports	Provide information at regular time intervals so that products, services, and work can be tracked, planned, and adjusted.
Progress (500–503)	Initial, interim, completion, and follow-up reports; activity, campaign, project, grant, or status reports	Give a supervisor, client, or administrator details on how a project or job is going and what needs to be done.
Trip or Call (504–507)	Conference, convention, customer service, evaluation, field, inspection, installation, maintenance, repair, sales trip or call, seminar, visit, or workshop reports	Share results of activities that happened away from the workplace or of contacts with people outside the company.
Major Reports (Chapter 31)		
(509–520)	Long, formal, or annual reports; position, white, or trade papers	Explore a major issue of serious concern to the company, particularly its long-term goals.
Proposals (Chapter 32)		
(521–538)	Suggestions, concept proposals; justification, grant, research, sales, or troubleshooting proposals; major bids	Analyze a problem and recommend a solution; identify a need and show how to meet it.

CHECKLIST for Report and Proposal Basics

Use this seven-traits checklist to review, revise, and refine any report or proposal that you write.

- ☐ **Ideas.** The report or proposal has a clearly spelled-out purpose; provides complete, accurate data; offers conclusions and recommendations that follow logically from the information; and uses graphics to communicate information.
- ☐ **Organization.** The report or proposal is ordered logically into three parts.
    - An introduction explains the problem or issue, purpose, and background; previews what the report or proposal covers; and summarizes main points when being direct.
    - Findings give readers the detailed explanations they need, organized according to logical patterns (e.g., time, space, order of importance, category, problem/solution, cause/effect, compare/contrast).
    - Closing key points summarize, conclude, and recommend, as appropriate.
- ☐ **Voice.** The tone is matter-of-fact and measured, but energetic and positive.
- ☐ **Words.** The report or proposal uses words that fit primary and secondary readers—the right level of formality and technical complexity.
- ☐ **Sentences.** Constructions are generally short, but varied.
- ☐ **Copy.** The report or proposal uses correct grammar, punctuation, spelling, and mechanics; it is free of typing errors.
- ☐ **Design.** The report or proposal has a format and presentation that follow company guidelines; it uses white space, boldface, lists, and graphics effectively.

CRITICAL THINKING Activities

1. Examine an essay or report that you have written for a college course. Which features of a workplace report does it share? How are workplace reports and proposals distinct from academic papers?
2. Find a sample workplace report or proposal—check the Internet, contact an office on your campus, or consider a business where you have worked. Using information in this chapter, determine the report or proposal's type and assess its strengths and weaknesses.

WRITING Activities

1. Take a college essay or report that you have written and rework it into a workplace report using the strategies offered in this chapter.
2. Using the basic strategies in this chapter, write a report to your instructor explaining what you have learned thus far in the course and projecting what you still hope to accomplish.

In this chapter

Guidelines for Incident Reports 486
Guidelines for Investigative Reports 490
Guidelines for Periodic Reports 496
Guidelines for Progress Reports 500
Guidelines for Trip or Call Reports 504
Checklist for Short Reports 508
Critical Thinking Activities 508
Writing Activities 508

CHAPTER

30

Writing Short Reports

The verb *report* (coming from *re,* meaning "back," and *portare,* meaning "carry") literally means "to carry back." The noun *report* refers to "a statement, account, analysis, record, summary, or recommendation." In the workplace, your job when writing a report is to (1) study the topic, (2) analyze it, and (3) "carry back" to your readers the details of your analysis. Regardless of which type of report you're writing, your job is always the same: study, analyze, and carry back.

By far, the most common type of report that you will write is the short report. Such documents carry back to readers the basic information they need to keep track of work and keep it moving. Short reports range from two to ten pages and vary in form from memos and letters to filled-out forms and online documents. Whatever their final form, short reports help readers with tasks such as the following:

- Creating a record for routine and special activities
- Documenting details related to an incident, project, or study
- Answering basic questions about what's happening in the company
- Checking results of work, plans, production, and so on
- Evaluating options and making decisions

Work through this chapter carefully, looking for strategies that will help you to study a topic, analyze it, and carry back to your readers the details of your analysis. When you finish the chapter, you will be much better prepared to write the short reports that the workplace requires.

GUIDELINES FOR INCIDENT REPORTS

Your goal when writing an incident report is to provide accurate details about and clear analysis of an accident, breakdown, or similar situation.

1 Prewrite
Rhetorical analysis: Consider your purpose, your audience, and the context.
- Which of these aims must be fulfilled: to describe the incident, analyze causes, recommend corrective action, and/or support preventive measures?
- Who are your potential readers—the people involved in the incident, managers, clients, inspectors, insurance agents, union leaders, government officials, lawyers, and the media? What do primary and secondary readers need?
- Is background information important for understanding this incident? What short-term and long-term effects might result?

2 Draft
Organize the report with the reader(s) in mind.
[opening] Identify the incident and summarize key points.
[middle] Fully describe and analyze the incident.
[closing] Provide conclusions and recommendations.

3 Revise
Review your report's ideas, organization, and voice.
- Have you included clear, relevant information and accurate analysis?
- Have you used graphics (such as photographs) to support and visualize your discussion?
- Have you used the straight facts and stayed within the limits of your expertise? (If appropriate, consider a legal review of the report.)
- Have you structured the report to feature the most important points related to the incident?
- Have you used an objective tone throughout the report?

4 Refine
Check your report line by line for the following:
- Accurate, neutral phrasing and definitions of key terms related to the event
- Clear, readable sentences of short to medium length with strong transitional words indicating sequence and cause/effect links
- Correct grammar, punctuation, usage, and spelling
- Effective format (headings, white space, graphics)

Organizing Incident Reports

[opening] Introduction
- **Label** the report with a title or subject line indicating the type of incident, your name, the reader's name, and the date.
- **Introduce** the report by stating its purpose, providing background information on the incident, and previewing the report's content.
- **Summarize** the incident if you need to be direct: Address what happened, what caused it, and what should be done. To be indirect, state these points only in the closing.

[middle] Findings
- **Describe** the incident completely and accurately, answering *who, what, when, where,* and *why* questions with precise evidence: What happened? If similar incidents happened earlier, how are all incidents related? Who was involved? When, where, and why did it happen?
- **Organize** the details using a time pattern:
   1. What important events and conditions preceded the event?
   2. What happened during the incident—step-by-step?
   3. What happened after the incident? What were the effects?
   4. What measures or steps were taken afterward? By whom?
   5. What still needs to be done? How, when, by whom, and at what cost?
- **Use graphics** such as drawings, diagrams, photographs, or charts to help the reader understand the incident.

[closing] Key Points
- **Summarize** the facts of the incident.
- **Provide conclusions** (if required) on the incident's cause, carefully phrased and clearly following from the facts described.
- **Offer recommendations** (if required) to fix the problem, address the effects, and prevent future incidents.

> "The incident report is the link between the crime and the court case—maybe a year later."
>
> —Dan Altena, County Sheriff

F.Y.I.
Attach copies of statements, interviews, and other documentation referred to in the report.

Incident Report

Rhetorical situation

Purpose: To provide details about a near-accident; to recommend preventive measures

Audience: General manager and safety committee chair; interested in productivity, bottom line, and safe workplace environment

Format: Memo (creates formal weight and paper record)

---

1 of 2

MAINTAINER CORPORATION OF NEW MEXICO

Date: March 25, 2005

To: Alice Jenkins, General Manager
Roger Smythe, Safety Committee Chair

From: Gwen Vos, Supervisor
Truck Finishing Department

Subject: Undercoating Safety Incident on March 20, 2005

[opening]
Identify the type of incident.
Preview the report.

This report details a recent event in the undercoating bay. You will find (A) a description of the incident, (B) conclusions about the causes, and (C) recommendations for fixing the problem.

[middle]
Divide the report into logical sections with clear headings.

Provide context, describe the incident, and explain what followed.

List and number the events in the order they happened.

A. The Incident: Tangled Air Hose

During a routine inspection of work on Thursday, March 20, at 10:45 a.m., I found undercoater Bob Irving struggling to breathe underneath the truck he was working on. While spraying liquid-rubber sealant on the undercarriage, he had rolled his dolly over his air hose, cutting off the air supply. I immediately pulled him out, untangled him, and took the following steps:

1. I checked Bob for injuries and determined that he was unharmed.
2. I asked him what had happened. He explained that he couldn't free himself because (a) he became tangled in the spray-gun cords, (b) his air hose was locked into his oxygen suit, and (c) he was lying down in a cramped space.
3. I discussed the incident with the undercoating crew. They confirmed that similar problems had developed before, but they hadn't filed reports because no one had actually been injured.
4. I inspected thoroughly all undercoating equipment.
5. As a short-term solution, I bought airhorn alarms to attach to the undercoaters' dollies.

B. Conclusions: Probable Causes

Presently, undercoaters maneuver under trucks and spray liquid-rubber sealant on the undercarriage while lying on their backs. Maintainer provides oxygen suits to protect their skin and oxygen supply from this sealant that (a) produces noxious fumes, (b) causes choking if swallowed, and (c) injures skin upon contact. However, this incident shows that our safeguards are inadequate:

- Ten-year-old oxygen-suit meters and air tubes frequently malfunction. Masks and hoses are beginning to crack.
- The practice of lying on a dolly while spraying can cause undercoaters to become tangled in cords and hoses or roll over their air hoses.
- Spraying from a prone position allows liquid rubber to drip onto undercoaters' masks. This dripping obscures vision and makes it more likely that workers will become entangled and more difficult for them to get untangled.

C. Recommendations: New Safety Measures

To further protect undercoaters from these hazards, I recommend the following actions:

1. Replace oxygen suits and equipment to meet the 2004 OSHA oxygen-safety standards (air-hose locks with emergency-release latches).
2. Put trucks on lifts so that undercoaters can work standing up.
3. Have two undercoaters work together on the same truck to monitor each other.
4. Purchase No-Drip Sealant Applicators to eliminate dripping liquid rubber.

With these measures, undercoating incidents such as the one with Bob Irving should not happen again. Please contact me at extension 2422 with any questions and with your response to these recommendations.

[middle]
Develop clear cause/effect thinking.

Use strong transitional words.

Be precise and objective.

[closing]
Suggest solutions that clearly match the problem.

Be realistic.

Stress benefits of action.

F.Y.I.

Incident reports may also be called accident, breakdown, error, occurrence, spoilage, stoppage, or trouble reports.

GUIDELINES FOR INVESTIGATIVE REPORTS

Your goal when writing an investigative report is to find, analyze, synthesize, evaluate, and report facts that will help your readers understand an issue and perhaps take effective action.

1 Prewrite

Rhetorical analysis: Consider your purpose, your audience, and the context.
- ☐ Why are you writing this report? How is the information important?
- ☐ For whom are you writing this report? How are they likely to use it? Who are potential secondary readers?
- ☐ What important background lies behind this investigation? Where could the investigation eventually lead?

Clarify, plan, and complete your research.
- ☐ Review the issue and discuss it with those involved.
- ☐ Reflect on any ethical challenges related to the issue and research methods.
- ☐ Formulate research questions: Why is this problem happening? What are the effects? Will this idea work? How can we avoid this problem?
- ☐ Develop testing instruments—observations, interviews, surveys, and so on.
- ☐ Do your research. See Chapter 7, "Conducting Research for Business Writing."
- ☐ Formulate conclusions, test them, and develop recommendations.

2 Draft

Organize the report with the reader(s) in mind.

[opening] State the purpose and focus of the investigation; summarize results (if appropriate).

[middle] Provide findings (from methods to results). Present your data in a way that makes them accessible to readers.

[closing] Summarize results, state conclusions, and offer recommendations.

3 Revise

Review your report's ideas, organization, and voice.
- ☐ Have you used clear, accurate information that's right for readers—whether specialists needing all the proof or nonspecialists looking for the bottom line?
- ☐ Are your conclusions convincing and your recommendations workable?
- ☐ Did you use an objective, stick-to-the-facts tone?

4 Refine

Check your writing line by line for the following:
- ☐ Precise nouns (with key terms defined when necessary) and strong verbs
- ☐ Varied sentences with logical transitions and effective parallelism
- ☐ Correct grammar, punctuation, usage, spelling, and mechanics (especially in the presentation of numbers, abbreviations, and tables)
- ☐ A fitting format (e.g., standard experiment report or field report)
- ☐ Strong design (white space, headings, lists, graphics)

Organizing Investigative Reports

[opening] Introduction
- **Label** the report with a title or subject line indicating the nature of the investigation, the group authorizing the research, your name, the reader's name, and the investigation's time frame.
- **Introduce** the report by stating its purpose, providing background information on the topic, and previewing the report's contents.
- To be direct (bottom line first), **summarize** your investigation's key results, conclusions, and recommendations. To be indirect, state these points only in the closing.

[middle] Findings
- **Describe** the investigation.
  - Explain research equipment and materials.
  - Outline the research method, perhaps using flowcharts or diagrams.
  - Introduce research results.
- **Organize** the findings using a logical pattern that fits your research.
  - List, describe, and evaluate possible solutions to the problem using criteria against which to measure the options.
  - Use a compare/contrast pattern to weigh choices against each other.
  - Offer hypotheses (educated guesses) about the answer to your research question. Then present evidence supporting or disproving each hypothesis.
  - Categorize the data. Address one part of the puzzle at a time.
- **Present** key evidence in tables, charts, diagrams, and lists.

[closing] Key Points
- **Summarize** your research results.
- **Provide conclusions** that answer the key research question.
- **Offer recommendations** (if required) to solve the problem or meet the need.

> "This report, by its very length, defends itself against the risk of being read."
> —Winston Churchill

F.Y.I.
If helpful for your readers, attach primary documents, tables containing data, and other key pieces of evidence.

Investigative Report

Rhetorical situation

Purpose: To study a cockroach infestation, research reliable solutions, and recommend an action plan

Audience: Primarily the vice president in charge of tenant relations; secondarily other members of management and tenants themselves

Format: Memo (appropriate for in-house report)

---

1 of 4

SOMMERVILLE DEVELOPMENT CORPORATION

Date: September 19, 2005

To: Bert Richardson, VP of Tenant Relations

From: Hue Nguyan, Cherryhill Complex Manager
Sandra Kao, Building Superintendent
Roger Primgarr, Tenant Relations
Juan Alexander, Tenant Representative

Subject: **Investigation of Cockroach Infestation at 5690 Cherryhill**

[opening]
Specify the nature of the research.

During the month of July 2005, 26 tenants of the 400-unit building at 5690 Cherryhill informed the building superintendent that they had found cockroaches in their units. On August 6, the management–tenant committee authorized us to investigate these questions:

Provide the research goals.

1. How extensive is the cockroach infestation?
2. How can the cockroach population best be controlled?

Preview the report's content.

We monitored this problem from August 9 to September 8, 2005. This report contains a summary, an overview of our research methods, findings, conclusions, and recommendations.

Summary

Summarize the outcome.

The 5690 Cherryhill building has a moderate infestation of German cockroaches. Only an integrated control program can manage this infestation. Pesticide fumigations address only the symptoms, not the causes. We recommend that Sommerville adopt a comprehensive program that includes (1) education, (2) cooperation, (3) habitat modification, (4) treatment, and (5) ongoing monitoring.

Research Methods and Findings

Overview of Research
We researched the problem in the following ways:

1. Contacted the Department of Agriculture, the Ecology Action Center, and Ecological Agriculture Projects.
2. Consulted three exterminators.
3. Inspected 5690 Cherryhill building, from ground to roof.
4. Placed pheromone traps in all units to monitor cockroach population.

The Cockroach Population
Pheromone traps revealed German cockroaches, a common variety. Of the 400 units, 112 units (28 percent) showed roaches. Based on the numbers, the infestation is rated moderate.

The German Cockroach
Research shows that these roaches thrive in apartment buildings.

- Populations thrive when food, water, shelter, and migration routes are available. They prefer dark, humid conditions near food sources.
- The cockroach seeks shelter in spaces that allow its back and underside to remain in constant contact with a solid surface.

Methods of Control
Sources we consulted stressed the need for an integrated program of cockroach control involving sanitation, habitat modification, and nontoxic treatments that attack causes. Here are the facts:

- The German cockroach is immune to many chemicals.
- Roaches detect most pesticides before direct contact.
- Spot-spraying simply causes roaches to move to unsprayed units.
- Habitat modification through (1) eliminating food and water sources, (2) caulking cracks and crevices, (3) lowering humidity, and (4) increasing light and airflow makes life difficult for cockroaches.

[middle]
Detail research methods to show thoroughness.

Categorize findings logically.

Use subheadings and lists.

State findings concisely, clearly, and confidently.

[closing]
State conclusions that follow logically from the findings.

Offer conclusions and recommendations positively, confidently.

Focus on usable results of the research.

Clearly spell out all needed steps.

Guide readers with strong transitions.

Conclusions

Based on our findings, we conclude the following:

1. A single method of treatment, especially chemical, will be ineffective.
2. A comprehensive program of sanitation, habitat modification, and nontoxic treatments will eliminate the German cockroach.

Recommendations

We recommend that Sommerville Development adopt an Integrated Program of Cockroach Prevention and Control for its 5690 Cherryhill building. Management would assign the following tasks to appropriate personnel:

Education. (1) Give tenants information on sanitation, prevention, and home remedies; and (2) hold tenant meetings to answer questions.

Cooperation. To ensure 100 percent compliance, revise the current lease to include program requirements and management's access to all units (offering tenants advance notice and accurate information). Plan treatments to minimize inconvenience and ensure safety.

Habitat Modification. Revise the maintenance program and renovation schedule to give priority to the following:

- Apply residual insecticides before sealing cracks.
- Caulk cracks and crevices (baseboards, cupboards, pipes, sinks). Insert steel wool in large cavities (plumbing, electrical columns).
- Repair leaking pipes and faucets. Insulate pipes to eliminate condensation.
- Schedule weekly cleaning of common garbage areas.

Treatment. In addition to improving sanitation and prevention through education, attack the roach population through these methods:

- Use home remedies, traps, and hotels.
- Place diatomaceous earth in dead spaces to repel and kill cockroaches.
- Use borax or boric acid powder formulations as residual, relatively nontoxic pesticides.
- Use chemical controls on an emergency basis, no more than once per year.
- Ensure safety by arranging for a Health Department representative to make unannounced visits to the building.

F.Y.I.

Investigative reports may also be called appraisal, audit, evaluation, experiment, feasibility, lab, impact, inspection, interview, research, or test reports.

Monitoring and Evaluating. Monitor cockroach population and evaluate the program's effectiveness:

- Every six months, use traps to check on cockroach activity in all units.
- Keep good records on the degree of occurrence, population density, and control methods used.
- Adjust the program as needed in response to the data.

We believe that this comprehensive program will solve the cockroach infestation problem. We recommend that Sommerville adopt this program for 5690 Cherryhill and consider implementing it in all its buildings.

[closing] Stress the value and benefits of the research.

WRITE Connection

In many investigative reports, you need to carefully integrate and document your research to build your credibility and avoid plagiarism. For help, turn to pages 152–166.

GUIDELINES FOR PERIODIC REPORTS

Your goal when you write a periodic report is to record accurately and promptly the activities and outcomes of a particular time period. Such reports (whether filed daily, weekly, monthly, quarterly, or annually) might offer data and analysis with respect to production, sales, services, consumption, profits, losses—virtually any work-related topic that needs to be tracked.

1 Prewrite

Rhetorical analysis: Consider your purpose, your audience, and the context.
- ☐ What information and conclusions must you share? To what end?
- ☐ Who are the primary and secondary readers? How can you help them understand the period's developments and make decisions for the future? With this topic, what do your readers need or value?
- ☐ Do you need to take into account background and special conditions?

Prepare to draft.
- ☐ Identify the report period—its beginning and ending.
- ☐ Review the prior report, meeting minutes, and similar documents.
- ☐ Address what has developed, continued, and ended.
- ☐ Explore what should happen next.
- ☐ Gather data and study them to determine key points and conclusions. Analyze trends, including their causes and effects.
- ☐ Anticipate issues to address in the next report.

2 Draft

Organize the report into three parts.

[opening] Identify the period and purpose of the report, and summarize key points (if appropriate).
[middle] Include all essential information: achievements, problems, plans.
[closing] Offer conclusions and make recommendations.

3 Revise

Review your draft's ideas, organization, and voice.
- ☐ Have you included accurate, complete coverage of the period?
- ☐ Have you connected what has happened with what may follow?
- ☐ Have you used logical reasoning and an objective tone?

4 Refine

Check your writing line by line for the following:
- ☐ Precise terms and concise phrasing—a condensed, factual style
- ☐ Correct grammar, punctuation, usage, and spelling
- ☐ A format that makes information accessible and meets company standards
- ☐ Helpful headings, white space, and graphics (e.g., tables, charts)

Organizing Periodic Reports

[opening] Introduction
- **Label** the report with the title or subject, time period covered, any reference numbers, your name, and the reader's name.
- **Introduce** the report by stating its purpose, providing background information, and previewing its content. If your report goes to the same reader regularly, keep this material to a minimum.
- To be direct (bottom line first), **summarize** key developments during the period, and provide your conclusions and recommendations. To be indirect, state these points only in the closing.

[middle] Findings
- **Provide** the following information in detail:
  - Goals achieved and not achieved
  - Routine activities
  - Problems encountered
  - Plans for the coming period
- **Organize** the information using a logical pattern.
  - Time sequence (the order in which developments happened)
  - Category types (activities, events)
  - Order of importance (most to least important)
  - Alphabetical (using first letters in key words)
- **Present** your data with the help of headings, lists, tables, spreadsheets, and other graphics.

[closing] Key Points
- **Summarize** totals, major changes, events, and dates.
- **Provide conclusions** about trends, developments, and significant events in the time period. Where are things headed?
- **Offer recommendations** (if required) to solve problems and guide future action.

> "The trouble with facts is that there are so many of them."
> —Samuel Crothers

Periodic Report

Rhetorical situation

Purpose: To review the year's production accomplishments and challenges
Audience: Senior management of the parent company, a plastics manufacturer; interested in efficiency, productivity, the bottom line
Format: Formal, titled report (fitting for audience and situation)

---

1 of 2

STEWART PLASTICS, INC.
JOLIET ANNUAL REPORT

[opening] — Identify the report, the time period, the writer, and the reader. State the purpose of the report and what it covers.

Date Submitted: November 15, 2005
Period Covered: November 1, 2004, to October 31, 2005
Prepared by: Denzel Irving
Prepared for: Senior Management

The following report reviews 2004–2005 activities at Stewart Plastics' Joliet facility. Topics covered include manufacturing, safety, and quality. Based on this review, the report projects sales and production needs for 2005–2006.

Provide main points in a summary. (The summary may be placed at the end.)

Summary

Major projects involved plastite screws for Fimco, spray-on graphics for Newland, Kelch steel clamps on EcoLab molds, and soda blasting to clean molds. Safety remained a priority through consultation with Liberty Mutual. First-Time Quality improved from 96.8 to 97.5 percent. Anticipated sales should be $5.1 million. To reach maximum output of $6 million, purchase a 5-Axis Router and hire a Process Engineer.

Year-End Operations Report

[middle] — Divide information into logical categories with clear headings. Use a "telegraphic" style (short clauses).

Manufacturing and Process Engineering

- Evaluated using thread-forming plastite screws to replace aluminum T-inserts on standard spot tank. Submitted samples to Fimco for testing.
- Completed testing of spray-on graphics and submitted samples to Newland. Although test parts looked good, the masking process proved time-consuming and costly. Alternative technology (Mark-It Company's post-molding graphic application) looked promising.
- Tested stainless steel clamps from Kelch on EcoLab molds. Clamps reduced maintenance downtime by 50 percent.

Safety and Maintenance

- Tessa Swann, Loss Prevention Consultant from Liberty Mutual, continued to identify safety issues and to implement solutions.
- Safety Committee facilitated improvements.
- Crews completed annual maintenance on all machines.

Quality

- Thermo King Quality System assessment gave us a 53.7 rating.
- Awarded Newland Quality Award in 2004.
- First-Time Quality for 2004 was 97.5 percent, up from 96.8 percent.

2005–2006 Goals

Sales Overview

Anticipated production output is $5.1 million, based on the following:

- Newland: anticipate sales to go from $1.4 to $1.6 million.
- Fermont: release of 5, 10, 15, 30, and 60KW parts for production should mean $150,000 increase in sales.
- Fimco: anticipate sales to go from $950,000 to $1.1 million.

Manufacturing Goals and Production Needs

- Combine the 800 and 160 work areas into a cell and analyze the effects on throughput and overall machine efficiencies.
- Consider purchasing a 5-Axis Router for custom-job applications requiring exact-trim procedures.
- Hire a Process Engineer by March 2006: crucial to success of custom-job applications and implementing 5-Axis Router technology.

Conclusions and Recommendations

The Joliet plant is capable of $6 million in sales. Roadblocks include the lack of a 5-Axis Router for custom-trimmed products and training for the use of this router. Therefore, I recommend the following:

1. Purchase a 5-Axis Router.
2. Hire a Process Engineer who can use the 5-Axis Router and conduct training.

F.Y.I.

Periodic reports may also be called activity, compliance, control, department, evaluation, monitor, operations, or status reports.

GUIDELINES FOR PROGRESS REPORTS

Your goal when writing a progress report is to inform readers about how a project is progressing—how far you've come, where you are, and what's left to be done. Such a report might also aim to identify and resolve problems or difficulties in the project.

1 Prewrite

Rhetorical analysis: Consider your purpose, your audience, and the context.
- ☐ What measures of progress must you share?
- ☐ Who is your reader—a supervisor, a client, a funder? Why does he or she care about the project? What does the reader already know about it? What does he or she need to know about its current status?
- ☐ How does your report relate to the whole project? If there is only one progress report, write it near the project's midpoint and estimate the project's completion date. If this is the first of many reports, provide background information and create a template. In interim reports, cover developments since the previous report and anticipate next steps.

Gather information.
- ☐ Collect details from a variety of sources: project records, planning charts, previous progress reports, observations, and interviews.
- ☐ Answer these questions: What has been accomplished since the last report? Have problems occurred? How will they be fixed, when, and by whom? Is the project on schedule, on target, under or over budget?

2 Draft

Organize the report with the reader(s) in mind.
[opening] Clarify the project and time period, summarizing progress.
[middle] Discuss developments in detail.
[closing] Give conclusions about the project's status, and point to the next stage.

3 Revise

Review your report's ideas, organization, and voice.
- ☐ Have you given clear, accurate, and complete project details?
- ☐ Is the discussion ordered sensibly—chronologically, for example?
- ☐ Have you used a tactful tone that explains objectively any problems that have occurred in the project and that focuses on solutions?

4 Refine

Check your writing line by line for the following:
- ☐ Accurate terminology that the reader will understand
- ☐ Clear sentences with strong transitions that signal time
- ☐ Correct grammar, punctuation, usage, and spelling
- ☐ Strong format and attractive appearance (white space, headings, graphics)

Organizing Progress Reports

[opening] Introduction

- **Label** the report with the project name, a reference number, the deadline, the time frame of the current period, your name, and the reader's name.
- **Introduce** the report by providing any background information needed and previewing what's to follow.
- To be direct (bottom line first), **summarize** key achievements and problems, as well as your conclusions and recommendations about future work, schedules, and budget. To be indirect, state these points only in the conclusion.

[middle] Findings

- **Provide** detailed information on work planned for the period. Identify which tasks are completed, in-process, or unfinished. Focus on schedule issues.
- **Organize** the details using one or more of the following patterns:
  - Time: stages of the project completed during the report period
  - Categories: progress of each part of the project
  - Problem/solution: problems and steps taken to fix them
- **Present** the report primarily in narrative form, but attach cost analyses, inspections, schedules, and other evidence such as photos or drawings.

[closing] Key Points

- **Summarize** the project's present status, major achievements, and problems during the report period.
- **Provide conclusions** about the project's success, schedule for completion, budget, and main goals for the next stage.
- **Offer recommendations**, if necessary, to get or keep the project on track, or to change the original plan.

> There is a profound difference between information and meaning.
>
> —Warren G. Bennis

F.Y.I.

If you are writing several reports for a project, use consistent headings and numbering systems to help readers find information quickly in each report.

Progress Report

Rhetorical situation

Purpose: To report progress on a project for improving services to minorities
Audience: Officer from funding body for project—a representative familiar with the original project proposal and the previous year's work on the project
Format: Signed letter on organization's letterhead (makes report official, weighty)

---

1 of 2

Hope Services *Development Center*
199 Myrtle Avenue, Reading, PA 19606-5464 • Telephone 610-555-6577

July 16, 2005

Mr. Anthony Jenson
Contract Compliance Officer
Community Planning and Development
473 Maple Street
Reading, PA 19608-3361

Dear Anthony:

Subject: Hope Services Annual Progress Report (CDBG 2368-04)

Please accept this Annual Progress Report concerning Hope Services' work with minorities for fiscal year July 1, 2004, through June 30, 2005.

I have included these statistics: (1) total number of minority persons assisted, (2) the number of households and their ethnic origin, and (3) their status as low- or moderate-income households. In addition, I have included a narrative describing highlights of culturally specific services for the past fiscal year.

Client Numbers (July 1, 2004–June 30, 2005)
The following is client information for the minority households served by the Hope Services' staff through the Cultural Diversity Program at the shelter:

1. 178 minorities served, including 102 children.
2. 76 households served, 100 percent female-headed (36 African American, 10 Asian, 23 Hispanic, 7 Native American).
3. 96 percent of households served had incomes below the poverty level, while the remaining 4 percent of households were at low-income levels.

[opening]
Give a title and a reference number if appropriate.

Clarify the period and preview the report.

[middle]
Whenever possible, provide numerical data related to progress.

Outreach Highlights of the Cultural Diversity Program

In addition to the previous statistics, here are two illustrations of our progress on cultural-diversity issues:

- In January, Jasmine Michaels joined Hope Services to develop the Cultural Diversity Program, including (a) services for victims of sexual or domestic assault and (b) community outreach to minority populations.
- In April, representatives from the following organizations formed Project SART (Sexual Abuse Response Team): Hope Services, Reading Hospital, Berks County Attorney's Office, Reading Police Department, and Penn State University.

Conclusions and Projections

Hope Services continues to improve its services to minority clients and communities in Reading.

- Numbers indicate that Hope Services is helping its target clientele (low-income minority households headed by women).
- Hiring a Cultural Diversity Specialist has given Hope Services a strong presence in the community.
- In the coming year, Hope Services will focus on strengthening its outreach to minority communities and increasing its training of staff and volunteers in cultural-diversity issues.

Thank you for supporting our work with minorities through Hope Services. If you need additional information, please contact me at (610) 555-6577, ext. 427.

Yours sincerely,

Melissa S. Drummond

Melissa S. Drummond
Resource Development Director

[middle]
Explain key developments—milestones related to project goals.

Use lists where appropriate.

[closing]
Summarize the project's status, and look forward to the next stages.

Anticipate further contact.

F.Y.I.

Progress reports may be scheduled as initial, interim, completion, or follow-up reports, and they may be called activity, project, or status reports.

GUIDELINES FOR TRIP OR CALL REPORTS

Your goal when writing a trip or call report is to record the purpose of the visit or conversation and to state what was accomplished: contacts made, sales generated, repairs done, strategies learned, or other appropriate details.

1 Prewrite
Rhetorical analysis: Consider your purpose, your audience, and the context.
- ☐ What aspects of the trip or call should you share?
- ☐ Who will read this report, and why? What will the reader do with the information? What secondary readers might see your report? How might they use the information?
- ☐ Is there important background information you should consider, such as previous trips or calls? What follow-up should you anticipate or initiate?

Plan and complete the trip or call.
- ☐ Gather materials and equipment (e.g., recorder, camera, and log).
- ☐ Collect necessary information (recordings, receipts, log entries).
- ☐ Think about your reader's questions: Where did you go, when, how, why, and with whom? What did you accomplish? Did you have any problems? If so, what steps did you take to solve them?

2 Draft
Organize the report with the reader(s) in mind.
[opening] Identify the trip or call and summarize results, if appropriate.
[middle] Provide all the details and analysis your reader needs.
[closing] Summarize the trip or call and point to follow-up.

3 Revise
Review your report's ideas, organization, and voice.
- ☐ Have you included a clear account of events, developments, and results?
- ☐ Does the organization help the reader understand what you did?
- ☐ Have you used a positive but honest and objective tone?

4 Refine
Check your writing line by line for the following:
- ☐ Accurate terminology that the reader will understand
- ☐ Smooth sentence transitions that signal time and location
- ☐ Correct grammar, punctuation, usage, and spelling
- ☐ Strong format and attractive appearance (white space, headings, graphics)

Organizing Trip or Call Reports

[opening] Introduction

- **Label** the report with a title, your name, the reader's name, the date and identifying details about the trip or call. List who went where, when, and for how long; or who was called, when, and why. Include specific project names, client names, and reference numbers.
- **Introduce** the report by providing background information for the trip or call, especially its purpose and goals, and by previewing what's to come.
- To be direct (bottom line first), **summarize** main achievements and concerns, as well as your conclusions and recommendations. To be indirect, state these points only in the conclusion.

[middle] Findings

- **Provide** details on events, developments, and contracts, including details on the work planned, started, completed, and left unfinished.
- **Organize** details using a pattern or patterns that fit the situation, such as the following:
  - Time: break the work done into stages
  - Place: for several sites, treat each separately
  - Categories: deal separately with different types of trip or call activities (installations, repairs, service calls, sales calls)
- **Present** findings in paragraph form, but to make data accessible, include graphics such as tables, maps, charts, and photos. Attach copies of work orders, expense receipts, and so on.

[closing] Key Points

- **Summarize** highlights and achievements. What goals were met? What targets were missed?
- **Provide conclusions** about benefits and costs involved (if appropriate).
- **Offer recommendations** for follow-up—future visits or calls, other contacts to be made, changes to implement, meetings to hold, material to share with others.

> "Everything should be made as simple as possible, but not simpler."
> —Albert Einstein

WRITE Connection

If you file trip or call reports frequently, create efficiency by developing a template or a report form (see Chapter 33, "Designing Report Forms").

Trip Report

Rhetorical situation

Purpose: To accurately describe a repair trip—the problem, the fix, the cost, and so on

Audience: District supervisor for natural gas pipeline company; concerned about efficiency, smooth operations, and safety

Format: Printed report (follows company's standard template for repair report; creates easy access to information plus consistent recordkeeping)

---

1 of 2

PACIFIC PIPELINE CORPORATION
REPAIR REPORT

Date:	February 3, 2005
To:	Ralph Arnoldson Pasco District Supervisor
From:	Chris Waterford, Crew Chief
Repair:	Leak Clamp Installation
Location:	Camas Eugene Lateral near the city of Mollala, Oregon
Repair Date:	January 30, 2005
Crew:	#3 (Brad Drenton, Lena Harold, John Baldritch, Laura Postit, Jill Reynaldo, Chris Waterford)

[opening] — Identify the job.

Crew #3 and I (Chris Waterford, Crew Chief) responded to a call from the Eugene District crew asking for help on repairing a leak. Based on their request, we took the emergency trailer and a 24-inch Plidco clamp. Our response time was 6.5 hours (2 hours loading, 4.5 hours driving). We arrived about 2:30 p.m.

Provide trip background.

Assessment of Problem

The Eugene crew had exposed the leak area on the pipe. Then Crew #3 and I assessed the damage and conditions:

- Mud covered the work area and the ramps in the ditch.
- Water in the ditch came up to the bottom of the pipe.
- Rain was falling.
- The leak came from two quarter-inch cracks on a seam (at 10 o'clock looking south).
- A power generator was on site.
- Three bolt-on leak clamps and one hydraulic leak clamp were on hand.
- Space for a hydro crane and an emergency trailer was limited.

[middle] — Divide trip activities into logical categories. Use headings, subheadings, and lists.

Repair Plan and Decisions

With the conditions in mind, we considered three issues: the possibility of more leaks, clamp selection, and safety precautions.

More Leaks?

Eugene crew members probed three-foot sections of pipe on each side of the two quarter-inch cracks and found no more leaks.

Best Clamp?

Laura indicated that the clamp would not need to be welded, so we decided to use the 24-inch bolt-on clamp.

Safety Precautions?

We decided not to use the air systems for these reasons:

- The gas leak was minor (detectable only with a detecting agent).
- Safer installation of the clamp in daylight would be delayed by using air systems.
- Suits and breathing systems limit visibility, add weight, and create fatigue that could cause errors and injuries.

Based on these considerations, we installed the 24-inch bolt-on clamp in 1.5 hours. The Eugene crew took responsibility for site cleanup. We packed up, returned home in 4.5 hours, and unloaded in 1 hour.

Summary

The January 30, 2005, repair trip for Crew #3 aimed to help the Eugene District crew repair two leaks on the Camas Eugene Lateral near Mollala. Based on the small leak size, the muddy site, and the poor weather (rain), we decided to repair the leak without using protective air systems while installing a 24-inch bolt-on clamp. As a result, we completed the repair with these benefits:

- A less-expensive clamp was used.
- A time savings of approximately 2 hours was realized.

Total time spent on this repair was 13.5 hours, including travel, repair, loading, and unloading.

[middle]
Condense key trip events and issues.

Highlight decisions and developments.

[closing]
Summarize work done.

Stress costs and benefits.

F.Y.I.

Trip or call reports may also be called conference, convention, customer service, expense, field, inspection, installation, maintenance, repair, seminar, or workshop reports.

CHECKLIST for Short Reports

Use this checklist to benchmark and revise your short reports.

- ☐ **Ideas**
  - The report pursues a clear purpose: to answer questions, monitor work, move something forward, create a record, and so on.
  - It spells out the purpose and provides complete and accurate data.
  - It offers conclusions and recommendations based on solid information.
  - It uses lists and graphics to communicate information clearly.

- ☐ **Organization**
  - The report is structured logically into three parts: (1) introduction, (2) findings, and (3) key points (conclusions and recommendations).
  - It arranges findings in a pattern: order of importance, time or space, cause/effect, problem/solution, comparison/contrast, and so on.
  - It has informative, parallel headings to signal the contents of sections.

- ☐ **Voice.** The tone is conversational and objective, but positive.

- ☐ **Words.** Phrasing fits the reader in terms of level of formality and technical complexity.

- ☐ **Sentences.** Constructions are generally short to medium length; read smoothly; use brief clauses or phrases (telegraphic style); and follow parallel structure in lists.

- ☐ **Copy.** Grammar, punctuation, usage, spelling, and mechanics are correct.

- ☐ **Design.** Format, page layout, and typography choices follow company guidelines. The document uses white space effectively.

CRITICAL THINKING Activities

1. Reflect on the profession for which you are currently training. What routine reporting activities would this career involve? If necessary, interview someone in this profession to get an inside view.
2. Track down a sample short report by checking your college's offices or Web site or by contacting an area social agency or business organization. Then (a) identify the type of report; (b) describe its topic, readers, purpose, and information; and (c) assess the report's quality.

WRITING Activities

1. Based on your experiences in a workplace, develop one of the following short reports: a report on an incident that occurred, an investigative report on a key issue, a periodic report on regular activities, a progress report on a project, a trip report related to field work.
2. For one week, track and review key activities and issues happening on your campus. Based on what you discover, write a short report to the student government or an appropriate administrator in student life.

> **In this chapter**
> Guidelines for Major Reports 510
> Checklist for Major Reports 520
> Critical Thinking Activities 520
> Writing Activity 520

CHAPTER

31

Writing Major Reports

Major reports (also called long reports, formal reports, annual reports, major-research reports, position papers, or white papers) are both similar to and different from short reports. For example, major and short reports are similar in the sense that your writing task for both is to (1) study the topic, (2) analyze it, and (3) carry back to your readers the details of your analysis. Another similarity is that you should write both types in an objective, impersonal (yet readable) style.

However, major reports also differ from short reports in several ways:

- They are typically longer than ten pages (sometimes more than one hundred).
- They are formal, bound, titled documents with many parts.
- They focus on topics important to a company's long-term development (an expansion, a major industry change, an operations review).
- They are written by teams of writers that include—or are working closely with—upper management.

When you write a major report, you also have additional responsibility for organizing the text. In a sense, you're like an architect designing a major building. The architect must create a floor plan with distinct areas, and then design signs that help people move through the building and use those areas. Similarly, you must create a format with distinct sections of information, and then design "signs" (such as tables of contents, headings, and indexes) that help readers move through the document and use information.

Studying this chapter will help you develop the skills needed for writing a well-designed, information-rich major report.

GUIDELINES FOR MAJOR REPORTS

Your goal when writing a major report is to generate a thoroughly researched, well-organized study that helps readers understand and make decisions about a big issue.

1 Prewrite

Rhetorical analysis: Consider your purpose, your audience, and the context.
- What outcome do you want? What key issues are you addressing? Must you supply information only, or conclusions and recommendations as well?
- Who are your primary and secondary readers? What are their needs, concerns, and values? Why is the issue or topic important to them?
- How does this report relate to your organization's long-term plans?

Create a project plan and do reliable research.
- With your team, clarify assignments, key goals, and important resources. (See Chapter 3, "Teamwork on Writing Projects.")
- Create a schedule, monitor progress, map out design features, establish style guidelines, and construct a working outline.
- Conduct research using company and external sources of information. (See Chapter 7, "Conducting Research for Business Writing.")

2 Draft

Draft the report in manageable sections.

[opening] Write the executive summary and introduction after you have completed the middle findings and closing points.

[middle] Start with the middle findings, drafting chunks in separate sittings or according to team assignments.

[closing] Develop conclusions (and recommendations).

3 Revise

Review your draft's ideas, organization, and voice.
- Is each section informative and accessible?
- Are all points supported with reliable, documented evidence?
- Have you included introduction, transition, and summary paragraphs?
- Are the headings and subheadings parallel?
- Do the graphics clarify information?
- Do you use a professional and confident tone?

4 Refine

Check your writing line by line for the following:
- Accurate terminology that readers will understand
- Smooth sentences with logical transitions—an accessible, not dense style
- Consistent style from section to section (especially in team-written reports)
- Correct grammar, punctuation, usage, and spelling
- Effective presentation and integration of graphics
- Overall appearance—quality paper, clean typography, effective binding
- Margins and page-number placement that accommodate the binding

Organizing Major Reports

Front Matter

To help readers access important information and specific parts of the report, **provide** some of the following elements:

- A title page (full report title; names, titles, and affiliations of writers and readers; submission date)
- A transmittal letter (optional) reviewing the assignment, summarizing results, acknowledging help, and promoting follow-up
- A table of contents, list of tables, and list of figures

[opening] Introduction

- **Summarize** the report's main points, conclusions, and recommendations in an executive summary (roughly 10 percent of the report's length).
- **Introduce** the report's theme and direction:
  - State the issue or problem that the report addresses.
  - Explain the purpose, goals, key questions, and expected results.
  - Provide background information and the big picture.
  - Preview the report's content, and relate the report's scope.

[middle] Findings

Provide a full discussion of findings, information gathered, and analysis developed.

- **Organize** the findings into manageable chunks using compare/contrast, time, space, order-of-importance, categories, cause/effect, or problem/solution patterns.
- **Present** the findings using a heading system (with lettering or numbering) as well as introductions, transitions, and summaries.

[closing] Key Points

- **Summarize** the main points of the findings and key facts about the problem, issue, or need.
- **Provide conclusions** if required—the main results of your analysis of the topic or issue.
- **Offer recommendations** (if required) explaining actions to take or not take if required.

Back Matter

At the end of your report, include supporting material:

- A bibliography or references page
- A glossary of key terms
- Appendices (tables, charts, diagrams, survey copies)

Major Report: Title Page

Rhetorical situation

Purpose: To develop an in-depth, strategic marketing plan for farm herbicide sales, a plan that meets current and anticipated farm economy conditions

Audience: Primarily the vice president of the agriculture division; secondarily employees throughout the same division, and potentially management in other divisions

Format: Printed, titled, bound document (formalizes report; limits distribution, sharing, and forwarding of sensitive company document)

[purpose]
Include the report's purpose in the title.

[layout]
Place identifying details in four blocks.

[design]
Present information attractively, on quality paper.

Do not number the title page.

**THE NORTH AMERICAN SMALL-GRAINS MARKET:
2005–2006 SITUATION ANALYSIS
FOR HERBICIDE SALES**

Prepared for

Robert Kellogg
Vice President
Agricultural Products Division

Prepared by

The Small-Grains Market Management Team
Gene Bottoms, Gail Yakamoto, Syd Tindale,
Selena Martin, and Aaron Buchwald

Rankin Chemical Company
2311 Industrial Rd.
Fargo, ND 58103-0682

June 12, 2005

Major Report: Table of Contents Page

CONTENTS

Executive Summary iii
- I. Introduction 1
  - A. Background on the Small-Grains Market 1
  - B. Present Market Position 3
  - C. The Small-Grains Herbicide Market Challenge 5
- II. General Market Conditions for Small Grains 7
- III. Market Segments, Sizes, and Priorities 9
  - A. North-Central States 9
  - B. Western Canada 14
  - C. Pacific Northwest 16
  - D. South-Central States 18
- IV. Product Portfolio: Meeting Growers' Needs 20
  - A. Chase Herbicide 20
  - B. Ucopian Herbicide 26
  - C. Victor Herbicide 29
- V. Comparisons with Competition 34
  - A. Advance Herbicide 34
  - B. Binder Herbicide 36
  - C. Defeat Herbicide 37
  - D. Knockout Herbicide 39
  - E. Prevent Herbicide 41
- VI. Keys to Marketing Success 42
  - A. Manufacturing Issues 42
  - B. Distribution Strategies 45
  - C. Third-Party Support for Portfolio 47
  - D. Product Registrations 52
  - E. Current Public Issues 54
- VII. Conclusion 56

Acknowledgments 58

References 58

ii

[spacing] Space items for easy reading.

[headings] Show two or three levels of headings.

[numbering] Indicate both section numbers and page numbers.

[setting off] Use leaders (periods) between titles and page numbers.

[pagination] For front matter, use lowercase roman numerals.

Major Report: Executive Summary

[key points]
Present key points in the same order as in the body of the report.

Use lists, numbers, and white space for easy reading.

[key facts]
Highlight key facts from the report.

Present numbers consistently and clearly.

[word choice]
Avoid overly technical word choice.

EXECUTIVE SUMMARY

North American Small Grains is a high-priority herbicide market for Rankin Chemicals in 2005–2006:

1. The market size is 115 million acres.
2. Rankin can effectively draw on its European division's resources for the North American market.
3. The available product portfolio meets the varied needs of small-grain growers.
4. U.S. grain growers are economically healthy—experiencing reduced debt in 2004, decreased interest rates, improved land prices, and increased availability of money.
5. Canadian grain growers' economic health remains steady. A weakened Canadian dollar has meant exporting strength for small-grain growers.

In order of priority, market segments for Rankin herbicides are (1) the North-Central States, (2) Western Canada, (3) the Pacific Northwest, and (4) the South-Central States.

In the North-Central States, a 1.9 million acre potential exists for Chase Herbicide and an undefined market share for Ucopian. Some opportunity may also exist for a herbicide formulation containing a small amount of Victor to control kochia. In the Western Canada market segment, a 1.6 million acre potential exists for Chase Herbicide, a 0.5 million acre opportunity for Ucopian, and approximately 3.5 million acres for Victor. In the U.S. Pacific Northwest, Chase sales are projected at 314,000 acres, with a small, undefined niche for Victor. In the U.S. South-Central States, Ucopian Herbicide has the largest potential at 1.1 million acres.

Rankin products targeted for this small-grains market fall into two groups:

1. Chase and Ucopian Herbicides work on Canada thistle and wild buckwheat on irrigated land or in the subhumid or humid, moisture-deficient regions.

iii

2. Victor attacks field bindweed and wild buckwheat in the semi-arid and subhumid, moisture-deficient regions.

Comparisons with competing herbicides show that Rankin's portfolio has strengths in spectrum of coverage, crop safety, cost, and tank-mix flexibility. Weaknesses include a narrow window for application and short residual for Victor Herbicide.

Grain producers have responded well to Rankin's product portfolio, but the Small-Grains Market Management Team must manage the following business issues to ensure a profit in this market:

1. **Manufacturing Issues.** Priority must be given to streamlining the pentamine process, operating the Ucopian plant at capacity, and accurately forecasting demands for Victor.

2. **Distribution.** If distributors and dealers can maintain an attractive profit margin, they will push a product's sales potential to its maximum. To ensure these profit margins, the Small-Grains MMT is implementing a margin-maintenance program.

3. **Third-Party Support.** In Canada, ECOW gave Rankin products favorable recommendation status for use in cereals. In the United States, ten significant land-grant universities have recommended all three herbicides in the portfolio.

4. **Product Registrations.** During 2004, Rankin received approvals in the United States for Victor, plus several label amendments for Chase and Ucopian. The Small-Grains MMT submitted applications to Agriculture Canada for using Victor in cereals, with approval pending.

5. **Public Issues.** The Small-Grains MMT's 2004 Strategy Report noted that the EPA might classify Victor Herbicide as a "Restricted-Use Pesticide" and require a groundwater statement on the label. The EPA decided to classify Victor as a "General-Use Pesticide" and require a groundwater statement in the "Use Precautions" section of the label.

iv

[voice]
Use an objective but energetic tone in crisp sentences.

[bottom line]
Provide bottom-line information to help decision makers.

[context]
Reference related reports.

Major Report: Introduction

[opening]
Put the report in context with background information, purpose, and scope details.

I. INTRODUCTION

A. Background on the Small-Grains Market

In 1998, Rankin's North American Herbicide Business Team selected small grains, primarily wheat and barley, as its top priority market. While basing this decision on several factors, the main reason was the size of the potential market. Globally, growers devote the largest percentage of available crop land to wheat and barley. Each year, these two crops account for 570 million acres, followed by 356 million acres for rice, 304 million acres for corn, 120 million acres for soybeans, and 59 million acres for cotton.

Of wheat and barley's total global acreage, 115 million acres are planted in North America. . . .

[headings]
Use headings and subheadings to indicate subjects and key issues.

B. Present Market Position

Because of the size of the global cereals market, a major portion of Rankin International's herbicide synthesis program is targeted toward small grains. Most of this effort supports European markets. . . .

[transitions]
Use strong transitions between sections and paragraphs.

C. The Small-Grains Herbicide Market Challenge

Given Rankin's global position, product portfolio, and past performance, these are the challenges addressed by this report for the coming year in the North American small-grains market:
(1) . . . (2) . . . (3) . . .

Major Report: Findings

II. GENERAL MARKET CONDITIONS FOR SMALL GRAINS

On the whole, market conditions for 2005–2006 are favorable. This judgment is based on production estimates, price trends, debt and loan conditions, government action, and climate projections.

Small-grains production looks to remain steady. The International Wheat Council estimates world wheat production for 2005–2006 at . . .

[preview] Preview the section's content in an introductory paragraph.

Level 1

III. MARKET SEGMENTS, SIZES, AND PRIORITIES

The North American small-grains market contains four main segments, listed here in order of priority: the North-Central States, Western Canada, the Pacific Northwest, and the South-Central States (Figure 1). Order of priority is based on grain types planted, total acreage planted, major broadleaf weeds and grasses found in the segment, and regional economic conditions.

Level 2

A. North-Central States

The North-Central States segment of the United States provides the greatest market potential for Rankin. In this area, growers plant 36,557,000 acres in small grains, 10,670M acres in winter wheat, 16,939M acres in spring wheat, and 8,948M acres in barley (Table 1). The total dollar value for herbicides in this market segment is $75MM.

1. Weed Spectrum for Region
   The weed spectrum for the region is . . .   *Level 3*

[headings] Show heading levels by shifting from center to side, all capital letters to uppercase and lowercase, bold to normal typeface, roman type to italics.

B. Western Canada

The Western Canada market segment contains 43,100,000 acres of small grains. Although it has more total grain acres than the North-Central States, it is ranked second in potential sales for these reasons: (1) . . . , and (2) . . .

[numbering] Designate sections with a letter and number system.

Major Report: Key Points

VI. KEYS TO MARKETING SUCCESS

Earlier sections of this report show that Rankin's product portfolio (Chase, Ucopian, and Victor) compares favorably with the competition, that we have clear strategies developed for all four market segments (including niches), and that the 2005–2006 conditions for the small-grains market look positive.

However, Rankin's success also depends on marketing. This section reports on both progress and work-to-be-done in manufacturing, distribution, third-party support, product registrations, and public issues. . . .

[summary]
Summarize previous sections, and introduce what is coming with transition paragraphs.

VII. CONCLUSION

This report has explored the 2005–2006 small-grains market's potential for herbicide sales: key business issues, the competition, the product portfolio's strengths and weaknesses, strategies for different market segments, and anticipated market conditions.

Given this discussion, acreage statistics, and breakeven analyses, the Small-Grains MMT concludes the following about the 2005–2006 market situation. . . .

Recap the key findings.

Major Report: Back Matter

ACKNOWLEDGMENTS

The Small-Grains Market Management Team wishes to express their gratitude to the following Rankin contributors to this report:

Chris Pratt, HMS, Wichita, KS
Patricia Fenwick, HMS, Saskatoon, SK
Hubert Pullman, HMS, Spokane, WA
Julio Franco, TS&D Specialist, Omaha, NE
Michelle Barton, Account Salesperson, Lubbock, TX
Bill Baldwin, Field Sales Rep., Fargo, ND
Lucy Shaw, TS&D Specialist, Kansas City, KS

REFERENCES

Branson Consultants. (2004). *Innovations in agri-business: Distribution systems* (Project No. 99-144565). Wichita, KS: Author.

National Association of Wheat Growers. (2005). *Report from Washington: 2005 issues.* Washington, DC: Author.

Nowlan, M. (2005, March 8). Small grains futures. *Agri-Business Reports, 5*(12), 93–97. Retrieved May 16, 2005, from http://www.jdgreen.com/news

Sternum, A. R., Cook, A., Lanham, R., & Zezulka, J. J. (2004). *Ag-products management team situation analysis.* St. Louis, MI: Agri-Biz Analysis.

[sources]
Acknowledge people who supplied information and feedback.

List sources used and referred to in the report.

F.Y.I.

Often, major reports can include "boilerplate"—material written for previous reports and other documents. Use boilerplate material carefully, making sure that the style and content fit well with your report. Integrate and modify boilerplate as needed.

CHECKLIST for Major Reports

Use this checklist to benchmark and revise your major reports.

- ☐ **Ideas.** The report supplies clear, accurate, well-researched, and relevant information—a reliable foundation for making decisions about key issues.
- ☐ **Organization.** As shown in the outline on page 511, the report has a reader-friendly structure that includes front matter, an introduction, findings, conclusions, and back matter.
- ☐ **Voice.** The report uses a professional and confident, but objective tone.
- ☐ **Words.** The report avoids contractions, uses personal pronouns sparingly, and defines technical terms in the text or in a glossary.
- ☐ **Sentences.** The report contains sentences that are generally short to medium in length, are varied in structure, and read well aloud.
- ☐ **Copy.** The report uses correct grammar, usage, spelling, and mechanics. Headings and lists are parallel in phrasing.
- ☐ **Design.** The report meets company specifications for format; is printed on quality paper; is effectively bound; and uses white space, margins, headers or footers, page numbering, and graphics to make the report accessible and attractive.

CRITICAL THINKING Activities

1. Working with a classmate, analyze the guidelines, model pages, and checklist in this chapter to assess how a major report is both similar to and different from a short report. List the similarities and differences. Assess how the characteristics of the two documents reflect the function of each.
2. Find a sample major report by contacting an organization, checking with an office on your campus, or searching Web sites (e.g., government agencies, think tanks, business sites). Study the report carefully and develop a presentation for classmates in which you (a) describe the report's contents and use and (b) assess its quality based on the instructions in this chapter.

WRITING Activity

Consider a business sector, industry, or field closely associated with your career or profession. What is the current state of that sector? What challenges does it face? What opportunities are developing? Write a major report for your fellow students in which you describe that sector, industry, or field and provide conclusions about its history, current state, and future.

WRITE Connection

For related research activities, see page 168.

In this chapter

Guidelines for Proposals **522**
Operational Improvement Proposals **524**
Sales or Client Proposals **530**
Grant and Research Proposals **533**
Checklist for Proposals **538**
Critical Thinking Activities **538**
Writing Activity **538**

CHAPTER

32

Writing Proposals

The verb *proposal* comes from *pro*, meaning "forth," and *ponere*, meaning "to put or place." The noun *proposal* means "plan." During your career, you will write many proposals—or "plans" that you "put forth" to convince others how to address specific issues.

Some of your proposals may be as simple as suggestion-box memos; others may be as complex as book-length bids. Regardless of the document's type or length, your writing goal will always be the same: (1) to study the issue, (2) to analyze the need or problem, and (3) to lay out a convincing plan that meets the need or solves the problem. When you achieve this goal, your proposal is a force for positive change.

In your career, what needs or problems will you address in a proposal? That depends on your specific situation. For ideas, consider these examples of positive change:

- Fixing inefficient operating practices
- Winning contracts or bids
- Selling products or services
- Developing new markets, products, or services
- Improving current products or services
- Gaining approval or funding for projects
- Advancing pure or applied research
- Meeting legal and ethical requirements

This chapter includes guidelines, a checklist, and seven model proposals arranged in three categories. The end-of-chapter activities will hone your ability to study an issue, analyze a problem, and lay out a convincing plan that brings about positive change.

GUIDELINES FOR PROPOSALS

When you write a proposal, your goal is to persuade others that you have a workable solution or plan that solves a problem or meets a need. Such a proposal might improve operations, win business, or gain approval and/or funding for a project.

1 Prewrite
Rhetorical analysis: Consider your purpose, your audience, and the context.
- ☐ Exactly what are you proposing? Why? What measurable, testable outcome do you want to result from your proposal?
- ☐ To whom are you making this proposal? What are your reader's needs, attitudes, and concerns in relation to the issue? To what degree might he or she resist change?
- ☐ How is this proposal related to your company's goals, mission, and policies? Is it affected by broader cultural, social, or global forces?

Study the need or problem and possible solutions.
- ☐ Research the problem's background and history.
- ☐ Break the problem into parts, investigating causes and effects.
- ☐ Review solutions attempted in the past, noting successes and failures.
- ☐ Identify and study all possible solutions with the aim of choosing the best one—the one that most fully meets the need or solves the problem.
- ☐ Consider ethical dimensions of both the problem and your proposal.

2 Draft
Organize your proposal into three parts.
[opening] Provide context as well as a summary, if appropriate.
[middle] Present the problem or need and your solution.
- ☐ Explain what the problem is and why it should be corrected; or outline the specific need and its significance.
- ☐ Map out the solution and stress its value.

[closing] Summarize your conclusions and recommendations.

3 Revise
Review your draft's ideas, organization, and voice.
- ☐ Have you provided all the details and analysis that your readers require to understand and embrace the problem or need?
- ☐ Does the proposal address alternatives, stress benefits, consider ripple effects (who will be affected and how), and show your ability to implement the solution?
- ☐ Is your tone confident and positive, but not aggressive?

4 Refine
Check your writing line by line for the following:
- ☐ Precise words, easy-to-read sentences, and strong transitions
- ☐ Correct grammar, punctuation, usage, mechanics, and spelling
- ☐ Document design that follows specifications, creates a good impression, and makes information accessible

Organizing Proposals

[opening] Introduction

- **Label** your proposal with a title or a subject line promising productive change, your name, your reader's name, the date, and reference numbers.
- **Introduce** your proposal with background and the theme—the need to be met, problem to be solved, and benefits to be gained.
- **Summarize** your proposal if you want to be direct. To be indirect, do not include the summary.

[middle] Problem and Solution

- **Define** the problem or need. Explain its importance, limits, causes, effects, history, and connection with larger issues. Review any past attempts to solve the problem, noting their successes and failures. If the reader is aware of the need or problem, be brief and informative. If the reader is unaware or resistant, build a persuasive case about the problem or need's existence and its importance.
- **List** criteria for a solution. What should a solution accomplish?
- **Compare and contrast** alternative solutions.
- **Promote** the best solution—your essential proposal—by stressing how it best meets the need or solves the problem. Refer to the solution criteria and explicit benefits.
- **Prove** the solution's practical workability by highlighting the following:
  - Outcomes of the solution, plan, or project
  - Requirements (e.g., facilities, equipment, material, personnel)
  - Schedules for start-up, stages, finishing dates, and follow-up
  - Budget or cost breakdowns (e.g., services, equipment, materials, travel)
  - Methods of monitoring costs and quality
  - Your qualifications for undertaking the task

[closing] Key Points

- **Summarize** the problem or need and alternative solutions.
- **Provide conclusions** about the best solution—results and benefits.
- **Review your recommendations** for implementing the solution.
- **Request approval,** solicit feedback, and anticipate a response.

> "In good writing, words become one with things."
> —Ralph Waldo Emerson

WRITE Connection

For additional discussions of persuasive strategies, check "The Art of Persuasion" on pages 452–453, "Persuasion Ethics" on pages 174–177, and Chapter 10, "Trait 1: Strong Ideas," especially making and supporting claims on pages 188–191.

OPERATIONAL IMPROVEMENT PROPOSALS

Many proposals seek to strengthen a company's operations by making practices more effective, efficient, and safe, thereby improving productivity and profits. Such proposals can be concept, troubleshooting, or justification proposals.

Concept Proposals or Suggestions

A concept proposal simply puts an idea on the table for discussion. Unlike a full proposal, it doesn't give a detailed analysis or solution.

Rhetorical situation

Purpose: To convince reader to consider customer relations training for staff
Audience: Director of marketing; interested in sales, customer satisfaction
Format: Memo (gets thoughts on paper for further discussion)

Date: July 6, 2005

To: Maria Estavez
Training Supervisor

From: Jamie Tobias
Director of Marketing

Subject: Concept Proposal for Speaking/Listening Training

[opening] — *State the subject positively, identify the problem, and give background.*

Maria, I'm writing with an idea for training that will improve customer relations and retention. Specifically, I have noticed that our sales and marketing staff could sharpen their skills in speaking and listening to clients.

[middle] — *Explain the problem in a neutral tone. Stress its impact on the reader and/or company.*

The Need: Connecting with and Keeping Clients
Clients must feel that their needs are being heard and acted upon. Through recent client feedback, I've learned about some problems, including our staff's poor phone manners and brusque responses to questions and complaints.

This behavior clearly affects our sales: Potential clients choose competitors, and current clients are tempted to go elsewhere. In addition, the behavior hurts relationships between staff by creating tension and interfering with teamwork.

Offer your solution, focusing on how it will solve the problem. Discuss costs and benefits.

Possible Solution: Training in Speaking and Listening
I suggest that we hire a trainer who can lead communication workshops with our sales and marketing staff. This training would address issues such as communicating with respect, interviewing clients to discover their needs, observing telephone etiquette, and making persuasive presentations. If well done, the training will help staff (1) present our products more effectively, (2) strengthen our relationships with clients, and (3) improve internal communication.

[closing] — *Invite feedback.*

Please let me know when you're able to discuss this matter.

Troubleshooting Proposal

A troubleshooting proposal seeks to fix a problem, error, waste, or danger that is impeding effective operations and productivity. Although usually an internal proposal, it can also be a study solicited from an external consultant.

Rhetorical situation

Purpose: Convince reader that phasing in electric lifts will solve carbon monoxide problems in a cost-effective manner
Audience: Plant manager responsible for both efficient, profitable operation and employee health and safety; secondary readers potentially include OSHA officials
Format: Formal document with memo heading (fitting for in-house communication)

---

1 of 3

Rankin Manufacturing
PROPOSAL

Date: June 19, 2005

To: John Cameron

From: Nick Jeffries

Subject: Reducing Carbon Monoxide Levels

As you requested, I have investigated the high levels of carbon monoxide in the main warehouse. The following proposal (A) explains the source of the problem, (B) proposes a solution, and (C) details an implementation plan.

A. Problem: Emissions from Lift Trucks

From November 2004 through March 2005, Rankin has been registering high carbon monoxide (CO) levels in Area 3 of the warehouse. General CO levels in the area have exceeded 35 ppm, and many office spaces show levels of 40–80 ppm (OSHA recommends 25 ppm).

These CO levels are a concern for three reasons:

1. High CO levels cause sickness and lower productivity.
2. Using summer exhaust fans in winter to reduce CO results in low humidity that shrinks wood used for manufacturing.
3. High CO levels can result in a substantial OSHA fine.

To determine the cause of the high CO levels, I investigated all sources of combustion in the warehouse. I concluded that the excess CO was caused by lift trucks operating in Area 3.

[opening]
Focus on positive change in the subject line and introduction.

[middle]
Define the problem and detail its impact.

Be objective.

I then checked all lift trucks. They were in good working condition and were being properly used and maintained.

B. Proposal: Phase Out Internal-Combustion Lifts

To correct the CO emissions problem, the ideal solution should accomplish the following in a timely and cost-effective manner:

1. Bring CO levels within OSHA limits.
2. Maintain relative humidity to ensure product quality.

To meet these targets, Rankin could continue using the exhaust fans and install humidifying equipment at a cost of $32,000. (See attached estimate.) Rankin could also replace all internal-combustion lift trucks in Area 3 with electric lift trucks for $195,000.

Instead, I propose gradual replacement of the internal-combustion lifts in shipping with electric lift trucks, for these reasons:

- Shipping (Area 3) has 14 lift trucks that operate almost 24 hours per day.
- The shipping area is well suited for electric lift trucks (no long-distance travel or use of ramps is required).
- While electric lifts cost more initially, they have lower operating and maintenance costs. A five-year cost analysis shows that the costs of operating the two types of lift trucks are similar. (See attachment.) In fact, after five years, electric lift trucks save money.

C. Implementation: Phase in Electric Lifts

Because Rankin buys an average of four lift trucks annually, the plans below will complete the changeover in the shipping area within the next four years.

1. When an area requests replacement of an existing lift truck, management approves purchase of a new electric lift truck.
2. The new electric lift truck goes to the Shipping Department.
3. The newest internal-combustion lift truck in Shipping is transferred to the area that requested a new lift truck.

By beginning this plan in January 2006, we could replace all internal-combustion lift trucks in Shipping by December 31, 2009.

Conclusion

Gradual replacement of internal-combustion lift trucks in Shipping with electric lift trucks will involve a higher initial cost but will reap two important benefits: (1) CO levels will fall below OSHA's 25 ppm standard, enhancing the safety of Rankin employees and guarding product quality; and (2) the electric lift trucks will prove less costly to own and operate in the long run.

Therefore, I recommend that Rankin management approve this plan, and phase it in over the next four years as outlined. The improved safety, increased product quality, and lower total long-term operating costs outweigh the higher initial costs.

If you wish to discuss this proposal, please call me at extension 1449.

Attachments: Renovation Estimate
Five-Year Cost Analysis

[closing]
Restate your solution and its benefits.

Justification Proposal

A justification proposal—usually internal—argues for an expenditure or an expansion, whether to purchase equipment or develop a new product. Such a proposal supports an investment in the company's operations.

Rhetorical situation

Purpose: To convince readers that purchasing a power washer will improve cleanup while saving time and money
Audience: Vet partners interested in medicine, customer relations, productivity, profits
Format: Memo (fitting for unsolicited proposal from subordinate)

---

1 of 2

Heartland Veterinary Services, Inc.
MEMO

Date: March 14, 2005

To: Drs. Hom-Kuh Kao, Arnold Shaffer, and Adrie Markus

From: Chris Tanyel

Subject: Improving Cleanup with a Power Washer

[opening] *Present your idea and stress its benefits.*

I'm writing to recommend that Heartland buy a power washer for the cattle-processing department. A power washer will (1) save time, (2) cut costs, (3) clean effectively, and (4) improve the work environment.

Current Cleaning Practices

The current cleanup procedure in cattle processing is labor-intensive, costly, ineffective, and unsanitary.

[middle] *Give the reason for your proposal. Review the present situation in detail.*

Presently, we use a large fire hose, a garden hose, and a hand brush to clean the manure, dirt, and grease off equipment in the cattle-processing room. Cleanup involves spraying an area with water, letting the water soak into the floor, scrubbing with Rocal D, and rinsing. Here are some of the problems with this method:

1. Doing the process thoroughly takes about two hours. (If time is limited, things get missed.)
2. This procedure uses 1/2 gallon of Rocal D ($48.50 per gallon).
3. The fire hose washes the floor effectively, but lacks the high pressure needed for walls and equipment. It uses 600 gallons per cleanup.
4. Metal surfaces, especially those in hard-to-reach areas, often appear dirty and unsanitary even after the process.

Power Washer Benefits

Replacing the current cleaning equipment with a power washer would supply pressure to remove even the driest material from both flat and hard-to-reach surfaces, would eliminate brush streaks, and would cut costs:

- Cleanup time would be reduced to 30 minutes.
- A 2000 psi washer would use 75–80 gallons of water per cleanup.
- Only 1 pint of Rocal D disinfectant would be needed per cleanup.

Savings realized by a power washer would pay for the washer in less than two months. While the current method costs $53.15 per day, a power washer would reduce that figure to $13.05 per day. (See attached cost analysis.)

Purchase Options and Recommendation

After checking with several retailers, I recommend that Heartland purchase a Douser power washer. The Douser by Hancock is available for $1,499 (including delivery and installation). It has a pressure rating of 2200 psi and comes with 100 feet of pressure hose and three nozzles.

The Douser power washer will keep the cattle-processing room cleaner and will recoup the initial investment in less than two months through savings on water, labor, disinfectant, and repairs.

If you need more information or wish to discuss this proposal, please call me at ext. 366.

Attachment 1

Part 5 Reports and Proposals

SALES OR CLIENT PROPOSALS

Whether solicited or unsolicited, a sales proposal seeks to win a potential client's business. Often, such a proposal is a response to a company's request for proposals (RFP).

Bid Form

With its standardized format, a bid form lays out precisely what a company promises to deliver at what cost to a potential client.

Rhetorical situation

Purpose: To convince a client that the project bid is competitive and that the work will be done well on schedule
Audience: School superintendent concerned with restricted education budgets
Format: Standardized form (offers accurate estimate and contractual weight)

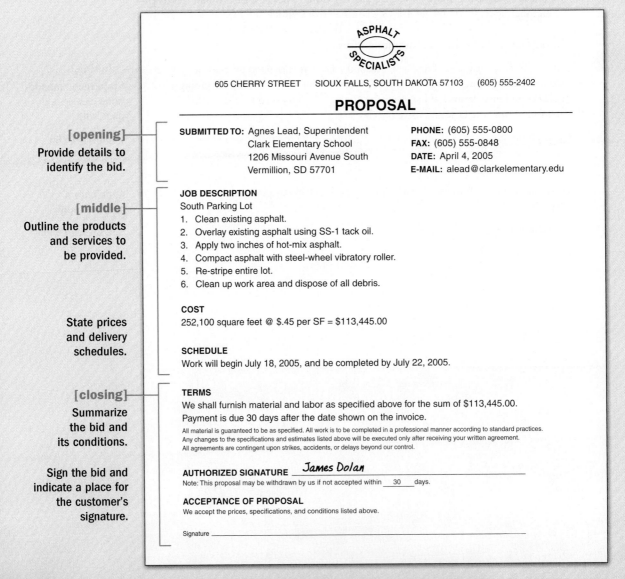

[opening] Provide details to identify the bid.

[middle] Outline the products and services to be provided.

State prices and delivery schedules.

[closing] Summarize the bid and its conditions.

Sign the bid and indicate a place for the customer's signature.

Sales Proposal (Letter)

While a letter may contain many of the same details as a bid form, the letter includes more discussion and analysis of the product or service. This discussion aims to persuade the reader of the product's or service's quality and value, as well as of the company's credibility and trustworthiness.

Rhetorical situation

Purpose: To convince the reader to accept the bid for waste removal services
Audience: Pharmaceutical plant maintenance supervisor; interested in efficient, cost-effective maintenance that adheres to local, state, and federal regulations
Format: Formal, signed letter on company letterhead (makes bid official)

BONIFACE SANITATION, INC.
846 Watson Way Tallahassee, FL 32308
302·555·2356 www.bonif.com

February 25, 2005

Ms. Agnes Grey
Millwood Pharmaceuticals
2211 Green Valley Road
Tallahassee, FL 32303-5122

Dear Ms. Grey:

Thank you for the opportunity to bid on Millwood's waste removal and recycling needs. We would be very pleased to take care of your refuse and recycling. — **[opening]** Be positive and polite.

Based on the bid requirements, we are submitting the following proposal:
- One eight-cubic-yard container for regular refuse, serviced twice a week.
- One eight-cubic-yard container for cardboard, serviced once a week.
- Total cost per month: $169.00, with extra pick-ups at $30.00 per trip.

We would place the containers to guard your premises' appearance, use new or like-new containers, and empty your containers at times best for you. In addition, we can dispose of harmful or hazardous materials at competitive prices while meeting all state and federal regulations.

Our goal is serving you well. As Tallahassee's leading waste collector, Boniface serves more than 300 organizations like yours. I've attached references, crew lists, and equipment brochures for your review.

— **[middle]** Provide a precise bid with the necessary details.

Sell your bid's value—benefits for the reader.

Sell your company.

I look forward, Ms. Grey, to your response. Until then, I'm at your "disposal" to answer questions. — **[closing]** Anticipate a positive reply.

Sincerely,

Robert Estavez

Robert Estavez

Enclosures 3

Major Sales Proposal or Bid

A major bid is usually a response to a request for proposal (RFP) published by a company or a government agency. Such a bid is typically a long, titled, and bound document that carefully follows the RFP specifications for content, organization, format, and so on. Besides observing these specifications, consider these tips:

Preface your bid with some or all of the following "front matter"
- A title page with the title, writer, reader, and submission date
- A cover letter that introduces the proposal, sells its strengths, notes the key players, and thanks the reader
- A copy of the RFP or the letter of authorization
- A table of contents and a list of illustrations
- An executive summary in nontechnical language

Build a strong case for your product or service
- Review the reader's need as indicated in the RFP; then brainstorm ways that your product or service matches up with or meets that need.
- Present the solution—your product or service—through accurate, vivid description. Link features of your product or service with concrete reader benefits.
- Prove that you can deliver the product or service by focusing on practical details of implementation:
  - The precise list of products and services included
  - Methods of delivery and a schedule of delivery
  - Costs, fees, budget breakdowns
  - Evaluation plans for checking progress and results
  - Personnel requirements
  - A statement of responsibilities (yours and the company's)
- Stress the bid's benefits for the client: results, efficiency, reliability, cost, value, flexibility, growth, and so on.
- Offer an executive summary for busy decision makers. Capture in a nutshell the reader's need, your solution, the results you can deliver, and the advantages.

Build credibility—your own and your company's
- Describe your company and its resources; indicate memberships and awards.
- List relevant past and current jobs; provide work samples, if appropriate.
- Provide references, testimonial letters, and résumés of key personnel.

WEB Link

To study a sample major bid, visit this book's Web site at http://college.hmco.com/english.

GRANT AND RESEARCH PROPOSALS

Grant proposals and research proposals ask readers to approve or fund a project. When writing such a proposal, your goal is to demonstrate a need, show how the project will meet that need, and prove that the project will happen. A persuasive proposal—one that your readers will approve—is creative and ambitious, but also logical and practical.

Grant Proposal

Rhetorical situation

Purpose: To win a matching grant from the state to fund road paving for industrial expansion in a small town
Audience: Government officials at the state department of transportation
Format: Official format required by funder, both sections and attachments

---

1 of 3

**OKLAHOMA DEPARTMENT OF TRANSPORTATION
APPLICATION FOR LOCAL DEVELOPMENT (LD) FUNDING**

Introduction
The City of Ada requests a Local Development matching grant of $104,635 to pave 1,106 ft. of 20th St NW. This grant will contribute to a total capital investment of $6,502,949, support the growth and workforce expansion of a successful local business (Ada Automation), and open 65 acres for new industrial development. (Formal details are included in Attachment A, the Application Form.)

[opening]
Summarize the project and its benefits.

The Transportation Need
This application for an LD matching grant involves an industrial expansion project and related roadway improvements. The grant will be used to hard-surface 1,106 ft. of 20th St NW, which is in a new industrial park in the northwest corner of Ada.

This roadway improvement will serve Ada Automation, Inc.'s growth and expansion. AA is a 42-year-old Oklahoma company that currently employs 145 workers. Beginning in 1961 as an agricultural-implement repair shop, AA has grown into a multifaceted firm that manufactures feed mixers, mixer-replacement parts, silage-bagging machines, service-truck bodies, and truck cranes. For the fiscal year ending February 28, 2004, AA's annual corporate sales were $22.3 million.

Ada Automation is currently in the first phase of a three-phase expansion project. Within five years, the company anticipates sales growth from $22 million to $70 million. To achieve this growth, AA has the following expansion plan:

1. In 2002, AA purchased 40 acres west of Highway 377 on the southern side of 20th St NW to relocate and consolidate its operations out of the downtown area.
2. In 2004, AA completed the first phase of its expansion by constructing a 100,000 sq. ft. facility on the 40-acre site. Total project cost for this phase was approximately $5.4 million.

[middle]
Describe the need objectively but persuasively.

Supply background that creates a project "story" focusing on people.

Include concrete facts and details

Typically, a grant proposal seeks funding from a government agency (e.g., a state small-business program) or a private philanthropic organization (e.g., the Pew Charitable Trust). To write a successful funding proposal, follow these tips:

- Research granting bodies to find a good match, and establish a relationship with someone at the agency. Learn what motivates the funder to give, and address those values in your application.
- Follow the grant instructions perfectly: Don't let a technical slip disqualify you.
- Be honest, straightforward, and objective. Clearly state what you want from the funder, and promise only what you can deliver. By doing so, you practice

---

2 of 3

[middle]

Use numbered and bulleted lists to highlight points and make information accessible.

Refer readers to attachments.

3. In 2006, AA plans to complete phases 2 and 3 by expanding the facility to 284,000 sq. ft. (See Attachment B.) The project will allow the company to handle the following on site:
   - All of its paint preparation, painting, and drying.
   - The full assembly of all its agricultural and service-body products.
   - All incoming freight and acquisition of product.
   - All loading for distribution of completed goods and parts.

Total project cost for phases 2 and 3 is estimated to be $5.4 million.

Stress the need's importance by creating a "before" picture.

The improvement to 20th St NW is needed to support heavy truck-and-trailer traffic to and from the new facility. Currently, the road is graveled, which diminishes AA's ability to provide easy access to vendors and attract clientele. Current traffic studies show that approximately 75 trucks use the road daily, and that number is expected to triple in the next five years. Because the company relies on heavy trucks, its future expansion cannot happen efficiently without the improvements.

The Paving Project

Describe the project so that the reader sees how it addresses the need.

Provide realistic projections.

With a firm commitment for economic development from Ada Automation (see Attachment C), the City of Ada proposes to pave 1,106 ft. of 20th St NW.

Total Cost:	$209,270
Matching Funds (assured by City of Ada):	$104,635
LD Matching Grant:	$104,635

Please see Attachment D, a cost estimate by the engineering firm of Grant Rihbold & Associates, for itemized construction costs; see also the site map (Attachment E) for technical details. We anticipate grading and paving to take place in July 2005.

Use positive, energetic, and clear phrasing—not jargon.

This paving project is consistent with the City of Ada's economic and transportation planning. In 2000, the City of Ada completed a comprehensive plan for the community that designated future industrial areas and road improvements. The current project fits with that plan by efficiently serving current and future development. (See Attachment F, excerpt from the comprehensive plan.) The project has the full support of the city in terms of zoning, water, sewage, drainage, electrical, and gas service. (See Attachment G, Local Commitment and Initiative Information.)

- sound ethics and avoid the need for future excuses and problems that could break the relationship with the funder.
- Show that you know your own organization, and that the project has had widespread input and will enjoy widespread support. Include management statements, results of focus groups, memberships of committees, survey results, and so on—evidence of backing and participation.
- Develop a strong, realistic project budget, getting expert help as needed.
- Submit a neat, error-free document that is easy to navigate for readers.
- Submit your grant proposal early so that it's more likely to be studied carefully and viewed positively.

Ada has a local development organization that works closely with the city to attract businesses to the community. In 1970, there were fewer than 100 manufacturing jobs in Ada. Today, 15 Ada companies offer more than 2,000 manufacturing jobs. The proposed paving project will continue this growth process.

Outcomes and Benefits
The paving of 20th St NW will have both immediate and long-term benefits:

1. Ada Automation will be able to proceed with phases 2 and 3 of its expansion. Specifically, the road construction would contribute $209,270 to a total capital investment of $6,502,949. (See Attachment H, Sources of Capital Investment.) Such an investment would strengthen the community's economic stability.
2. As part of its growth, AA would expand its workforce from 145 to 225 employees, with all hires anticipated to be Oklahoma residents.
3. The 65 acres on the north side of 20th St NW will become available for economic development. The city will strongly pursue this industrial development once the paving project is approved.

Conclusion
The hard paving of 20th St NW will allow an existing industry to expand and will attract new development. As a result, the City of Ada will continue to flourish and its residents will enjoy greater access to well-paying manufacturing jobs.

In light of these benefits to the City of Ada, area businesses, and the State of Oklahoma, the City of Ada requests a matching grant of $104,635 to hard-surface 1,106 ft. of 20th St NW in July 2005.

Attachments: A Application Form
B Ada Automation Expansion Plan
C Letter of Commitment from Ada Automation
D Paving Project Itemized Cost Estimate
E Site Map and Sketch Plan
F Excerpt from City of Ada Comprehensive Plan
G Local Commitment and Initiative Information
H Sources of Capital Investment

[middle]
Show that the project is supported by all interested parties.

List concrete benefits.

While referring to numbers, keep the focus on people by creating an "after" picture.

[closing]
Summarize the need, the project, and its benefits.

Restate the formal request for support.

List attachments.

Research Proposal

While a grant proposal can seek funding for a wide range of projects, a research proposal more specifically seeks approval and possibly resources for a research project. Your goal in such a proposal is to demonstrate that the research is both valid (makes good scientific sense) and valuable (will lead to significant knowledge and perhaps to practical applications).

Rhetorical situation

Purpose: To convince a supervisor that an experimental study of testosterone in chicks is workable and valuable
Audience: Biology professor familiar with the writer's classroom and lab work
Format: Memo (creates hard copy of plan for discussion and approval)

---

1 of 2

To:	Dr. Adam Pomme, St. George College Biology Department
From:	Tricia Walton
Date:	11 March 2005
Subject:	Independent Study 370 Course: The Effects of Testosterone on Developing Chicks

[opening] — *Identify your research project's title. Summarize what you propose to study and what you hope to learn: your hypothesis, working thesis, or key research question.*

Purpose and Project Description

For my Independent Study 370 project next fall, I propose to study the effects of the hormone testosterone on the bodies and brains of developing chicks. Specifically, I would like to explore the correlation between different doses of testosterone and factors such as comb length, width, and height, as well as body weight. In addition, I would like to gauge the effect of the hormone on the chicks' brains. My hypothesis is that the testosterone should affect both physical appearance and the nervous system by making them measurably more masculine.

[middle] — *Map out how you intend to carry out the research. Indicate both secondary and primary research. Provide details demonstrating that you have a good grasp of fitting research methods.*

Project Plan

My plan is to work through the following stages of the experiment. These stages will be spelled out in detail after consultation with you:

1. Review appropriate science literature on testosterone.
2. Divide a group of eighteen chicks approximately five days old into six groups of three and inject them with testosterone every other day for a period of four weeks.

   Female Groups:
   #1 Control group receiving no testosterone
   #2 1 mg/mL injections
   #3 2 mg/mL injections
   #4 3 mg/mL injections
   Male Groups:
   #1 Control group receiving no testosterone
   #2 3 mg/mL injections

3. On days chicks do not receive injections, weigh chicks and measure comb length, width, and height.

4. After the four-week period, continue weighing and measuring the chicks for one week. Then tabulate and analyze the data.
5. Kill the chicks and for each make samples of the brain into stained microscope or histology slides.
6. Analyze the slides and make qualitative measurements of the diameter of the two individual brain halves.
7. On the basis of analyzing the measurement data and the brain slides, determine whether testosterone has the effect hypothesized.
8. Complete a report and give an oral presentation to the Biology Forum.

Timetable for Project

I have set the following target dates for my project during the fall semester:

1. Begin the experiment on September 9 and complete injections on October 7.
2. Complete the bulk of library research by September 13.
3. Meet biweekly with you, and submit progress reports on October 2 and November 9.
4. Kill chicks on October 21, prepare histology slides by November 1, and analyze data by November 11.
5. Complete the project report by November 22 and turn it in to you as the project supervisor.
6. Present the results to the Biology Forum on December 4; submit a proposal to present the report to the Arizona Academy of Science; revise the report and submit it for publication in a science journal.

Project Outcome

This project should have the following outcomes:

1. I will become more familiar with effective measuring techniques, making slides, graphing data, and analyzing data.
2. The results of the experiment should show whether the injections of testosterone do affect the size of a chick's brain. Normally, in females the cerebral hemispheres are roughly equal in size. In males, the right hemisphere is distinctly larger than the left. I hypothesize that the repeated testosterone injections will "masculinize" the brains of the developing chicks to the point that the right hemisphere will be distinctly larger than the left. This experiment would show that the effects of such injections are not limited to comb size but also affect at least one internal organ.

Project Approval

Please approve my project or suggest needed changes. You can reach me at 555-8884 or twalton@stgeorge.edu.

[middle]
Focus on data to be gathered and analyzed.

Depending on the nature of the project, supply a working bibliography indicating your preliminary survey of possible resources (not shown).

Map out a realistic research schedule indicating major milestones in the project.

Indicate methods of sharing research results.

[closing]
Summarize the project's outcomes: benefits of the research for you and others.

Request both input and approval from the reader, as well as funding or other resources, if appropriate.

CHECKLIST for Proposals

Use this checklist to benchmark and revise your proposals.

☐ **Ideas.** The proposal shows a detailed, logical, and accurate understanding of the problem, alternative solutions, the reader's perspective, and your own resources.

☐ **Organization.** As shown in the outline on page 523, the proposal effectively moves from problem to solution to implementation.

☐ **Voice.** The proposal has a positive and confident, but objective tone that shows careful attention to the reader's perspective.

☐ **Words.** The proposal uses language at an appropriate level of formality. It uses technical terms carefully, defining them when necessary.

☐ **Sentences.** The proposal's sentences read smoothly, are linked by logical transitions, and present information clearly.

☐ **Copy.** The proposal contains no grammatical errors.

☐ **Design.** The proposal follows the company's style guide; uses consistent, parallel, informative headings; uses graphics effectively; and includes white space, underlining, boldface, and other layout features.

CRITICAL THINKING Activities

1. Working with a classmate, study this chapter's guidelines for writing proposals, along with the three categories of proposal (operational improvement, sales or client, and grant and research). List six to ten key characteristics of each type, noting which traits the models have in common and which they do not. Discuss your findings with the class.
2. Ask an area business to give you copies of two or three proposals written in that organization. Identify their type and analyze their quality using this chapter's guidelines, models, and checklist as benchmarks.

WRITING Activity

Doing the appropriate research and investigation, write a proposal for one of the following situations:

College: What problems or needs do you see at your college? Consider issues such as old or lacking equipment, community relations, internship opportunities, a student-life issue, a new course or program, an environmental issue, a facilities issue.

Workplace 1: Consider places you've worked in the past or where you presently work. Is there some change that would improve that workplace?

Workplace 2: Imagine that you work in a student-run enterprise, either a business or a nonprofit service group. Write a sales proposal to an appropriate client.

In this chapter

Guidelines for Designing Report Forms **540**
Checklist for Designing Report Forms **544**
Critical Thinking Activity **544**
Writing Activities **544**

CHAPTER
33

Designing Report Forms

In some ways, a report form is a combination survey *and* report. The report form is like a survey in the sense that the person who designs the form wants to get information quickly from the person who fills out the form. The designer uses strategies such as fill-in-the-blank, multiple-choice, and short-answer formats. The respondent, or report writer, fills in the appropriate information to create the report itself. A report form, then, is like a report in the sense that the form is designed to share findings with readers who use the information to understand issues, make decisions, or take action.

A well-designed report form yields efficiency for both the writer and the reader of the report. To develop a form that gathers information quickly and shares it efficiently, the designer must do the following:

- Make recording information easy.
- Help the respondent provide all necessary information.
- Help readers access and understand the information quickly.

Because report forms are designed to fill out quickly, they are most useful for gathering and sharing basic facts such as financial data, employment records, itemized expenses, and details of work completed. Report forms are less useful when the readers also need analysis and explanations of the facts.

GUIDELINES FOR DESIGNING REPORT FORMS

Your goal when designing a report form is to create a clear, user-friendly form that efficiently gathers, stores, and shares information. Good report forms almost fill themselves out, helping report writers supply readers with the right information. Poor forms are hard to fill out and read, frustrating both report writers and readers.

1 Prewrite
Rhetorical analysis: Consider the report purpose, the report writer, the report reader, and the report context.
- ☐ Why is a form a good choice for this reporting situation? How will the form save time and advance work?
- ☐ Who will fill out the form, using what technology—pen, pencil, typewriter, or computer? How does the report form relate to this user's job and tasks?
- ☐ Who will read the report, and why? How will he or she use the information and analysis to understand, decide, or act?
- ☐ Should the format be paper or electronic? Which medium will work best for completing, studying, and storing the report?

2 Draft
Organize the form into three parts.
[opening] Provide a title and instructions (if necessary).
[middle] Create distinct sections with easy-to-complete features.
[closing] Cover practical details about processing the report.

3 Revise
Review your draft's ideas, organization, and voice.
- ☐ Is there enough space to supply necessary information?
- ☐ Does the form use easy-to-complete strategies such as options to circle or check?
- ☐ Are there clear instructions and labels for what to do?
- ☐ Is all of the information requested actually needed, and does the form avoid asking for the same thing twice?

4 Refine
Check your form line by line for the following:
- ☐ Words, phrases, and sentences that are clear, precise, and understandable to all those who complete the form and read the report
- ☐ A design that effectively uses white space, typography (especially large enough print), boxes, and numbers

F.Y.I.
Test the form with potential users (recorders and readers) before sending it to production. Then periodically review and revise the form. Note the revision date on the form itself, and inform users of any changes made.

Organizing Report Forms

[opening]

- **Label** the report form with a clear title and the company name centered at the top. You may include a reference number as well.
- **Introduce** the purpose of the form. If appropriate, add instructions for completing the form and a note about confidentiality.

[middle]

- **Organize** sections of the form to fit the subject and the information being reported. Sequence sections so that they flow from left to right and top to bottom.
  - Gather basic identifying information (e.g., writer's name, address, identification number, the date).
  - Collect major facts using a logical pattern—time or sequence, space, order of importance, categories, alphabetical, compare/contrast, cause/effect, problem/solution, and so on.
  - Signal the order with boxes, numbers, labels, and other graphical elements.
- **Use** a variety of techniques to minimize the writer's work of organizing and recording the information:
  - Write clear headings and questions (in readable type size) to get the right details.
  - Allow adequate room for the writer's answers on blanks or in boxed spaces.
  - Provide options to circle, check, or click.
  - Furnish white space to make writing and reading easy.

[closing]

- **Leave** room for necessary signatures.
- **List** attachments that should be included with the report, and explain how to submit them.
- **Provide** details for transmitting, copying, routing, filing, reading, and responding to the report.

WRITE Connection

For more help with effectively designing report forms, check out Chapter 16, "Trait 7: Reader-Friendly Design."

Report Form: Paper

Rhetorical situation

Purpose: To accurately record veterinary treatment details for "farm" calls
Audience: Primarily office manager using the report to maintain records and do billings
Format: Paper form (gives veterinarians flexibility for field use)

[opening]
Provide a clear title for the form and brief instructions.

[middle]
Collect important identifying information.

Divide the report into sections that flow logically.

Use boxes, lines, and other features to make completing and reading the form easy.

[closing]
Indicate how to use and process the report.

Heartland Veterinary Services, Inc.
Farm Visit and Inspection Report

Complete the visit and treatment information noted below. Then submit your report to the office manager within 24 hours of the visit for billing. The manager will complete the bottom portion of the form.

Visit and Client Information:

Date of Visit: *May 3, 2005* Time of Visit: *9:30 p.m.*
Client's Name: *Cliff Bailey* Length of Visit: *1 hour*
Client's Address: *886 182nd Ave., Belleview, SD* Client's Phone: *555-4954*
Visiting Doctor: *Pauline Kao* Type of Visit: *Urgent*

Background Information:
Cliff had given penicillin (6) and antihistamine (Recovr) for two days before my visit. The heifer had begun to eat less, and her eyes were sunken and listless. Heifer's wt. 450.

Diagnosis:
- *Heifer's temperature registered at 104.0.*
- *Breathing pattern was accelerated and lungs sounded congested.*
- *Tongue was swollen.*
- *Heifer was suffering from diphtheria, a common form of necro bacillosis.*

Treatment and Services:
Administered intravenously 22.5 cc of gentamicin sulfate and 5 cc dexamethasone. Administered intramuscularly (neck area) 30 cc of antihistamine IB-complex vitamins.

Comments, Recommendations, Follow-up:
The heifer should be given 8 cc of dexamethasone intravenously along with 20 cc of antihistamine IB-complex vitamins for the next three (3) days.

Billing (to be completed by the office manager):

Service charge	*45.00*
Gentomicin sulfate	*12.00*
Antihistamine IB-complex	*9.50*
Tripelennamine hydrochloride	*5.75*
Total Charges	*72.25*

Note: Process this report into electronic billing, and place a copy in the customer's file.

Report Form: Electronic

Rhetorical situation

Purpose: To accurately track media contacts so as to maintain a consistent organizational public-relations message
Audience: Resource Development Director for nonprofit agency: coordinates public relations
Format: Electronic form (makes timely, convenient reporting and follow-up possible)

HOPE SERVICES
Media Contact Form

Note:
Whenever possible, please notify the Resource Development Director before you are interviewed by someone from the media. That way, we will better convey a clear, accurate, and consistent message about Hope Services to the media and, in turn, to the public.

Your Name: Jasmine Michaels

Date You Are Filling Out This Form: April 11, 2005

Date of Contact by or to Media: April 10, 2005

Media Contact Was Made by: (Click one)   Myself ❑   Media ☒

Medium: (Click one)   Television ❑   Radio ❑   Newspaper ☒
                              Other:

Name of Medium: (If television or radio station, please include station number.)
Berks County Examiner

Name of Contact Person(s) from Media: Robert McKennit

Subject of Contact: Start-up of Project SARA (Sexual Abuse Response Team)

When, Where, and How Coverage Will Take Place (date, time, section, etc.):

Sunday, April 24, 2005, edition; Section B, City News

Brief Description of What Was Discussed or Answered:

Bob asked for an interview to discuss Hope Services' involvement in Project SARA. He was especially interested in discussing how we saw the project improving life in local communities. The interview is scheduled for Tuesday, April 19, at 9:00 a.m.

To submit your report to the Resource Development Director, click below.

[**SUBMIT**]

[opening]
Identify the report form with a title and offer instructions.

[middle]
Organize the form so that parts flow logically from basic identifying information to key report details.

Make completing the form easy with text boxes and items that simply need to be clicked.

[closing]
Indicate how to submit the report.

CHECKLIST for Designing Report Forms

Use this checklist to benchmark and revise your report forms.

- ☐ **Ideas.** The form focuses on gathering, sharing, and storing the required information.
- ☐ **Organization.** As shown on page 541, the form has an easy-to-complete structure.
- ☐ **Voice.** The tone in all instructions and questions is neutral but inviting.
- ☐ **Words.** The form includes clear labels, accurate terminology, strong action verbs, and language understood by readers.
- ☐ **Sentences.** The form uses short, clear instructions with command verbs.
- ☐ **Copy.** The form uses correct grammar, punctuation, spelling, capitalization, and mechanics. It includes no abbreviations unfamiliar to the reader, and it is free of typographical errors.
- ☐ **Design.** The form offers a format that allows easy access for both the writer and the reader; provides adequate space for all answers or responses; and includes labels, reference numbers, a title, headings, boxes, colors, icons, lines, and other devices to make the form easy to use.

CRITICAL THINKING Activity

Working with a classmate, get report forms from three different businesses. (Consider obtaining both paper and electronic formats.) Compare and contrast the documents in terms of their purpose and design. Use this chapter's guidelines and checklist to assess the quality of each form.

WRITING Activities

1. Identify an organization (e.g., business, nonprofit agency, college club) that might use a form to generate reports from its employees, clients, or members. Interview an individual responsible for gathering this information, show him or her the model forms in this text, and offer to write a report form designed to gather the information that he or she needs. Then use this chapter's instructions to write the document.
2. Reflect on your past work, internship, or volunteer experiences. Identify situations when a report form would have helped workers efficiently record and share information about activities. (For example, consider short reports such as periodic, progress, investigative, incident, and trip reports; see Chapter 30, "Writing Short Reports.") Develop a report form for one of the situations.

PART 6

Special Forms of Writing

34 Public-Relations Writing

35 Writing Instructions

36 Writing for the Web

In this chapter

Guidelines for News Releases **548**
Guidelines for Flyers and Brochures **552**
Guidelines for Newsletters **560**
Checklist for Public-Relations Writing **564**
Critical Thinking Activities **564**
Writing Activities **564**

CHAPTER

34

Public-Relations Writing

Any organization needs a team of employees who understand its goals and support its work. To develop such a team, an organization must share information about its plans, products, or policies; show employees that they're valued and important; and spell out the benefits of successful teamwork. Workplaces try to do these things by producing *in-house* documents such as a brochure on health care benefits, a flyer on a new product, or a newsletter article about an employee. Documents such as these can develop employees' understanding of and appreciation for the business.

Of course, an organization also needs good relationships with its customers and other external groups. Such relationships are based on information about what the organization is, what it does, and what principles or guidelines distinguish its work. Customers need this kind of information to understand the organization's products or services, be convinced of their value, and purchase them. Other groups (e.g., community members or members of a profession) need the information to conduct their business with the organization. To communicate such information, organizations produce *public* documents such as a brochure on a new product, a news release about a new program, or a newsletter article about the organization's plans to expand.

Whether the writing is *in-house* material for readers inside the organization or *public* material for readers on the outside, the overall goals are similar: to build good relations between the organization and its readers. At some point in your career, you will write public-relations documents. In fact, if you are a department manager or work in the Public-Relations Department or Marketing Department, you will write these documents often.

This chapter will help you. In it you will find guidelines, tips, and models that will enable your organization to build the relationships that make it strong.

Part 6 Special Forms of Writing

GUIDELINES FOR NEWS RELEASES

News releases share information (a plant expansion, product recall) that organizations want the news media to present to the public. Your goals when writing a news release are (1) to present your organization positively and (2) to develop a newsworthy angle so that editors will publish the story.

1 Prewrite
Rhetorical analysis: Consider your purpose, your audience, and the context.
- What do your readers know about your organization and the topic?
- What angle or features could get the story published?

Gather details that will interest readers and tell the story.
- Interview knowledgeable people and tour a worksite.
- Read relevant documents (sales reports, résumés, product descriptions).
- Develop eye-catching graphics (photographs, logos, drawings).

2 Draft
Organize the release into three parts.

[opening] Introduce the story.
- Use an eye-catching lead: a strong quotation, question, or fact.
- Answer the five W's and H: Who? What? When? Where? Why? How?

[middle] Present information from most to least important.
- Introduce the history of the event, product, or person.
- Use quotations, anecdotes, or engaging data.
- Emphasize the quality, design, and function of new products or procedures.
- Explain how this information affects employees or customers.

[closing] Give extra details, provide background, or forecast future stories.

3 Revise
Review your draft's ideas, organization, and voice.
- Is the story clear, complete, and engaging?
- Is the organization logical and are the transitions between parts clear?
- Is the voice objective and engaging?

4 Refine
Check the writing for the following:
- Correct spelling, punctuation, grammar, and usage
- Double-spacing on letterhead stationery
- Contact names, phone and fax numbers, and e-mail addresses
- If necessary, a note stating when the story may be published
- The word "MORE" at the bottom of each page (except the last one) for multiple-page releases

Topics for News Releases

Organizations use news releases to inform the public about themselves. Tables 34.1 and 34.2 offer details on common topics for stories.

Tips for Distributing News Releases

News agencies publish only those stories that their editors believe are newsworthy. To increase your chances of having stories published, follow these tips:

- Be punctual, polite, and professional in all correspondence.
- Get to know each news organization's editors, audience, style, and news release guidelines—including the preferred format and medium for stories and photos.
- Cast a story from a fresh angle that editors will find engaging, and be thorough and correct in an article's content and format.
- Send only newsworthy stories and (when appropriate) discretely suggest why they are worthy.
- Deliver materials to the agency's designated recipient and occasionally offer exclusive coverage.

Table 34.1 Topics for Good-News Releases

Personnel	Policy	Product	Event
Management changes	New return/exchange policy	New products	Branch openings, grand openings
Promotions, advancements	New waste-disposal policy	Plant expansion	Company anniversaries
Awards	Improved employee benefits package	Mergers	Awards, contracts, achievements
Hiring announcements	Equal opportunity employer	Guarantees	Sponsor social or cultural programs

Table 34.2 Topics for Bad-News Releases

Personnel	Policy	Product	Event
Strikes	Reduced customer-services policy	Product recall	Plant closings or cutbacks
Death of an employee	Reduced public hours	Discontinuation of product or service	On-site accident or incident
Employee layoffs	More limited returns policy	Price increases	Response to legal action
Employees criticize company	Less public access to facilities	Product losses	Response to bad report or penalty
Key replacement or firing	Seek materials from foreign markets	Product delays	Move factory to another country

Release (Good-News)

Rhetorical situation

Purpose: To describe positively the Donovans' gift to the American Cancer Society
Audience: The American Cancer Society, area businesses, and the public
Format: Identifying information followed by the title and story double-spaced

---

News Release

Four Seasons Health Club
153 South Fairway Drive
Virginia City, Nevada 89440

Contact: Robert or Maria Donovan
(775) 555-1492

LOCAL FITNESS CLUB OWNERS SPONSOR RELAY-FOR-LIFE TEAM

[opening] — Double-space the title and story for easy reading. Introduce the story with a strong lead and identifying information.

Virginia City, Nevada—Four Seasons Health Club owners Robert and Maria Donovan will again sponsor a Relay-for-Life team to benefit the American Cancer Society. The Relay for Life will be held on Saturday, July 9, 2005, at the Virginia City High School track.

[middle] — Present details, organizing information from most to least important.

The Donovans challenge other area businesses and individuals to join in this important event by sponsoring their own team of 8–12 participants, or by pledging to support existing teams. For further information or to make a contribution, call (775) 555-1492.

[closing] — Because editors may shorten stories, give secondary details last.

"The Relay for Life celebrates the lives of cancer patients and raises money for cancer research," says Maria Donovan. "As the mother of a cancer survivor, I know first-hand how important the work of the American Cancer Society is." In the past 10 years, the Four Seasons Health Club and other area businesses, churches, and schools have contributed more than $300,000 to the American Cancer Society. Every year since its opening 27 years ago, the Four Seasons Health Club has sponsored health or community-related fundraisers.

Release (Bad-News)

Rhetorical situation

Purpose: To report the strike and related details objectively and fairly
Audience: Atlantic City Airport's employees, customers, and the public
Format: Identifying information followed by the title and story double-spaced

---

News Release

Atlantic City Airport
4567 Emerald Way
Atlantic City, New Jersey 08082

FOR IMMEDIATE RELEASE

Contact: Samuel DellaSanta
Phone: 609-555-9054

ATLANTIC CITY, January 10, 2005—A strike against the Atlantic City Airport Tuesday by the Aircraft Mechanics and Service Technicians Association (AMSTA) failed to convince other airport personnel to walk off the job. "The majority of our staff, including union and non-union pilots, flight attendants, and ticket agents crossed the picket line," reported Tim Redland, Human Resources Director for the airport.

[opening] **Begin with an attention-getting sentence and key details.**

Scheduled flights have been unaffected by the strike. Replacement mechanics and service technicians are maintaining and repairing aircraft. Redland stressed that customer safety remains the highest priority. Contract talks ended last Saturday when AMSTA turned down an airport offer related to health care issues. Redland said, "Atlantic City's current contract, based on independent surveys, offers some of the highest wages in the Eastern Seaboard aviation system." The point of disagreement, says Redland, is AMSTA's demand for fully funded health care. AMSTA members now pay 30 percent of the policy.

[middle] **Organize details from most to least important.**

Atlantic City Airport has served the Eastern Seaboard for 30 years with direct flights to New York, Chicago, and Dallas. The airport employs more than 1,500 aviation personnel.

[closing] **Offer background information.**

GUIDELINES FOR FLYERS AND BROCHURES

A flyer contains time-sensitive information often printed on one side of a single page. In contrast, a brochure contains longer-life information and may vary in form from a single folded page with four panels (or sides) to a brief booklet. Your goals when writing these documents are (1) to get your reader's attention, (2) to describe your organization, product, service, or event, and (3) to help readers respond to your message.

1 Prewrite
Rhetorical analysis: Consider your purpose, your audience, and the context.
- ☐ What overall "story" do you want to tell, and what is its main idea or "theme"?
- ☐ What information (data, quotations, details) and visuals (pictures, graphs) will grab the reader's attention, tell the story, and develop the theme?

Prepare to draft.
- ☐ Gather information about the organization (mission, motto, logo, products), product (specifications, features), service (description, value, and provider's qualifications), or event (e.g., time, place, featured speakers).
- ☐ Consider design issues: type (flyer or brochure), form (length, size, and color), content (balance of visuals versus text), and distribution (e.g., mailed, displayed).

2 Draft
Organize the message into three parts.

[opening] Use the top of a flyer or the cover panel of a brochure to
- ☐ Catch the reader's attention with a strong title, graphic, and/or organizational logo.
- ☐ Introduce the story by naming the organization, product, service, or event.

[middle] Use the middle of a flyer or center panels of a brochure to
- ☐ Describe your topic in brief sentences with clear nouns and strong verbs.
- ☐ Give details in bulleted lists or a question-and-answer format.
- ☐ List supporting information (e.g., special features or customers' quotes).

[closing] Use the bottom of a flyer or the back panel of a brochure to restate the theme and give contact information or an easy-to-complete response form.

3 Revise
Check your draft's ideas, organization, and voice.
- ☐ Are the message and main idea clear, logically ordered, correct, and complete?
- ☐ Is the tone engaging, convincing, and professional?

4 Refine
Check your writing and page layout for the following:
- ☐ Correct titles, dates, prices, spelling, grammar, and punctuation
- ☐ Attractive layout (no overlap on folds), fonts, white space, and color
- ☐ Appropriate balance of visuals versus text

Flyer

Rhetorical situation

Purpose: To promote a public forum on building a civic center
Audience: Riverside City Council, area businesses, and the public
Format: Large title in blue; place, time, and date in red; program details in list; campaign theme in blue; contact information last

---

ARKANSAS

A New Civic Center for Riverside?

Come learn about plans for a new civic center and share your ideas on the project!

Public Forum on New Civic Center
Wednesday, December 7, 2005
7:30 to 9:30 p.m.
Riverside Community Center

Program:
- City planners introduce the project.
- Vision Architects, Inc., presents concept drawings and a model.
- Riverside Orchestra director describes use of the new building.
- Neighborhood property owners and businesspeople offer their views.
- The public shares their ideas.

Riverside—a city that's building its future!

Civic Center Planning Committee
Eloise Denchenko, Chairperson
Phone: 870-555-8000, ext. 604
E-mail: edenchenko2@earthlink.net

[opening]
A bold heading set off by white space introduces the topic.

[middle]
A brief statement explains what is happening when and where.

Details are stated in brief sentences with clear nouns and strong verbs.

[closing]
The theme is restated and highlighted.

Contact information is given.

Part 6 Special Forms of Writing

Brochure (Front Cover)

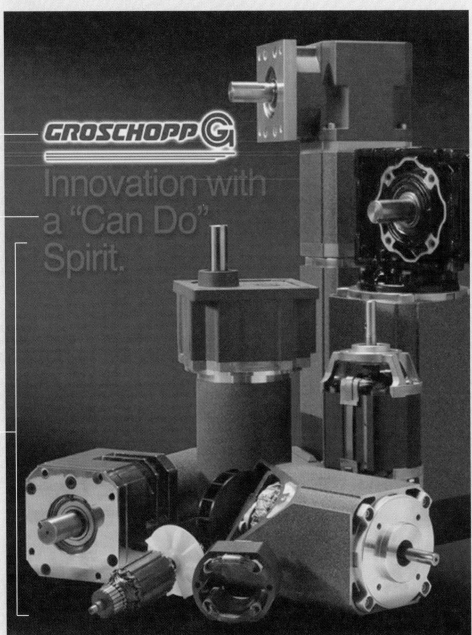

Company name and logo

Company motto

Pictures of products

Chapter 34 Public-Relations Writing 555

Brochure (Inside Cover)

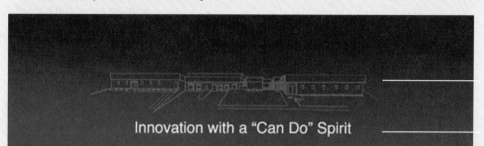

Innovation with a "Can Do" Spirit

For nearly seventy years, Groschopp, Inc. has served as a standard of excellence for fractional horsepower motor drives. We have a well-earned industry reputation for innovation in products and services that meet demanding needs in an ever-expanding market. In addition, the company presents an unsurpassed reputation for integrity, a "can do" attitude and good value - all with high quality and on-time delivery.

Our Sioux Center location in Northwestern Iowa benefits from a long-standing Dutch heritage with a strong work ethic resulting in unmatched overall customer service. Added to our design capability and state-of-the-art manufacturing systems is a strong vendor base, ISO 9001 recognition, and a proactive sales team. Groschopp continues to be a progressive leader in fractional motor innovation.

We want to work closely with you, our valued customer, to provide the cost advantage of our broad range of standard configurations, modifications and innovative new designs to satisfy exacting application needs. We understand that each order must be completed to your product specifications and deadlines. In addition to full OEM sample and pre-production services, Groschopp also offers *FASTTrack*™, a 48-hour delivery service on standards which includes many typical modifications.

Our industry-leading design service, MotorTec℠, is custom software developed in-house that enables our engineers and technical sales force to give you prompt and accurate product suggestions. In a matter of seconds, speed/torque, efficiency and other essential to function design variables and our product recommendations are made available.

Your Groschopp team is fully committed to producing fractional horsepower motor products at a good value that do the job right the first time.

Gerald Van Roekel
GERALD VAN ROEKEL
President & General Manager

FastTrack™
48-Hour Delivery

Differentiated
Services

MotorTec™ Assisted
Design Solutions

- Line drawing of local plant
- Company motto
- Purpose or mission
- Pictures of services

Brochure (Inside Back Cover)

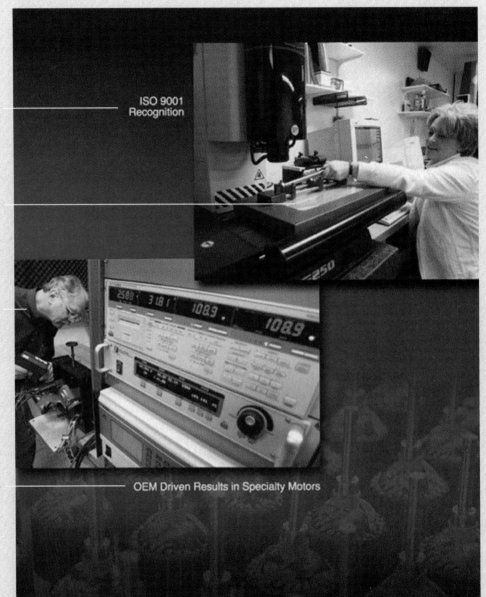

Market advantage

Testing procedures

Pictures of personnel

Sample products

Chapter 34 Public-Relations Writing

Brochure (Back Cover)

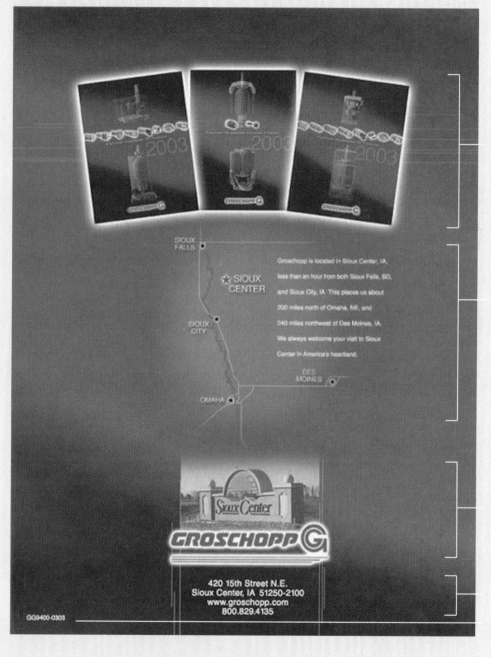

Similarly designed brochures advertising products

Factory location

Company name and logo

Contact information

Publication number

Part 6 Special Forms of Writing

Six-Panel Brochure Advertising Products

The previous pages show Groschopp's four-page brochure advertising the company, a business that produces electric motors. Figures 34.1, 34.2, 34.3, and 34.4 show Groschopp's six-panel brochure advertising its products.

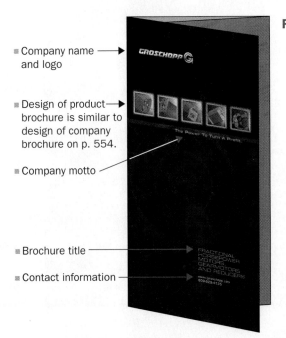

Figure 34.1 Brochure: Front Cover

- Company name and logo
- Design of product brochure is similar to design of company brochure on p. 554.
- Company motto
- Brochure title
- Contact information

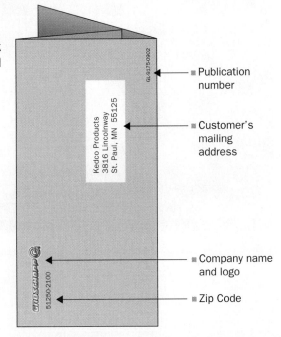

Figure 34.2 Brochure: Back (or Address) Panel

- Publication number
- Customer's mailing address
- Company name and logo
- Zip Code

Figure 34.3 Brochure: Outside Panels

The brochure is 5 1/2" × 11" folded and 16 3/4" × 11" unfolded.

- **Left Panel:** Product design and description of services

- **Middle Panel:** Customer's mailing address and company's Zip Code

- **Right Panel:** Front cover with brochure title and company's name, logo, and contact information

Figure 34.4 Brochure: Inside Panels

- **Left Panel:** AC motors and performance data

- **Middle Panel:** DC motors and performance data

- **Right Panel:** Reducers, product features, and product options

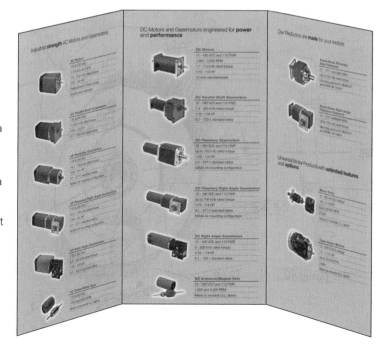

GUIDELINES FOR NEWSLETTERS

In-house newsletters are intended for employees, and public newsletters are designed for customers and constituents. Your goal when writing a newsletter is to share information that helps readers to support your organization, and to understand and use its products and services.

1 Prewrite

Rhetorical analysis: Consider your purpose, your audience, and the context.
- ☐ **In-House Newsletter:** What topics (e.g., introducing new employees or products, explaining changes in programs) would inform employees about the organization, boost their morale, and inspire their support?
- ☐ **Public Newsletter:** What topics (e.g., sales report, new building project, new vice president, new products) would inform customers, investors, and the public about the company and engender their support?

Plan your production schedule, content, and design.
- ☐ Set due dates for choosing topics, writing, editing, printing, and distributing.
- ☐ Choose a design (e.g., regular features, overall format).

Gather information (including photos, quotes, and other details).
- ☐ Interview people.
- ☐ Read reports and minutes.
- ☐ Visit work sites.
- ☐ Examine new products and procedures.

2 Draft

Organize each newsletter article into three parts.

[opening] Introduce the topic with an eye-catching title, and then answer the questions Who? What? When? Where? Why? and How?

[middle] Add details to help readers understand how the topic relates to their and the organization's welfare.
- ☐ Use common words in short sentences.
- ☐ Develop the story with quotes, data, and pictures.

[closing] Close with a brief paragraph that restates a key point or ties this article to related articles.

3 Revise

Review your draft's ideas, organization, and voice.
- ☐ Is the main point clear and supported with interesting details?
- ☐ Are sentences smooth, transitions clear, and words precise?

4 Refine

Check your writing line by line for the following:
- ☐ Correct details, spelling, punctuation, usage, and grammar
- ☐ Correct format and layout as explained on the next page

Newsletter Layout

Successful newsletters build on recurring elements in every issue. Although the elements may look simple, they are surprisingly complex.

Distribution

Design a newsletter with distribution in mind. Will the newsletter be sent full size or folded, as a self-mailer or in an envelope?

Nameplate

Your nameplate should provide immediate identification and explain the purpose of your newsletter.

Publication Information

Clearly identify the source of the newsletter and include contact information such as your address and phone numbers. Give the volume number and issue date.

Columns

Spacing between columns should be proportionate to type size.

- Larger type calls for more space between columns.
- "Rivers" of white space in justified text can be distracting.
- Excessive hyphenation in narrow columns of type should be corrected by reducing type size, increasing column width, or choosing a ragged-right alignment.

Text

For long feature articles, choose a smaller type size in multiple columns. For shorter articles, choose wider columns and larger type.

- Use only as many different type styles as are necessary to organize your text. Strive for maximum contrast when using more than one typeface. Each variation in typeface, size, or weight slows down the reader.
- Offer visual relief with lead-ins between headlines and text, and pull quotes.
- Start text the same distance from the top of each page throughout a document.
- Make tabs proportional to column widths.
- Eliminate awkward lines or line breaks by editing text.

Headlines

Choose a type font that is noticeably larger and bolder than the font in the body of text.

- Avoid parallel headlines, subheads, and quotations in adjacent columns.
- Avoid tombstoned headlines that can overwhelm a page.
- Place headlines closer to the text they introduce than to the preceding text.

Artwork and Captions

Use photographs or illustrations to strengthen your message. Allow space for captions and make sure names are spelled correctly.

Newsletter (In-House)

Organizations produce in-house newsletters to communicate with employees, the board of directors, and others who are inside or part of the business.

State the newsletter's name, host company, and purpose.

Give the volume, issue, and date.

ROYALE REVIEW

A newsletter by, for, and about employees of Royale Telephone Company

Royale TELEPHONE

VOL. 2, ISSUE 3   OCTOBER 2005

Use clear, descriptive titles.

■ RTC Harvest Picnic: October 15

Plan to attend the annual RTC Harvest Picnic at noon on Saturday, October 15, 2005, in Berthesen Park. RTC managers will grill hamburgers and brats. Dan's Deli will provide salads and beverages. After the picnic, the RTC Raiders will take on any brave souls in a softball game at Berthesen Field!

■ Trick or Treat

The annual Halloween party for cancer patients at St. Luke's Children's Hospital will take place October 29. Grab a costume and help us entertain children who wouldn't otherwise be able to participate in trick-or-treat festivities. If you can join us, meet at the nurses' station on the third floor at 5:30 p.m.

Use a distinct font and graphic to distinguish section heads.

■ Business Briefs

Security Committee Reminder: RTC employees use passwords to keep the computer network secure. Therefore, you must not share logins and passwords with anyone. Change your password regularly and avoid using obvious passwords. Finally, remember that most software used at RTC belongs to the company and may not be copied by users.

■ Congratulations

Congratulations to Mickey and Rose Paletto on the birth of their first child—Angela Rose, a 7 pound 11 ounce baby girl born Friday, September 16.

■ Sports Beat

Walleyball • Once again, it's time for co-ed walleyball—a combination of volleyball and racquetball. Because of last year's success, we're adding a second league. Play begins November 3 on Thursday nights through February. Interested players should contact Mari Sanchez (ext. 1899) before October 14. The fee is $45 per person.

Basketball • Get in shape and enjoy a great game of hoops! Our three basketball teams (men's, women's, and co-ed) are gearing up for another sweaty, fun-filled season. Practices are held twice a week (games once a week) at the Tuscaloosa Community Center. Sign up on the cafeteria bulletin board by Friday, October 14.

Explain how readers can contribute to the newsletter.

Royale Review depends on news and views contributed by all employees of RTC. Submit written items, ideas, questions, or comments to the Review mail slot, or e-mail the editor directly at reviewrtc@busi.net.

Newsletter (Public)

Organizations produce public newsletters to share information with customers, clients, the community, and others interested in the business.

April 2005 • Volume 11, Issue 6

CONNECTIONS
A Publication of Royale Telephone Company

Royale TELEPHONE

List the date, volume, and issue.

Give the names of the company and newsletter.

Relay for Customers with Hearing Impairments
A new service called Relay now enables people with hearing impairments to communicate via telephone with people who can hear and speak (and who do not own a teletypewriter [TTY]).

Since 1987, RTC has provided teletypewriters (TTY) and telecommunication devices for the deaf (TDD). These typewriter-like machines allow customers with hearing impairments to communicate over phone lines with other persons with hearing impairments.

Relay calls are initiated by dialing the local Relay number in the telephone directory. As the hearing person speaks, the Relay operator types the message on his or her TTY. The person with the hearing impairment reads the message on his or her TTY machine and responds vocally or uses the TTY to type a response. If the person with the hearing impairment uses the TTY, the Relay operator reads the message to the hearing person.

RTC is pleased to provide another avenue of communication for our customers. After all, helping people communicate is our business.

Call Before You Dig
Spring is in the air! No matter what your construction project, whether planting new trees or installing a fence, if it involves digging, call Utility One (1-800-555-0000). Utility One then notifies local utilities such as Royale Telephone so that we can locate underground lines and save everyone a lot of trouble.

Open House for 20th Anniversary
Royale Telephone Company has provided outstanding communication services for this community for 20 years. Join us as we celebrate our anniversary at an open house at our main office in downtown Tuscaloosa on Tuesday, May 16, from 9 a.m. to 4 p.m.

Extended Hours
Royale Telephone is now open on Saturdays from 9 a.m. to noon.

CONNECTIONS is published quarterly for the customers of Royale Telephone Company. Established in 1981, Royale Telephone Company is a local, independent telephone company serving Tuscaloosa and the surrounding area.

Royale Telephone Company
378 11th Avenue NE
Tuscaloosa, AL 35401
(205) 555-4444

Use an appropriate font style and size to accent titles of feature articles.

Write about topics relevant to readers.

Describe the newsletter's purpose and contact information.

CHECKLIST for Public-Relations Writing

Use the seven traits to check your document and then revise as needed.

- ☐ **Ideas.** In all public-relations forms, the information helps readers understand and appreciate the business, its products, and services. In flyers and brochures, the information helps readers understand an organization and encourages them to purchase its products or services. In newsletters, the information nurtures understanding of and support for the company.

- ☐ **Organization.** In all public-relations forms, the document identifies the business, introduces the topic, makes the main point, and follows with supporting details organized from most to least important. In news releases and newsletter articles, key concepts are presented first so that the message will be clear even if the last part of the article is cut.

- ☐ **Voice.** The tone is objective, positive, and trustworthy.

- ☐ **Words.** The words are precise but common, and technical terms are defined.

- ☐ **Sentences.** In all public-relations forms, the sentences are brief, clear, and easy to read. In flyers and brochures, information may be presented in phrases or abbreviated sentences arranged in bulleted lists.

- ☐ **Copy.** The document or article accurately cites names, policies, products, and prices and includes no errors in mechanics, usage, spelling, or grammar.

- ☐ **Design.** The design follows guidelines and formats in this chapter. The types of visuals used and the balance of visuals versus text fit the document's purpose.

CRITICAL THINKING Activities

1. Visit the public-relations department of an area business and ask for two drafts of a recent news release: (a) the draft submitted to newspapers and (b) the draft published by newspapers. Study both drafts, noting what information in the original draft was changed in the published draft. Analyze how the changes affect the story.
2. Compare and contrast an organization's in-house newsletter with its public newsletter. What similarities or differences do you find in the (a) writing (topics, titles, supporting details, voice); (b) design (page size, layout, fonts; use of photos, graphics, and color); and (c) distribution. Explain how the qualities of each newsletter may reflect the organization's goals for the newsletter.

WRITING Activities

1. Research an organization for which you would like to work, and identify a product- or service-related topic that warrants an article in the area newspaper. Write a news release on that topic compatible with the length and format of the paper's similar articles.
2. Research a product or service of the organization described in Writing activity 1. Write a flyer or brochure that advertises the product or service.

In this chapter

Types of Instructions **566**

Tips for Writing Instructions **567**

Guidelines for Instructions **568**

Checklist for Instructions **574**

Critical Thinking Activities **574**

Writing Activity **574**

CHAPTER

35

Writing Instructions

Instructions are how-to documents that both business organizations and their customers or clients need. Organizations need instructions that help employees produce goods and services successfully, efficiently, and safely. For example, instructions for operating a sheer help a factory worker cut sheet metal successfully; instructions for delivering documents help a courier complete her route efficiently; and instructions for waste disposal help a lab technician handle toxic chemicals safely. Regardless of the type of work, well-written instructions, when used properly, can help employees do their jobs.

Customers and clients also need instructions. For example, a customer buying a multifunction printer may need the manufacturer's instructions for using the fax option. A client preparing for a tax conference may need the accountant's instructions for completing the pre-conference worksheets. In either case, well-written instructions, when followed carefully, can help the person use the product or service purchased.

Because written instructions play such an important role in the workplace, writing them well is essential. This chapter will help you write effective instructions, regardless of your writing situation. First, pages 566–567 describe the types of instructions written in the workplace and provide style tips for writing the documents. Second, guidelines explain how to work step-by-step through the process of prewriting, writing, revising, and refining. Third, models demonstrate helpful strategies such as listing materials, organizing steps, or using photographs, drawings, or signal terms. Finally, the checklist helps you polish your document.

TYPES OF INSTRUCTIONS

Instructions are practical how-to documents that are written in many forms (from brief statements on product packaging to book-length manuals) and for many purposes (from operating an electric can opener to operating a dialysis machine). For an overview of the types of instructions produced by businesses, read the lists below and check the forms that you may use or write.

Instructions for putting things together

Assembling a child's playhouse
Assembling a picnic table or chair
Assembling a prefabricated building
Assembling a tool rack
Connecting a PC, printer, and fax
Installing a dishwasher
Installing computer chips
Replacing a lavatory faucet
Replacing an auto's water pump
Replacing windshield wipers
Setting up a tent
Wiring an electric armature

Instructions for operating equipment or machines

Connecting a camera to a computer
Connecting a tractor and backhoe
Lighting a furnace
Operating a snow blower
Ordering a plane ticket online
Replacing a print cartridge
Reviewing online bank records
Scanning a document
Starting a lawn mower
Testing a battery
Using a bread maker
Using a gas-powered hedge trimmer
Using a heart defibulator
Using a scanner
Using a Web browser
Using a wire welder
Using an adding machine
X-raying a tooth

Instructions for implementing a process or procedure

Changing a car's antifreeze
Closing down a cash register
Completing UPS mailing instructions
Consuming medication
Disarming a surveillance system
Disciplining a student
Drawing and labeling a blood sample
Filling out tax forms
Installing floor tile
Interrogating a crime suspect
Interviewing a job applicant
Preparing an annual report
Processing a delinquent account
Releasing in-house information
Reporting an auto accident
Reporting an injury in a care facility
Reporting signs of child abuse
Reporting workplace discrimination
Requesting a job transfer
Requesting a leave-of-absence
Routing a document for review
Scheduling a board meeting
Scheduling a social worker's visit
Submitting an article for publication
Submitting travel expenses
Using stain remover

TIPS FOR WRITING INSTRUCTIONS

Follow Safety-Information Guidelines

When writing instructions, your primary goal is to help readers complete the task safely. To do so, use safety-information guidelines produced by the American National Standards Institute (ANSI), the International Organization for Standardization (ISO), or the U.S. military (MILSPEC).

Choose the authority most closely aligned with your readers: ANSI for U.S. readers other than the military, ISO for international readers, and MILSPEC for the U.S. military. Be aware that each authority may offer slightly different guidelines. For example, in its "Product Safety Signs and Labels" standard (Z535.4-2002), ANSI advocates using the three signal words *Danger*, *Warning*, and *Caution* as follows:

- *Danger* indicates an imminently hazardous situation that, if not avoided, will result in death or serious injury. This signal word is to be limited to the most extreme situations.
- *Warning* indicates a potentially hazardous situation that, if not avoided, could result in death or serious injury.
- *Caution* indicates a potentially hazardous situation or practice that, if not avoided, may result in minor or moderate injury.

ISO and MILSPEC may not use these words in exactly the same way. In addition, ANSI, ISO, and MILSPEC may offer different guidelines regarding the use of colors or symbols in safety-related information. To ensure workers' safety and to comply with current standards, check the organization's guidelines.

Use Command Verbs

In instructions, use precise command verbs that readers will understand:

address	dig	insert	pour	scan	tip
align	download	inspect	press	scroll	total
blow	drag	lift	print	select	transect
boot up	drain	load in	pull	send	trim
call up	drill	lock	push	shift	turn
change	drop	loosen	raise	shut off	twist
check	ease	lower	remove	slide	type
choose	enter	make	replace	start	unhook
clean	fasten	measure	reply	state	unplug
click	fill	move	review	switch	use
clip	find	notify	rinse	take	ventilate
close	flip	oil	roll	tear	verify
connect	follow	open	rotate	test	wash
cut	identify	place	save	tighten	wipe
delete	include	plug	saw	tilt	wire

GUIDELINES FOR INSTRUCTIONS

When writing instructions, your goal is to break down a task into logical steps and describe those steps so clearly that a reader can do the task. To achieve this goal, follow the guidelines below, using those directives appropriate for your writing situation.

1 Prewrite

Rhetorical analysis: Consider your purpose, your audience, and the context.
- ☐ Who will use the instructions, in what setting, and for what outcome?
- ☐ What are the user's probable reading skills?
- ☐ How familiar is the reader with the process and related terminology?
- ☐ What are the user's process-related skills, including skills with needed tools?
- ☐ What government, industry, and company standards must the instructions follow, and where can I find information about (or help with) these standards?
- ☐ What potential damage or safety issues are involved, and where can I find information about (or help with) addressing these issues?

Think about and practice the task.
- ☐ What is the core task? How well do I know the task? What should I review, learn, and practice?
- ☐ What materials and tools are needed, and what safety issues are related to each?
- ☐ How many people are required, and what are their needed skills?
- ☐ What steps are required? In what order?
- ☐ How are the steps related or linked?
- ☐ How long should each step take? How long should the full process take?

Think about the document's content and form.
- ☐ How long must the document be (one page, four pages, a manual)?
- ☐ What visuals (photos, drawings, illustrations) will help clarify the task?
- ☐ Is a document the best form for the instructions, or would another form (e.g., a CD with video and audio support) better fit the task, audience, and situation?

2 Draft

Organize the message into three parts.

[opening] Introduce the process.
- ☐ Identify the task in the title.
- ☐ Describe the task briefly, explaining its importance and goals.
- ☐ Point out issues related to potential damage, personal injury, adherence to code, or legal requirements such as necessary certification or training.
- ☐ List and, if necessary, picture required materials, equipment, and tools.
- ☐ State the approximate time that the activity will take.

[middle] Tell the reader what to do.
- ☐ Give step-by-step instructions in succinct sentences.
- ☐ Number steps chronologically.

- Use command verbs in the present tense:

    Loosen the 1/2" nut in the center of the blade.
    Remove the blade from the driveshaft.

- Address the reader directly with personal pronouns like *you* and *your*.
- Limit instruction to one basic action per sentence.
- Use precise terms for materials, tools, and measurements.
- Keep subjects, verbs, and objects close together for easy reading.

    POOR: Install (being careful that the red arrow on the blade is turning clockwise) the replacement blade on the driveshaft.

    GOOD: Install the replacement blade on the driveshaft.
    Caution! Be sure that the red arrow on the blade is turning clockwise.

- Place adjectives and adverbs close to the words they modify.

    POOR: Tighten the 1/2" bolt to the driveshaft *securely*.
    GOOD: *Securely* tighten the 1/2" bolt to the driveshaft.

- State clear signal words where needed: **WARNING** (potential for danger or injury), **Caution** (possible error or damage to equipment), and *Note* (tips or clarification on how to do the task). For current, legal standards for the use of signal words, check the guidelines of professional organizations such as ANSI, ISO, and MILSPEC (see page 567).
- Illustrate the action with clear, well-placed photos, drawings, or diagrams.

[closing] When appropriate, give additional details:

- The desired outcome of the task or any reminders (e.g., the task's relevance)
- A summary of the task or note regarding its relationship to other procedures

3 Revise
Carefully check your draft for ideas, organization, and voice.

- Are the lists of materials and tools complete and correct?
- Is each step in the process clear, complete, and supported by needed visuals?
- Are steps in the correct sequence, and are they correctly numbered?
- Is each measurement (e.g., weight, distance, length) correct?
- Test the instructions by reading them to a listener who completes the task by doing *precisely* what you read. Afterward, discuss the instructions' clarity, completeness, and correctness, and then revise as needed.
- For activity that could cause damage or injury, test the instructions more than once and with a knowledgeable observer.

4 Refine
Check your writing line by line for the following:

- Terms are used in a consistent, correct manner.
- **WARNINGS** are in boldface and caps, and **Cautions** are in boldface.
- Spelling, labels, numbers, and terms are accurate.
- Graphics, photos, and drawings are clear and placed near their respective steps.

> "When writing instructions, don't assume anything. If the reader knew what to do, he or she wouldn't need instructions."
>
> —Dennis Walstre, Plumbing Contractor

Instructions with a List of Materials

Rhetorical situation

Purpose: To explain to employees how to close off a cash register
Audience: All sales personnel who use a cash register
Format: Highlighted title, list of materials, and numbered steps listed chronologically

---

[opening]
Identify and introduce the task.

List materials or tools needed.

[middle]
Present steps in chronological order.

Use short sentences with command verbs.

Describe each action precisely and succinctly.

Place **WARNINGS** in boldface and capital letters; place **Cautions** in boldface only.

State closely related actions in a single step.

[closing]
Review a key point.

Instructions for Closing Off a Cash Register

Use the instructions below (1) to close off the cash register and (2) to account for the day's receipts.

Materials Needed

- Daily Account Form
- Deposit bag
- Adding machine
- Pen

Steps

1. **WARNING: MAKE SURE THE STORE DOOR IS LOCKED.** Take the cash tray out of the register drawer and place it on the counter. Leave the empty drawer open to deter thieves.

2. Turn the cash register key to the X setting and press the X key. The machine will print the X reading: the total amount of receipts (cash, credit card slips, and checks) taken in for the day.

3. Turn the key to the Z setting and press the Z key. The machine will print the Z reading: itemized, department-by-department subtotals (camera sales, film sales, film processing, etc.).

4. Count out $100 and place it in the envelope marked FLOAT. (The FLOAT is the $100.00 of cash placed in each cash register when the store opens.) The remaining cash, checks, and credit card slips make up the day's receipts. **Caution: Do not place the float back in the drawer.**

5. Using the adding machine, total the day's receipts and check them against the X reading. If the totals differ, count the receipts a second and third time if necessary. If the totals still do not agree, make a note of the difference and attach it to the receipts.

6. Fill out the Daily Account Form by entering these totals: the X and Z readings and the day's receipts (credit card slips, checks, and cash). Place the day's receipts in the deposit bag.

7. Place the following in the safe: (1) the deposit bag, (2) the Daily Account Form, (3) the X and Z readings, and (4) the envelope marked FLOAT.

8. **Caution: Double-check the safe door to make sure it is locked.**

Chapter 35   Writing Instructions   571

Instructions for a Procedure

Rhetorical situation

Purpose: To explain to caregivers when and how to wash their hands
Audience: All employees who provide care for patients
Format: Highlighted title, introduction, caution, numbered steps, and final note

---

Regis City Hospital
HAND-WASHING PROCEDURE

Perform this procedure whenever you report for duty and before and after providing care for a patient. Also use this procedure after bathroom use; after eating, coughing, or sneezing; and before and after using sterile gloves, gowns, and masks. Whenever in doubt, wash your hands.

Caution: Rings and other jewelry harbor bacteria and are difficult to clean. Before reporting for work, remove all jewelry.

Steps

1. Remove the first paper towel, and place that towel in the wastebasket.
2. Take the next paper towel and use it to turn the water on to a comfortable temperature. **WARNING: DO NOT TOUCH THE CONTROLS WITH YOUR HANDS.**
3. Put your hands and wrists under the running water, keeping your fingertips pointed downward. Allow the water to flow gently.
4. Once your hands and wrists are completely wet, apply antiseptic solution.
5. Bring your hands together and create a heavy lather. Wash at least three inches above the wrists, and get soap under your fingernails and between your fingers. **Caution: Wash well for one full minute.**
6. With the fingertips of your opposite hand, circle each finger on the other hand with a rotary motion from base to tip. Pay careful attention to the area between your fingers, around nail beds, and under your fingernails.
7. Rinse your hands well under running water. **Caution: Hold your hands down** so that the direction of the water flow is from the wrist to your fingertips.
8. Pat your hands dry with a clean paper towel, and turn off the water with the towel. Discard the paper towel into the wastebasket.

Note: If your hands touch the sink, faucet, or spout, repeat the entire procedure.

[opening]
Identify the task.

Explain when to do the procedure and how to prepare to do it.

[middle]
List steps in chronological order.

State **WARNINGS** in boldface and capital letters.

State **Cautions** in boldface without all capital letters.

[closing]
End with a final point if needed.

F.Y.I.

If two or more steps should be done simultaneously, insert an appropriate signal word (*Note*, **Caution**, or **DANGER**), and then clearly state which of the following steps must be done simultaneously.

Part 6 Special Forms of Writing

Instructions Containing Photographs

Rhetorical situation

Purpose: To explain how to download photos from the MC-150 digital camera
Audience: All camera sales personnel and users
Format: Highlighted title, note, and numbered steps illustrated by photos

---

[opening]
Use a descriptive title. Note or list materials.

[middle]
Give steps and photos in chronological order.

Add graphics (such as the arrow) to create a quick visual cue.

Boldface words that need special attention.

To show an object's size, use a reference (such as the fingers).

Use only well-focused photographs.

[closing]
Note common problems that are easily solved.

Downloading Photographs from the MC-150 Digital Camera

Note: MC-150 software must be loaded on your computer to download photographs from the camera.

1. Turn your computer on.
2. Plug the camera's USB cable into your computer.
3. Turn the camera's mode dial to the **data transfer setting** (Figure 1).
4. Open the camera's flash-card door and plug the other end of the USB cable into the **camera port** (Figure 2).
5. Select USB transfer from the camera screen menu. The MC-150 software will then launch onto your computer.
6. Follow the instructions on the computer screen to download all of your photos or specific photos.
7. When your download is complete, turn the camera off and unplug the USB cable from the camera and the computer.

Figure 1: Data Transfer Setting

Figure 2: Camera Port

Note: If MC-150 software doesn't launch, disconnect the camera (step 7), restart the computer, and continue on from step 2.

Instructions with Drawings

Instructions for Replacing Auger Bearing on Bale-Press 64-D

Use the instructions below to replace the right, rear auger bearing on the Bale-Press 64-D.

Parts Needed

Item No.	Quantity	Part No.	Description
1	1	RRAB20024	right, rear auger bearing
2	1	RRASP20007	right, rear auger-support post
3	2	BLT5/8X3.5HX	⅝" × 3 ½" hex bolt
4	2	NUT5/8HX	⅝" hex nut
5	2	CP1/8X2	⅛" × 2" cotter pin

WARNING: SHUT OFF ENGINE AND REMOVE KEY!

Steps

1. Take out the existing bearing by removing the two ⅛" cotter pins, two ⅝" hex nuts, and two ⅝" hex bolts. (See Figure 1.)

2. Insert the new bearing (item #1) into the 4 ⅜" hole on the post (item #2).

3. Fasten the bearing to the post with two ⅝" hex bolts (item #3), two ⅝" hex nuts (item #4), and two ⅛" cotter pins (item #5).

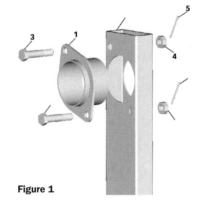

Figure 1

Note: For ease of installation, insert both bolts before tightening the nuts.

Caution: For safer installation, fasten the replacement bearing with the new bolts, nuts, and cotter pins included with the replacement kit.

[opening]

Identify topic.

List the parts needed in a table format. If helpful, add illustrations.

[middle]

Indicate danger with a **WARNING** message in capital letters and in boldface print. Use an icon, if helpful.

Give steps in chronological order.

Use the same terms in the steps as were used in the parts list.

Place the illustration next to the steps.

[closing]

Use **Cautions** (boldface without capital letters) to warn of damage.

CHECKLIST for Instructions

Use the seven traits to check your document and then revise as needed.

- ☐ **Ideas.** The task and all necessary parts, materials, and tools are clearly identified. Steps (what to do and how to do it) are fully explained. Directives with accurate measurements (e.g., time, distance, length) are cited. Clear and appropriate warnings, cautions, and notes are given.

- ☐ **Organization.** The instructions have a clear *opening* (title or introduction) that identifies the procedure and sets the context; a distinct *middle* that states numbered steps clearly in chronological order; and a *closing* that presents the last step and/or a summary or clarification.

- ☐ **Voice.** The tone is knowledgeable, confident, forthright, and sensitive.

- ☐ **Words.** Words (including technical terms) *match* the task, situation, and reader; *identify* parts correctly and uniformly; and *state* notes, warnings, and cautions objectively and without alarm.

- ☐ **Sentences.** The sentences are succinct but complete. Steps are stated with command verbs and clear transitions.

- ☐ **Copy.** The document accurately cites specifications, names, part numbers, and signal words. The copy includes no errors in grammar, punctuation, capitalization, and spelling.

- ☐ **Design.** The document effectively uses headings, numbers, white space, and graphics to identify and organize information. Illustrations are labeled, easy to read, and located next to their corresponding steps.

CRITICAL THINKING Activities

1. Working with a classmate, find a set of instructions. Test the document by having one person read it while the other person does the actions. Then use the guidelines on pages 568–569 and the checklist above to evaluate the document's form and content. Report your findings to the class.
2. Research the signal terms advocated by one of the following organizations: American National Standards Institute (ANSI), International Organization for Standardization (ISO), or the U.S. military (MILSPEC). Report the following to the class: (1) which signal terms the organization advocates, (2) how it defines the terms, and (3) which symbol or sign is used along with each term.

WRITING Activity

Working with a classmate, choose a product that comes with confusing or incomplete assembling instructions. Develop your own instructions for the product, including needed drawings, photos, and signal terms. Show both documents to the class and explain why—based on the checklist above—your document is stronger.

In this chapter

Web Page Elements and Functions **576**
Strategies for Developing a Web Site **578**
Sample Web Sites and Pages **582**
Checklist for Developing Web Pages and Sites **586**
Critical Thinking Activities **586**
Writing Activities **586**

CHAPTER

36

Writing for the Web

Whether large or small, nationwide or local, today's businesses rely on Web sites to market their products and services, recruit employees, service customers, share information internally, procure supplies, conduct market research, and plan for the future.

A successful Web site depends on well-crafted and organized content. Such success requires planning, involves many tasks, and calls for collaboration between people in various business departments. In addition, people creating Web content must have some understanding of the technical workings—and limitations—of Web browsers, pages, sites, and servers.

When you're writing for the Web, remember that—above all else—Web content should be concise, focused, and visually appealing. People don't *read* Web sites so much as they *scan* them, so information should be presented in short chunks of text. Web pages should also be brief, designed so that readers don't have to scroll down more than two screens before going on to the next page. Online, people can be impatient readers, so content must always remain focused on the page's purpose, audience, and topic. Many sites have distinct menus and pages for different audiences (investors, suppliers, job seekers), for different purposes (ordering, serving customers, conducting research), and for different topics (products, services, company history). Whether you choose to build your own site or pay someone else to do so, this chapter will help you craft a Web site that meets your business's goals.

WEB PAGE ELEMENTS AND FUNCTIONS

To design a strong Web site and develop strong Web pages, you need to start with a basic understanding of Web page elements and functions. Web pages use the same elements as printed pages, so many of the same design principles apply. However, unlike printed pages, Web pages can include both elements and functions.

Page Elements

On the Web, page elements, such as text and images, can appear in many formats, sizes, and colors. However, the use of too many page elements—or poor combinations of elements—may distract or annoy readers, meaning your message gets lost. Whenever you use the following elements on a Web page, make sure that they serve the purpose of your site.

Headlines

Headlines (also called headers) come in six sizes and are used to separate different sections and subsections of Web documents.

Body Text

As you might guess, body text defines the text format, also called the font. Body text is organized into chunks, called *paragraphs*, which are separated by white space. The body text's *font size* should be large enough to read, but small enough to fit a sufficient amount of text on each page. The *font color* should be dark enough to remain readable against any background.

Background Color

Pages can be displayed in almost any color, but color should be used sparingly. Any background color must work with the font color and images on the page.

Lists

Web pages typically use three types of lists. *Ordered lists* are numbered, *unordered lists* are bulleted, and *definition lists* present pairs of information—usually terms alongside their definitions, which are indented.

Images

Images can include photographs, clip art, graphs, line drawings, cartoon figures, icons, and animations. Because the Web is considered a visual medium, Web pages typically incorporate several visual elements, whereas others—such as library catalogs—limit visuals to present text more efficiently. *Icons* are small symbols that denote a category of information, such as the "flying letter" symbol ✉ for e-mail. An icon is useful only when its meaning is commonly understood. To review the principles of using images, see Chapter 5, "Using Graphics in Business Documents."

Tables

Tables (grids made up of rows and columns) are a basic tool for Web page layout.

Page Functions

Web page functions set electronic pages apart from printed pages. On a Web page, readers can browse pages in almost any order, send and receive e-mail, send messages and files, post messages, and join live "chat" sessions. In short, readers can interact with Web pages in ways they cannot with printed pages, and readers can do these things everywhere and all the time. Like any Web page element, each Web page function you employ should somehow serve the purpose of your site.

Hyperlinks

Hyperlinks—*links* for short—are strings of specially formatted text that enable readers to jump to another section of the existing Web page or another Web page on your site. These are called internal links. External links lead to pages on other Web sites. *Mailto links* allow readers to address e-mail to recipients, such as your company's director of customer service or a salesperson.

Menus

Menus are a structured list of links that operate like a Web site's table of contents. They are typically presented in a column or row at the edge of a Web page. Good Web sites include a standard site menu on every page so that readers won't get lost.

Forms

Forms enable the host of a Web site to interact with the site's readers. Web forms can function as questionnaires, suggestion boxes, complaint forms, job-application forms, service-request forms, or other business forms for gathering information.

Given all of the enticing features and functions available for Web site development, it's easy for authors to get carried away. To create successful Web sites, however, writers must stay focused on the rhetorical principles that strengthen all good business writing: purpose, audience, and topic.

If you find yourself distracted by the many bells and whistles of the Web, remember the "KISS" maxim of information technology professionals: Keep It Short and Simple. It's better to have a simple Web site that works well and is effective at presenting information than a complex site that does not.

> ### WRITE Connection
> Check the following chapters and section for supporting information related to Web page elements and functions:
> - Chapter 16, "Trait 7: Reader-Friendly Design."
> - Chapter 5, "Using Graphics in Business Documents."
> - "Lists," page 224.

> ### WEB Links
> To study these elements and functions at work, check this book's Web site at http://college.hmco.com/english. For a second worthwhile example, we also recommend www.upwritepress.com.

STRATEGIES FOR DEVELOPING A WEB SITE

Regardless of the purpose, topic, and audience of your Web site, you can develop it by following these essential steps: Get focused, establish your central message, create a site map, study the competition, gather and prioritize content, think about support materials, develop and design individual pages, and test your site.

> **WRITE Connection**
>
> Developing a Web site typically requires collaboration. For help, see Chapter 3, "Teamwork on Writing Projects."

Get Focused

The first step in developing a Web site is getting an overview of the project—the purpose, the audience, and the topic. Answering the questions below will help you get focused and develop fitting content for the site.

- **What is the primary purpose of the Web site?** Am I going to present company information and announcements? Am I trying to promote a specific product line or service? Am I creating a library of documents that my audience will reference?

- **Who is the site's audience?** What kind of people will seek out this site? Why? What do they need? How often will they visit the site, and how often should it be updated? How comfortable are they using computers and Web sites? What level of formality is appropriate for the language? What kind of graphics and design will appeal to them?

- **What is the site's central topic?** What do I already know about the topic? What do I need to learn, and where can I find the information? What will my audience want to know about the topic? How can I divide the information into brief segments? What visual elements would help me present my message? What other Web sites address this topic, and should my Web site link to them?

Establish Your Central Message

After you've made decisions about your purpose, audience, and topic, it's a good idea to write out the main idea you want to communicate. You might call this the theme or "mission statement" of your Web site. Begin with the simple statement "The purpose of this Web site is," and then add a phrase like *to explain, to provide, to promote,* or *to inform*. Finish by inserting your topic, along with the main mission of your Web site.

> The purpose of this Web site is to *inform the general public* about the *potential dangers of static electricity* when people refuel their vehicles at filling stations.

Consider posting this mission statement on the wall as you develop the Web site. Posting the mission statement in plain sight may help you stay on target with the project, and you may want to modify your goals as the site is developed, or add secondary goals for the site.

Create a Site Map

After focusing your task, create a site map using these principles:

- **No one will read your entire site.** Provide only the content that readers need.
- **Your site will have many small audiences—not one big one.** A site's audience is not limited to a conference room or mailing list but may include anyone with a computer, an Internet connection, and an interest in your company. As you organize your site, keep all potential readers in mind.
- **Web sites are not linear.** Few Web sites proceed linearly (opening, middle, closing). Instead, a single "home page" or "splash page" introduces the site, which branches out like tree limbs into varied content.
- **You may need to build the site in phases.** You can add pages to a Web site after it has been published, so be careful that your site organization does not limit future additions.

Web site maps can be hierarchical: the organization used for family trees and company organizational charts (see Figure 36.1). A menu for a simple site might include only four items—home page, page "A," page "B," and page "C"—allowing users to "jump" between any of the secondary-level pages or back to the home page.

A more complex Web site typically needs more levels. Likewise, its menu—or menus—will offer more navigation choices. Related pages on secondary levels might be connected with links—as represented by the dotted lines in Figure 36.2. These links function as cross references in a printed text.

As you begin creating your Web site plan, consider developing multiple menus for the same set of content so that different readers can navigate according to their different needs.

Figure 36.1 Hierarchical Web Site Map

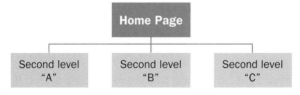

Figure 36.2 Complex Web Site Map

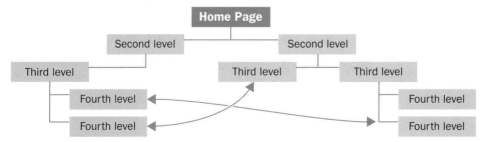

Study the Competition

One of the best ways to design a Web site is to review and evaluate successful sites. The seven traits can supply helpful benchmarks:

- Does the site present and support strong ideas?
- Are they organized?
- Is the language conversational?
- Does the site use the best words possible?
- Are the sentences smooth?
- Are there any distracting errors?
- Are the pages user-friendly?

Gather and Prioritize Content

Using your Web site's mission statement and map, brainstorm and research the actual content, with the goal of creating an outline for the site. How many topics will your site address? How wide will your coverage of a topic be? How deep? Your outline may also serve as the Web site's table of contents, which can be used to create a menu for the site. Also, seek input from other departments of your company, because a company Web site should represent the entire organization.

Using your outline as a guide, begin to gather and prioritize the information. Based on your research, discussions with others in the company, the budget, and the amount of time available for the project, select which content, features, and functions your site will offer.

Think About Support Materials

List the types of company documents—brochures, instructions, manuals, reports—that will be presented on your site, and decide whether they will be displayed as Web pages, made available for readers to download, or both. You may want to construct a grid like Table 36.1 to keep track of how documents will be used on the Web site.

List graphics, technical drawings, graphs, logos, photographs, and animations that can help readers grasp the meaning of your message. Review Table 36.2 for electronic files that may be appropriate to your topic, audience, and purpose.

Table 36.1 Tracking Document Use on a Web Site

Purpose	Document	Displayed	Downloadable	Both
Promotional	Brochure #11	☑	☐	☐
Informational	Report #26	☐	☑	☐

Table 36.2 Electronic Files for a Web Site

Images	Audio	Video
Cartoons, drawings	Music	Animations
Charts, graphs	Sound effects	Film clips
Photographs	Spoken text	Presentations
Tables		Webcasts

Design and Develop Individual Pages

Most Web pages—and the pages of most other publications—are designed on *grids*. Look at any newspaper or magazine page, and you should be able to draw horizontal and vertical lines denoting columns and rows of content. Some rows may span multiple columns, while some columns may overrun several rows.

Another fundamental design concept is *balance*. You may balance light elements with dark ones, open space with crowded space, text with images, and so on. The balance of your page design should be driven by the purpose of your Web site, its audience, and its topic.

Business Web sites may contain a variety of pages—each tailored to a different purpose, audience, and topic—to present some combination of informational and promotional content. Even sites devoted to sheer entertainment may promote a product or service or present information related to it. Use the purpose of each page to guide your decisions about page design elements and functions to include.

Many books describe graphic Web design, and this chapter cannot begin to cover more than some elemental aspects. Review Chapter 16, "Trait 7: Reader-Friendly Design," and keep these six design principles in mind as you begin to mock up pages:

1. See readers reading.
2. Choose fitting design features.
3. Make pages open and balanced.
4. Go with the flow.
5. Develop consistent visual "cues."
6. Stress connections and contrasts.

Test Your Site

Most Web sites are developed through the collaborative effort of writers, graphic designers, and programmers. In such a collaborative environment, many content and layout ideas may be considered, rejected, reformulated, or developed to produce and launch the site. Nevertheless, it is still the external audience—the customer—who determines a Web site's success or failure.

After a site has been launched, its success may be measured by the amount of traffic on the site, feedback submitted by users, and—if transactions can be completed online—the amount of business the site generates. A Web development team may test a site "mock-up" by writing a questionnaire and compiling the questionnaire responses or by organizing a focus group to discuss the proposed site. See "Testing Documents with Readers," pages 68–69.

WRITE Connection

If you need information on using the Web as a research resource, see pages 145–150.

SAMPLE WEB SITES AND PAGES

On the next four pages, you'll find sample Web pages from four typical Web sites: a manufacturing company, a retailer, a nonprofit organization, and a government agency. Study each model and the description of the rhetorical situation for insights into what makes for strong Web content and design.

Manufacturing Company Web Site

The KeyTronic Corporation designs and makes keyboards, laser printers, telecommunication satellite units, multimedia touch panels, digital control panels, and medical devices for varied companies. Figure 36.3 shows a page from its Web site.

Rhetorical situation

Purpose: This site aims to generate business, whether as an investment in the company's stock or as an order for its manufacturing services. Notice that some menu options lead readers to make a transaction ("Shop Online") or to contact a salesperson (the e-mail link at the bottom of the page).

Audience: This page targets specific audiences and anticipates their needs with three different menus.

- Investors may be interested in the "Corporate Profile" or "News" menu buttons on the menu at the top of the page.
- Potential customers may want to see the "Manufacturing Portfolio" or "Outsourcing" information in the left-column menu.
- Existing customers may "Login" or take the "Customer Survey."

Format: The page's menus address both service topics—"Engineering Services"—and manufacturing topics—"Custom Molding & Tooling." A third topic—the company itself—is the object of several menu buttons. Note the simple layout of the page: two horizontal menus, one vertical, and a large field of text. The page includes only two images (company logo and photograph), but bold colors create visual impact, while the highly textual presentation and the businesslike language lets readers know this Web site "means business."

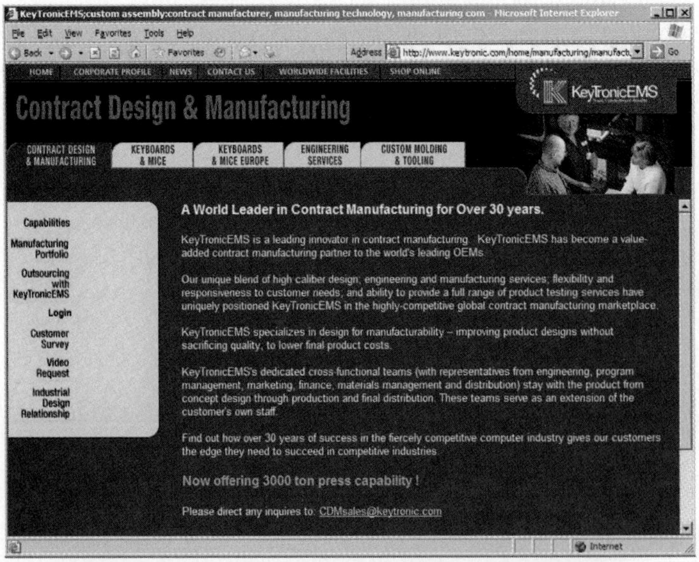

Figure 36.3 Web Site: Manufacturing Company

Retail Business Web Site

As the Karmaloop Web site tells it, the company was founded in 1999 "to battle the evil forces of McFashion." In short, Karmaloop is one of many companies in the competitive market of youth-oriented fashion. Figure 36.4 shows a page from its Web site.

Rhetorical situation

Purpose: The site juggles two main purposes: to sell clothing that is considered "cool" by the target audience and to show that Karmaloop is a "cool" company. Note the prominent shopping cart, photos of fashionably dressed youths, and the product logos. The site's "About Us" page claims "Every member of Karmaloop's staff is in their 20's."

Audience: The site's obvious audience is young people—like those pictured on the home page—who want to buy clothing. Notice how the page's title (in the blue bar at the top of the screen) includes descriptive search words that define Karmaloop's target market: cool, urban, streetwear, hip-hop, and skate. The home page is probably updated with new merchandise offers frequently to capture its readers.

Format: The primary focus of Karmaloop's site is its product catalog, which may be searched or browsed in several ways. Notice how the drop-down menus divide the catalog into product type (brands) and audience (women, men, and even bargain hunters—see the "$5 Off," "Sale!," and "Free Shipping" links). Secondary topics on the site, including the company itself, can be accessed via the horizontal menu above the photographs.

In summary, Karmaloop's page design is divided into three simple columns, but its six horizontal rows—coupled with three photos, multiple text fonts, assorted background colors, and colorful logos—add some complexity. As an online retailer, Karmaloop is highly dependent on its Web presence to generate business transactions, and its Web pages are designed to draw readers in and quickly direct them toward purchasing products.

Figure 36.4 Web Site: Retail Business

Nonprofit Agency Web Site

The Oklahoma Trucking Association (OTA) is a nonprofit trade association serving the interests of the state's trucking industry. Founded in 1932, the group provides services to trucking firms, promotes trucking safety, and lobbies state and federal legislators. Figure 36.5 shows a page from its Web site.

Rhetorical situation

Purpose: The site aims to inform a very specific audience about—as the horizontal menu shows—the association's services, events, press (news releases), history, and safety programs. The site also aims to attract new members.

Audience: From this Web page, we can surmise that OTA's audience is narrow. Its members would visit the site to find news or information specific to the trucking industry. We can also speculate that the site was created to be user-friendly to people who may not be expert Web surfers. Note that the horizontal menu includes only seven entries and that nearly one-third of the screen is devoted to a single graphic.

Format: The two main topics of this site—the association itself and news related to the state's trucking industry—are presented both as textual content on this page and as menu links that lead to other pages on the site. In other words, the design of the page is deliberately simple. Three large rows divide the screen visually, while one wide column displays textual content, and one narrow column lists "News" links. Notice that more than half of the page is used to describe OTA and present contact information.

Figure 36.5   Web Site: Nonprofit Agency

Government Agency Web Site

The Oregon Labor Market Information System (OLMIS) is a Web site that provides economic information to employers and job seekers. The site was also created to distribute information produced by the state's employment department to labor researchers, policy analysts, and others. Figure 36.6 shows a page from this Web site.

Rhetorical situation

Purpose: Because OLMIS serves many purposes, offers multiple tools, and presents different types of content, it may be considered a portal or an entryway to a large grouping of distinct but related content and tools. Note how the site's creators named it a "system" rather than a "site" and how they've created a tour of online tools and content.

Audience: The audience of OLMIS is obviously diverse, as illustrated by the complexity of the home page design. The left-column menu anticipates the needs of OLMIS's audience by including seventeen links and a search field. Notice that a Spanish version of the site is also available. The designers of the site expect its readers to visit often, so news stories in the right column are date-stamped to show they are current.

Format: Because Oregon employment news takes up two-thirds of the screen, we can conclude that this news is the primary topic of the home page, while the menu options enable the reader to pursue many other topics. So many topical menu links are listed, in fact, that they have been organized into three groupings in the left column. The page design is, therefore, defined by three clear-cut columns: a menu, newsworthy features, and other news links. Because the page presents so much text—as both links and as content—little room is left for photos or graphics. Too many visual elements would make the page appear cluttered, so the photos are kept small and the background colors are soft. Compare this design to the bold colors of the KeyTronic home page in Figure 36.3 (page 582).

Figure 36.6   Web Site: Government Agency

CHECKLIST for Developing Web Pages and Sites

Use this checklist to review and revise your site and its pages.

- ☐ The **purpose** of the Web site is evident on the home page and elsewhere.
- ☐ The **page elements**—headlines, body text, colors, lists, images, and tables—work together and are suited to the page's audience, topic, and purpose.
- ☐ The **page functions,** including navigation menus, are logically presented and enable readers to find what they need quickly.
- ☐ The **content** collected for the site (from brochures to reports) and **support materials** (images, audio, and video) are available, approved, accurate, and in the correct format for presentation or download.
- ☐ The **site plan** allows information to be presented and cross-referenced in logical and efficient pathways.
- ☐ The **page design** incorporates the six design principles on page 292.
- ☐ The **informational** and promotional aspects suit the audience, topic, and purpose.

CRITICAL THINKING Activities

1. Visit a familiar Web site. Take a few minutes to study its home page, or another page, in terms of the site's purpose, audience, and topic. Analyze and evaluate the page's content and design.
2. Working with a classmate, evaluate your company's (or college's) Web site. Assess how well the site addresses its purpose, audience, and topic.

WRITING Activities

1. Choose a company that has a Web site and produces a product or service you use or have used. Analyze the Web site in terms of its content and design. Create a PowerPoint presentation to share your findings and recommendations for improvement.
2. Working with classmates, compare and contrast Web sites from two competing companies. Develop a report in which you analyze and evaluate how these competitors use the Web similarly and differently.
3. Working with classmates, brainstorm and investigate campus clubs, associations, and other groups that currently do not have a Web site (or at least not a quality Web site). Using the guidelines from this chapter, develop a Web site. As another option, consider a nonprofit organization in your area. Collaborate and research as needed.

PART 7

Management and Management Writing

37 Managing Your Time and Manners

38 Managing Effectively

39 Management Writing

40 Human Resources Writing

"Before we begin our Time Management Seminar, did everyone get one of these 36-hour wrist watches?"

In this chapter

Managing Your Time **590**

Evaluating Your Time-Management Skills **592**

Practicing Workplace Etiquette **594**

Polishing Your Etiquette **596**

Eating and Drinking **597**

Checklist for Appropriate Dress **598**

Critical Thinking Activity **598**

Writing Activity **598**

CHAPTER

37

Managing Your Time and Manners

In nearly any job, your success will largely depend on how well you do the basics—writing, reading, speaking, listening, thinking, creating a spreadsheet, and working with technology. But to thrive in the workplace, you need other skills as well.

For example, you must manage your time effectively and be able to show supervisors how you have managed it. You must know basic etiquette such as how to introduce yourself, introduce others, or be introduced. And you must handle yourself with decorum in both the workplace and social settings. Practicing decorum includes knowing how to dress, participate in a meal, and carry on a good conversation. This chapter will help you do these things.

> You can either lay in bed all day and moan about what doesn't work, or you can get up and use what does.
>
> —Grandma Besse's principle of time management

MANAGING YOUR TIME

To offer products or services at competitive prices, every business needs employees who manage their time well. Why? Poorly used time slows production, adds unnecessary cost, and erodes employee morale. You can improve your time-management skills by adhering to four practices.

> "Managing your time means managing yourself."
> —Jack D. Ferner, *Successful Time Management*

Periodically review what your employer expects regarding your use of time. Read documents like the following and highlight task- and time-management details:

- Your contract and job description, particularly duties for planning and goal setting
- Policies regarding arrival and departure times, conducting personal business in the workplace, and documenting time use
- Job-review forms, particularly time-management issues

Determine your available work hours per week. Calculate these hours by using the following equation:

168 (total hours in a week) − committed nonwork hours = available work hours

Add committed nonwork hours per week:

Personal activities (eat, sleep, exercise, home chores)	54 hours
Family time	32 hours
Social and professional organizations, church	16 hours
Recreation and leisure	18 hours
Professional and personal development	10 hours
Total committed nonwork hours	130 hours

Subtract

168 (hours in a week) − 130 (committed nonwork hours) = 38 available work hours

Consider how you can satisfy your employer's expectations during the number of hours that you have available. Start by developing a clear, reasonable schedule.

- Each Friday, review the week's activities, noting finished work, unfinished work, and related issues. Then review prescheduled activities for the next week and revise the schedule as needed. (To help you plan, use printed and electronic time-management tools like those described on the next page.)
- At the end of each day, review the day's activities and note finished and unfinished tasks. Then list and prioritize tasks for the next day.
- At the beginning of each day, review and update your schedule and to-do lists.

Printed time-management tools

- **Calendars** in which you list items such as appointments, events, deadlines; names, addresses, phone/fax numbers; flight schedules, hotels, car rentals; birthdays, holidays, and special nonbusiness events
- **Weekly planning guides** in which you write (1) main activities, (2) key projects, and (3) important people involved in these activities or projects
- **To-do lists** in which you record the day's tasks and prioritize each as a 1, 2, or 3:
    1. Urgent—must be done today.
    2. Priority—should be done today.
    3. Important—but may be postponed.

Electronic time-management tools

- **Personal information-manager (PIM) software** helps you (1) maintain contact information (e.g., names, phone numbers, e-mail addresses), (2) create and update your schedule in a daily, weekly, or monthly format, and (3) keep track of projects, including work status, dates, and parties involved.
- **Personal digital assistants (PDAs)** are hand-held computers that can do all the functions of PIM software, plus download or upload information from a PIM program, take digital pictures, and more.
- **Project-management software** helps you plan and supervise a project by breaking it into tasks and subtasks, scheduling work, setting deadlines, monitoring progress, and showing progress in chart formats.
- **Government Web sites,** such as the Small Business Administration's www.sba.gov, have information on topics such as goal setting, developing a business plan, managing a project, and budgeting resources such as time.

Make time saving a personal goal. Begin with these basic activities:

- Take time each day to review past work and to plan and prioritize tasks.
- Avoid indecision or procrastination, particularly on major projects.
- Prepare and use form letters and messages (see Chapter 28, "Writing Form Messages").
- Block out quiet time for tasks such as drafting major writing assignments.
- Prepare lists of (1) business partners and contacts; (2) office-equipment personnel, along with related warranties, receipts, contracts, and notes; (3) business-supply companies; (4) cleaning services; (5) plumbing or electrical services; (6) airline and car-rental services; (7) hotels and catering services; and (8) temporary-employment agencies and freelance employees.
- Avoid hoarding versus delegating tasks (see page 601).
- Keep an organized desk and office, and use orderly filing practices.
- Use technology to achieve both efficiency and quality.
- Periodically review the amount of time used in your workday to address personal (versus work-related) issues; develop strategies to correct the problem.

EVALUATING YOUR TIME-MANAGEMENT SKILLS

Once or twice a year, record your activities over a one-week period and evaluate both how and how well you manage time. To record your activities, make a table like the one described below and shown on the following page.

Prepare a Spreadsheet

Using your computer, create a table.

- At the top of the page, provide spaces for the date, your name, and job title.
- Label the far-left column, "Time"; below that, list your workday in 15-minute segments.
- Label the far-right column, "Other Tasks."
- Label each remaining column with the name of one of your common tasks.
- Make a copy of the spreadsheet for each workday in the week.

Fill in Details

Each morning, fill in the date, along with your name and title. Then two or three times that day, place check marks in the grid indicating which tasks you did when.

- Place a check in the time slot during which you do each task.
- If a task is listed multiple times under "Other Tasks," give it its own column.
- Design a legend with symbols that distinguish whatever additional information you want to record. For examples, see the legend on the next page.

Analyze the Data

After you have recorded a week of activity, evaluate the data by asking and answering questions related to your assessment goal. For example, if you want to learn how you can manage your time more efficiently, ask questions like these:

- Which tasks occur most often, and why?
- Which tasks took longer than expected? Why? What is the effect?
- Which time slots are busiest? Most productive? Least productive?
- When do most interruptions occur? What causes them? Why?
- What tasks could be rescheduled? Assigned to another person or department?
- When and how often were personal tasks done? Why?
- What tasks were initiated by people outside your department or organization?

Other Spreadsheet Applications

You can also design questions for these purposes:

- To show your need for additional staff, training, or equipment
- To identify tasks to be reassigned
- To document information supporting an evaluation, proposal, or promotion
- To generate information for a periodic report, grant, or news story

Chapter 37 Managing Your Time and Manners

Date: January 11, 2005
Name and Title: Kathy Missoula, Associate Designer

Time	Design Multimedia	Design Graphics	Design Interactive Media	Get Feedback on Designs	Refine Interactive Scripts	Prepare Program Files	Photograph Images	Plan/Review Schedule	Update Program Files	Respond to Writer's Drafts	Print Materials	Project Meetings	Production Meetings	Department Meetings	Other Tasks
8:00–8:15								√							
8:15–8:30	√														
8:30–8:45	√														
8:45–9:00				√											
9:00–9:15	√														
9:15–9:30								√							
9:30–9:45										√ RR					
9:45–10:00										√ RR					
10:00–10:15													√		
10:15–10:30													√		
10:30–10:45													√		
10:45–11:00													√		
11:00–11:15													√		
11:15–11:30													√		
11:30–11:45			√												
11:45–12:00			√												
12:00–12:15															PB
12:15–12:30															PB
12:30–12:45			√												
12:45–1:00					√										
1:00–1:15					√										
1:15–1:30												√			
1:30–1:45		√													
1:45–2:00		√													
2:00–2:15														√	
2:15–2:30														√	
2:30–2:45														√	
2:45–3:00														√	
3:00–3:15							√								
3:15–3:30							√								
3:30–3:45							√								
3:45–4:00				√											
4:00–4:15												√ ND			
4:15–4:30												√ ND			
4:30–4:45											√ ND				
4:45–5:00								√							

Legend

DI Department Interruption   **RR** Rush Request   **PB** Personal Business
OI Outside Interruption   **ND** Next-Day Delivery   **BP** Boss's Personal Business

PRACTICING WORKPLACE ETIQUETTE

Proper etiquette (on the job or in social settings) helps you develop confidence, earn others' respect, and build strong business relationships. Use the guidelines that follow.

Make Polite, Helpful Introductions

An effective introduction includes the following elements:

- The person's name and title stated clearly and pronounced correctly
- Appropriate details that help individuals initiate a conversation
- Correct order of names, when you are introducing others

Introducing Yourself

Smile, look in the person's eyes, extend your right hand, and do as follows:

1. Offer your greeting, name, and helpful details.

   Hello, I'm Jean Olms from Nova Industries, and I'm here for your presentation.

2. Shake hands firmly while looking at and listening to the person.
3. Think about his or her name, its pronunciation, and other details that he or she says. Then use the name and details in your follow-up conversation.

Being Introduced

If you are seated, stand, *smile*, shake hands firmly, and look and listen carefully.

1. Note the person's name, job, company, related details, and facial expression.
2. Use the name in your follow-up greeting and parting comment.
3. Use details from what you've seen or heard to develop the conversation; or ask a question on a topic like the occasion, sports, hobbies, or news items (avoid politics, religion, or personal issues in most settings).

Introducing Others

State each person's name clearly and in the correct order.

- Name the person of higher rank first, regardless of gender or age.

  President Barker, I'd like you to meet Hadassa Marquart, our new sales representative for Tampa—and our department's new computer whiz!

- If you're not sure of differences in rank, first name the person from outside your organization, group, or conversation.

  Dr. Ortiz, I'd like you to meet President Barker, another Notre Dame alumnus.

- If you're introducing someone from outside your organization to someone inside, name the client or guest first.

  Ms. Sandau, here is the person who founded our company, President Barker.

F.Y.I.

The standard custom in North America is a firm, two- or three-pump handshake.

Introducing a Speaker

Your goal when introducing a speaker is to establish a setting in which (1) the speaker feels welcome, and (2) the audience is prepared to listen to the message. Using no more than two minutes, proceed as follows.

1. State the speaker's key qualifications, topic, purpose, and basic theme.
2. Inspire interest in the presentation by highlighting the title, occasion, or key details in the content; telling some little-known fact about the speaker or topic; quoting from the speaker's writing; or noting how the speech relates to the group's goals, work, or welfare.
3. Deliver the information in a positive, engaging tone.
4. After the presentation, show your and the group's appreciation by briefly thanking the speaker, noting the speech's importance, and facilitating follow-up activities such as a question-and-answer session.

Speak and Listen Well

While all oral communication situations share some characteristics (e.g., someone speaks and someone listens), each situation is also distinct, requiring participants to learn relevant skills, protocol, and etiquette. Below is a list of common situations, along with page numbers where you will find helpful information on each topic.

- Conducting one-on-one conversations (see pages 648–649)
- Participating in small-group discussions (see pages 655–664)
- Sending and receiving messages (see page 649)
- Giving and taking instructions (see page 650)
- Preparing and delivering speeches (see pages 677–694)
- Preparing and delivering multimedia presentations (see pages 688–693)
- Participating in business meetings (see pages 665–676)

Develop a Positive Personality

The personality that you cultivate throughout your life is the core out of which you communicate. That core influences countless conscious and unconscious communication choices. Each choice (from word choice to sentence structure, from facial expression to gesture, and from verbal inflection to nonverbal sound) sends a message that's usually loud and clear. To control the messages, take time to refine your core personality:

- Seek to practice thoughtful, polite, and even-tempered behavior.
- Identify and change brooding, brusque, or hot-tempered behavior.
- When you succeed, congratulate yourself. When you fail, apologize, and keep trying!

F.Y.I.

While changing a personality trait is challenging, the task is doable. For help, check with your human resources department.

POLISHING YOUR ETIQUETTE

While social customs vary between cultures, the following guideline is universally appropriate: *Practice behavior that inspires your colleagues' trust, confidence, cooperation, and mutual respect.*

Using the Workplace

- Respect others' **workspaces** by keeping your own work area clean and orderly.
- Respect others' **preferences** by not using tobacco, strong perfume, cologne, or air fresheners.
- Respect others' **work atmosphere** by avoiding annoying habits such as loud speech, loud laughter, gum chewing, nose blowing, coughing, or belching.
- Respect others' **investments** by not using the employer's office space, equipment, supplies, or vehicles for your own personal business.

Using Time

- Respect others' **time** by being on time for all appointments, meetings, conference calls, or group work.
- Respect others' **responsibilities** by meeting all deadlines for assignments, requests for information, or special requests for time-sensitive chores.
- Respect others' **needs** by refraining from idle chatter, distracting humor, personal stories, or self-congratulatory anecdotes.
- Respect others' **plans** by informing individuals early when you cannot meet a deadline or are unable to complete a task.

Using Manners

- Respect others' **feelings** by expressing appropriate courtesies (please, thank you, excuse me, forgive me).
- Respect others' **goodwill** by sending appropriate and timely correspondence (thank you, congratulations, sympathy, best wishes, RSVP).
- Respect others' **peace of mind** by not using unnecessary technical jargon, fancy words, or unfamiliar language.
- Respect others' **contributions** by documenting sources used in your writing and by acknowledging sources used in your speech.
- Respect others' **self-respect** by not criticizing a person in the presence of others, divulging his or her personal information, or treating a personal file carelessly.
- Respect others' **trust** by being open, honest, and genuine (no talking down to another person, pulling rank, or relaying lies, half-truths, or gossip).
- Respect others' **dignity** by avoiding foul language, racist or sexist humor, demeaning gestures or innuendo, and prying personal questions.
- Respect others' **integrity** by not asking for favors, cover-ups, or rule bending.
- Respect others' **rights** by not sharing personal information and by shredding unneeded personal documents.

EATING AND DRINKING

Working lunches, postinterview dinners, midconference hors d'oeuvres, and guest receptions are just a few occasions when your job will require that you eat, drink, and do business all at the same time. For tips on how to proceed at meals, look below.

Guest Duties

Before the meal
- Arrive a few minutes early to get your bearings and visit the restroom.
- Watch the host for cues regarding when and where to sit.
- If the host chooses alcohol, you may do so, too—but drink *little* and *slowly!*
- When ordering food, choose entrées that you can eat with confidence and grace; order nothing more expensive than your host's choice.
- Place your napkin on your lap before consuming anything.

During the meal
- Use utensils working from the outside (salad fork) in.
- After all have been served, eat with modest bites and swallow before speaking.
- If others progress faster than you, ask questions and listen as you eat.
- Follow the host's lead regarding ordering dessert or after-dinner coffee.

After the meal
- Unless the host offers to pay for your meal, you should do so.
- Thank the host, reaffirm follow-up business contacts, say farewells, and leave.
- Promptly send the host a thank-you note or e-mail.

Host Duties

Before the meal
- Reserve space at a restaurant appropriate for the occasion and guests.
- Send invitations along with clarification of its purpose, RSVPs, arrival time, directions, meeting place, and a discreet indication regarding who will pay for the meal.
- Arrive early to confirm your space, number of guests, and payment plan.
- Greet guests, note and meet their special needs, request that your party be seated, show individuals where to sit, and introduce them to one another.

During the meal
- Nurture conversation on appropriate topics (the occasion, sports, hobbies, and news items are good; avoid politics, religion, or personal issues).
- If guests order dessert or coffee, do so, too.

After the meal
- Request the check and quietly pay it (plus at least a 15 percent gratuity).
- Thank guests, reaffirm follow-up business contacts, and say farewells.

CHECKLIST for Appropriate Dress

Use this checklist to select clothing for formal business settings.

General guidelines: My clothing is

- ☐ Clean, pressed, and well maintained.
- ☐ Practical, fits well, and makes me feel comfortable and confident.
- ☐ Within the range of styles worn by other employees, particularly supervisors.

Tips for men: My clothing includes

- ☐ Suits in dark colors (blues, browns, grays), muted patterns, and current styles.
- ☐ Shirts in white or conservative colors that match the suit. No short sleeves.
- ☐ Ties that complement (not match) the colors, textures, and patterns of the suit and shirt. The tie tip extends to the middle or bottom of the belt.
- ☐ Socks in solid colors or simple patterns. Colors match the trousers.
- ☐ Shoes that are attractive and well maintained. The color complements the trousers.
- ☐ Jewelry that is complementary and modest.

Tips for women: I select my apparel by

- ☐ Choosing solid, neutral colors rather than bold prints or trendy colors—choices that give me a classic look which I can spice up with accessories.
- ☐ Building a wardrobe around complementary pants, blouses, suits, and dresses that I can mix and match, creating a variety of choices on a limited budget.
- ☐ Avoiding clothes that are too tight, too short, or too low-cut—anything that may be uncomfortable or considered provocative.
- ☐ Wearing comfortable, well-maintained shoes with low heels and closed toes. (*Exception:* In some offices, casual-dress rules do allow sandals in the summer.)

CRITICAL THINKING Activity

Review the first two practices given in "Managing Your Time" (page 590). Use that information to identify (1) the time that you spend in class and (2) the time that you have available for out-of-class study. Assess your efficiency.

WRITING Activity

Using your preferred time-management tool, create your schedule for the next day, week, or month.

In this chapter

Managing Writing Tasks **600**
Delegating Work **601**
Solving Problems **602**
Sustaining a Supportive Work Climate **603**
Developing Successful Employees **604**
Dealing with Discrimination **607**
Checklist for Managing Effectively **608**
Critical Thinking Activities **608**
Writing Activities **608**

CHAPTER
38

Managing Effectively

Regardless of whether you're a CEO or an entry-level employee, doing a good job in your organization requires management skills. Those skills will help you effectively use your own resources (such as your skills and training) as well as your employer's resources (such as your time on the job, your workspace, or others' time).

While developing management skills is a lengthy process often requiring time spent on the job, you can start the process right now by reading this chapter. In it you will find helpful tips on issues including managing writing skills, delegating work, solving problems, maintaining a positive workplace, and developing your own skills as well as the skills of those with whom you work.

Later chapters will build on the lessons presented here. They will show how the broad concepts described in this chapter apply to the specific tasks of writing management-oriented documents such as mission statements, company profiles, job descriptions, employee evaluations, and instructions for procedures. After studying this material, you will be well on your way to developing the management skills that you will need in your career.

MANAGING WRITING TASKS

In a competitive work world, efficiency and effectiveness go hand-in-hand. Use the tips below to complete your reading and writing tasks well and on time.

Remember the 80–20 Rule

Eighty percent of your accomplishments will come from 20 percent of your efforts. Keep a work log to identify when you read and write most efficiently, and apply these practices to all your writing tasks.

Use This Handbook's Writing Shortcuts

- Writing process tips and strategies (see pages 13–44)
- List of guidelines and checklists (see inside back cover)
- Editing strategies (see pages 32–36)
- Writing form messages (pages 473–478)
- Writing report forms (pages 539–544)
- Table of contents and index

Use Your Organization's Writing Resources

- Engage colleagues in brainstorming ideas, evaluating arguments, or editing documents.
- Review company documents periodically, and discuss how to improve them.
- Bring in writing specialists, or send employees to workshops.
- Use templates for often-used documents.
- Communicate frequently requested information by sending a brief cover letter, along with a brochure, flyer, or other document that includes the information.
- Arrange your workspace so that you have easy access to writing tools and supplies.
- Periodically purge computer files, traditional files, and bookshelves of unused materials.
- Select software that helps you and your colleagues produce and share writing.

Follow Practices That Help You Work Efficiently

- Learn and use the writing process (see pages 13–44).
- Delegate writing tasks that can be done by others.
- Delegate lengthy reading tasks, and ask for written summaries.
- When appropriate, assign others to research, outline, and draft documents for you.
- Maintain to-do lists that schedule blocks of time for writing tasks.

DELEGATING WORK

Managing a workplace includes delegating work orally and in writing. Whichever method you use, do so in a manner that presents assignments clearly and treats employees fairly. The guidelines below will help you assess—and possibly revise—how you delegate work.

Assignment Issues

- Whether you give assignments orally or in writing, explain these aspects of the job:
    - What must be done
    - Why it is important
    - When the work must be finished
    - How the results should be recorded (e.g., oral or written report, work log, billing notation)
- Make assignment decisions based on criteria that you share with your staff, and keep written records of potentially sensitive or legal issues.
- Review job descriptions periodically so that assignments match the tasks listed.
- Revise job descriptions to conform with your assignment practices and your organization's procedures for such changes.
- Delegate work evenly so that employees can see that others share the workload.
- For tips on giving assignments, see "Giving and Taking Instructions," page 650.

People Concerns

- Know people's training, knowledge, skills, schedules, interests, and job descriptions.
- Delegate tasks to individuals whose skills, knowledge, and interests most closely match those required for the work.
- Avoid making assignment decisions based on favoritism, your self-interest, or intimidation from others.
- Request oral or written feedback from employees about their assignments and the assignment process.
- Offer appropriate praise or suggestions for improvement. (For help, see "Giving and Taking Criticism," page 651.)

WRITE Connection

To delegate work within an organization, use memos (see Chapter 22, "Writing Memos") and e-mail (see Chapter 23, "Writing E-mail Messages and Sending Faxes"). To delegate work to part-time, home-office employees, or subcontractors, you may also use letters (see Chapter 24, "Writing Letters").

SOLVING PROBLEMS

Whatever your job, solving problems is an important task. Use the steps below to save time and develop strong solutions.

1. Define the problem in writing by answering these questions:
   - Is there a problem? Who says so? What is the evidence? Is it reliable?
   - Has the problem occurred before? When? Where? How was it addressed? With what success?
   - Who and what are affected by the problem? How?
   - How can information about the problem be gathered and possible solutions be found?
   - Which organizational documents (e.g., mission statement, policies, procedures) can guide the work?
   - Which legal issues are relevant, and why?
   - Which resources (time, money, materials) are available and warrant use?
2. Identify possible solutions by answering these questions:
   - Which solutions were successful in the past, and why?
   - How could they be improved?
   - Which solution is easiest? Most economical? Longest-lasting?
   - Which new approach is worth trying?
   - Which technology could improve each solution?
3. Choose the best solution.
   - Reject solutions that conflict with organizational guidelines.
   - List remaining solutions from strongest to weakest.
4. Implement the best solution.
   - Write out a timetable to implement and test the solution.
   - Commit needed resources (time, people, materials, and money).
   - Keep all relevant parties informed, and solicit regular feedback.
   - Record details needed for the final report.
5. Measure, evaluate, and record results.
   - Identify the steps taken, and evaluate the quality of the work done.
   - Note how well the solution succeeded, and how it could be improved.
   - Assess how the lessons learned could help solve other problems.
   - Write a report, share it, and file it.

WRITE Connection

To recommend a solution or problem-solving procedure, write a proposal (see Chapter 32, "Writing Proposals").

SUSTAINING A SUPPORTIVE WORK CLIMATE

Workplace conflicts can hurt an organization by poisoning employees' relationships, impeding efficiency, adding production costs, and damaging customer relations. While a supportive work climate cannot preempt all conflicts (see pages 652–653), it can help employees solve most problems before they become conflicts. Improve your work climate by using the two lists below (1) to prepare a survey (see pages 134–135) that gathers information about a problem, (2) to guide a group discussion (see Chapter 42, "Communicating in a Group") that educates employees, or (3) to write a proposal (see Chapter 32, "Writing Proposals") that calls for change.

Supportive climate	Nonsupportive climate
▪ Encourages understanding and tolerance of religious and cultural differences	▪ Tolerates prejudiced comments and behavior
▪ Encourages respect for both sexes	▪ Tolerates sexism
▪ Values people with disabilities	▪ Demeans people with disabilities
▪ Promotes respect for all individuals	▪ Bases respect on rank
▪ Encourages employees to share ideas without the fear of being wrong	▪ Discourages ideas that challenge the status quo or call for change
▪ Welcomes experimentation and sensible risk taking	▪ Stifles creativity, fears failure, and discourages innovation
▪ Urges sound, researched judgment	▪ Settles for weak research and reasoning
▪ Offers training programs and supports skill development	▪ Expects employees to find and pay for their own training
▪ Looks for decisions based on the group's welfare	▪ Encourages self-promotion and self-interest
▪ Treats employees as valued individuals	▪ Treats employees as commodities
▪ Invites honesty, trust, and mutual support	▪ Tolerates dishonesty if it leads to winning
▪ Encourages healthy cooperation	▪ Permits destructive competition
▪ Promotes accurate, clear, and fair communication	▪ Relies on manipulative, misleading, and dishonest communication

F.Y.I.

Any strong relationship between two people (e.g., husband/wife, sister/brother, or friend/friend) succeeds only when both parties invest the effort needed to sustain that relationship. Sustaining a supportive work climate requires a similarly steady investment from both employers and employees.

DEVELOPING SUCCESSFUL EMPLOYEES

According to the U.S. Labor Department's *Learning a Living: SCANS Report,* your organization needs employees who have—or develop—both the foundational skills outlined below and the five competencies described on the next page. The skills help workers achieve the competencies. For example, to make good use of resources such as "time" (the first *competency*), a person must design a work schedule. However, to design the schedule, he or she needs these foundational skills: (1) basic skills in *mathematics,* (2) thinking skills in *problem solving,* and (3) personal qualities such as *self-management.*

Foundational Skills

Basic Skills

Reads, writes, performs mathematical operations, listens, and speaks.

- **Reading:** Locates, understands, and interprets written information in prose and in documents such as manuals, graphs, and schedules.
- **Writing:** Communicates ideas and information in writing; creates documents such as letters, directions, manuals, reports, graphs, and flowcharts.
- **Arithmetic/mathematics:** Performs basic computations and solves practical problems by using appropriate mathematical techniques.
- **Listening:** Interprets and responds to verbal and nonverbal messages.
- **Speaking:** Organizes ideas and communicates orally.

Thinking Skills

Thinks creatively and logically, makes decisions, solves problems, and learns well.

- **Creative thinking:** Generates new ideas.
- **Decision making:** Specifies goals and constraints, generates alternatives, considers risks, and evaluates and chooses the best alternative.
- **Problem solving:** Recognizes problems; devises and implements solutions.
- **Seeing things in the mind's eye:** Organizes and processes symbols, pictures, graphs, objects, and other information.
- **Knowing how to learn:** Uses efficient learning techniques to acquire and apply knowledge and skills.
- **Reasoning:** Discovers a rule or principle underlying the relationship between two or more objects and applies the rule.

Personal Qualities

Displays responsibility, self-esteem, sociability, self-management, integrity, and honesty.

- **Responsibility:** Exerts a strong, committed effort toward meaningful goals.
- **Self-esteem:** Demonstrates understanding, empathy, and politeness.
- **Self-management:** Sets goals, monitors progress, and exhibits self-control.
- **Integrity/honesty:** Chooses and pursues ethical courses of action.

Five Competencies

The *SCANS Report* argues that success in business requires *competence* in five basic activities: using resources, using interpersonal skills, using information, using systems, and using technology. Doing each activity requires a combination of the *basic skills* discussed on the previous page.

- **Resources:** Identifies, organizes, plans, and allocates resources.
  - **Time:** Selects and ranks goal-related activities; prepares and follows schedules.
  - **Money:** Uses or prepares budgets, makes forecasts, keeps records, and makes adjustments to meet objectives.
  - **Material and facilities:** Acquires and uses materials and space efficiently.
  - **Human resources:** Assesses skills and distributes work accordingly; evaluates performance and provides feedback.
- **Interpersonal:** Works with others.
  - **Participates as a member of a team:** Contributes to the group effort.
  - **Teaches others:** Shares knowledge and skills.
  - **Serves clients/customers:** Works to satisfy customers' expectations.
  - **Exercises leadership:** Effectively analyzes issues and engages others' support.
  - **Negotiates:** Works toward agreements; resolves divergent interests.
  - **Works with diversity:** Works well with people from diverse backgrounds.
- **Information:** Acquires and uses information.
  - Acquires and evaluates information.
  - Organizes and maintains information.
  - Interprets and communicates information.
  - Uses computers to process information.
- **Systems:** Understands complex interrelationships.
  - **Understands systems:** Effectively uses social, organizational, and technical systems.
  - **Monitors and corrects performance:** Distinguishes trends, predicts effects on systems operations, and diagnoses and corrects problems.
  - **Recommends or designs systems:** Suggests changes to existing systems and develops new or alternative systems to improve performance.
- **Technology:** Works with a variety of technologies.
  - **Selects technology:** Chooses appropriate procedures, tools, or equipment.
  - **Applies technology to task:** Understands and uses technical equipment.
  - **Maintains and troubleshoots equipment:** Prevents or solves technical problems.

Using SCANS Skills and Competencies

Whether you are an employer or an employee, the SCANS lists on pages 604–605 can help you achieve your staff-development goals.

Employer (Management or Human Resources)

Use the lists to evaluate and develop human resources strategies, programs, and policies.

- Study the *SCANS Reports* (or similar materials) to learn what the SCANS qualities are and how they can enhance your organization's effectiveness.
- Study your mission statement to learn whether and how these qualities are addressed in your organization's purpose, mission, and goals.
- Study your employee handbook, job descriptions, and related policies and procedures to assess what you expect from employees regarding these qualities.
- Survey employees regarding how and why the SCANS qualities would strengthen their divisions, departments, or work teams.
- Examine your staff-development programs to identify where, how, and how well you test for and develop the SCANS qualities.
- Examine your employee-evaluation forms to learn how you assess performance.
- Examine your criteria for bonuses and promotions to see how you promote the SCANS qualities and reward employees who exemplify them.
- Write about these qualities in the employee newsletter, showing their relevance for the success of individuals, teams, departments, and the entire organization.
- Discuss the SCANS qualities with employees; gain consensus regarding programs and policies that strengthen performance throughout your organization.

Employees

Use the lists to improve your own effectiveness and that of your team or department.

- Learn how the SCANS qualities are relevant for you. Note how they are addressed in your employee handbook, job description, or self-assessment report.
- Use what you found in the activity above to revise your goals, or to request training, materials, or equipment.
- Cite skills shown in your work that warrant a raise, promotion, new assignment, or sabbatical.
- Discuss the SCANS qualities with your team or department, analyzing how they would improve the group's performance on specific projects or activities.
- Set goals for the group regarding how you will improve and assess your progress.
- Use your findings in the preceding activities to write a proposal for training, equipment, or a group retreat.

DEALING WITH DISCRIMINATION

Discrimination involves observing differences between people or things, and making judgments based on those observations. When it produces an informed, fair decision, discrimination is constructive. However, when discrimination results in people being treated unfairly, it is both destructive and illegal. For example, Title VII of the 1964 Civil Rights Act prohibits discrimination that is based on the following:

- National origin
- Race
- Skin color
- Age
- Gender
- Religion
- Mental disability
- Physical disability
- Pregnancy

Addressing Discrimination

Employers should do the following (seek legal advice as necessary):

1. Write a policy (see pages 614–615) stating the organization's definition of and response to discrimination.
2. Write procedures (see pages 618–619) for the following:
   - Steps to be taken by discrimination victims
   - Steps that supervisors should take in response to an incident
3. Publish the information in an employee handbook.
4. Enforce the policy.

Employees who are discriminated against should do the following:

1. Check the organization's policy to confirm that the incident is discriminatory. Then follow the related procedure.
2. If the organization lacks a policy, report the problem to the personnel manager.
3. If he or she does not help, explain the problem to the personnel manager's supervisor.
4. If these steps do not resolve the problem, file a complaint with the Equal Employment Opportunity Commission (EEOC).
5. Throughout the process, keep written notes that objectively record the following information:
   - The discriminatory incident or incidents, including precisely what the accused party said and did, what you said and did, dates, places, times, and people present
   - Things said and done in response to the problem
   - Action taken by those to whom you reported the problem

F.Y.I.

Employers and employees have both the personal and communal responsibility to correct discriminatory practices; such behavior hurts the victim, the perpetrator, and the organization.

CHECKLIST for Managing Effectively

Use this checklist to assess how well you handle management tasks.

- ☐ **Writing Tasks.** I use the writing process, along with appropriate writing tools such as guidelines, models, checklists, templates, and online writing resources.
- ☐ **Delegating Work.** I assign tasks courteously and clearly, asking others to do work that is within the range of their job descriptions, skills, experiences, and time limits.
- ☐ **Solving Problems.** I address problems effectively by carefully analyzing the causes, implementing effective solutions, and testing the results.
- ☐ **Sustaining a Supportive Work Climate.** I promote behavior, policies, and programs that make the workplace a positive, supportive environment.
- ☐ **Developing Employees.** I encourage employees to measure their performances, understand how to improve, and develop the necessary knowledge and skills.
- ☐ **Developing My Skills and Competencies.** I periodically measure and refine my own skills and competencies.
- ☐ **Dealing with Discrimination.** I promote behavior and policies that value all people equally, regardless of nationality, race, age, gender, religion, or ability.

CRITICAL THINKING Activities

1. Use "Managing Writing Tasks" (page 600) to assess how well you manage your writing. Summarize your ideas and identify strategies for improvement.
2. Review "Delegating Work" (page 601), "Solving Problems" (page 602), and "Sustaining a Supportive Work Climate" (page 603). Explain why the ideas on these pages are or are not relevant for work in your career.
3. Review "Developing Successful Employees" (pages 604–606). Explain which competencies and skills you either have developed or need to develop.

WRITING Activities

1. Use what you learned in Critical Thinking activity 3 to evaluate (and possibly revise) your career plan (pages 312–314).
2. Write an incident report (see pages 486–489) in which you (a) describe a workplace experience in which you saw someone discriminated against, and (b) recommend strategies that would have helped the victim address the problem. For ideas, review "Dealing with Discrimination" (page 607).

In this chapter

Guidelines for Mission Statements **610**
Guidelines for Position Statements **612**
Guidelines for Policy Statements **614**
Guidelines for Procedures **618**
Guidelines for a Company Profile (or Fact Sheet) **620**
Checklist for Management Writing **624**
Critical Thinking Activities **624**
Writing Activities **624**

CHAPTER
39

Management Writing

To be productive, an organization needs a vision regarding what it wants to be as well as clear strategies for achieving that vision. Management writing helps the organization articulate both a vision and strategies.

For example, a mission statement is a visionary document describing the organization's overall purpose or identity. The organization uses the document to clarify what the company is and what it wants to do—such as be a social-service agency that treats clients respectfully. Then, based on broad concepts in its mission statement, the organization writes other forms of management writing—like a confidentiality policy—to explain how employees should address a specific issue, such as protecting a client's confidential information.

This chapter will help you write both types of documents—the broad visionary statements as well as the specific applications of that vision. In addition, the chapter will help you understand how the various application documents (position statements, policy statements, procedures, and company profiles) are related.

GUIDELINES FOR MISSION STATEMENTS

Your goal when writing a mission statement (also called a corporate plan, statement of purpose, or vision and goals document) is to spell out what the organization is and what it wants to do. The organization then uses the document as a guide for tasks such as developing policies (see pages 614–617) and procedures (see pages 618–619).

> Good writing is concise. A sentence should contain no unnecessary words, a paragraph no unnecessary sentences, for the same reason a drawing should have no unnecessary lines and a machine no unnecessary parts.
> —William Strunk

1 Prewrite

Rhetorical analysis: Consider your purpose, your audience, and the context.
- ☐ What are your organization's purpose, goals, and key practices?
- ☐ How can you compress that information into a clear, tidy, usable statement?
- ☐ Who will read the statement and in what situations?

Prepare to draft.
- ☐ Review documents that present the organization's philosophy, goals, practices, strengths, weaknesses, annual sales and expenditures, and scope of operations.
- ☐ Interview top management about the organization's function and mission.

2 Draft

Organize the message into three parts.

[opening] Give the organization's name, and state briefly what the organization is.

[middle] In complete, unified sentences, explain
- ☐ Why the company exists,
- ☐ What it wants to do, and
- ☐ How it plans to do it.

[closing] Conclude by briefly summarizing or restating the company's vision.

   Note: Divide a longer mission statement (one or two pages) into sections using appropriate headings like *Purpose, Values, Principles,* and *Company Strategy*.

3 Revise

With help from management, check the ideas, organization, and voice.
- ☐ Does the writing clearly state the organization's goals, products, or services?
- ☐ Are paragraphs complete, linked, and logically organized?
- ☐ Is the voice consistent, confident, and goal oriented?

4 Refine

Check your writing line by line for the following:
- ☐ Smooth sentences free of clichés, generalizations, and jargon
- ☐ Correct names, dates, titles, grammar, mechanics, and spelling
- ☐ Clear headings, proper formatting, and attractive graphics

Mission Statements

Rhetorical situation

Purpose: To inform readers that Abix Technologies is an engineering company that provides technological solutions to environmental problems
Readers: Board of directors, shareholders, customers, employees, and the public
Format: Brief paragraph (can be printed by itself or as part of other documents)

Abix Technologies' Mission

Abix Technologies, Inc., is an innovative international company that provides technological solutions to environmental problems. Our primary commitment is to develop UV-based technologies for environmental solutions through focused research, innovative engineering, and quality manufacturing. We seek to market these long-term solutions globally to communities and industries. We foster growth through the development of new products, synergistic technologies, and strategic business alliances. Our goal is to provide excellent value to our customers, rewarding career opportunities for our employees, and a superior, long-term return on investment for our shareholders.

[opening] Name the organization and describe it briefly.

[middle] Explain what the organization does, what it plans to do, and how.

[closing] Restate the vision.

Rhetorical situation

Purpose: To inform readers that Helping Hands is a nonprofit, social-service organization dedicated to eliminating poverty housing worldwide
Readers: Board of directors, donors, clients, customers, employees, and the public
Format: Brief paragraph (can be printed by itself or as part of other documents)

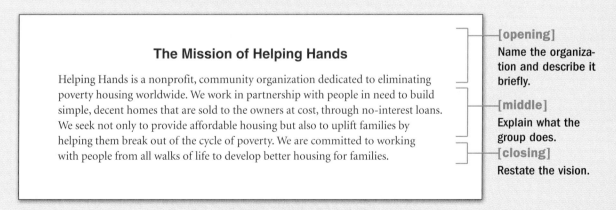

The Mission of Helping Hands

Helping Hands is a nonprofit, community organization dedicated to eliminating poverty housing worldwide. We work in partnership with people in need to build simple, decent homes that are sold to the owners at cost, through no-interest loans. We seek not only to provide affordable housing but also to uplift families by helping them break out of the cycle of poverty. We are committed to working with people from all walks of life to develop better housing for families.

[opening] Name the organization and describe it briefly.

[middle] Explain what the group does.

[closing] Restate the vision.

F.Y.I.
Mission statements are often included in documents such as annual reports, press releases, client proposals, and brochures.

GUIDELINES FOR POSITION STATEMENTS

Your goal when writing a position statement (or "white paper") is to explain your organization's stand on a specific issue or problem. The writing may take the form of a letter, memo, or report.

1 Prewrite

Rhetorical analysis: Consider your purpose, your audience, and the context.
- What is the topic, and how is it related to your organization's purpose, goals, community, products, services, and clients?
- Who are your readers, and how are they affected by the topic? What do they know about it, and what should they know?

Prepare to draft.
- Research the topic thoroughly, particularly its history, legal or ethical issues, and relationship to your organization.
- Review documents and interview people (including legal staff) to understand your organization's past experience with—or position on—the topic.
- Research how organizations like yours deal with the topic.

2 Draft

Organize the message into three parts.

[opening] Introduce the topic, explain its relevance to your organization, and describe what will follow.

[middle] Spell out your position in a well-organized argument that has a logical main point, clear supporting points, and relevant supporting details.

[closing] Conclude by briefly summarizing or restating your position.

In most cases, use direct organization by stating your position early on. For controversial or sensitive topics, consider using indirect organization.

3 Revise

Check your draft for ideas, organization, and voice.
- Are related documents (e.g., mission statements, policies, and procedures) clearly cited?
- Is the argument clear, logically organized, and well supported?
- Are main points in distinct paragraphs linked with clear transitions?
- Is the tone objective, respectful, and forthright?

4 Refine

Check your writing line by line for the following:
- Smooth sentences and word choice free of jargon or clichés
- Correct names, titles, quotations, and data
- Correct spelling, grammar, punctuation, and usage
- Clear headings, proper formatting, and attractive layout

> "Read over your compositions and, when you meet a passage which you think is particularly fine, strike it out."
> —Samuel Johnson

Position Statement

Rhetorical situation

Purpose: To explain why the company supports building a recreation center
Audience: Employees, community leaders, and the public
Format: Business letter (full block and printed on company stationery)

Kendal Laboratories, Inc.

1206 Stevens Road
Burnside, Minnesota 51342
Phone: 612-555-3000
Fax: 612-555-3248

August 5, 2005

Mr. Theodore Snappers, Mayor
Burnside City Offices
310 Main Street NW
Burnside, MN 51342

Dear Mayor Snappers:

On behalf of the Management Team at Kendal Laboratories, I am writing to support the City Council's proposal to build a $9,500,000 Sports and Recreation Center. This complex, which will include an indoor pool, outdoor pool, and indoor ice arena, will give our community a facility that it needs.

Twelve years ago, Kendal Laboratories moved to Burnside because of the city's fine assets such as strong schools, a good college, a fine hospital, bike trails, and Burnside Lake. These assets enhance the quality of life in Burnside, thereby helping Kendal Laboratories recruit the specialists that it needs to compete in the pharmaceutical industry. The new Center will strengthen that quality of life.

During the next five years, Kendal Laboratories will hire an additional 120 employees. The jobs we create pay well, add money to the local economy, and contribute substantial taxes to the city and state. The new Sports and Recreation Center will help us fill those jobs, benefiting both Burnside and Minnesota.

Sincerely yours,

Dr. Gabriel Benjamin

Dr. Gabriel Benjamin, D.V.M.
Director of Operations

cc: Nancy Zachary (Minnesota Development Council)

[opening]
Introduce the topic and state your position.

[middle]
Give details showing that your position is logical, fair, and supported by facts.

[closing]
Restate your position, along with additional support.

Close politely.

GUIDELINES FOR POLICY STATEMENTS

Your goal when writing a policy statement is to explain your organization's (1) understanding of an issue and (2) method of dealing with it.

1 Prewrite

Rhetorical analysis: Consider your purpose, your audience, and the context.
- ☐ What is the issue (including its history), and how does it relate to concepts in the mission statement?
- ☐ What is the organization's definition of the issue and method of addressing it?
- ☐ What procedural guidelines (see pages 618–619) will carry out the policy?

Prepare to draft.
- ☐ Review the mission statement, related policy statements, minutes, the employee handbook, and other relevant documents.
- ☐ Interview people (management, legal council, and employees) knowledgeable about the topic or related issues.
- ☐ Study other organizations' policies on this or similar topics.

2 Draft

Organize the message into three parts.

[opening] Introduce the policy by giving the following information:
- ☐ Its name and main idea
- ☐ Reasons why the policy was adopted

[middle] Explain what the policy means and how it will be implemented:
- ☐ What employees must do, including when, where, and how
- ☐ What the organization will do, including when, where, and how

[closing] List documents used as references, along with the dates the policy was adopted or revised

3 Revise

With help from management (and possibly an attorney), check the ideas, organization, and voice.
- ☐ Is the policy correctly cited, clearly explained, and sufficiently supported?
- ☐ Is the organization clear and logical?
- ☐ Is the voice informed, reasoned, fair, and concerned?

4 Refine

Check your writing line by line for the following:
- ☐ Correct names, titles, and dates; spelling, mechanics, usage, and grammar
- ☐ Formatting choices (e.g., headings, numbering of sections, white space) that provide easy access and use

> "Managers don't have problems; managers solve problems."
> —Bernard Schulman

F.Y.I.
A procedure gives specific instructions on how to implement a policy.

Policy Statement on Quality Control

Rhetorical situation

Purpose: To spell out the company's "Quality Policy" clearly and succinctly
Audience: Management, employees, customers, investors, and the public
Format: Brief statement (single spacing between lines; double spacing between paragraphs)

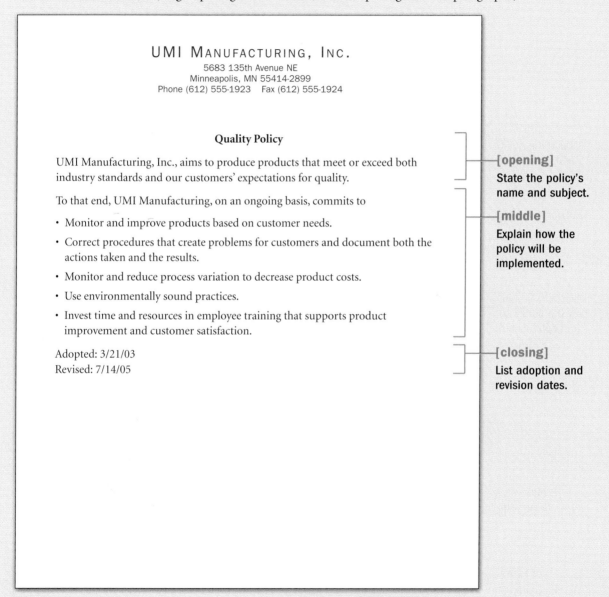

UMI MANUFACTURING, INC.
5683 135th Avenue NE
Minneapolis, MN 55414-2899
Phone (612) 555-1923   Fax (612) 555-1924

Quality Policy

UMI Manufacturing, Inc., aims to produce products that meet or exceed both industry standards and our customers' expectations for quality.

To that end, UMI Manufacturing, on an ongoing basis, commits to

- Monitor and improve products based on customer needs.
- Correct procedures that create problems for customers and document both the actions taken and the results.
- Monitor and reduce process variation to decrease product costs.
- Use environmentally sound practices.
- Invest time and resources in employee training that supports product improvement and customer satisfaction.

Adopted: 3/21/03
Revised: 7/14/05

[opening] State the policy's name and subject.

[middle] Explain how the policy will be implemented.

[closing] List adoption and revision dates.

F.Y.I.

The policy above (1) defines what the company means by "quality" and (2) asserts the organization's commitment to achieving quality. The company then writes quality-control procedures to spell out how it implements this policy.

Policy Statement on Reimbursing Expenses

Rhetorical situation

Purpose: To explain the university's health policy and how it functions
Audience: Administration, board of directors, and employees
Format: Statement (organized in question-and-answer format)

[opening]
Name the policy and describe it.

[middle]
Explain the policy.

Give details about how the policy will be implemented.

Explain what readers must do.

[closing]
List adoption and revision dates.

Middleburg University's Dental/Optometry/Audiometry Reimbursement Policy

What is this policy?
The Dental/Optometry/Audiometry Reimbursement (DOAR) policy is a Middleburg University plan that reimburses employees for dental, visual, and hearing-related expenses not covered by the medical coverage.

Who is covered by the DOAR policy?
The policy covers full-time employees, their spouses, and dependents. A dependent is a person who is

- Single,
- Younger than 18 years of age or a full-time student, and
- Receiving 50 percent of his or her financial support from the employee.

What expenses does the policy cover?
Each benefit year, the policy reimburses an employee up to $750 on $1,000 worth of paid expenses.

- The first $500 is reimbursed dollar-for-dollar.
- Expenses from $501 to $1,000 are reimbursed at 50 percent.
- Expenses over $1,000 are not reimbursed.

What is the benefit year?
The benefit year runs from August 1 to July 31.

- Unused funds do not accumulate from year to year.
- Expenses submitted by the 15th of each month will be paid by the 20th of that month.

How does an employee receive reimbursement?
The employee submits the following items to the Business Office:

- A DOAR form describing the expense, and
- A receipt indicating that the amount has been paid.

Adopted: August 6, 2003
Revised: September 12, 2005

Chapter 39 Management Writing

Policy Statement on Confidentiality
Rhetorical situation

Purpose: To define a confidential staff–client relationship and explain how the organization will promote such relationships
Audience: Board of directors, employees, and clients
Format: Statement (single-spaced lines, numbered items, concluding signature and dates)

Family and Children Coalition
384 Nickler Street, Queenstown, Maryland 21658 PHONE 301-555-6271
contact@aspenstatebank.com

Confidentiality Policy

The clients of Family and Children Coalition, Inc. (FCC), have the right to confidentiality. Any conversations between the client and the staff will be treated with respect and will not be communicated to others without the consent of the client, except in the case of a life-threatening emergency. To protect client confidentiality, FCC requires the following:

1. No information will be provided to a third party as a general release of information (a release that does not specifically name the client).
2. The written consent of the client or legal guardian is required for all disclosures of personally identifying information concerning clients.
3. Clients will be informed that if they choose not to release such information, FCC will still provide all services that it has offered the clients.
4. The Consent to Release form will specify to whom the information is released and what information will be shared.
5. To be valid, all releases of information must be dated and signed by the client or legal guardian and the FCC counselor.
6. The original consent form is to remain in the client's clinical file with a copy to accompany the disclosed information.

I understand the Family and Children Coalition Confidentiality Policy and will comply with it.

_____   _____
    Staff Signature   Date

Policy approved: April 5, 2005

[opening]
Give the policy's name and main idea.

[middle]
Explain what the policy means.

Explain how the policy affects employees' work.

[closing]
Give compliance information, staff's signature and date, and the policy's approval/revision date.

GUIDELINES FOR PROCEDURES

Your goal is to instruct readers (both employees and management) how to carry out an organization's policy (pages 614–617). A clear procedure makes a policy workable.

1 Prewrite
Rhetorical analysis: Consider your purpose, your audience, and the context.
- ☐ What must readers know to implement the policy?
- ☐ What can they expect others to know and do?

Prepare to write.
- ☐ Study relevant documents such as related policies and procedures, as well as the minutes of the committee that adopted the policy.
- ☐ Interview people familiar with the issues addressed in the policies.
- ☐ Identify each person who will carry out the policy.
- ☐ List each person's tasks and due dates.

2 Draft
Organize the message into three parts.

[opening] Introduce the procedure.
- ☐ Identify the policy that the procedure implements.
- ☐ Briefly state or summarize the policy.

[middle] In clear, step-by-step instructions, explain what must be done.
- ☐ Spell out which forms, reports, or proposals to submit, which people to contact, and which tasks to complete.
- ☐ Cite who must do each activity and by when.

[closing] State the outcome of the procedure, including any follow-up action that the reader must perform.

3 Revise
Review your writing for ideas, organization, and voice.
- ☐ Does the procedure accurately implement the policy?
- ☐ Are safety, security, and legal issues correctly addressed?
- ☐ Are steps, assignments, and due dates clear and correct?

4 Refine
Check your writing line by line for the following:
- ☐ Correct names, titles, due dates, and procedural terms
- ☐ Action verbs, numbered steps, and attractive format
- ☐ Correct spelling, punctuation, and grammar

F.Y.I.
Because a procedure implements a policy, write or revise both documents at the same time.

Instructions for a Policy Procedure

Rhetorical situation

Purpose: To explain how to implement the confidentiality policy on page 617
Audience: Board of directors, employees, and clients
Format: Statement with a clear title, section headings, and numbered items

---

Family and Children Coalition Confidentiality Procedures

The Family and Children Coalition Confidentiality Policy states that all clients have the right to confidentiality. Conduct your work in keeping with this policy by following the procedures below.

[opening] Identify the organization and topic; summarize the policy.

Client Intakes

During the client intake, the FCC counselor should discuss the conditions of confidentiality with the client. These conditions include the following:

1. Information will never be shared unless the client has given written permission using the Consent to Release Form.
2. Confidentiality may be limited or canceled if FCC staff have serious concerns about child abuse or neglect, or if the client is a danger to herself or himself, or to others.

[middle] Use clear headings throughout.

Outside Requests for Information

FCC counselors will handle outside requests for client information.

1. No client information will be shared without the client's written permission.
2. Clients will be notified of any outside requests for information. If the client gives permission, he or she must sign the Consent to Release Form and specifically indicate what information may be released and to whom. An FCC counselor must also sign the release form.

List steps clearly, including which documents to complete and submit.

Breaches of Confidentiality

If a client believes that FCC staff have not observed the confidentiality policy and procedures, the client should be directed to follow the Client-Grievance Process.

By carefully following the procedures above, the FCC staff can help clients while also respecting their rights to confidentiality.

[closing] Close with a fitting summary or restatement.

GUIDELINES FOR A COMPANY PROFILE (OR FACT SHEET)

When writing a company profile, your goal is to describe what the organization is and what it does. Note that a profile is rarely used by itself; rather, it is inserted into other documents (such as an annual report, news release, or catalog of products) to support their messages.

> "You can be a little ungrammatical if you come from the right part of the country."
> —Robert Frost

1 Prewrite

Rhetorical analysis: Consider your purpose, your audience, and the context.
- ☐ What or who are your organization's purpose, products, services, customers, facilities, mission, and long-term goals?
- ☐ What format would feature details and serve as an insert into other documents?

Prepare to draft.
- ☐ Gather information by reviewing documents such as mission statements, annual reports, sales reports, and catalogs of products or services.
- ☐ Review details (locations, contact information, staff, and focus of work) that distinguish each company facility or office.
- ☐ Find appropriate graphics (e.g., logo, pictures, or line drawings).

2 Draft

Organize the message into three parts.

[opening] Briefly state the organization's name, location, and function—what it does.

[middle] Describe the organization's basic goals, products or services, offices and plants, personnel, major projects, marketing practices, and awards.
- ☐ In a short document, address key subjects only and present the message in a list or in a few brief paragraphs.
- ☐ In a longer document, give a more detailed picture and organize the message in separate sections with clear headings.
- ☐ In any profile, organize subjects from broader to narrower or from more important to less important.

[closing] Conclude by giving the organization's contact information.

3 Revise

Check your draft for ideas, organization, and voice.
- ☐ Is the business (including all offices) described clearly and in sufficient detail?
- ☐ Are paragraphs unified, organized logically, and linked by clear transitions?
- ☐ Is the voice engaging but objective, informed but not inflated?

4 Refine

Check your writing and page layout for the following:
- ☐ Correct names, titles, numbers, spelling, punctuation, and grammar
- ☐ Consistent headings, correct graphics (e.g., logo, graphs, pictures), and attractive layout

Company Profile (Brief)

Rhetorical situation

Purpose: To present basic information about what the organization is and does
Audience: Job applicants, company personnel, investors, customers, and the public
Format: Statement (heading and title followed by succinct paragraphs and contact information)

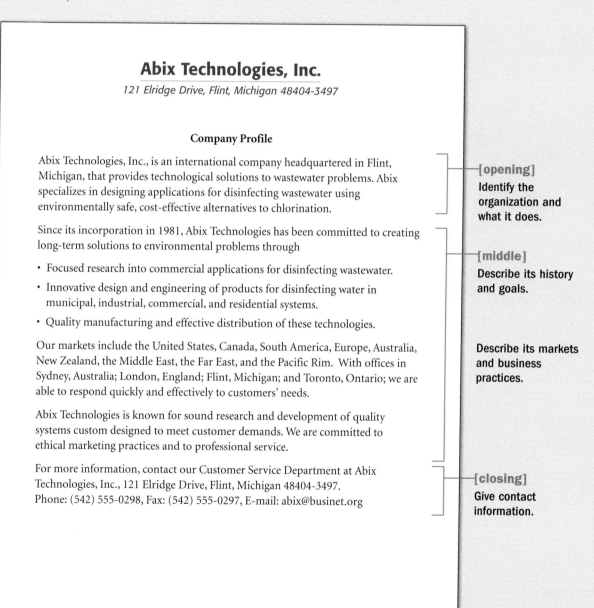

Abix Technologies, Inc.
121 Elridge Drive, Flint, Michigan 48404-3497

Company Profile

Abix Technologies, Inc., is an international company headquartered in Flint, Michigan, that provides technological solutions to wastewater problems. Abix specializes in designing applications for disinfecting wastewater using environmentally safe, cost-effective alternatives to chlorination.

— [opening] Identify the organization and what it does.

Since its incorporation in 1981, Abix Technologies has been committed to creating long-term solutions to environmental problems through

- Focused research into commercial applications for disinfecting wastewater.
- Innovative design and engineering of products for disinfecting water in municipal, industrial, commercial, and residential systems.
- Quality manufacturing and effective distribution of these technologies.

— [middle] Describe its history and goals.

Our markets include the United States, Canada, South America, Europe, Australia, New Zealand, the Middle East, the Far East, and the Pacific Rim. With offices in Sydney, Australia; London, England; Flint, Michigan; and Toronto, Ontario; we are able to respond quickly and effectively to customers' needs.

Describe its markets and business practices.

Abix Technologies is known for sound research and development of quality systems custom designed to meet customer demands. We are committed to ethical marketing practices and to professional service.

For more information, contact our Customer Service Department at Abix Technologies, Inc., 121 Elridge Drive, Flint, Michigan 48404-3497.
Phone: (542) 555-0298, Fax: (542) 555-0297, E-mail: abix@businet.org

— [closing] Give contact information.

Long Company Profile (Fact Sheet)

Rhetorical situation

Purpose: To present detailed information about what the organization is and does
Audience: Job applicants, company personnel, investors, customers, and the public
Format: Statement (heading, title, section heads, succinct paragraphs, and contact information)

---

1 of 2

Abix Technologies, Inc.
121 Elridge Drive, Flint, Michigan 48404-3497

Company Profile

[opening] — *Give the organization's name, location, and function.*

Abix Technologies, Inc., is an international company headquartered in Flint, Michigan, that provides technological solutions to environmental wastewater problems. Abix provides lasting solutions through research, product development, marketing, manufacturing, and service.

[middle] — *Describe key products and services.*

Research and Development

Our team of chemical engineers and microbiologists has developed a comprehensive range of quality disinfection systems, from two-gallon-per-minute residential applications to municipal systems that treat millions of gallons per day. Our experts are continually researching new applications, developing more efficient and advanced systems, and providing in-house consultations and test services for our clients.

Marketing

Describe the company's scope and influence.

Our professional network of more than 90 representatives in the United States, Canada, Europe, South America, the Middle East, the Far East, and the Pacific Rim has been thoroughly trained in the sale and service of our products and responds quickly and effectively to customers' needs.

Abix shares its expertise with regulatory agencies and other professionals through published articles in trade journals. In addition, factory seminars and visits to operational Abix systems are arranged for potential customers.

Manufacturing

Use headings to highlight information.

At Abix's modern facilities, we design quality, operator-friendly systems. In 1985, Abix pioneered and patented the modular Abix 2000 System. Our Abix 3000 System, a computer-controlled system, was introduced in 1991, continuing our leadership role in state-of-the-art technology. Our latest in this series of disinfection systems, the Abix 4000 System, was introduced in 2004. This system is designed specifically to meet the needs of high-volume treatment facilities. All our systems are built to rigid factory specifications, with final assembly and inspection under Abix's direct supervision. More than 2,000 Abix disinfection systems are in operation worldwide. In industry, Abix has provided systems to government agencies, bottling plants, transportation companies, and computer businesses. In addition, smaller Abix systems provide bacteria-free drinking water to 100,000 households worldwide.

Service
Abix systems require nominal maintenance. Periodic cleaning, part replacement, and other servicing are provided on a local or regional basis.

At facilities in Sydney, Australia; London, England; Flint, Michigan; and Toronto, Ontario; Abix maintains fully stocked warehouses and factory-trained personnel to service Abix equipment and address customer needs.

Awards
Abix Technologies is known for sound research and development of quality systems custom designed to meet customer demands. Our success has resulted in numerous awards, including, most recently, the Flint Chamber of Commerce Outstanding Business Achievement Award for 2004.

For more information, contact our Customer Service Department at Abix Technologies, Inc., 121 Elridge Drive, Flint, Michigan 48404-3497. Phone: (542) 555-0298, Fax: (542) 555-0297, E-mail: abix@businet.org

[middle]
Give details about key products.

Explain the type and quality of services.

Cite important awards.

[closing]
Invite requests for information and give contact details.

CHECKLIST for Management Writing

Use the seven traits to check your document and then revise as needed.

- ☐ **Ideas.** A mission statement points out what the organization wants to be. A policy statement clarifies the organization's practices concerning a key issue. A procedure explains how to implement a policy—what, when, where, how. A profile describes what an organization is and what it does.

- ☐ **Organization.** A mission statement, policy, or profile begins with the main point and follows it with supporting details. A procedure identifies the policy and chronologically lists the steps needed.

- ☐ **Voice.** The tone is informed, objective, confident, and sensitive.

- ☐ **Words.** In a mission statement, position statement, or profile, the words are common terms understood by a broad audience. In a policy or procedure, the words might be more precise and include work-related terminology understood only by people familiar with specific situations or activities.

- ☐ **Sentences.** In a procedure, sentences state steps in a succinct telegraphic style that includes crisp phrases with command verbs. In other forms, the sentences are usually longer; they flow smoothly with clear statements and transitions.

- ☐ **Copy.** Accurate details are provided with no errors in grammar or mechanics.

- ☐ **Design.** The document's appearance is formal and attractive with appropriate fonts, headings, and graphics.

CRITICAL THINKING Activities

1. Choose an organization, research it, read its mission statement, and analyze whether the statement accurately describes the organization.
2. Review two or three of the organization's policies, and analyze whether they do or do not reflect the organization's mission.

WRITING Activities

1. Revise the mission statement that you analyzed in Critical Thinking activity 1.
2. Choose an issue that the organization above has not addressed in a position statement. Research the topic carefully, and write a position statement.
3. Choose a personnel or problem that the same organization may have. Write a policy stating how the organization will address the problem.
4. Write a procedure giving employees step-by-step instructions on how to implement the policy created in Writing activity 3.
5. Write a profile statement for the organization addressed in Writing activities 1–4.

In this chapter

Guidelines for Job Descriptions **626**
Guidelines for Job Advertisements **630**
Guidelines for Employer's Follow-up Letters **632**
Guidelines for Employee Evaluations **636**
Guidelines for Employee Recommendations **640**
Checklist for Human Resources Writing **644**
Critical Thinking Activities **644**
Writing Activities **644**

CHAPTER

40

Human Resources Writing

Wise organizations understand that to succeed, they must recruit, hire, and retain skilled and motivated employees. The employees are the organization's "human resources," its most valuable asset.

Human resources (HR) writing involves preparing documents that develop this asset. For example, to find and hire strong employees, an organization (usually a department manager or someone in the HR Department) writes job descriptions, job advertisements, interview follow-up letters, and job-offer letters. Then after the organization hires the best applicants, it tries to develop and retain them by writing documents such as evaluation forms, evaluation letters, and recommendations for promotion.

At some point in your career, you will write documents like these. If you work in an HR Department, become a department manager, or start your own business, you will write HR forms many times. However, no matter how often you write the documents, your writing must always be strong. Why? Because *your* success at HR writing will affect your *organization's* success in recruiting and retaining its most valuable asset—employees.

GUIDELINES FOR JOB DESCRIPTIONS

Your goals when writing a job description are (1) to inform—not persuade; (2) to describe the job's duties and compensation; and (3) to be brief—writing only one or two pages. Employees use job descriptions to understand their responsibilities and compensation, and employers use the documents to recruit and evaluate staff.

1 Prewrite
Rhetorical analysis: Consider your purpose, your audience, and the context.
- What does each reader (employer, new recruit, present employee) need to know about the job? Why?
- How can you present the information clearly to all readers?

Prepare the draft.
- Gather information about the job's responsibilities, duties, and opportunities, especially by interviewing people familiar with the job.
- Review the content and format of other job descriptions.

2 Draft
Organize the description into three parts.

[opening] Use the headings below to present the big picture first:
- Job Title (sometimes called Position Title)
- Division (or department)
- Reports to (name of supervisor)
- Salary (included when readers need this information)

[middle] Give additional information introduced by these headings:
- Job Summary (brief description of the job)
- Major Duties (list of main tasks)
- Qualifications (education, experience, special licenses, and skills)
- Working Conditions (work site, travel and physical demands)
- Career Ladder (optional)

[closing] Conclude with the following headings and details:
- Approved by (person who approves this document)
- Date (date that it was approved—or revised)

3 Revise
Check your draft for ideas, organization, and voice.
- Is information correct, complete, and logically organized?
- Have you presented duties rather than specific tasks (*prepares press releases* versus *sends PSAs to WGN*)?

4 Refine
Check your writing line by line for the following:
- Correct spelling, punctuation, capitalization, and grammar
- Effective layout, headings, and listing techniques

> "To err is human, but when the eraser wears out ahead of the pencil, you're overdoing it."
> —J. Jenkins

Job Description (Brief)

Rhetorical situation

Purpose: To inform readers about the job's duties and compensation
Audience: Management, job applicants, and employees doing the job
Format: Document title, section heads, and succinct, bulleted statements

---

CARLYLE INTERNATIONAL, INC.
JOB DESCRIPTION

Job Title: Administrative Assistant
Department: Carlyle International—Upper Midwest Region
Reports to: Regional Director
Salary Status: Hourly, Level 1

Job Summary
The Administrative Assistant manages the administrative tasks of the Regional Center and produces the regional newsletter.

Major Duties

- Coordinate production of the regional newsletter.
- Answer telephone and respond to or redirect inquiries.
- Open mail and answer inquiries or direct for response.
- Plan training events, and state, regional, and district meetings.
- Expedite production, duplication, and mailing of affiliate materials.
- Coordinate annual reporting for the Regional Center.
- Oversee maintenance of office files and databases.
- Complete Regional Center accounting requirements.
- Coordinate schedules of Regional and Associate Directors.
- Complete other duties as assigned by the Regional Director.

Qualifications

- Associate's/bachelor's degree or three years of administrative experience
- Strong interpersonal, communication, and organizational skills
- Basic bookkeeping skills
- Excellent keyboarding skills (50 wpm minimum)
- Knowledge of office equipment, including Macintosh computers

Working Conditions
Some off-premises meetings or conferences may be required.

Approved: Martin Matthews
Date of Approval: January 27, 2005

[opening] Give the organization's name and basic details about the job.

[middle] Summarize job tasks.

Use headings to organize information.

Use lists to present details quickly.

State qualifications clearly but briefly.

[closing] List who approved the document and when.

Job Description (Extended)

Rhetorical situation

Purpose: To inform readers about the job's duties and compensation
Audience: Management, job applicants, and employees doing the job
Format: Document title, section heads, and succinct, bulleted statements

---

[opening] *Give the organization's name and basic details about the job.*

**FAMILY AND CHILDREN COALITION, INC.
JOB DESCRIPTION**

Job Title: Sexual Assault Specialist
Division: Crisis Services
Reports to: Crisis Services Coordinator
Salary Status: Salaried, Level 2

[middle] *Summarize job tasks.*

Job Summary

The Sexual Assault Specialist provides direct services to families or individuals affected by domestic abuse/sexual assault and offers community education and staff training about sexual assault and rape prevention.

Use headings to organize information.

Specific Duties:

Use lists to present details quickly.

- Help culturally diverse clients identify their needs, set goals, and implement intervention strategies to reach their goals.
- Provide crisis counseling.
- Lead women's individual and group counseling at the Family Shelter.
- Make appropriate referrals for clients with needs beyond the scope of the agency.
- Provide training and consultation on cultural diversity and sexual assault for Family and Children Coalition staff, other human-services professionals, emergency-room personnel, and community professionals.
- Make educational presentations to agencies and organizations.
- Arrange legal advocacy and follow-up for victims of sexual assault.
- Maintain communication with the South Carolina Partnership Against Sexual Assault and use its resources.
- Serve on the Crisis Services Leadership Team.
- Maintain case records and reports, and compile statistics.
- Attend Family and Children Coalition monthly staff meetings and other required meetings.
- Complete other tasks requested by the Crisis Services Coordinator.

State duties using strong verbs and precise nouns and modifiers.

Qualifications

- Bachelor of arts degree in a human-services field; or an associate's degree and two years' full-time employment in domestic violence, rape/sexual assault, and/or cultural diversity.
- One year of experience (paid or volunteer) in human services or a related field.
- Knowledge of counseling techniques, confidentiality, observation/recording techniques, informed choice, and functional limitations.
- Ability to schedule, observe, record, and interpret behaviors.
- Ability to instruct.
- Ability to organize, prioritize, and meet deadlines.
- Strong written and oral communication skills.
- Excellent interpersonal and problem-solving skills.
- Strong listening skills.

Working Conditions

There are some evening and weekend on-call duties. A valid driver's license is required, and some driving is needed.

Approved: Sheila Bauer
Date Approved: July 15, 2005

[middle]
If possible, begin each new page with a new heading.

State the duties and skills.

Describe special working conditions.

[closing]
List approval details.

GUIDELINES FOR JOB ADVERTISEMENTS

Your goals when writing a job advertisement are to describe the job accurately, to attract strong applicants, and to show that your organization is a good place to work.

1 Prewrite

Rhetorical analysis: Consider your purpose, your audience, and the context.
- ☐ What may readers already know about the job and your organization, and what should they know?
- ☐ What details will encourage strong candidates to apply?

Prepare to draft.
Gather information about the job (special tasks and qualifications):
- ☐ Review the job description and documents that discuss the position.
- ☐ Interview individuals or departments who are seeking to hire the employee.
- ☐ Read documents describing the department's projects and plans.

2 Draft

Present your message clearly.
[opening] Name your organization and briefly describe what it does. You may include a brief description of its location, products, and services.
[middle] Describe the job.
- ☐ Give the title (using terms common in the profession).
- ☐ List primary duties (worded positively, but honestly).
- ☐ Include requirements (licenses, experience, and other job-specific qualifications).
- ☐ Mention salary and benefits (optional).

[closing] Note additional details and help the reader respond.
- ☐ Note any legal issues (e.g., equal-opportunity employer).
- ☐ List the person and address to which application letters and résumés should be sent (for anonymity, use a post office box).

3 Revise

Check your draft for ideas, organization, and voice.
- ☐ Are details related to the job and institution correct and complete?
- ☐ Is the organization logical and clear?
- ☐ Is the voice positive, professional, and encouraging?

4 Refine

Check your writing line by line for the following:
- ☐ Clear, concise wording
- ☐ Correct grammar, punctuation, and spelling
- ☐ A format that is appropriate for the job, your organization, and the publication

Job Advertisements

Rhetorical situation

Purpose: To describe the job and workplace accurately and attract strong applicants
Audience: Job seekers, management, and employees seeking a new position
Format: Job ads may be posted in newspapers, journals, private employment agencies, and Web sites. The first ad below is written for a newspaper's classified section; the second is designed for display in an employment agency.

Full-Time Administrative Specialist position available at Unicorp International. Organized, detail-oriented person needed to arrange meetings, conferences, travel, schedules, and materials for Unicorp's Operations Support Services office. Prefer business school/college-level courses and minimum of three years of secretarial/administrative experience. Typing speed of 55–65 wpm and proficiency in office software required. Interested applicants send résumé to Staffing Services, 5300 Eucalyptus Road, Sacramento, California 95823. EEO.

[opening]
Name job, organization.

[middle]
Give details.

[closing]
Give contact information.

ADMINISTRATIVE SPECIALIST

Unicorp International seeks a full-time Administrative Specialist for its Operations Support Services office. The successful candidate will be responsible for the following tasks:

- Design graphic presentations.
- Plan project meetings, organize schedules, and prepare materials.
- Use desktop publishing to compose letters, reports, and reviews.
- Maintain confidential department files and records.
- Maintain manager/executive's calendar.
- Arrange meetings, conferences, and travel itineraries.
- Screen phone calls and visitors and make appropriate referrals.

Applicants must have the following:

- High school diploma/equivalent.
- Business school/college-level courses preferred.
- Three years of secretarial/administrative experience.
- Typing speed of 55–65 wpm.
- Advanced writing, reading, and math skills.
- Ability to respond to top-level management inquiries.
- Strong customer-service skills.
- Initiative and analytical ability to research and prepare reports.
- Proficiency with office software.

Unicorp International is an equal-opportunity employer. Send résumé to Staffing Services, 5300 Eucalyptus Road, Sacramento, CA 95823.

[opening]
Name the job and organization.

[middle]
List tasks and responsibilities.

List requirements, including special training, skills, and certification.

[closing]
List special notes and contact information.

GUIDELINES FOR EMPLOYER'S FOLLOW-UP LETTERS

Your goals when responding to an applicant are (1) to express your appreciation for the application, (2) to show that the application process is fair, and (3) to keep strong candidates interested.

1 Prewrite

Rhetorical analysis: Consider your purpose, your audience, and the context.
- ☐ What are the applicant's strengths and weaknesses, and how well does he or she match the job requirements?
- ☐ How does this applicant compare with other applicants?

Gather details and prepare to write.
- ☐ Reread the job description and note details.
- ☐ Review the application for key (or missing) information.
- ☐ Note the person's current job situation, availability, and contact information.

2 Draft

Organize the letter into three parts.

[opening] Use a formal but conversational voice.
- ☐ Express appreciation for the person's interest in the job and organization.
- ☐ Mention when and how you received the application.

[middle] Explain the selection process.
- ☐ State how and when applications will be evaluated.
- ☐ For strong candidates, list additional information needed and due dates; list the time and place for interviews; and note when and how the reader will learn of the outcome. For weak candidates, state objectively—but courteously—why the person does not meet the job requirements.

[closing] Thank the applicant.
- ☐ Restate your appreciation for the application.
- ☐ Encourage strong candidates to complete the process.

3 Revise

Review your draft's ideas, organization, and voice.
- ☐ Are details related to the applicant, job, and organization correct and complete?
- ☐ Are paragraphs linked with clear transitions?
- ☐ Is the voice informed, confident, professional, and positive?

4 Refine

Check your writing line by line for the following:
- ☐ Clear wording and correct spelling, punctuation, and usage
- ☐ Effective format (e.g., parts, spacing, typography)

> "If any man wishes to write a clear style, let him first be clear in his thinking."
> —Wolfgang Von Goethe

Employer's Follow-up Letter (Strong Candidate)

Rhetorical situation

Purpose: To show appreciation for the application and encourage further interest
Audience: A strong applicant whom the company wants to encourage
Format: Letter with company logo (full block)

TRIPLEX TECHNOLOGIES, INC.
3020 Gore Road • Imlay City, MI 48444-0981 • 616-555-8069

June 24, 2005

Mr. Marvin Greenfield
1554 Bastian Street
Lapeer, MI 48446-1601

Dear Mr. Greenfield:

Thank you for submitting your application and résumé for the position of Chief Microbiologist at Triplex Technologies, Inc. Your academic research on environmental hazards with the University of Michigan's Biology Department and your subsequent work as a microbiologist for the EPA indicate that you have much to offer our company.

Evaluation of all applications is nearly complete, and we will begin interviewing in two weeks. To help us gain a clearer view of your qualifications, please send us a detailed summary of your current work responsibilities.

Thank you again for your interest in Triplex Technologies. Your résumé and credentials show you to be a deserving candidate for the position of Chief Microbiologist. Please send the requested materials to our Human Resources Department by Monday, July 11.

Sincerely,

Keith Ryster
Human Resources Manager

[opening]
Greet the reader and acknowledge the application.

[middle]
Describe the status of the process and (if needed) request information.

[closing]
Express thanks and encourage the applicant.

Employer's Follow-up Letter (Unqualified Candidate)

Rhetorical situation

Purpose: To show appreciation for the application and to diplomatically explain why the reader does not meet the job requirements
Audience: An unqualified applicant
Format: Letter (full block, printed on company stationery)

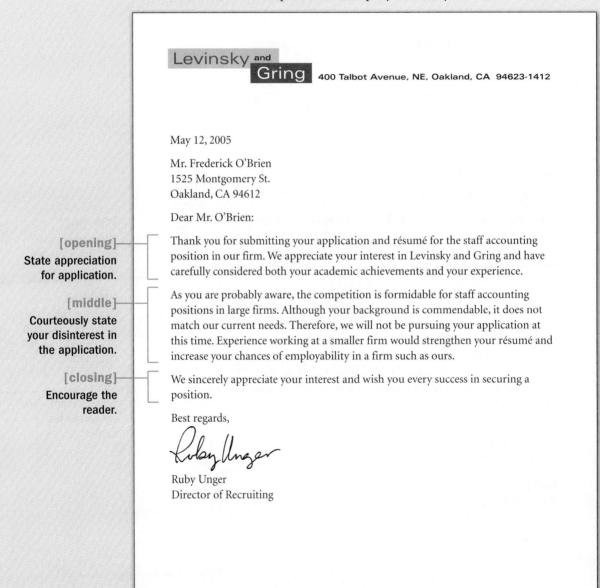

Levinsky and Gring 400 Talbot Avenue, NE, Oakland, CA 94623-1412

May 12, 2005

Mr. Frederick O'Brien
1525 Montgomery St.
Oakland, CA 94612

Dear Mr. O'Brien:

[opening] — State appreciation for application.

Thank you for submitting your application and résumé for the staff accounting position in our firm. We appreciate your interest in Levinsky and Gring and have carefully considered both your academic achievements and your experience.

[middle] — Courteously state your disinterest in the application.

As you are probably aware, the competition is formidable for staff accounting positions in large firms. Although your background is commendable, it does not match our current needs. Therefore, we will not be pursuing your application at this time. Experience working at a smaller firm would strengthen your résumé and increase your chances of employability in a firm such as ours.

[closing] — Encourage the reader.

We sincerely appreciate your interest and wish you every success in securing a position.

Best regards,

Ruby Unger
Director of Recruiting

Job-Offer Letter

Rhetorical situation

Purpose: To offer the job, cite key details, and encourage acceptance
Audience: A strong applicant whom the company wants to hire
Format: Letter with company logo (full block)

TRIPLEX TECHNOLOGIES, INC.
3020 Gore Road • Imlay City, MI 48444-0981 • 616-555-8069

July 26, 2005

Mr. Marvin Greenfeld
1554 Bastian Street
Lapeer, MI 48446-1601

Dear Mr. Greenfeld:

I am pleased to offer to you the position of Chief Microbiologist at Triplex Technologies, Inc. Our selection committee noted that your enthusiasm and range of experience set you apart from the other applicants whom we considered. — **[opening]** Offer congratulations and affirm the person's strengths.

Following are further details regarding this position: — **[middle]** Cite key contract details.

1. Your remuneration will be $85,370 per annum with a salary review based upon the anniversary-performance appraisal.
2. The starting date is Monday, August 22, 2005, with the usual three-month probationary period.
3. An overview of employee benefits is enclosed. If you have questions about the benefits, please call Judy Owen, our Human Resources Assistant (616-555-8911), and she will give you that information.

Mr. Greenfeld, we sincerely hope that you will accept our offer. We are eager to work with you and believe that our relationship will be mutually beneficial. Together we can continue Triplex's leadership role in state-of-the-art technology. — **[closing]** Encourage acceptance.

Please inform me of your decision by August 2, 2005, by calling my office at (616) 555-2376. Thank you again for your interest in Triplex Technologies, and I look forward to your reply. — Note the date for the reader's decision.

Sincerely,

Keith Ryster

Keith Ryster
Human Resources Manager

Encl.: Employee benefits

GUIDELINES FOR EMPLOYEE EVALUATIONS

Your goals when writing performance evaluations are to objectively evaluate the person's strengths, to build on the strengths, and to correct the weaknesses.

1 Prewrite

Rhetorical analysis: Consider your purpose, your audience, and the context.
- What is the person's job, training, and work record, including past evaluations?
- What evidence indicates his or her present performance, effort, and attitude?
- How can you help the person build on success and correct weaknesses?

Gather details and prepare to write.
- Review the job description and relevant sections of the employee handbook.
- Review the employee's self-assessment.
- Note measurements of the person's work (awards, sales records, assessments by employees or clients), and analyze reasons for success or failure.
- Note signs of personal responsibility (e.g., attendance, punctuality, organization).

2 Draft

Organize the letter into three parts.

[opening] Establish a positive, objective tone.
- Address the employee using his or her title.
- Show appreciation for the person's employment.
- State the purpose for writing.

[middle] Present your evaluation.
- For a positive critique, use direct organization: State the good news followed by specific supporting details.
- For a negative critique, use indirect organization: Report (1) what led to your assessment, (2) the assessment itself, and (3) attainable, corrective steps along with dates for achieving each.

[closing] Conclude with forward-looking comments.
- Anticipate continued good work from strong employees.
- Anticipate specific corrective action from weak employees.

3 Revise

Check your draft's ideas, organization, and voice.
- Are the main points clear, correct, well organized, and supported by solid evidence?
- Is the tone objective, fair, genuine, and concerned?

4 Refine

Check your writing line by line for the following:
- Clear wording and correct grammar, spelling, punctuation, and usage
- Correct format for evaluations (company evaluation form or business letter)

> "I do not choose the right word. I get rid of the wrong one."
> —A. E. Houseman

Employee Evaluation (Short Form)
Rhetorical situation

Purpose: To assess the person's performance objectively and fairly
Audience: The employee, director of human resources, and management
Format: Memo (formalizes request and creates written record)

Empire Estates Employee Evaluation

Employee: Larry Mott Supervisor: Sarah Iverson
Job Title: Mortgage Specialist II Title: Mortgage Specialist Supervisor
Hire Date: April 17, 2000 Evaluation Date: December 3, 2004

Using the scale below, rate the employee's performance. (If an area does not apply to this position, write "NA.") Support your rating with details.

Rating Scale:
5 – Excellent: Performance frequently exceeds requirements and expectations. Outstanding work.
4 – Good: Performance is above average. Employee is reliable and independent.
3 – Acceptable: Performance meets requirements and expectations. Goals met with normal supervision.
2 – Marginal: Performance needs improvement to meet job requirements and expectations.
1 – Unacceptable: Performance fails to meet job requirements and expectations. Not reliable, unable to work independently. Poor understanding of position.

__5__ Quality of Work
Comments: *Larry is innovative, is hard working, and regularly attends seminars to refine his skills.*

__5__ Analytical and Problem-Solving Skills
Comments: *Other employees rely on Larry as the troubleshooter for their technical problems.*

__4__ Job Knowledge
Comments: *Larry understands/applies information, instructions, and procedures. He has learned to do cash reservations, lockboxes, and CPI system balancing.*

__5__ Communication Skills
Comments: *Larry explains technical information to other workers and provides excellent customer service.*

__5__ Interpersonal Skills
Comments: *Larry is well liked by co-workers. He is patient and considerate—an excellent ally to colleagues.*

[opening]
State the form's title and basic details describing the job and person.

[middle]
Define the grading scale clearly.

Provide space for grades and comments.

When completing the form, state grades fairly, along with supporting details.

[closing]
The closing can include a signature and date as shown on page 639.

Employee Evaluation (Long Form)

Rhetorical situation

Purpose: To assess the person's performance objectively and fairly
Audience: The employee, director of human resources, and management
Format: Memo (formalizes request and creates written record)

[opening] — State the organization, person, and key details about the job.

[middle] — List the job's responsibilities in order of importance.

In the first column, succinctly respond to the topics listed.

In the second column, state additional details in a fair, objective voice.

---

1 of 2

Empire Estates Incorporated
Employee Evaluation

Employee: Mary Lamont
Job Title: Mortgage Specialist II
Hire Date: April 16, 2003

Supervisor: Sam Everly
Title: Mortgage Specialist Supervisor
Evaluation Date: September 7, 2005

Part A **Job Responsibilities:** (List 3–5 key roles from job description)
1. Reconciling A/A and S/S accounts
2. Reconciling FHLMC P & I accounts
3. CPI System balancing
4. Troubleshooting and technical adviser/supervisor

Part B **Job Characteristics:** Rank the employee's performance (E for Excellent, G for Good, A for Acceptable, and U for Unacceptable) and specify areas of strength and areas for improvement.

Job Characteristics	Evaluation/Comments
1. Quality of work: Produces excellent work.	E Innovative, outstanding work.
2. Productivity: Produces an expected quantity of work on time.	G Production levels good, some delays due to troubleshooting.
3. Job Knowledge: Understands, retains, and applies information, instructions, and procedures.	E Excellent analytical skills, quick learner.
4. Communication Skills: Communicates effectively. Writes clearly and concisely in language appropriate for the reader.	E Works well with customers, accommodating.
5. Decision-Making Skills: Identifies problems and implements solutions effectively.	E Solves problems on her own, troubleshoots for others.
6. Interpersonal Skills: Works respectfully and cooperatively with co-workers and clients.	E Works well with both administrators and customers.

Part C **Performance Evaluation Summary:** Review Parts A & B; then summarize the demonstrated strengths and areas for improvement.

Demonstrated Strengths
Mary's excellent analytical skills help her serve as the department's primary problem solver. She's a quick learner of new systems. In addition, Mary's excellent interpersonal and communication skills enable her to work well with colleagues and customers.

Areas for Improvement
Mary should receive further training to strengthen her knowledge of Monarch and Access. She also needs additional supervisory training to expand her skills in troubleshooting and problem solving.

Part D Individual Development Plan
Establish an individual plan to further develop or improve skills. Plans may include additional training, seminars, on-the-job assignments, etc.

1. Fall 2005:   Additional training in Monarch and Access
2. Spring 2006: Leadership seminars to strengthen supervisory skills

Part E   **Employee Response:** I appreciate this fair evaluation and look forward to growing professionally through the suggested training and seminars.

Prepared by: Sam Everly, Mortgage Specialist Supervisor
Date: September 9, 2005
Reviewed by: Dan Levine, Mortgage Department Supervisor
Date: September 13, 2005
Employee Signature: *Mary Lamont*
Date: *September 15, 2005*

Note: The employee's signature above does not necessarily constitute agreement with this evaluation.

[middle]
Name the person's strengths and give details.

Cite weaknesses precisely and clearly.

List activities and dates.

Respond honestly but politely.

[closing]
Sign and date as requested.

GUIDELINES FOR EMPLOYEE RECOMMENDATIONS

Your goals are to evaluate the person's qualifications objectively and to state your recommendation clearly.

1 Prewrite

Rhetorical analysis: Consider your purpose, your audience, and the context.
- ☐ What is the job applied for, and what are its requirements?
- ☐ What are the employee's training, skills, and work record, and how can you state them clearly and fairly?
- ☐ What are the person's plans and needs?

Prepare to draft.
- ☐ Collect measurements (awards, sales records, evaluations) of the person's work.
- ☐ Read the person's past evaluation, response, and self-improvement plan.
- ☐ Record signs of work habits (attendance, cooperation, initiative).
- ☐ List details about the job being applied for, and note how they match the person's training, skills, and personal qualities.

2 Draft

Organize the message into three parts.

[opening] Establish a positive, professional tone.
- ☐ Cite the person you are recommending, your relationship (e.g., supervisor, colleague), and its length.
- ☐ Note the position applied for.

[middle] State your assessment clearly and objectively.
- ☐ For recommendations without reservations, cite strengths (accenting those you know the reader is seeking), plus supporting details.
- ☐ For recommendations with reservations, cite strengths, and note relevant weaknesses, along with supporting details where needed.
- ☐ For a person who was laid off or fired, tactfully explain why.

[closing] Conclude in a focused, business-like manner.
- ☐ For strong applicants, restate the recommendation and offer help.
- ☐ For weak applicants, close politely.

3 Revise

Review your draft for ideas, organization, and voice.
- ☐ Are your main point and supporting details clear and complete?
- ☐ Is the organization logical and the word choice fair?

4 Refine

Check your writing line by line for the following:
- ☐ Precise, sensitive wording
- ☐ Correct names, dates, punctuation, grammar, and spelling
- ☐ Correct, attractive format

> "When you know a thing, to hold that you know it; and when you do not know a thing, to allow that you do not know it—this is knowledge."
> —Confucius

Chapter 40 Human Resources Writing

Recommendation (Promotion)

Rhetorical situation

Purpose: To convince the reader that the promotion is warranted and feasible
Audience: The director of human resources and management
Format: Memorandum

Memorandum

Date: October 14, 2005

To: Lizette Luna

From: Samuel Everly

Subject: Recommendation for Mary Lamont

During last Friday's management meeting, you mentioned your plan to add an accounting team to serve new and existing accounts. Via the office grapevine, Mary Lamont heard about the plan and asked that I recommend her for team leader. Because she has shown the accounting skills, dedication, and communication skills needed by leaders, I agreed to write this recommendation. — **[opening]** Name the person and position; describe your relationship.

Mary joined Empire Estates in April 2003 and was assigned to my team in June 2004 as a Mortgage Specialist II. Her excellent skills at reconciling A/A and S/S accounts, as well as FHLMC P & I accounts, enabled Mary to make an immediate, strong contribution. Moreover, last January when our team needed help with CPI System balancing, Mary took a night class on the topic and voluntarily added CPI System balancing to her previous assignment. That training further helped our team. — **[middle]** Objectively cite strengths that the reader wants in an applicant.

While Mary has excellent accounting and information systems skills, she also works well with colleagues. In fact, within six months after she arrived, team members were looking to her as the group's troubleshooter. They respected her ability to listen, think through a problem, and solve it.

In summary, Mary Lamont has shown the skills and dedication that Empire Estates wants in its leaders. As you look for someone to lead the new accounting team, please consider Mary. The assignment would require that she be promoted to Mortgage Specialist Supervisor, but as this memo shows, Mary has earned the promotion, and she can handle the work. — **[closing]** Restate your recommendation and offer help.

If you want more information, please call me at ext. 6029.

F.Y.I.
The writer pulls details from Mary Lamont's job evaluation shown on pages 638–639.

Recommendation (Strong Candidate)

Purpose: To convince the reader that the applicant can do the job well
Audience: The director of human resources and others assessing the applicant
Format: Letter (full block and printed on company stationery)

ABC Mortgage
3555 Beverly Avenue, Gainesville, TX 76240

June 3, 2005

Mr. Daniel Chiu
Empire Estates Incorporated
1697 Nebraska Street SE
Sioux Falls, SD 57102

Dear Mr. Chiu:

[opening] — *State your role, the applicant's name, job applied for, and your recommendation.*

As manager of the reconciliation group at ABC Mortgage, I am pleased to recommend Janice O'Rourke for the position of Mortgage Account Manager for your company. Janice came to ABC Mortgage as an independent consultant in January 2002 and proved to be an asset at a crucial time.

[middle] — *Give details supporting your point.*

We had recently expanded our Texas Servicing Center and had a backlog of tasks. Janice worked on a payment-clearing account. From start to finish, she worked diligently to help us manage this transition. Janice identified procedural issues, made recommendations to avoid future difficulties, and trained a permanent employee to manage the account.

Throughout her seven months on our staff, Janice consistently maintained a professional demeanor and work ethic. Her strong people, organizational, and management skills helped her work well with co-workers and senior management.

[closing] — *Restate your point and offer additional information.*

Without reservation, I recommend Janice O'Rourke for any accounting/management position. If you would like to discuss her qualifications in more detail, please call me at (214) 555-5000.

Sincerely,

Sue Atheron

Sue Atheron
Manager

Recommendation (with Reservations)

Purpose: To convince the reader that the applicant can do the job well
Audience: The director of human resources and others assessing the applicant
Format: Letter (full block and printed on company stationery)

ABC Mortgage
3555 Beverly Avenue, Gainesville, TX 76240

June 3, 2005

Mr. Daniel Chiu
Empire Estates Incorporated
1697 Nebraska Street SE
Sioux Falls, SD 57102

Dear Mr. Chiu:

I am writing to recommend Gail Dubois for the position of Mortgage Account Specialist for your company. Gail came to work at ABC Mortgage in March 2003. We had earlier expanded our Texas Servicing Center, and she was one of several employees whom we hired to manage the additional work resulting from our expansion. — **[opening]** State the person's name, job applied for, and your recommendation.

Gail was a diligent, well-organized worker whose skills were particularly important in her work on a payment-clearing account. She worked closely with senior management, consultants, and subordinates and her interactions were always accommodating and capable. Though Gail had some difficulty supervising her own assistants, I believe she was improving in this area. — **[middle]** Illustrate the applicant's strengths and diplomatically note documented weaknesses.

Gail was a quick learner and receptive to constructive criticism. She knew her limitations and was eager to improve. When inaccuracies in her work were pointed out, she worked hard to resolve the problem. To that end, she attended work-related seminars and workshops.

I found Gail to be an intelligent, personable, and motivated worker. If I can be of further service, please contact me at (214) 555-5000. — **[closing]** Restate your point; offer more information.

Sincerely,

Sue Atheron

Sue Atheron
Manager

CHECKLIST for Human Resources Writing

Use the seven traits to check your document and then revise as needed.

- ☐ **Ideas.** The job description accurately presents the job's tasks and responsibilities. The job advertisement clearly describes the job and gives contact information. The letter to the applicant gives details needed to pursue the application process. The evaluation/recommendation measures the employee's performance according to standards in documents such as a job description, policy, or employee handbook.

- ☐ **Organization.** The job descriptions, advertisement, or evaluation follows organizational tips in this book. The letters include the main point, details, additional help, and a polite conclusion.

- ☐ **Voice.** The tone is knowledgeable, forthright, respectful, and sensitive.

- ☐ **Words.** In evaluations and recommendations, the words state details objectively and may include technical terminology used by those who do the job. Similarly, in job descriptions, advertisements, evaluations, and letters to an applicant, the words are precise terms understood by readers familiar with the job. In recommendations, the words are accessible to nonspecialists.

- ☐ **Sentences.** In letters, memos, and e-mails, sentences are complete, smooth, and clear. In documents such as form evaluations or job ads in newspapers, ideas are stated in phrases (often bulleted) and in sentences in telegraphic form.

- ☐ **Copy.** The copy includes correct names, dates, titles, tasks, and performance criteria. It lacks any errors in grammar, punctuation, capitalization, or spelling.

- ☐ **Design.** The document follows formats shown in this chapter.

CRITICAL THINKING Activities

1. Find five job advertisements in your field: two from newspapers and three on the Web. Evaluate the effectiveness of each ad by assessing its (a) clarity, (b) strategies to interest applicants, (c) completeness, and (d) overall appeal.
2. Choose the job advertisement that you find weak, research the organization, and revise the ad using the guidelines in this chapter.

WRITING Activities

1. Research a job and organization at which you want to work, and write a detailed advertisement (for posting at an employment agency) for the job.
2. Imagine that you are hiring someone for the position described in Writing activity 1 and that two people applied—one weak candidate and one strong candidate. Write each person a follow-up letter.
3. Imagine that you hired the strong applicant and that he or she has done good work. Write a recommendation for promotion.

PART 8

Speaking, Listening, and Giving Presentations

41 Communication Basics

42 Communicating in a Group

43 Communicating in Meetings

44 Writing and Giving Presentations

In this chapter

Speaking Effectively **648**

Listening Effectively **649**

Giving and Taking Instructions **650**

Giving and Taking Criticism **651**

Understanding Conflicts **652**

Resolving Conflicts **653**

Checklist for Communication Basics **654**

Critical Thinking Activity **654**

Writing Activity **654**

CHAPTER
41

Communication Basics

Of course you need good speaking and listening skills to thrive in your career. But what are effective speaking and listening? For example, which strategies will help you present an oral report, and how do these strategies differ from those needed to participate in a group discussion or to give instructions? What is effective listening, and should you listen to a speech in the same way that you listen to a sales proposal or criticism?

This chapter describes effective speaking and listening skills. It explains how to use these skills in situations such as giving instructions, taking criticism, and resolving conflicts. Other chapters in Part 8 explain how to use these skills in other situations, such as participating in group discussions or communicating in business meetings.

You can improve your speaking and listening skills significantly by reading the instructions in Part 8 and doing the end-of-chapter activities. However, you will learn far more by practicing your oral communication skills not just in class but outside of class as well. Every time you join a group discussion or give instructions, you have an opportunity to refine these important skills that will enhance your career.

SPEAKING EFFECTIVELY

Whether you're presenting a sales proposal or participating in an interview, you must prepare, look, listen, and think.

Prepare to Speak by Considering Your Subject, Purpose, and Audience

- **Subject:** Facts and background information. For proposals, anticipate objections. For problems, suggest solutions.
- **Purpose:** Why you must speak. Determining the purpose helps you shape the type of message (statement, question, or command), form (discussion, brainstorming session, or interview), and medium (face-to-face talk, phone conversation, or voice-mail).
- **Audience:** Including the listener's personality, experience, and familiarity with the topic and terminology.

Look at Your Listener

- Establish eye contact to show that you are paying attention.
- Watch for responses and adjust the message accordingly. If the person looks bored, ask a question or offer additional information.

Listen to Your Listener

- Invite dialogue. Ask questions and weigh the listener's responses to determine whether more information is needed.
- Answer questions precisely. Don't ramble.

Think While You Speak

- Focus on your prepared main points, but always be ready to revise your approach.
- Remember that nonverbal messages (facial expressions, posture, gestures) communicate as much as words. If you're describing a project with enthusiasm—and then yawn—the listener may decide that the yawn reflects your real attitude.
- Remember that individuals from different cultures may have different nonverbal messages. For example, a Japanese listener who nods as you speak may be indicating that she hears you, but doesn't necessarily agree with your position.

F.Y.I.

Note that the four verbs in the headings above are all in active voice: *prepare, look, listen,* and *think*. The point is obvious—to be an effective speaker, you must be an engaged, *active* listener.

LISTENING EFFECTIVELY

Whereas hearing is a passive activity, listening takes effort and skill. To do it well requires that you do at least three things: prepare, focus, and reflect.

Prepare to Listen

- Decide what you want to gain: acquire facts, understand a process, or learn an idea.
- Research the topic and prepare questions beforehand.
- Acknowledge your own biases about the subject, speaker, and situation.
- Withhold judgment about the message until you've heard it and thought about it.
- Have note-taking materials and, if appropriate, a tape recorder.

Focus on the Message

- Concentrate on hearing all the words and on picking up nonverbal cues: gestures, facial expressions, and vocal tones.
- Determine the speaker's purpose: to motivate, to explain, or to inform?
- Listen for major points and supporting details; think about the relationships between those points.
- Listen for transitions (*next, second, more importantly, as a result*) that indicate how points are related in terms of order, importance, or cause-and-effect.
- Listen for bias or prejudice: Is the speaker being fair and objective?
- Identify why the speaker uses emotion and humor—to aid understanding or to manipulate the audience?
- Take notes carefully and objectively. Jot down main points, key evidence, conclusions, and questions.
- If relevant (e.g., field report or project report), take notes on aspects of the setting that impact what was said or heard.

Reflect on the Message

- Review your notes as soon after the presentation as possible and think about the message.
- Summarize the message in one sentence.
- Ask "How does this relate to me?" and "How can I use this information?"
- Discuss the message with others.
- Create a record of your response—write it out.

GIVING AND TAKING INSTRUCTIONS

Giving and taking instructions orally are two of the most common—and most important—forms of communication that you will use in the workplace. For example, most of your workdays will begin with instructions regarding the day's tasks. During the day, you will exchange additional instructions that help you and others solve problems related to those tasks. At the end of the day, you will likely exchange further instructions as you assess your progress, analyze remaining tasks, and plan for the next day's activities.

To improve your ability to give and take instructions well, follow these guidelines.

Giving Instructions

1. **Know the task.** Understand the entire task, including (1) safety issues, (2) goals or outcomes, (3) materials needed, and (4) each step in the process.
2. **Summarize.** Give an overview of the task (safety issues, purpose, and outcome).
3. **Identify materials.** List the materials needed.
4. **Describe the task.** Present the task in a logical, step-by-step sequence. Include the names of parts, along with measurements (distances, weights). Clarify the task with drawings, gestures, or illustrations.
5. **Ask for questions.** Answer each question clearly.
6. **Review.** Have the listener explain the task to you; make corrections as needed.
7. **Rehearse.** For important or complex tasks, do a practice run.

Taking Instructions

1. **Listen to the summary.** Try to understand the entire task (from purpose to outcome).
2. **Note the steps.** Listen for details about the materials needed, what to do, and the sequence of steps.
3. **Ask for clarification.** Ask questions about precisely what to do, or repeat what you heard and ask for confirmation.
4. **Follow the steps.** Do the steps in order, and pay attention to safety issues.
5. **Ask for help.** If you have problems, don't risk a major error—ask for help.

WRITE Connection

For help putting instructions in writing, see Chapter 35, "Writing Instructions."

GIVING AND TAKING CRITICISM

A person who can't give criticism well has a limited ability to help others, and a person who can't take criticism well loses access to others' insights. If you want to develop your skills as an employee or a manager, you'll need to learn to both give and take criticism. Use the guidelines below to strengthen both skills.

Giving Criticism

1. Establish your goal. Decide what you want to happen after you speak.
2. Ponder the problem. Limit criticism to a problem that the listener can understand and fix. Think about what the problem is, what caused it, and how the listener can solve it.
3. Consider the listener. How will the person benefit from the criticism? Does he or she have special needs that you must consider? Is he or she able to take criticism at this time?
4. Craft the message. State the message so the listener knows that you seek to correct the *problem* and not the *person*.
5. Think about the setting. Is this the best time and place to share this criticism? For example, will others overhear your conversation, or will competing activities distract the discussion?
6. Rehearse the message. Say the criticism aloud to yourself.
7. Deliver the message. Speak in a positive tone and determine whether the listener
   - Understands the problem and takes ownership of it,
   - Is committed to working toward a solution, and
   - Feels your support in achieving the solution.

Taking Criticism

1. **Listen for the message.** Ask yourself, "What is the problem?" and "How will this fix it?"
2. **Respect the speaker.** Focus on the person's ideas rather than how the criticism is delivered or how you will defend yourself.
3. **Value criticism.** Let criticism help you understand that a problem exists, why it exists, and how it can be resolved. Remember that a person unable to take criticism loses the opportunity to benefit from that criticism.
4. **Respect yourself.** Concentrate on taking criticism calmly, evaluating it wisely, using the good, and politely disregarding the rest.

UNDERSTANDING CONFLICTS

Constructive conflicts help people identify a problem and solve it, whereas destructive conflicts hurt people and hinder their work. You can make most conflicts constructive by working toward a win-win situation.

Win-Win Situations

- The focus is on solving the problem rather than defeating either party.
- Both parties give something up, but gain something as well.

  A landscape designer disagrees with his supervisor when she asks him to spend more time in the office and less time on job sites. Finally, he suggests that they hire a crew manager so he has more time to meet with customers. Both accept the plan.

Win-Lose Situations

- One party may refuse to work toward a reasonable solution.
- The final solution benefits only one party.

  A research chemist regularly demeans lab technicians, and they file complaints about her behavior. The supervisor cautions the chemist to stop this activity. When the chemist persists, the supervisor first reprimands her, and then he fires her. The chemist loses her job, but the lab technicians gain better working conditions.

Lose-Lose Situations

- The focus is on defeating each other rather than finding a shared solution.
- Both parties lose things that they care about and are unhappy with the solution.

  The manager of a real estate agency asks two realtors to meet with a client who wants to purchase a commercial property. Rather than work together, the realtors offer competing suggestions and criticize each other's ideas. The client leaves without purchasing anything, and the manager becomes angry with both employees.

RESOLVING CONFLICTS

These guidelines will help you resolve most conflicts constructively. Practice the guidelines daily for your own sake and for the sake of your organization.

Listen carefully and think clearly. Weigh competing arguments and then think through the information to gain a win-win solution.

Be honest, clear, and direct. Build trust so that others will invest the effort needed to find a constructive solution.

Avoid harsh or slanted language. Your use of derogatory language will diminish other people's respect for you.

Avoid spiraling conflicts. Conflicts that spiral out of control become extremely destructive as both parties abandon hope for a constructive solution and seek victory at all costs.

Show goodwill. Even if your position is rejected, show goodwill while working toward the best solution. Your good demeanor will earn the respect that you want and need for the future.

Be sensitive to the circumstances.

Defer in the following situations:
- You learn that you are wrong.
- The issue is more important to the other person than it is to you.
- You want the person to learn from a mistake.
- The cost of winning isn't worth it.

Compromise in the following situations:
- There is not enough time to seek a win-win outcome.
- The issues don't warrant lengthy negotiations.
- The person isn't willing to seek a win-win outcome.

Compete in the following situations:
- The outcome is important.
- The other person will exploit a noncompetitive approach.

Cooperate in the following situations:
- The issue is too important for a compromise.
- Your long-term relationship with the other person is important.
- The other person is willing to cooperate.

CHECKLIST for Communication Basics

To communicate well, you must respect (1) yourself, (2) the listener, and (3) your organization. Test your level of respect with this checklist.

I respect myself by

- ☐ Believing that my own thoughts and feelings have value.
- ☐ Taking responsibility for what I think and feel.
- ☐ Expecting others to respect what I think and feel.
- ☐ Being ready to present my ideas clearly and to listen to others carefully.
- ☐ Wearing appropriate dress and using polite manners.

I respect the listener by

- ☐ Learning who he or she is (personality, experiences, and culture).
- ☐ Speaking courteously—no profanity, vulgarity, racism, or sexism.
- ☐ Learning what information the individual needs and why.
- ☐ Using words and illustrations that he or she understands.
- ☐ Getting to the point and not wasting time.
- ☐ Being forthright—no hidden agenda or manipulation.
- ☐ Qualifying my statements or, when necessary, admitting that I don't know.
- ☐ Anticipating questions and having answers.
- ☐ Following up with necessary phone calls, memos, and other tasks.

I respect my organization by

- ☐ Assessing how my speaking/listening practices affect co-workers.
- ☐ Periodically upgrading my communication skills.
- ☐ Helping others critique and develop their skills.
- ☐ Using my skills for my organization's benefit as well as my own.

CRITICAL THINKING Activity

Review the instructions in "Speaking Effectively" and "Listening Effectively" (pages 648–649). Based on those guidelines, evaluate the quality of a group discussion on a TV news program. Report your findings to the class.

WRITING Activity

Review "Understanding Conflicts" and "Resolving Conflicts" (pages 652–653). For a week, practice these instructions each time you are involved in a conflict. After each incident, describe what happened and why.

In this chapter

Beginning a Group **656**
Working in a Group **657**
Making Decisions **659**
Listening in a Group **660**
Responding in a Group **661**
Roles in a Group **662**
Disagreeing in a Group **663**
Checklist for Communicating in a Group **664**
Critical Thinking Activity **664**
Writing Activities **664**

CHAPTER
42

Communicating in a Group

As research like the U.S. Labor Department's *SCANS Reports* has shown, businesses need employees who can work well in groups. But what if you or your colleagues have problems communicating, cooperating, making decisions, and working toward shared goals? Would such problems hinder your work? How, why, and what can you do to correct the problems?

This chapter has answers to these questions. It begins by explaining what an effective group is and how it functions. Next, it describes specific skills that help individuals participate in and strengthen a group. Some of those skills include brainstorming, problem solving, making decisions, listening, responding, and disagreeing.

The chapter also distinguishes common roles that people play when participating in groups. Some of those roles—like the ones that help a group do its tasks—include a diagnoser, direction-giver, energizer, opinion-seeker, or summarizer. Other positive roles include an empathetic listener, harmonizer, or praise-giver. Negative roles include an attacker, blocker, or joker.

You can use information in this chapter to assess your own group communication skills, along with the roles that you play as a group member. You can also use the information to polish your skills and to take on new, more challenging roles. While improving your skills and role-playing abilities will take effort, the *SCANS Reports* promises that the benefits will last a lifetime!

BEGINNING A GROUP

A collection of individuals is not a group, but it may become one. For example, seven people who happen to step into an elevator at the same time are not a group. However, if that elevator gets stuck between floors, the individuals will probably form a group as they respond to their common problem. In other words, a *group* is a number of people who get together, communicate, do something—such as solve a problem—and affect one another.

Stages of Group Building

To form a group, individuals work through several stages of group development. By understanding these stages, you can assess how well a group is taking shape, and you can help the process along.

Groping. During this stage, individuals usually aren't clear about the purpose of the group, the direction of the group, or the role each individual will play in the group.

Griping. Individuals get frustrated and complain as they adjust to the group tasks and roles.

Grasping. People communicate and begin to understand their purpose and direction.

Grouping. People work together by sharing ideas, influencing one another, and accomplishing something together that they could not do alone. Individuals identify with the group.

Group-Building Tips

Early in your group's history, do the following:

Clarify your group's purpose.
- Discuss what you intend to do and why.
- Limit your focus to those issues directly linked to the problems you face.
- Begin work on a purpose statement that clarifies who and what you are.

Select probable participants.
- Write guidelines to help you identify and select participants.
- List individuals whose abilities and interests match those guidelines.
- Avoid "groupthink" (members all think alike) by selecting participants whose backgrounds and skills strengthen the group.

Based on your purpose, determine how you will function.
- How often and where will the group meet?
- What group officers are needed, and how and when will they be chosen?
- What rules (e.g., *Robert's Rules of Order*) will govern your proceedings?
- How will you develop an agenda and make decisions? Why?
- How will you record and publish your agenda, minutes, and reports?
- How will you finance the group's activities?

WORKING IN A GROUP

To do its work, a group usually concentrates on *brainstorming, problem solving,* or *informing,* often moving from one activity to another. For example, if a group gets stuck problem solving, it may take ten minutes to brainstorm new ideas before returning to problem solving.

Brainstorming

This activity generates ideas quickly. Groups brainstorm for three reasons:

- To produce new ideas
- To list problems that the group must address
- To collect possible solutions to problems

Brainstorming usually includes three steps: *generating, discarding,* and *evaluating.*

Generating

Group members have one objective—to generate as many good ideas as possible.

- The leader states the discussion topic or problem.
- Someone records the ideas on a board or flipchart.
- Participants pitch ideas quickly as they come to mind; ideas are welcome.
- Nothing is allowed to slow or block new ideas:
  - No one judges—all ideas are equally valuable.
  - No one may make negative or evaluating comments such as "That's not possible" or "Wouldn't this cost too much?"

Discarding

Group members discard duplicate ideas.

Evaluating

Group members discuss the value of each idea, weeding out the weaker ones, keeping stronger ones, and perhaps combining several ideas.

F.Y.I.

Some practices may impede your group's ability to generate ideas.

Negative treatment of people
- Destructive criticism of group members, including leaders
- Impolite or unnecessary interrupting of a speaker
- Forming cliques or polarizing subgroups

Negative treatment of ideas
- Judgmental comments, gestures, or facial expressions
- Unprofessional attitude (yawns, careless word choice)
- Lack of commitment (unwillingness to reach for success and try hard)
- Immediate criticism of an idea—a "that will never work" attitude

Problem Solving

Group discussions are an effective method of problem solving because they enable individuals to listen to ideas, evaluate them, and reach a mutually acceptable solution. Problem-solving discussions should work through these steps:

1. Define the problem.
   - What is it?
2. Analyze the problem.
   - Why is it important?
   - What are its causes?
3. Set standards for choosing the best solution.
   - What should a solution accomplish?
   - Will the solution create other problems?
   - What will the solution cost (in terms of money, time, effort)?
4. Identify possible solutions.
   - What solutions have been tried and how have they worked?
   - What new solution would avoid past failures?
   - What solutions are now available?
5. Select the best solution.
   - Which solution best matches our standards and resources?
6. Decide how to evaluate the solution.
   - How will you test the solution after you implement it?
7. Implement the solution.
   - Set guidelines for how and when work must be done.
   - Assign people to put the solution into effect.
   - Assign people to evaluate the solution.
   - Set guidelines and dates for the evaluation.

Presenting Information

Group presentations are common instructional activities used to inform a group about a topic. Informing usually requires that the presenter does the following:

1. Introduce the topic, explain why it's important, and provide an overview that sets the direction of the presentation.
2. Present information on the topic—often with the help of handouts, props, or computer-generated displays.
3. Lead the group in a discussion during which individuals ask questions and give their opinions.
4. Conclude the discussion when group members clearly understand the information presented.

MAKING DECISIONS

To resolve issues, groups may choose one of these decision-making methods: *authority, minority, majority,* or *consensus.*

Authority

The group discusses issues and may make a recommendation; however, one authority figure (a leader or an invited expert) makes the decision. Although this method is efficient, members might feel that it usurps their authority.

Minority

A vocal or powerful minority may make a decision that the majority disagrees with, yet feels forced to accept. This "minority rule" often leaves the majority feeling resentful or uncooperative.

Majority

The group votes, and the side with the majority (most votes) wins. However, a majority is not necessarily more than half the votes.

- If a group has ten members and they choose between two options, a majority is six votes (more than half of ten).
- If the same group chooses between three options, option 1 might get three votes, option 2 might get three votes, and option 3 might get four votes. In this case, the majority is four (fewer than half of ten).

While majority rule is common and efficient, it has problems:

- If the group has two choices, and the vote is close (49 percent to 51 percent), the majority might not be able to gain the minority's cooperation.
- If the group has three choices and the vote is close (e.g., 32 percent, 33 percent, and 35 percent), the two losing minorities (with 32 percent and 33 percent) have 65 percent of the vote, and they might rebel against the majority, which has only 35 percent of the vote.

Consider ways to avoid majority-rule problems:

- A group might decide *before it votes* that it will not accept a choice that does not receive a large majority (e.g., 60 percent).
- In a close multiple-choice vote, indicate that the group must vote twice: first to determine the two most popular choices, and second to choose one of these two options.

Consensus

Group members agree to support a solution even though some individuals have reservations about it. While a consensus decision takes time, members might support it more easily than a majority decision.

LISTENING IN A GROUP

Listening skills help group members understand one another and work together. Here are some tips that could help your group communicate more effectively.

Active Listening

When people in a group practice active listening, everyone feels valued and understood.

- Listen for a speaker's main point and supporting arguments.
- Interpret gestures and facial expressions.
- Take notes.
- Consider how the message relates to what others have said.
- Concentrate on the speaker rather than on your response.
- Think about how the message relates to the group's task.

How and When to Interrupt

It's rarely appropriate to interrupt a speaker (usually the group leader does it). If you must interrupt, do it politely and for the right reasons.

Right reasons

- To clarify: "Excuse me, Vic, can you explain your logo . . . ?"
- To steer discussion: "Dee, let's table that discussion until later."
- To keep someone from dominating discussion: "Thanks, Aaron, but I think Melissa has an idea."
- To set a schedule: "Excuse me, Adam. Our next session begins in five minutes, so would you please summarize . . . ?"

Wrong reasons

- You weren't listening carefully.
- You disagree with the speaker's statement.
- You are losing an argument or you are angry.

Responding to Interruptions

If a person asks you to repeat something, do so. If the person asks again and is not paying attention, you could reply, "Yes, Jimmy, but please let me finish this point." Always respond politely—even if the person interrupting you is not being polite.

F.Y.I.

Before you speak up in a group, be sure your comment passes the following tests (if it doesn't, don't speak):

- Is my comment related to the topic being discussed?
- Is it appropriate at this time—does it clarify a point or answer a question?
- Does it add something new without simply repeating?

RESPONDING IN A GROUP

For a group to work as a team, each member must respond to the needs of other individuals and to the needs of the group.

Respond to the Needs of Individuals

- Be sensitive to people who need more information. If you're reporting on a new patient-monitoring system and two student nurses are in the group, consider sharing additional details about the system.
- Help individuals who lack needed skills. If you're presenting a flowchart and an individual seems confused, you may ask, "Is this information clear, Matt?"
- Include everyone in the conversation. If you're explaining a software program and someone appears bored, you may say, "Jennifer, you've used this program. What do you think?"

Respond to the Needs of the Group

- Clarify confusing issues or terminology. If you're describing a retirement program, define special terms such as "deferred annuity" or "limited trust."
- Assist groups struggling with language. If you're presenting a policy to a group who speak English as a second language, consider displaying the policy as you read it aloud.
- Encourage discussion. If you're presenting guidelines to sales personnel who appear frustrated, ask, "How could these help us?"

Improving Group Effectiveness

As an individual

- Understand the roles that you play in a group.
- Identify the positive contribution that you make.
- Learn how to improve your contribution.
- Recognize the contributions made by others.
- Respond effectively to others.

As a group

- Analyze how your group functions.
- Understand each person's contribution.
- Evaluate whether group members are playing all necessary roles.
- Ask individuals to play unfilled roles.
- Assess the value of each individual's contribution.
- Help those playing dysfunctional roles to fill more constructive roles.

ROLES IN A GROUP

Throughout group discussions, individuals play various roles: task roles, maintenance roles, or dysfunctional roles. Recognizing these roles will help you improve your own skills and help others improve theirs.

Task Roles

These roles distinguish activities that help the group.

Diagnoser: Observes how the group is working. ("We've discussed the problem, but we need solutions.")

Direction-giver: Gives directions or instructions.

Energizer: Encourages members to work hard.

Gatekeeper: Enables all members to participate.

Information-giver: Offers helpful facts, evidence, and so on.

Information-seeker: Asks others for information.

Opinion-giver: States personal opinions, attitudes, and beliefs.

Opinion-seeker: Asks others for opinions, attitudes, and beliefs.

Reality-tester: Monitors whether group ideas are realistic.

Starter: Helps the group get started. ("We'd better begin.")

Summarizer: Reviews by pointing out what has been accomplished and what still needs doing.

Maintenance Roles

These roles help members of the group work well together.

Empathetic listener: Listens without evaluating or judging.

Evaluator of emotional climate: Senses and shares the tone of the group. ("I think we're all feeling a bit defensive now.")

Harmonizer: Helps members solve personal conflicts.

Participation-encourager: Encourages shy members to speak.

Praise-giver: Points out the group's good work.

Tension-reliever: Helps group members relax.

Dysfunctional Roles

Dysfunctional roles keep people from working well as a group.

Attacker: Questions others' abilities or motives.

Blocker: Prevents group progress by raising objections.

Joker: Distracts members by joking.

Recognition-seeker: Attracts attention by boasting, telling irrelevant experiences, or seeking sympathy.

Withdrawer: Refuses to respond openly or honestly.

DISAGREEING IN A GROUP

An effective group listens to the ideas of all its members, evaluates those ideas, and then uses the best ones. In the process, individuals should be able to disagree constructively, without feeling threatened or getting defensive. To improve your ability to disagree constructively, follow the tips below.

Be Tactful

Use "I" statements, not "you" statements. Report your thoughts and feelings without blaming others for not agreeing with you.

> **POOR:** "You're not getting the point, Caleb."
> **GOOD:** "Caleb, let me state that differently."

Disagree with Ideas or Work, Not with a Person

Phrase your comments to help the speaker know that even if you disagree with his or her idea, you do not think less of him or her.

> **POOR:** "I don't agree with you, Addison—your argument seems weak."
> **GOOD:** "I disagree with the proposal, Claire, because we need data from more than the last six months."

Communicate Ideas Without Emotional Distortion

- Keep quiet if you need to get control of your emotions.
- Clarify your feelings by answering these questions: "What's bothering me?" and "What can I do about it?"
- Write out your points and then read only what you've written.

Be Willing to Disagree, Even If It Is Difficult

Group members who can't voice their disagreement are victims of "group think"—a condition in which all members support an idea or policy not because they agree with it, but because they're unable or unwilling to disagree. In the process, the group loses access to valuable ideas.

Be Responsible

After the group has made its decision, move on.

CHECKLIST for Communicating in a Group

Use this checklist to evaluate the quality of your group communication skills.

- ☐ **Beginning a Group.** I recognize the stages of group development (groping, griping, grasping, grouping) and help the group work through those stages.
- ☐ **Brainstorming.** I help the group (1) generate ideas spontaneously, (2) discard weaker ideas, and (3) select the best ideas.
- ☐ **Problem Solving.** I help the group define and analyze the problem, select the best solution in accordance with clear standards, and implement the solution.
- ☐ **Presenting Information.** I introduce a topic and its relevance, present supporting information clearly, and lead the group in a probing discussion.
- ☐ **Making Decisions.** I understand methods of decision making (authority, minority, majority, and consensus) and help the group implement the one it chooses.
- ☐ **Listening.** I listen actively, focusing on the speaker, concentrating on the message, evaluating it thoughtfully, and avoiding unnecessary interruptions.
- ☐ **Responding.** I sensitively respond to members of the group by using forms of address and other words that are polite and professional.
- ☐ **Role Playing.** I understand the roles commonly played in group discussion, and I practice positive roles appropriate for the group and the situation.
- ☐ **Disagreeing.** When I disagree, I do so tactfully, focusing on issues (rather than people) and presenting my ideas in a thoughtful, measured manner.

CRITICAL THINKING Activity

Choose and record a group discussion on a TV program. Use the recording to assess which role (or roles) each member of the group plays. Analyze how the role (or roles) played by each group member affects his or her ability to facilitate the discussion.

WRITING Activities

1. Review the recording that you used for the Critical Thinking activity, and write a report in which you (a) analyze how well each participant voiced his or her disagreement with others' ideas, and (b) explain how the practices you observed enhanced or diminished the group's discussion.
2. Write a report in which you describe your contribution to a recent formal or informal group discussion. Include what you said, how you said it, the roles that you played, and the quality of your contribution.

In this chapter

Formal Versus Informal Meetings **666**
Formal Meetings **667**
Order of Business for a Meeting **668**
Making Motions **669**
Officers and Their Responsibilities **670**
Guidelines for Minutes **672**
Checklist for Meeting Minutes **676**
Critical Thinking Activity **676**
Writing Activity **676**

CHAPTER
43

Communicating in Meetings

If three "techies" in a Denver law firm sit in the lunchroom talking gigabytes, is that a business meeting? If four Texas State Troopers stand beside the highway discussing DUIs, is that a business meeting?

Yes, both gatherings are business meetings in the sense that people get together and talk in an effort to do their work. But neither is a formal meeting. Instead, both are the informal kind of get-togethers that happen most often in the workplace.

Formal meetings are no less useful or important than informal ones. The formal type are simply more structured, including elements such as a written agenda and more specific rules for opening the meeting, offering ideas, making motions, and making decisions.

To thrive in the workplace, you need the knowledge and skill to participate fully in both types of meetings. This chapter includes guidelines, tips, and rules that will help you do so.

FORMAL VERSUS INFORMAL MEETINGS

Most organizations conduct business meetings using a combination of informal and formal procedures. For example, while floor managers in a South Carolina furniture store meet weekly with no published agenda (informal), the shop manager keeps detailed minutes (formal). By contrast, while employees in a Wisconsin publishing house do use a written agenda in their weekly production meetings (formal), they make most decisions through group discussion and consensus (informal).

Knowing the difference between informal and formal procedures will help you understand which type is being used during a meeting, and how you can best participate.

Informal Procedures

- No prepared agenda
- No minutes taken during meeting
- Good for short conferences and making on-the-spot, daily decisions
- Usually unscheduled and held when needed
- Generally efficient for small groups
- Decisions made by authority, minority, majority, or consensus—whatever is appropriate at the time
- No elected or appointed leader
- People speak and make decisions using only rules of common courtesy

Formal Procedures

- Prepared agenda
- Detailed minutes record things such as individuals present or absent, topics discussed, key comments, all motions, people who make and second motions, and all decisions
- Good for making a decision when it's necessary to record both the decision and the dialogue that led to it
- Usually held on a regular, scheduled basis
- Generally efficient for larger groups
- Decisions made according to the rules in the group's constitution
- Leader is formally elected or appointed
- People speak and make decisions according to the rules of parliamentary procedure
- Officers (e.g., president, vice president, secretary, and treasurer) play the roles as described on pages 670–671.

FORMAL MEETINGS

Since its birth in the thirteenth century, Great Britain's Parliament has used a set of rules (called *parliamentary law*) to guide its proceedings. These rules became known as parliamentary procedure. In 1876, Henry M. Robert simplified the rules and wrote them in a handbook called *Robert's Rules of Order*. Today, business people follow the same rules to conduct their formal meetings.

Principles of Parliamentary Procedure

Parliamentary rules are designed to help a group work together. The guidelines are based on the following principles:

1. A group must work in a peaceful, orderly way.
2. The group makes decisions by "majority rule." They vote, and the side with the majority of votes wins. (For information on what constitutes a majority, see "Making Decisions," page 659.)
3. The minority must be treated fairly and respectfully.
4. The group chooses its own officers who must serve the needs of the group.
5. All members are equal and have the opportunity to do the following:
    - Attend meetings.
    - Be informed about what's going to happen in a meeting.
    - Have copies of the group's rules and policies.
    - Make motions.
    - Vote.
    - Nominate candidates for office.
    - Run for office.
    - Disagree with the group.
    - Inspect the official records of the group.
    - Resign.
    - Be treated according to the group's rules, even when disciplined by the group.

ORDER OF BUSINESS FOR A MEETING

The order of business is the sequence of events that should take place during a meeting. The sequence helps the group think about their present work in relation to what they did in past meetings and what they plan to do.

1. **Call to order.** The chairperson says something like, "This meeting will now come to order," or "I believe everyone is here, so let's begin."
2. **Approving the agenda.** The chair asks the group to look at the agenda and suggest necessary additions or changes.
3. **Reading the minutes.** The chair asks the secretary to read the minutes aloud. After the secretary has done so, the chair asks, "Are there any additions or corrections?" If members have corrections, those are made. Then the chair asks for a motion to approve the minutes. After the motion is approved, the chair says, "The minutes are approved." (At this point, the minutes become an official—and legal—record of the meeting.)
4. **Officers' reports.** Officers may present reports. If a report is detailed, an officer usually distributes printed copies.
5. **Committee reports.** Committees who have been given a task by the president or the group are invited to report. If a committee has nothing to present, the committee chairperson declines by saying, "No report at this time."
6. **Old business.** The chair asks, "Is there any old business that we need to discuss?" ("Old business" includes issues from previous meetings that the group has not finished discussing, and important updates.)
7. **New business.** The chair asks, "We'll now deal with the new business." (These items are printed in the agenda or have been added to the agenda in step 2.)
8. **Announcements.** The chair asks for announcements. Usually these include information about future meetings or issues related to the agenda.
9. **Adjournment.** The chair closes the meeting by asking for a motion to adjourn. After the motion is made and approved, the chair says, "This meeting is adjourned."

MAKING MOTIONS

How do things get done in formal meetings? You make a motion. A *motion* is a proposal asking the group to think about an idea and to act on it. In informal discussion, a motion sounds like a suggestion: "I think that Jean's logo looks good and that we should adopt it, along with the new fonts, of course." In a formal meeting, you present the idea more precisely: "I move that we adopt Jean's logo but postpone using it until we approve the new fonts."

Being precise is important because the secretary will record exactly what you say, and the group will discuss your statement and act on it. A carelessly worded motion leaves the group unclear about what you want them to do. As a result, the group wastes time trying to understand or change the motion.

To make a motion, go through the following steps:

1. "Address" (or get the attention of) the chair. Usually you just raise your hand.
2. The chair "recognizes" you (or invites you to speak) by saying, "Yes, Miriam . . ."
3. You state the motion, and the secretary writes exactly what you say.
4. The chair asks for a "second" by saying, "Is there support for the motion?" If someone says "Support" or "I second that," the motion is ready for discussion.
5. The chair states the motion to the group or asks the secretary to read it.
6. The chair asks for discussion, and members give their opinions.
7. The motion may be amended (changed) or tabled (action on it is postponed), but eventually the group votes to accept or reject the motion.

There are four kinds of motions: (1) main, (2) subsidiary, (3) incidental, and (4) privileged. Each kind is used to accomplish different tasks, and different rules apply to each type. For example, some motions need to be seconded and voted on, whereas others do not. For details, see Table 43.1, Common Parliamentary Procedures, on page 675.

OFFICERS AND THEIR RESPONSIBILITIES

Groups usually have at least four officers: president, vice president, secretary, and treasurer. The group gives its officers the authority to carry out specific actions on behalf of the group.

President

The president is the group's primary officer and is responsible for the tasks listed below.

Before the meeting

- Prepare an agenda that lists what you plan to do during the meeting.
- Arrange for a comfortable meeting place that is accessible to all members and has necessary furniture, audio-visual equipment, and related materials.
- Send (or ask the secretary to send) group members copies of the agenda and the following information:
  - Time and place of the meeting
  - Approximate length of the meeting
  - Locations of parking, registration desk, and meeting room
  - Materials to read before the meeting or take to the meeting
  - Arrangements for meals (if the meeting extends over mealtime)
  - Available office equipment (e.g., projectors or copy machines) at meeting sites that are not offices

During the meeting

- Open the meeting by calling it to order, welcoming members, and explaining the purpose of the meeting.
- Introduce each discussion topic clearly.
- Keep the discussion focused on the topic and the agenda.
- Encourage all members to participate.
- Discourage any members from monopolizing the discussion.
- Summarize the discussion and recall the agenda if the group gets stalled.
- Call for a consensus ("Do we agree that . . . ?") if an issue seems unclear.
- Keep the discussion focused on ideas, not on personalities.
- Resolve conflicts fairly.
- Don't push your personal agenda or manipulate the discussion.
- Honor the schedule, but don't make poor decisions for lack of time.
- Encourage politeness and mutual respect by modeling those traits in your own behavior.
- Bring the discussion to a close and adjourn the meeting.

Vice President
- Chair a meeting when the president is absent.
- Take the office of the president if he or she resigns or is permanently unable to do the work.
- Represent the president when he or she asks you to do so.
- Do any assignments that the group's constitution requires the vice president to do.
- Help the president by doing tasks that he or she requests.
- Do your work well and avoid competing with or usurping the authority of the president.

Secretary
- Write minutes—the official record of a meeting (see pages 672–674).
- Send correspondence when asked to do so by the group or president.
- Write clear, correct, properly formatted documents that facilitate business and reflect well on the organization.

Before the meeting
- Make copies of documents to be discussed during the meeting.
- Distribute the copies to group members.
- Take reference materials that the president or others may need during the meeting.

During the meeting
- Check attendance.
- Read minutes of the previous meeting.
- Give the secretary's report (mention correspondence the group has received or that you have sent).
- Search minutes (or reference materials) for information requested by members.

Treasurer
- Keep the group's financial records (income and expenses).
- Pay the group's bills.
- Send invoices to anyone who owes money to the group.
- Prepare a financial report showing (1) what was spent and taken in since the last meeting and (2) the current balance.
- Prepare periodic reports as called for by the group's constitution.
- Distribute financial reports and present the information.
- Uphold the organization's integrity by keeping records that are clear, correct, and in conformance with professional bookkeeping procedures.

GUIDELINES FOR MINUTES

Good minutes create an accurate record of what took place in a meeting. In fact, after the group approves the minutes, usually at the following meeting, the minutes become an official (and legal) record.

1 Prewrite

Rhetorical analysis: Consider your purpose, your audience, and the context.
- ☐ How does the organization use the minutes now, and how might it use them in the future? What are my social and legal responsibilities?
- ☐ What topics or documents will be discussed, and who will give presentations?
- ☐ What is the history of these topics, and what facts should I review?

Prepare to draft.
- ☐ Read the previous minutes, the meeting's agenda, and discussion materials.
- ☐ During the meeting, listen carefully and take accurate notes.

2 Draft

Organize your minutes into three parts.

[opening] List details identifying the meeting—group's name, date of meeting, location, time, people present (and absent), and the purpose of the meeting.

[middle] Describe what happened in the meeting:
- ☐ Record motions word-for-word, along with who made and seconded each, the number of votes for and against, and the result (passed or failed).
- ☐ Summarize the discussion, leaving out personal feelings, emotions, or comments about personalities.
- ☐ State follow-up action, including who is to do what by when.
- ☐ Number items clearly (see the numbering system on pages 673–674).

[closing] Conclude the minutes:
- ☐ List the next meeting's business, date, time, and location.
- ☐ Print and sign your name.

3 Revise

Review your draft's ideas, organization, and voice.
- ☐ Is each action stated clearly and correctly?
- ☐ Is the wording of each motion, argument, or quotation correct?
- ☐ Is the order of events stated correctly?
- ☐ Is the tone objective, respectful, and fair?

4 Refine

Check your writing line by line for the following:
- ☐ Correct spelling, punctuation, capitalization, and usage
- ☐ Proper form and correct numbering

Minutes

The minutes below record proceedings of the Management Committee of the non-profit Family and Children Coalition.

Rhetorical situation
Purpose: To create a clear and accurate record of what took place in the meeting
Audience: Board of directors, management, employees, sponsors, and clients of the Family and Children Coalition
Format: Organization's name; meeting time and place; list of those present or absent; followed by the group's actions, each item numbered and stated succinctly

---

1 of 2

Management Committee, Family and Children Coalition

February 2, 2005: Conference Room

Present: Susan Van Weelden, Christine Turpin, Ezra Milford, Gabriel Bozeman, Elijah Wahpeton, Jack Spencer

Absent: Zachary Livingston, Nathaniel Traverse, Benjamin Racine

0519   Christine called the meeting to order and the agenda was approved.

0520   The minutes for the January 5 meeting were read.
- Jack noted an error in minute 16, which stated that the conference he and Elijah were planning was behind schedule and may be postponed. Jack explained that while much work needed to be done, the conference would take place on May 20, 2005, as planned.
- Susan acknowledged the error and revised the minute to read, "Planning for the Parenting Conference is behind schedule, but Jack and Elijah believe that it will take place on May 20, 2005, as planned."
- Elijah then moved that the minutes be approved, Ezra seconded the motion, and the motion passed.

0521   Gabriel reported on the Children's Advocacy Conference that he attended in Charlotte. He distributed new state standards for counselors' caseloads (1/20), noting that FCC's caseloads average 1/17.

0522   Ezra described the January 10 contract-negotiation session and asked for feedback on the board's proposed dental-optometry plan. Christine suggested that Ezra e-mail the plan to all employees to get their assessment. Ezra agreed.

[opening]
The writer lists the group's name, meeting date, location, and members present or absent.

[middle]
Each minute is numbered showing the year (05) plus the minute's number (19).

Each minute (including corrections) is stated accurately and succinctly.

One topic is addressed in each minute; related actions are presented as bulleted points.

[middle]

The voice is objective.

0523 Jack and Elijah reviewed the plans for the conference that they are organizing for parents and other caregivers who experience stress with the challenges of caring for young children. Jack noted that during the conference, he will present a workshop for parents with twins and triplets. Elijah reported that he will lead a workshop on how to discipline children in ways that develop their sense of security and well-being.

0524 Susan presented the revised confidentiality procedure, noting the changes requested by the staff.

The number of votes cast and the resulting action are given.

- Ezra moved and Gabriel seconded a motion that the Management Committee approve the changes. The motion passed (6—Yes; 0—No).
- Susan then asked Christine to address the revised confidentiality procedure in her report to the board of directors on March 7, and Christine agreed to do so.

0525 Zachary will report on his meeting with Gerlene Peoria regarding AEA-5's ability to help clients complete GED programs.

[closing]

The next meeting's business, time, and place are listed. The writer's name is printed and signed.

0526 Next meeting: March 3, 2005, at 3:30 p.m. in the conference room.

Minutes submitted by Susan Van Weelden.

Susan Van Weelden

Susan Van Weelden

Chapter 43 Communicating in Meetings

Table 43.1 Common Parliamentary Procedures

Motion	Purpose	Subsidiary Needs Second	Debatable	Amendable	Required Vote	May Interrupt Speaker	Motion Applied
I. Original or Principal Motion							
1. Main motion (general) Main motion (specific)	To introduce business	Yes	Yes	Yes	Majority	No	Yes
a. To reconsider	To reconsider previous motion	Yes	When original motion is	No	Majority	Yes	No
b. To rescind	To nullify or wipe out previous action	Yes	Yes	Yes	Majority or two-thirds	No	No
c. To take from the table	To consider tabled motion	Yes	No	No	Majority	No	No
II. Subsidiary Motions							
2. To lay on the table	To defer action	Yes	No	No	Majority	No	No
3. To call for previous question	To close debate and force vote	Yes	No	No	Two-thirds	No	Yes
4. To limit or extend limits of debate	To control time of debate	Yes	No	Yes	Two-thirds	No	Yes
5. To postpone to a certain time	To defer action	Yes	Yes	Yes	Majority	No	Yes
6. To refer to a committee	To provide for special study	Yes	Yes	Yes	Majority	No	Yes
7. To amend	To modify a motion	Yes	When original motion is	Yes (once only)	Majority	No	Yes
8. To postpone indefinitely	To suppress action	Yes	Yes	No	Majority	No	Yes
III. Incidental Motions							
9. To raise a point of order	To correct error in procedure	No	No	No	Decision of chair	Yes	No
10. To appeal for a decision of chair	To change decision on procedure	Yes	If motion does not relate to indecorum	No	Majority or tie	Yes	No
11. To suspend rules	To alter existing rules and order of business	Yes	No	No	Two-thirds	No	No
12. To object to consideration	To suppress action	No	No	No	Two-thirds	Yes	No
13. To call for division of house	To secure a countable vote	No	No	No	Majority if chair desires	Yes	Yes
14. To close nominations	To stop nomination of officers	Yes	No	Yes	Two-thirds	No	Yes
15. To reopen nominations	To permit additional nominations	Yes	No	Yes	Majority	No	Yes
16. To withdraw a motion	To remove a motion	No	No	No	Majority	No	No
17. To divide a motion	To modify a motion	No	No	Yes	Majority	No	Yes
IV. Privileged Motions							
18. To fix time of next meeting	To set time of next meeting	Yes	No, if made when another question is before the assembly	Yes	Majority	No	Yes
19. To adjourn	To dismiss meeting	Yes	No	Yes	Majority	No	No
20. To take a recess	To dismiss meeting for specific time	Yes	No, if made when another question is before the assembly	Yes	Majority	No	Yes
21. To raise question of privilege	To make a request concerning rights of assembly	No	No	No	Decision of chair	Yes	No
22. To call for orders of the day	To keep assembly to order of business	No	No	No	None unless objection	Yes	No
23. To make a special order	To ensure consideration at specified time	Yes	Yes	Yes	Two-thirds	No	Yes

CHECKLIST for Meeting Minutes

Use the seven traits to check your document and then revise as needed.

- ☐ **Ideas.** The minutes accurately summarize the group's meeting, including details such as when, where, who attended, what was discussed, what was decided and by whom, what is planned for the following meeting, and when and where it will be held.

- ☐ **Organization.** The minutes begin by introducing the group's name, meeting date and place, and members present or absent. Each significant action of the group is then described in a separate statement or "minute," and the minutes are organized chronologically. The first minute of the first meeting each year is numbered "1" and all subsequent minutes are numbered sequentially.

- ☐ **Voice.** The tone is fair and thoughtful, reporting each action (including emotional comments or decisions) objectively.

- ☐ **Words.** The wording is spare, well chosen, specific, and correct. Quotations and motions are stated precisely as they were delivered in the meeting.

- ☐ **Sentences.** The structures are brief, direct, and easy to read. Multiple sentences within a minute are clearly related and linked with transitions. The punctuation of a quotation reflects the speaker's meaning, not the writer's opinion of the speaker or quotation.

- ☐ **Copy.** The document accurately cites details, and includes no errors in mechanics, usage, spelling, grammar, or punctuation. Each minute is correctly numbered.

- ☐ **Design.** Each minute is featured as a separate paragraph or unit of information. Fonts are traditional and businesslike, and boldfacing and underlining are used sparingly.

CRITICAL THINKING Activity

Working with a classmate, select a group (city council, school board, civic organization, or business) whose meeting you may attend—preferably a group that allows you to videotape the meeting. Tape the activity and take careful notes. Use the tape and notes to analyze how the proceedings match the parliamentary procedural guidelines for organizing a meeting, making motions, discussing ideas, making decisions, and using minutes.

WRITING Activity

Working with a classmate, ask permission to attend a meeting of a local business or community group. Request copies of the agenda and the previous meeting's minutes, and take notes during the meeting. Write a report in which you (a) describe the group's business-meeting practices, and (b) assess the quality of those practices.

In this chapter

Giving Presentations **678**
Planning Your Presentation **679**
Organizing Your Presentation **680**
Writing Your Presentation **683**
Writing with Style and Motivational Appeals **684**
Using Visual Support **690**

Developing Computer Presentations **691**
Practicing Your Delivery **692**
Overcoming Stage Fright **693**
Checklist for Writing Presentations **694**
Critical Thinking Activity **694**
Writing Activities **694**

CHAPTER

44

Writing and Giving Presentations

People in business regularly give presentations to demonstrate new products, introduce new programs, or report on projects. The challenge is to share ideas orally in a clear, concise, and effective manner. Unlike a piece of writing that the reader can review at a later date, a presentation must be so clear and memorable that the listener gets the message after hearing it only once.

This chapter will help you produce such presentations. Specifically, it includes information that will help you plan, write, and deliver a presentation, whether it's a one-minute impromptu talk or an hour-long report with presentation software. You'll find practical strategies for drafting an attention-grabbing introduction, organizing the body of your presentation, and developing a focused conclusion.

In addition, you'll find a model presentation in outline form, a model in manuscript form, help with developing computer presentations, and tips for overcoming stage fright. Read the chapter carefully and then polish your presentation skills by doing the end-of-chapter activities. In the process, you'll polish a key workplace skill: the ability to share your ideas orally in a clear, concise, and effective manner.

GIVING PRESENTATIONS

Regardless of the topic, form, or length of your presentation, you can follow the same basic steps to develop and present it.

Getting Started

The first step in preparing an oral presentation is getting an overview of the task. Begin by asking yourself some pertinent questions: What is my purpose? Who is my audience? What is my topic? Answering these questions will help you write the presentation and shape its delivery.

What is my purpose?
- Am I going to explain something?
- Am I trying to persuade or inspire my audience to do something?
- Am I hoping to teach my audience about something?

Who is my audience?
- Is it an in-house group or an outside group?
- How many people are in the group, and what are their ages, backgrounds, and interests?
- What will people already know about the topic, and what will they want or need to know?
- What will their attitude be toward the topic and toward me?

What is my topic?
- What do I already know about the topic?
- What do I need to learn, and where can I find that information?
- What support materials (displays, computer projections, handouts) would help me present my message?

> "It usually takes me more than three weeks to prepare a good impromptu speech."
> —Mark Twain

Stating Your Main Idea

After you've made decisions about your purpose, audience, and topic, it's a good idea to write out the main idea you want to communicate.

Begin with the simple statement "My purpose is . . ." and then add a phrase like *to explain*, *to persuade*, or *to inform*. Finish by inserting your audience and your topic, along with the main idea of the presentation.

> My purpose is to explain to the Board of Directors how our new confidentiality procedure will help our staff serve our clients better.

WRITE Connection

To see the outline that develops this idea, turn to pages 686–687.

PLANNING YOUR PRESENTATION

Drafting an Outline

After you've clarified your purpose, audience, and topic—and you've written your purpose statement—think about how to present your message. Begin by brainstorming points that you want to communicate, and then organize those points into a list or working outline. For a brief, informal presentation, this list or outline might be the only script you need.

For a longer, more challenging presentation, you might need to significantly revise and develop your outline as you research and write the script. This outline serves as your tool for gathering and organizing your thoughts. (See page 683.)

Gathering Information

Using your outline as a guide, gather the information you need. Begin by reviewing key documents, manuals, and company materials related to the topic. If necessary, read current articles, review videos, explore the Internet, and talk with other people. What you gather will depend on your purpose, topic, resources, and available time.

For help in finding and organizing information, see pages 119–151.

Thinking About Support Materials

As you gather information, keep a list of graphics, displays, and handouts that could make your presentation clearer and more interesting. For example, charts, tables, and graphs can help an audience grasp the meaning of complex data. Technical drawings or sketches can help listeners visualize a product or site. Demonstrations or video clips can help listeners better understand a process or connect with the people involved.

Review the list below for items appropriate to your topic, audience, and setting (including available equipment). Then, as you do your research, make a note about an item that you could use and how you could use it (as a display, projection, handout, and so forth).

Audio clip (music)	Key quotation
Bibliography	List of authorities
Brochure	Overhead
Cartoon	Photograph
Chart	Sample product
Company document	Sketch
Demonstration	Table
Graph	Technical drawing
Handheld prop	Video clip

ORGANIZING YOUR PRESENTATION

After you've gathered your information, you must organize and develop the message. Start by thinking about your presentation as having three distinct parts: (1) introduction, (2) body, and (3) conclusion. The guidelines on this page and the following two pages will help you integrate, organize, and refine all three parts so that they communicate the message and achieve your purpose.

Introduction

For any speaking situation, you should develop an introduction that does the following:

- Greets the audience and grabs their attention.
- Communicates your interest in them.
- Introduces your topic and main idea.
- Shows that you have something worthwhile to say.
- Establishes an appropriate tone.

You might greet the audience in many ways, including the following: introducing yourself; thanking people for coming; or making appropriate comments about the occasion, the individuals present, or the setting. After making these comments, introduce your topic and main idea as quickly and as clearly as you can. For example, you could open with one of these attention-grabbing strategies:

- A little-known fact or statistic
- A series of questions
- A humorous story or anecdote
- An appropriate quotation
- A description of a serious problem
- A cartoon, picture, or drawing
- A short demonstration
- A statement about the topic's importance
- An eye-catching prop or display
- A video or an audio clip
- A playful comment about the audience, previous speaker, or setting
- An insight into the history of the occasion, setting, or topic
- A provocative statement or thematic challenge

> "What this country needs is more free speech worth listening to."
> —Hansell B. Duckett

Body

The body of your presentation should deliver the message and supporting points so clearly that the audience understands the presentation after hearing it just once. The key to developing such a clear message is choosing an organizational pattern that fits your purpose statement.

Before you outline the body of your presentation, take a moment to review what you want it to do: Explain a problem? Promote an idea? Teach a process? Be sure the organizational pattern you select will help you do that. For example, if you want to teach a process, the outline should list the process steps in chronological order. Consider these potential organizational patterns:

Chronological order. Arrange information according to the time order in which events (steps in a process) take place.

Order of importance. Arrange information according to its importance: greatest to least, or least to greatest.

Comparison/contrast. Give information about subjects by comparing and contrasting them.

Cause and effect. Give information about a situation or problem by showing (1) the causes and (2) the effects.

Order of location. Arrange information about subjects according to where things are located in relation to each other.

Problem/solution. Describe a problem and then present a solution for it.

After deciding how to organize your message, write it out in either outline or manuscript form. For help, see the following tips and the models on pages 683, 686–689.

Body-Building Tips

- Build your presentation around several key ideas. (Don't try to cover too much ground.)
- Follow a logical pattern.
- Write with a personal, natural voice.
- Address all points forecast in the introduction.
- Support your main points with reliable facts and clear examples.
- Present your information in short, easy-to-follow segments.
- Build clear transitions between segments.
- Use positive, respectful language. (Avoid jargon.)
- Use graphic aids and handouts.

> "When high words confuse the talk, low words will untangle it."
>
> —Jobo proverb

Conclusion

A strong introduction and conclusion work like bookends supporting the body of the presentation. The introduction grabs the audience's attention, sets the tone, states the main idea, and identifies the key points of the message. Almost in reverse, the conclusion reviews those points, restates the main idea, reinforces the tone, and refocuses the audience on what it should think about or do. Together, those bookends emphasize and clarify the message so that the audience understands and remembers it.

Here are some strategies—which you can use alone or in combination—for concluding a presentation:

- Review your main idea and key points.
- Issue a personal challenge.
- Come "full circle." (State those arguments or details that back up your original point.)
- Recommend a plan of action.
- Forecast a time when the problem addressed will be resolved.
- Suggest additional sources of information.
- Recite a summarizing poem or lyric.
- Thank the audience and ask for questions.

Q & A Session

Following your presentation, you may want to invite your audience to ask questions. Very often, a question-and-answer (Q & A) session is the real payoff for participants. They can ask for clarification of points or ask how your message applies to their personal situations. Audience members may even offer their own insights or solutions to problems mentioned in the presentation. The following suggestions will help you lead a good Q & A session:

- Listen carefully and think about each part of the question.
- Focus on answering the question rather than defending yourself.
- Repeat or paraphrase questions for the benefit of the entire group.
- Answer the question concisely and clearly.
- If helpful, link your answer to points raised in the presentation.
- Respond honestly when you don't know the answer, and offer to find an answer.
- Ask for a follow-up question if someone looks confused after your answer.
- Look directly at the group when you answer.
- Be prepared to pose an important question or two if no one asks a question.
- Conclude by thanking the audience for their participation.

> "We cannot ignore tone of voice or attitude. These may be just as important as the words used."
> —Maurice S. Trotter

WRITING YOUR PRESENTATION

How much of your presentation you actually write out depends on your topic, audience, purpose, and—of course—your personal style. The three most common forms to use when making a presentation are a list, an outline, and a manuscript.

List

Use a list for a short, informal speech such as an after-dinner introduction. Think about your purpose and then list the following:

- Your opening sentence (or two)
- A summary phrase for each of your main points
- Your closing sentence

1. Opening sentence or two
2. Phrase 1
   Phrase 2
   Phrase 3
3. Closing sentence

Outline

Use an outline for a more complex or formal topic. You can organize your material in greater detail without tying yourself to a word-for-word presentation. Here's one way you can do it:

- Opening (complete sentences)
- All main points (sentences)
- Supporting points (phrases)
- Quotations (written out)
- All supporting technical details, statistics, and sources (listed)
- Closing (complete sentences)
- Notes on visual aids (in capital letters or boldface)

I. Opening statement
   A. Point with support
   B. Point (purpose or goal) [VISUAL 1]
II. Body (with three to five main points)
   A. Main point
      1. Supporting details
      2. Supporting details
   B. Main point
      1. Supporting details
      2. Supporting details
   C. Main point
      1. Supporting details
      2. Supporting details
III. Closing statement
   A. Point, including restatement of purpose
   B. Point, possibly a call to action [VISUAL 2]

Manuscript

Use the guidelines below if you plan to write out your presentation word for word as you plan to give it:

- Double-space pages (or cards).
- Number pages (or cards).
- Use complete sentences on a page (do not run sentences from one page to another).
- Mark difficult words for pronunciation.
- Mark the script for interpretation (see the symbols on page 692).

WRITING WITH STYLE AND MOTIVATIONAL APPEALS

Style in writing or speaking is the "music" in the message, and stylistic devices help create this music. Style lures a listener into focusing on the message and enjoying it. Similarly, emotional appeals lure listeners with emotional "bait"—an idea, a story, or a challenge that draws the listener into identifying with the speaker and topic.

Study the stylistic devices and emotional appeals that follow. Then read aloud the examples taken from the speeches of President John F. Kennedy. As you read, listen to how Kennedy's music (or style), together with his emotional bait, distinguish his point and lure you into supporting it.

Allusion. An allusion is a reference in a speech to a familiar person, place, or thing.

Appeal to the Democratic Principle

One hundred years of delay have passed since President Lincoln freed the slaves, yet their heirs, their grandsons, are not fully free. *(Radio and television address, 1963)*

Analogy. An analogy is a comparison of an unfamiliar idea to a simple, familiar one. The comparison is usually quite lengthy, suggesting several points of similarity. An analogy is especially useful when attempting to explain a difficult or complex idea.

Appeal to Common Sense

In our opinion the German people wish to have one united country. If the Soviet Union had lost the war, the Soviet people themselves would object to a line being drawn through Moscow and the entire country defeated in war. We wouldn't like to have a line drawn down the Mississippi River. . . . *(Interview, November 25, 1961)*

Anecdote. An anecdote is a short story told to illustrate a point.

Appeal to Pride and Commitment

Frank O'Connor, the Irish writer, tells in one of his books how as a boy, he and his friends would make their way across the countryside and when they came to an orchard wall that seemed too high and too doubtful to try and too difficult to permit their voyage to continue, they took off their hats and tossed them over the wall—and then they had no choice but to follow them. This nation has tossed its cap over the wall of space, and we have no choice but to follow it. Whatever the difficulties, they will be overcome. *(San Antonio address, November 21, 1963)*

Antithesis. Antithesis balances or contrasts one word or idea against another, usually in the same sentence.

Appeal to Common Sense and Commitment

Mankind must put an end to war, or war will put an end to mankind. *(Address to the United Nations, 1961)*

Irony. Irony is using a word or phrase to mean the exact opposite of its literal meaning, or to show a result that is the opposite of what would be expected or appropriate; an odd coincidence.

Appeal to Common Sense

They see no harm in paying those to whom they entrust the minds of their children a smaller wage than is paid to those to whom they entrust the care of their plumbing. *(Vanderbilt University, 1961)*

Negative definition. A negative definition describes something by telling what it is not, rather than, or in addition to, what it is.

Appeal for Commitment

[M]embers of this organization are committed by the Charter to promote and respect human rights. Those rights are not respected when a Buddhist priest is driven from his pagoda, when a synagogue is shut down, when a Protestant church cannot open a mission, when a cardinal is forced into hiding, or when a crowded church service is bombed. *(Address to the United Nations, September 20, 1963)*

Parallel structure. Parallel structuring is the repeating of phrases or sentences that are similar (parallel) in meaning and structure. *Repetition* is the repeating of the same word or phrase to create a sense of rhythm and emphasis.

Appeal for Commitment

Let every nation know, whether it wishes us well or ill, that we shall pay any price, bear any burden, meet any hardship, support any friend, oppose any foe, in order to assure the survival and the success of liberty. *(Inaugural address, 1961)*

Quotations. Quotations, especially of well-known individuals, can be effective in nearly any speech.

Appeal for Emulation or Affiliation

At the inauguration, Robert Frost read a poem which began "the land was ours before we were the land's"—meaning, in part, that this new land of ours sustained us before we were a nation. And although we are now the land's—a nation of people matched to a continent—we still draw our strength and sustenance . . . from the earth. *(Dedication speech, 1961)*

Rhetorical question. A rhetorical question is a question that is asked to emphasize a point, not to get an answer.

Appeal to Common Sense and Democratic Principle

"When a man's ways please the Lord," the Scriptures tell us, "he maketh even his enemies to be at peace with him." And is not peace, in the last analysis, basically a matter of human rights—the right to live out our lives without fear of devastation—the right to breathe air as nature provided it—the right of future generations to a healthy existence? *(Commencement address, 1963)*

Presentation in Outline Form

Report on FCC's Confidentiality Procedure

[opening]
The opening is written out word for word and placed in boldface.

I. Good afternoon, everyone. I appreciate this opportunity to report on our work here at the Family and Children Coalition. Today I want to focus on one topic that affects all of our work—the new confidentiality procedure that was adopted last week. We believe this procedure will help us serve our clients more effectively. We think it's an effective tool because

- It is based on sound policy,
- It will help the staff implement the policy in a uniform manner, and
- It will enable staff and clients to develop trusting relationships with one another.

Italics and brackets signal a speaker's prompt.

[*Identify the handout and read the policy.*]

[middle]
Main points are stated as full sentences (word for word).

II. First, let's examine four strengths of the policy on which the new procedure is based. Note how the policy

Supporting details are listed as phrases.

- Enables clients to control most personal information.
- Calls for written records of who receives case-related information.
- Satisfies legal requirements related to privacy issues.
- Helps staff and clients develop trust.

[*Identify the handout and read the procedure.*]

III. Second, the confidentiality procedure will help staff members deal with confidentiality issues in a consistent, uniform manner. Note how the procedure

The speaker uses the phrases as cues and comments on each point.

- Lists issues that counselors must explain to clients at intake.
- Sets guidelines for releasing information to outside parties.
- Explains a client's recourse to a staff member's breach of confidentiality.
- Promotes the uniform application of the confidentiality policy—particularly by new staff and student interns.

2

IV. Third, the confidentiality procedure will help staff and clients establish the trusting relationships FCC needs to provide its services.

- Clients, needing a confidential advocate, often come to intake sessions fearful and suspicious.
- Abused wife with child: "**How do I know he won't find out that we are here? He said if I say anything, he'll kill me.**"
- Pregnant teenager: "**Thanks for listening . . . I just had to tell somebody.**"
- Clients become less fearful because they view the policy and procedure as contracts—promises that the information collected will be kept in confidence.
- Clients become less fearful because they feel information will be kept private from outsiders—staff can say that the confidentiality procedure does not allow them to release information.

V. As you know, the work that we do at the Family and Children Coalition requires that our staff and clients have trusting, confidential relationships with one another. To build such relationships, and to satisfy legal requirements related to privacy issues, FCC has long had a confidentiality policy. However, the staff believes that the new confidentiality procedure will help them provide better service to clients because the procedure (1) is based on sound policy, (2) improves uniform application of that policy, and (3) helps staff members and clients develop trusting relationships.

Are there any questions?

[middle]
Entire text is spaced for easy reading.

Quotations are written out word for word in boldface.

[closing]
The closing is written out and placed in boldface.

The speaker recaps main points and asks for questions.

Presentation in Manuscript Form

[opening]
[SLIDE 1] The title is projected.

The speaker delivers the speech as it appears on the page.

[middle]
[SLIDE 2] Monument

[SLIDE 3] Inscription

[SLIDES 4–8] People mentioned

[SLIDE 9] Company headquarters

All points and supporting details are stated.

[SLIDE 1] **Abix Technologies: Finding the Right Solutions**

Good afternoon, everyone, and welcome to Abix Technologies! This is Lynn, your tour guide, and I'm Benjamin Allyn, Director of Public Relations at Abix. Lynn will soon be taking you on a walk through our reception center, a research lab, and a manufacturing facility. During that tour, she will give you a lot of information and answer all of your questions. However, before Lynn takes over, I want to personally welcome each of you to Abix. In addition, I'd like to introduce you to our company by saying a few things about who we are, what we produce, and the people we serve.

First, who are we? [SLIDE 2] The inscription on the monument that you passed when entering the building answers the question with these words: [SLIDE 3] "Abix Technologies is an international corporation that provides technological solutions to environmental wastewater problems." What does that mean? It means that at Abix Technologies, [SLIDE 4] we have scientists who research wastewater problems and propose solutions. [SLIDE 5] We have engineers who develop products to implement those solutions. [SLIDE 6] We have people who produce the products. [SLIDE 7] We have sales personnel who market our products around the world. [SLIDE 8] And finally, we have technicians who service what we sell, wherever we sell it.

[SLIDE 9] **Second, what do we produce?** While Abix Technologies makes a wide variety of products for many different applications, it specializes in technology that disinfects wastewater. These products are

- Well researched,
- Environmentally safe, and
- Cost-effective.

[SLIDE 10] For example, as you tour the laboratory today, you'll meet researchers who have been working on a particularly challenging problem for more than three years. Although they needed only ten months to find a solution, more than two years later they're still refining it. Why? Because at Abix, products must not only solve problems, but must do so in ways that are environmentally safe and cost-effective. [SLIDE 11]

Third, who uses our products? The short answer to this question is "Smart people around the world!" In fact, as you visit the Shipping Department today, you'll see crews packaging products that will be sent to sites on three continents.

The longer answer to the question about our customer base is that our markets include [SLIDE 12] the United States, [SLIDE 13] Canada, [SLIDE 14] South America, [SLIDE 15] Europe, [SLIDE 16] Australia, [SLIDE 17] New Zealand, [SLIDE 18] the Middle East, [SLIDE 19] and the Far East. While serving such a broad clientele is not easy, we do it well for two reasons: [SLIDE 20]

1. We carefully assess each customer's needs to make sure that the products we sell meet those needs. [SLIDE 21]
2. We have offices in Sydney, Australia; London, England; Flint, Michigan; and Toronto, Ontario. Each office has highly trained technicians who respond to our customers quickly and effectively. [SLIDE 22]

While I'd like to tell you more, Lynn will show you these things for yourselves. So once again, welcome to Abix Technologies! I'm glad that you're here, and I hope that you enjoy the tour!

[middle]
[SLIDE 10]
Researchers

[SLIDE 11]
Products

[SLIDES 12–19]
Markets mentioned

[SLIDE 20]
Statement 1

[SLIDE 21]
Statement 2

[SLIDE 22]
Monument

[closing]
The main point is restated and a polite closing is added.

USING VISUAL SUPPORT

Most people are visual learners in the sense that they understand a message more clearly and recall it more readily when the oral presentation is supported by visual aids such as outlines, drawings, videos, or demonstrations. For this reason you should consider using visuals for all oral presentations, particularly those that are lengthy, complex, or technical. The key is choosing appropriate visuals and using them well.

When used well, visuals add clarity and impact to a presentation. Conversely, when not used effectively, they can get in the way of the message. For help in deciding which visuals to use and how to use them, consider these factors.

Types of Visuals

- Overhead transparencies are colorful, inexpensive, and easy to make. You can write on them during a presentation.
- Flipcharts are also colorful, inexpensive, and easy to make. You can also write on them during a presentation.
- Posters can be set up in the room before you begin, or you can introduce them at appropriate times during your presentation.
- Models, samples, or handouts can help you demonstrate a product or process. They also help your audience visualize what you're talking about.
- Computer presentations can communicate information in many forms, including quotations, photographs, technical drawings, maps, cartoons, video clips, and animations.

Tips for Using Visuals

- No matter which type of visuals you use, make words large enough to read from the back of the room.
- On visuals other than handouts, use key words and lists rather than sentences.
- Use strong contrasting colors (e.g., black on yellow) to feature the information you want to communicate.
- Always practice until you can use the visuals effectively.
- Remember that writing on visuals during a presentation can slow your delivery.
- Check out all electrical equipment thoroughly; take nothing for granted.

> "The best impromptu speeches are the ones written well in advance."
>
> —Ruth Gordon

DEVELOPING COMPUTER PRESENTATIONS

Business people commonly use computers to deliver presentations because this multimedia approach can powerfully reinforce and clarify a message. To use presentation software effectively, follow these guidelines.

Develop a design. Be sure your graphic design fits your topic and your audience—businesslike for a serious topic, casual for a team meeting, and so on.

Create pages. If a main idea has several parts, present each one on its own page. Each click of the mouse button (or computer key) should reveal a new detail.

Use transitions. Dissolves, fades, wipes, and other transitional effects refine a computer presentation and keep the audience's attention (as long as these stylistic devices don't detract from the message).

Try animation. Text can be animated to appear from off screen at just the right moment. Graphics can be made to appear one element at a time, and illustrations can change before the viewer's eyes. Remember to use special effects, especially animation, wisely.

Add sound. Just as graphics and animation can enhance a presentation, so, too, can sound. Music can serve as an introduction or backdrop, and sound effects can add emphasis. Voice recordings can add authority and help drive home key points.

Fine-tune your presentation. Practice delivering your presentation while clicking through your pages. Try it with an audience of co-workers, if possible, and ask for their input.

Check for word choice and style. Make sure that the words on the screen are key words. Use these words as talking points—don't try to cover any point word for word. Also, check that transitions, animations, and sounds are smooth and not disruptive.

Edit the final version. Check spelling, punctuation, usage, and other mechanics. Remember: On-screen errors are glaringly obvious to everyone.

Rehearse. Practice running the equipment until you can use it with confidence.

Make a backup copy. Protect all the effort you have invested in your presentation.

> "When you say something, make sure you have said it. The chances of having said it are only fair."
>
> —E. B. White

PRACTICING YOUR DELIVERY

Research shows that less than 40 percent of your message is communicated by your words. In contrast, more than 60 percent is communicated by your delivery—your voice, body language, and attitude. For this reason, rehearsing the delivery of a presentation is at least as important as revising the script.

Rehearsing Your Presentation

Keep going over your presentation until you're comfortable with it. Ask a family member or a co-worker to listen to you and offer feedback, or use a video recorder so that you can see and hear yourself. Practice these things:

Maintain eye contact with your audience. Eye contact helps people feel that you care about them, and it helps you learn how people are responding to your message.

Speak loudly and clearly. Also speak at an appropriate speed.

Take your time. Glance at your notes when necessary.

Use your hands to communicate. Practice using natural, unforced gestures.

Maintain a comfortable, erect posture. Avoid the following:

- Folding your arms across your chest
- Clasping your hands behind you
- Keeping your hands on your hips
- Rocking back and forth
- Fidgeting with objects
- Chewing gum

Use your voice effectively. You can mark your copy for vocal variety by using the techniques described next.

Marking Your Script

- Inflection (arrow up) for a rise in pitch, and (arrow down) for a drop in pitch.
  Why is education so important? I'll tell you why.
- Emphasis (underlining or boldface) for additional <u>drive</u> or **force**.
  Because education expands who we **are,** what we can **do,** and what we can **become.**
- Color (curved line or italic) for additional feeling or *emotion*.
  In other words, meaningful learning prepares us for meaningful living.
- Pause (diagonal) for a short / or long // pause.
  So/why should you get an education? You may as well ask, // "Why live?"
- Directions (brackets) for movement [walk to chart] or use of visual aids [hold up chart].
- Pronunciation (parentheses) for phonetic (fō NE tic) spelling of difficult words.

OVERCOMING STAGE FRIGHT

While it's okay to feel a little nervous before a presentation (the emotion keeps you alert), stage fright can limit your ability to communicate. The remedy for stage fright is confidence—confidence in what you want to say and how you plan to say it. To develop that confidence, take the following steps.

Presentation Preparation

- Know your subject well.
- Rehearse the presentation thoroughly, including the use of visuals.
- Schedule your time carefully, making sure to arrive early.
- Relax before the presentation by stretching or doing deep-breathing exercises, remembering that your presentation can be successful without being perfect.

The Room and Equipment

Confirm that the room is clean, comfortable, and well lit.

- Make sure tables and chairs are set up and arranged correctly.
- Check that audio-visual equipment is in place and working.
- Test microphone volume.
- Position the screen and displays for good visibility.

Personal Details

- Check clothing and hair.
- Arrange for drinking water to be available.
- Put your script and handouts in place.
- Place all support items (e.g., props or the projector's remote control) in easy reach.

Speaking Strategies

- Greet individuals as they arrive for the presentation.
- Learn some people's names.
- Be confident, positive, and energetic.
- Provide for audience participation such as a quick survey of the audience: "How many of you . . . ?"
- Speak up and speak clearly—don't rush.
- Reword and clarify when necessary.
- After the presentation, ask for questions and answer them clearly.
- Thank the audience.

CHECKLIST for Writing Presentations

Use the seven traits to check your document and then revise as needed.

- ☐ **Ideas.** The presentation clarifies my purpose (to explain, persuades, teach); introduces my topic; distinguishes my position on it; supports that position with details; and motivates listeners to believe my message.
- ☐ **Organization.** The presentation includes an engaging introduction; a logically ordered body with clear main points and strong supporting details; and a memorable conclusion that reinforces my main point and helps listeners embrace it.
- ☐ **Voice.** The tone is sincere (not manipulative), knowledgeable, and convincing.
- ☐ **Words.** The presentation uses words that are precise and appropriate for the topic, audience, setting, and situation.
- ☐ **Sentences.** The presentation includes sentences that are clear, concise, and linked. Stylistic devices such as repetition, parallel structure, and rhetorical questions create rhythm and interest.
- ☐ **Copy.** The presentation—particularly visuals and handouts—correctly cites names, dates, and other details and includes no errors in spelling, punctuation, or grammar.
- ☐ **Design.** The outline or manuscript is easy to read. Visuals are attractive and clearly legible from the back of the room.

CRITICAL THINKING Activity

Use the Internet or library to obtain a videotape of an outline or impromptu speech. Analyze the structure of the speech, checking whether and how the introduction, body, and conclusion follow the instructions in "Organizing Your Presentation" (pages 680–682). Analyze the content, checking whether the main points are clear and convincing and whether the style is engaging and fitting. Evaluate the delivery, including the speaker's use of visuals.

WRITING Activities

1. Use the Internet and library to find a recent manuscript speech on a business-related topic. Analyze the speech's structure and content, paying close attention to the organization, word choice, voice, sentence structure, and stylistic or motivational strategies. Based on your study, write a memo to your instructor in which you (a) grade the speech (A, B, C, D, or F) and (b) explain why your grade is fair.
2. Choose an organization that produces a product or service that you believe has value. Research that organization and its product or service, looking for both information on the topic as well as pictures, graphics, or other visuals that would help you talk about the topic. Develop and present a PowerPoint presentation in which you aim to convince your classmates that the product or service is one that they should buy or support.

PART 9

Proofreader's Guide

45 Understanding Grammar

46 Constructing Sentences

47 Using Punctuation

48 Checking Mechanics

49 Using the Right Word

50 Addressing ESL Issues

In this chapter

Noun **697**	Adverb **711**
Pronoun **700**	Preposition **712**
Verb **704**	Conjunction **713**
Adjective **710**	Interjection **713**

CHAPTER
45

Understanding Grammar

Grammar is the study of the structure and features of the language, consisting of rules and standards that are to be followed to produce acceptable writing and speaking. **Parts of speech** refers to the eight different ways words are used in the English language—as *nouns, pronouns, verbs, adjectives, adverbs, prepositions, conjunctions,* or *interjections.*

NOUN

A **noun** is a word that names something: a person, a place, a thing, or an idea.

Toni Morrison [author]   Lone Star [film]   A Congress of Wonders [book]
UC-Davis [university]   Renaissance [era]

Classes of Nouns

All nouns are either *proper nouns* or *common nouns.* Nouns may also be classified as *individual* or *collective,* or *concrete* or *abstract.*

Proper Nouns

A **proper noun,** which is always capitalized, names a person, a place, a thing, or an idea.

Rembrandt, Bertrand Russell [people]
Stratford-upon-Avon, Tower of London [places]
The Night Watch, Rosetta Stone [things]
New Deal, Christianity [ideas]

 F.Y.I.

See pages 767–768 for information on count and noncount nouns.

Common Nouns

A **common noun** is a general name for a person, a place, a thing, or an idea. Common nouns are not capitalized.

    optimist, instructor [people]   cafeteria, park [places]
    computer, chair [things]   freedom, love [ideas]

Collective Nouns

A **collective noun** names a group or a unit.

    family   audience   crowd   committee   team   class

Concrete Nouns

A **concrete noun** names a thing that is tangible (can be seen, touched, heard, smelled, or tasted).

    child   Pearl Jam   gymnasium   village   microwave oven   pizza

Abstract Nouns

An **abstract noun** names an idea, a condition, or a feeling—in other words, something that cannot be seen, touched, heard, smelled, or tasted.

    beauty   Jungian psychology   anxiety   agoraphobia   trust

Forms of Nouns

Nouns are grouped according to their *number, gender,* and *case.*

Number of Nouns

Number indicates whether a noun is singular or plural.

A singular noun refers to one person, place, thing, or idea.

    student   laboratory   lecture   note   grade   result

A plural noun refers to more than one person, place, thing, or idea.

    students   laboratories   lectures   notes   grades   results

Gender of Nouns

Gender indicates whether a noun is masculine, feminine, neuter, or indefinite.

    MASCULINE: father   king   brother   men   colt   rooster
    FEMININE: mother   queen   sister   women   filly   hen
    NEUTER (WITHOUT SEX): notebook   monitor   car   printer
    INDEFINITE OR COMMON (MASCULINE OR FEMININE): professor   customer   children   doctor   people

Case of Nouns

The **case** of a noun tells how it is related to other words within a sentence. There are three cases: *nominative, possessive,* and *objective*.

Nominative case describes a noun used as a subject. The subject of a sentence tells who or what the sentence is about.

> **Dean Henning** manages the College of Arts and Communication.

A noun is also in the nominative case when it is used as a predicate noun (or predicate nominative). A predicate noun follows a form of the *be* verb (*am, is, are, was, were, be, being, been*) and repeats or renames the subject.

> Either Mr. Cassett or Ms. Yokum is the **person** to talk to about the college's impact in our community.

Possessive case describes a noun that shows possession or ownership.

> Our **president's** willingness to discuss concerns with students has boosted campus morale.

Objective case describes a noun used as an object of the preposition, a direct object, or an indirect object.

> To survive, institutions of higher **learning** sometimes cut **budgets** in spite of **protests** from **students** and **instructors**. [*Learning* is the object of the preposition *of*, *protests* is the object of the preposition *in spite of*, and *students* and *instructors* are the objects of the preposition *from*. In addition, *budgets* is the direct object of the verb *cut*.]

A CLOSER Look
at Direct and Indirect Objects

A **direct object** is a noun (or pronoun) that identifies what or who receives the action of the verb.

> Budget cutbacks reduced class **choices**. [*Choices* is the direct object of *reduced*.]

An **indirect object** is a noun (or pronoun) that identifies the person *to whom* or *for whom* something is done, or the thing *to which* or *for which* something is done. An indirect object is always accompanied by a direct object.

> Recent budget cuts have given **students** fewer class choices. [*Choices* is the direct object of *have given*; *students* is the indirect object.]

F.Y.I.

Not every verb can be followed by *both* a direct object and an indirect object. Both can, however, follow *give, send, show, tell, teach, find, sell, ask, offer, pay, pass,* and *hand*.

PRONOUN

A **pronoun** is a word that is used in place of a noun.

> Roger was the most interesting 10-year-old **I** ever taught. **He** was a good thinker and thus a good writer. **I** remember **his** paragraph about the cowboy hat **he** received from **his** grandparents. **It** was "too new looking." The brim was not rolled properly. But the hat's imperfections were not the main idea in Roger's writing. No, the main idea was about how **he** was fixing the hat **himself** by wearing **it** when **he** showered.

Antecedents

An **antecedent** is the noun that the pronoun refers to or replaces. Most pronouns have antecedents, but not all do. (See "Indefinite Pronouns" on the next page.)

> As the wellness **counselor** checked *her* chart, several **students** *who* were waiting *their* turns shifted uncomfortably. [*Counselor* is the antecedent of *her*; *students* is the antecedent of *who* and *their*.]

 F.Y.I.

Each pronoun must agree with its antecedent in number, person, and gender.

Classes of Pronouns

There are several classes of pronouns: *personal, reflexive and intensive, relative, indefinite, interrogative, demonstrative,* and *reciprocal.* See Table 45.1.

Personal Pronouns

A **personal pronoun** refers to a specific person or thing.

> *Marge* started **her** car; **she** drove the antique *convertible* to Monterey where **she** hoped to sell **it** at an auction.

Reflexive and Intensive Pronouns

A **reflexive pronoun** is formed by adding *-self* or *-selves* to a personal pronoun. A reflexive pronoun can act as a direct object or an indirect object of a verb, an object of a preposition, or a predicate nominative.

> Charles loves **himself**. [direct object of *loves*]
>
> Charles gives **himself** A's for fashion sense. [indirect object of *gives*]
>
> Charles smiles at **himself** in store windows. [object of preposition *at*]
>
> Charles can be **himself** anywhere. [predicate nominative]

An **intensive pronoun** intensifies, or emphasizes, the noun or pronoun to which it refers.

> Leo **himself** taught his children to invest their lives in others.
>
> The lesson was sometimes painful—but they learned it **themselves.**

Relative Pronouns

A **relative pronoun** relates an adjective clause to the noun or pronoun it modifies. (The noun is italicized in each example below; the relative pronoun is in bold.)

> *Freshmen* **who** believe they have a lot to learn are absolutely right.
>
> Just navigating this *campus,* **which** is huge, can be challenging.

F.Y.I.
Make sure you know when to use the relative pronouns *who* or *whom* and *that* or *which*.

Indefinite Pronouns

An **indefinite pronoun** refers to unnamed or unknown people, places, or things.

> **Everyone** seemed amused when I was searching for my classroom in the student center. [The antecedent of *everyone* is unnamed.]
>
> **Nothing** is more unnerving than rushing last minute into the wrong room for the wrong class. [The antecedent of *nothing* is unknown.]

F.Y.I.
Most indefinite pronouns are singular, so when they are used as subjects, they should have singular verbs.

Table 45.1 Classes of Pronouns

Personal
I, me, my, mine / we, us, our, ours / you, your, yours
they, them, their, theirs / he, him, his, she, her, hers, it, its

Reflexive and Intensive
myself, yourself, himself, herself, itself, ourselves, yourselves, themselves

Relative
who, whose, whom, which, that

Indefinite

all	anything	everybody	most	no one	some
another	both	everyone	much	nothing	somebody
any	each	everything	neither	one	someone
anybody	each one	few	nobody	other	something
anyone	either	many	none	several	such

Interrogative
who, whose, whom, which, what

Demonstrative
this, that, these, those

Reciprocal
each other, one another

Interrogative Pronouns

An **interrogative pronoun** asks a question.

> So **which** will it be—highlighting and attaching a campus map to the inside of your backpack, or being lost and late for the first two weeks?

Demonstrative Pronouns

A **demonstrative pronoun** points out people, places, or things.

> We advise **this:** bring along as many maps and schedules as you need.
>
> **Those** are useful tools. **That** is the solution.

F.Y.I.

When a demonstrative pronoun *modifies* a noun (instead of replacing it), the pronoun functions as an adjective: *this* teacher, *that* test.

Forms of Personal Pronouns

The **form** of a personal pronoun indicates its *number* (singular or plural), its *person* (first, second, or third), its *case* (nominative, possessive, or objective), and its *gender* (masculine, feminine, neuter, or indefinite). See Table 45.2.

Number of Pronouns

A **personal pronoun** is either singular (*I, you, he, she, it*) or plural (*we, you, they*) in number.

> **He** should have a budget and stick to it. [singular]
>
> **We** can help new students learn about budgeting. [plural]

Person of Pronouns

The **person** of a pronoun indicates whether the person is speaking (first person), is spoken to (second person), or is spoken about (third person).

First person is used to name the speaker(s).

> **I** know **I** need to handle **my** stress in a healthful way, especially during exam week; **my** usual chips-and-doughnuts binge isn't helping. [singular]
>
> **We** all decided to bike to the tennis court. [plural]

Second person is used to name the person(s) spoken to.

> Maria, **you** grab the rackets, okay? [singular]
>
> John and Tanya, can **you** find the water bottles? [plural]

Third person is used to name the person(s) or thing(s) spoken about.

> Today's students are interested in wellness issues. **They** are concerned about **their** health, fitness, and nutrition. [plural]
>
> Maria practices yoga and feels **she** is calmer for **her** choice. [singular]
>
> An advantage of regular exercise is that **it** raises one's energy level. [singular]

Case of Pronouns

The **case** of each pronoun tells how it is related to the other words within a sentence. There are three cases: *nominative, possessive,* and *objective.*

Nominative case describes a pronoun used as a subject. The following are nominative forms: *I, you, he, she, it, we, they.*

>**He** found an old map in the trunk.
>
>My friend and **I** went biking. [not *me*]

A pronoun is in the nominative case when it is used as a predicate noun (predicate nominative) following a form of the *be* verb (*am, is, are, was, were, be, being, been*).

>It was **he** who discovered electricity. [not *him*]

Possessive case describes a pronoun that shows possession or ownership: *my, mine, our, ours, his, her, hers, their, theirs, its, your, yours.*

>That coat is **hers.** This coat is **mine. Your** coat is lost.

Objective case describes a pronoun used as the direct object, indirect object, or object of a preposition: *me, you, him, her, it, us, them.*

>Professor Adler hired **her.** [*Her* is the direct object of the verb *hired.*]
>
>He showed Mary and **me** the language lab. [*Me* is the indirect object of the verb *showed.*]
>
>He introduced the three of **us**—Mary, Shavonn, and **me**—to the faculty. [*Us* is the object of the preposition *of*; *me* is part of the appositive of the object *us*.]

Gender of Pronouns

The **gender** of a pronoun indicates whether the pronoun is masculine, feminine, neuter, or indefinite.

>**MASCULINE:** he, him, his
>
>**FEMININE:** she, her, hers
>
>**NEUTER (WITHOUT SEX):** it, its
>
>**INDEFINITE (MASCULINE OR FEMININE):** they, them, their

Table 45.2 Number, Person, and Case of Personal Pronouns

	Nominative Case	Possessive Case	Objective Case
First Person Singular	I	my, mine	me
Second Person Singular	you	your, yours	you
Third Person Singular	he, she, it	his, her, hers, its	him, her, it
First Person Plural	we	our, ours	us
Second Person Plural	you	your, yours	you
Third Person Plural	they	their, theirs	them

VERB

A **verb** shows action (*pondered, grins*), links words (*is, seemed*), or accompanies another action verb as an auxiliary or helping verb (*can, does*).

> Harry **honked** the horn. [shows action]
> Harry **is** impatient. [links words]
> Harry **was** honking the truck's horn. [accompanies the verb *honking*]

Classes of Verbs

Verbs are classified as *action*, *auxiliary* (helping), or *linking* (state of being).

Action Verbs: Transitive and Intransitive

As its name implies, an **action verb** shows action. Some action verbs are *transitive;* others are *intransitive*. (The term *action* does not always refer to a physical activity.)

> Rain **splashed** the windshield. [transitive verb]
> Josie **drove** off the road. [intransitive verb]

Transitive verbs have direct objects that receive the action (see page 699).

> The health care industry **employs** more than 7 million **workers** in the United States. [*Workers* is the direct object of the action verb *employs*.]

Intransitive verbs communicate action that is complete in itself. They do not need an object to receive the action.

> My new college roommate **smiles** and **laughs** a lot.

Some verbs can be either transitive or intransitive.

> Ms. Hull **teaches** physiology and microbiology. [transitive]
> She **teaches** well. [intransitive]

Auxiliary (Helping) Verbs

Auxiliary verbs (helping verbs) help to form some of the *tenses* (page 706), the *mood* (page 707), and the *voice* (page 707) of the main verb. In the following example, the auxiliary verbs are in **bold,** and the main verbs are in *italics*.

> I *believe,* I **have** always *believed,* and I **will** always *believe* in private enterprise as the backbone of economic well-being in America. —Franklin D. Roosevelt

Common auxiliary verbs

am	been	could	does	have	might	should	will
are	being	did	had	is	must	was	would
be	can	do	has	may	shall	were	

> **F.Y.I.**
> "Be" auxiliary verbs are always followed by either a verb ending in *ing* or a past participle. Also see "Common Modal Auxiliary Verbs" (page 772).

Linking (State of Being) Verbs

A **linking verb** is a special form of intransitive verb that links the subject of a sentence to a noun, a pronoun, or an adjective in the predicate.

> The streets **are** flooded. [adjective]
>
> The streets **are** rivers! [noun]

Common linking verbs

am   are   be   become   been   being   is   was   were

Additional linking verbs

appear   feel   look   seem   sound   grow   remain   smell   taste

The verbs listed as "additional linking verbs" above function as linking verbs when they do not show actual action. An adjective usually follows these linking verbs.

> The thunder **sounded** ominous. [adjective]
>
> My little brother **grew** frightened. [adjective]

When these same words are used as action verbs, an adverb or a direct object might follow them.

> I **looked** carefully at him. [adverb]
>
> My little brother **grew** corn for a science project. [direct object]

Forms of Verbs

A verb's **form** differs depending on its *number* (singular, plural), *person* (first, second, third), *tense* (present, past, future, present perfect, past perfect, future perfect), *voice* (active, passive), and *mood* (indicative, imperative, subjunctive).

Number of a Verb

Number indicates whether a verb is singular or plural. The verb and its subject both must be singular, or they both must be plural. (See "Subject-Verb Agreement," page 780.)

> My college **enrolls** high schoolers in summer programs. [singular]
>
> Many colleges **enroll** high schoolers in summer courses. [plural]

Person of a Verb

Person indicates whether the subject of the verb is *first, second,* or *third person*. The verb and its subject must be in the same person. Verbs usually have a different form only in *third person singular of the present tense.*

	First person	Second person	Third person
Singular	I think	you think	he/she/it thinks
Plural	we think	you think	they think

Tense of a Verb

Tense indicates the time of an action or state of being. There are three basic tenses (*past, present,* and *future*) and three verbal aspects (*progressive, perfect,* and *perfect progressive*).

Present Tense Present tense expresses action that is happening at the present time or action that happens continually, regularly.

> In the United States, more than 75 percent of workers **hold** service jobs.

Present progressive tense also expresses action that is happening at the present time, but this tense is always formed by combining *am, are,* or *is* and the present participle (ending in *ing*) of the main verb.

> More women than ever before **are working** outside the home.

Present perfect tense expresses action that began in the past and has recently been completed or is continuing up to the present time.

> My sister **has taken** four years of swimming lessons.

Present perfect progressive tense also expresses an action that began in the past but stresses the continuing nature of the action. Like the present progressive tense, it is formed by combining auxiliary verbs (*have been* or *has been*) and present participles.

> She **has been taking** them since she was six years old.

Past Tense Past tense expresses action that is completed at a particular time in the past.

> A hundred years ago, more than 75 percent of laborers **worked** in agriculture.

Past progressive tense expresses past action that continued over an interval of time. It is formed by combining *was* or *were* with the present participle of the main verb.

> A century ago, my great-grandparents **were farming**.

Past perfect tense expresses an action in the past that occurs before another past action or an action that was completed by some specific past time.

> By dinnertime my cousins **had eaten** all the olives.

Past perfect progressive tense expresses a past action but stresses the continuing nature of the action. It is formed by using *had been* along with the present participle.

> They **had been eating** the olives since they arrived two hours earlier.

Future Tense Future tense expresses action that will take place in the future.

> Next summer I **will work** as a lifeguard.

Future progressive tense expresses a continuous or repeating future action.

> I **will be working** for the park district at North Beach.

Future perfect tense expresses action that will begin in the future and be completed by a specific time in the future.

> By 10:00 p.m., I **will have completed** my research project.

Future perfect progressive tense also expresses future action that will be completed by a specific time, but (as with other perfect progressive tenses) stresses the action's continuous nature. It is formed using *will have been* along with the present participle.

> I **will have been researching** the project for three weeks by the time it's due.

Voice of a Verb

Voice indicates whether the subject is acting or being acted upon.

Active voice indicates that the subject of the verb is doing something.

> People **update** their résumés on a regular basis. [The subject, *People,* is acting; *résumés* is the direct object.]

Passive voice indicates that the subject of the verb is being acted upon or is receiving the action. A passive verb combines a *be* verb with a past participle.

> Your résumé **should be updated** on a regular basis. [The subject, *résumé,* is receiving the action.]

Using Active Voice   Generally, use active voice rather than passive voice for more direct, energetic writing. To change your passive sentences to active ones, do the following: First, find the noun that is doing the action and make it the subject. Then find the word that had been the subject and use it as either a direct object or an indirect object.

> **PASSIVE:** The winning goal **was scored** by Eva. [The subject, *goal,* is not acting.]
>
> **ACTIVE:** Eva **scored** the winning goal. [The subject, *Eva,* is acting.]

Using Passive Voice   When you want to emphasize the receiver more than the doer—or when the doer is unknown—use the passive voice. Much technical and scientific writing regularly uses the passive voice.

Mood of a Verb

The **mood** of a verb indicates the tone or attitude with which a statement is made.

Indicative mood, the most common, is used to state a fact or to ask a question.

> **Can** any theme **capture** the essence of the complex 1960s culture? President John F. Kennedy's directive [stated below] **represents** one ideal popular during that decade.

Imperative mood is used to give a command. (The subject of an imperative sentence is *you,* which is usually understood and not stated in the sentence.)

> **Ask** not what your country can do for you—**ask** what you can do for your country.
>
> —John F. Kennedy

Subjunctive mood is used to express a wish, an impossibility or unlikely condition, a necessity, or a motion in a business meeting. The subjunctive mood is often used with *if* or *that*. The verb forms below create a nontypical subject-verb agreement, forming the subjunctive mood.

> If I **were** rich, I would travel for the rest of my life. [a wish]
>
> If each of your brain cells **were** one person, there would be enough people to populate 25 planets. [an impossibility]
>
> The English Department requires that every student **pass** a proficiency test. [a necessity]
>
> I move that the motion **be** accepted. [a motion]

Verbals

A **verbal** is a word that is made from a verb, but functions as a noun, an adjective, or an adverb. There are three types of verbals: *gerunds, infinitives,* and *participles.*

Gerunds

A **gerund** ends in *ing* and is used as a noun.

> **Waking** each morning is the first challenge. [subject]
> I start **moving** at about seven o'clock. [direct object]
> I work at **jump-starting** my weary system. [object of the preposition]
> As Woody Allen once said, "Eighty percent of life is **showing up.**" [predicate noun]

Infinitives

An **infinitive** is usually introduced by *to*; the infinitive may be used as a noun, an adjective, or an adverb.

> **To succeed** is not easy. [noun]
> That is the most important thing **to remember.** [adjective]
> Students are wise **to work** hard. [adverb]

F.Y.I.
It can be difficult to know whether a gerund or an infinitive should follow a verb. It's helpful to become familiar with lists of specific verbs that can be followed by one but not the other.

Participles

A **present participle** ends in *ing* and functions as an adjective. A **past participle** ends in *ed* (or another past tense form) and also functions as an adjective.

> The students **reading** those study-skill handouts are definitely **interested.**
> The prospect of **aced** tests and assignments must be **appealing.**
> [These participles function as adjectives: *reading* students, *interested* students, *aced* tests and assignments, and *appealing* prospect. Notice, however, that *reading* has a direct object: *handouts.* Verbals might have direct objects.]

Using Verbals

Make sure that you use verbals correctly; look carefully at the examples below.

> **VERBAL:** **Diving** is a popular Olympic sport.
> [*Diving* is a gerund used as a subject.]
> **Diving** gracefully, the Olympian hoped to get high marks.
> [*Diving* is a participle modifying *Olympian.*]
>
> **VERB:** The next competitor was **diving** in the practice pool.
> [Here, *diving* is a verb, not a verbal.]

Irregular Verbs

Irregular verbs can often be confusing. That's because the past tense and past participle of irregular verbs are formed by changing the word itself, not merely by adding *d* or *ed*. Table 45.3 contains the most troublesome irregular verbs.

Table 45.3 Common Irregular Verbs and Their Principal Parts

Present Tense	Past Tense	Past Participle	Present Tense	Past Tense	Past Participle	Present Tense	Past Tense	Past Participle
am, be	was, were	been	forget	forgot	forgotten, forgot	shake	shook	shaken
arise	arose	arisen				shine (light)	shone	shone
awake	awoke, awaked	awoken, awaked	freeze	froze	frozen	shine (polish)	shined	shined
			get	got	gotten	show	showed	shown
beat	beat	beaten	give	gave	given	shrink	shrank	shrunk
become	became	become	go	went	gone	sing	sang	sung
begin	began	begun	grow	grew	grown	sink	sank	sunk
bite	bit	bitten, bit	hang (execute)	hanged	hanged	sit	sat	sat
blow	blew	blown	hang (suspend)	hung	hung	sleep	slept	slept
break	broke	broken	have	had	had	speak	spoke	spoken
bring	brought	brought	hear	heard	heard	spend	spent	spent
build	built	built	hide	hid	hidden	spring	sprang	sprung
burn	burnt, burned	burnt, burned	hit	hit	hit	stand	stood	stood
			keep	kept	kept	steal	stole	stolen
burst	burst	burst	know	knew	known	strike	struck	struck, stricken
buy	bought	bought	lay	laid	laid			
catch	caught	caught	lead	led	led	strive	strove	striven
choose	chose	chosen	leave	left	left	swear	swore	sworn
come	came	come	lend	lent	lent	swim	swam	swum
cost	cost	cost	let	let	let	swing	swung	swung
cut	cut	cut	lie (deceive)	lied	lied	take	took	taken
dig	dug	dug	lie (recline)	lay	lain	teach	taught	taught
dive	dived, dove	dived	make	made	made	tear	tore	torn
do	did	done	mean	meant	meant	tell	told	told
draw	drew	drawn	meet	met	met	think	thought	thought
dream	dreamed, dreamt	dreamed, dreamt	pay	paid	paid	throw	threw	thrown
			prove	proved	proved, proven	wake	woke, waked	woken, waked
drink	drank	drunk						
drive	drove	driven	put	put	put	wear	wore	worn
eat	ate	eaten	read	read	read	weave	wove	woven
fall	fell	fallen	ride	rode	ridden	wind	wound	wound
feel	felt	felt	ring	rang	rung	wring	wrung	wrung
fight	fought	fought	rise	rose	risen	write	wrote	written
find	found	found	run	ran	run			
flee	fled	fled	see	saw	seen			
fly	flew	flown	set	set	set			

ADJECTIVE

An **adjective** describes or modifies a noun or pronoun. The articles *a*, *an*, and *the* are adjectives.

    Advertising is **a big** and **powerful** industry. [*A*, *big*, and *powerful* modify *industry*.]

Numbers are also adjectives.

    Fifty-three relatives came to my party.

Many demonstrative, indefinite, and interrogative forms may be used as either adjectives or pronouns (*that, these, many, some, whose,* and so on). These words are adjectives if they come before a noun and modify it; they are pronouns if they stand alone.

    Some advertisements are less than truthful. [*Some* modifies *advertisements* and is an adjective.]

    Many cause us to chuckle at their outrageous claims. [*Many* stands alone; it is a pronoun and replaces the noun *advertisements*.]

Proper Adjectives

Proper adjectives are created from proper nouns and are capitalized.

    English has been influenced by advertising slogans. [proper noun]

    The **English** language is constantly changing. [proper adjective]

Predicate Adjectives

A **predicate adjective** follows a form of the *be* verb (or other linking verb) and describes the subject.

    At its best, advertising is **useful**; at its worst, **deceptive**. [*Useful* and *deceptive* modify the noun *advertising*.]

Forms of Adjectives

Adjectives have three forms: *positive, comparative,* and *superlative.*

The **positive form** is the adjective in its regular form. It describes a noun or a pronoun without comparing it to anyone or anything else.

    Joysport walking shoes are **strong** and **comfortable**.

The **comparative form** (*er, more,* or *less*) compares two things. (*More* and *less* are used generally with adjectives of two or more syllables.)

    Air soles make Mile Eaters **stronger** and **more comfortable** than Joysports.

The **superlative form** (*est, most,* or *least*) compares three or more things. (*Most* and *least* are used most often with adjectives of two or more syllables.)

    My old Canvas Wonders are the **strongest, most comfortable** shoes of all!

F.Y.I.

Two or more adjectives before a noun should have a certain order when they do not modify the noun equally. See "Placing Adjectives" (page 776).

ADVERB

An **adverb** describes or modifies a verb, an adjective, another adverb, or a whole sentence. An adverb answers questions such as *how, when, where, why, how often,* or *how much.*

> The temperature fell **sharply.** [*Sharply* modifies the verb *fell.*]
>
> The temperature was **quite** low. [*Quite* modifies the adjective *low.*]
>
> The temperature dropped **very quickly.** [*Very* modifies the adverb *quickly,* which modifies the verb *dropped.*]
>
> **Unfortunately,** the temperature stayed cool. [*Unfortunately* modifies the whole sentence.]

Types of Adverbs

Adverbs can be grouped in four ways: *time, place, manner,* and *degree.*

Time (These adverbs tell *when, how often,* and *how long.*)
> today, yesterday   daily, weekly   briefly, eternally

Place (These adverbs tell *where, to where,* and *from where.*)
> here, there   nearby, beyond   backward, forward

Manner (These adverbs often end in *ly* and tell *how* something is done.)
> precisely   regularly   regally   smoothly   well

Degree (These adverbs tell *how much* or *how little.*)
> substantially   greatly   entirely   partly   too

Forms of Adverbs

Adverbs have three forms: *positive, comparative,* and *superlative.* See Table 45.4.

The **positive form** is the adverb in its regular form. It describes a verb, an adjective, or another adverb without comparing it to anyone or anything else.

> With Joysport shoes, you'll walk **fast.** They support your feet **well.**

The **comparative form** (*er, more,* or *less*) compares two things. (*More* and *less* are used generally with adverbs of two or more syllables.)

> Wear Jockos instead of Joysports, and you'll walk **faster.** Jockos' special soles support your feet **better** than the Roksports do.

The **superlative form** (*est, most,* or *least*) compares three or more things. (*Most* and *least* are used most often with adverbs of two or more syllables.)

> Really, I walk **fastest** wearing my old Canvas Wonders. They seem to support my feet, my knees, and my pocketbook **best** of all.

Table 45.4 **Adverb Forms**

Regular Adverbs			Irregular Adverbs		
Positive	**Comparative**	**Superlative**	**Positive**	**Comparative**	**Superlative**
fast	faster	fastest	well	better	best
effectively	more effectively	most effectively	badly	worse	worst

PREPOSITION

A **preposition** is a word (or group of words) that shows the relationship between its object (a noun or pronoun following the preposition) and another word in the sentence. See Table 45.5.

> **Regarding** your reasons **for** going **to** college, do they all hinge **on** getting a good job **after** graduation? [In this sentence, *reasons, going, college, getting,* and *graduation* are objects of their preceding prepositions *regarding, for, to, on,* and *after.*]

Prepositional Phrases

A **prepositional phrase** includes the preposition, the object of the preposition, and the modifiers of the object. A prepositional phrase may function as an adverb or adjective.

> A broader knowledge **of the world** is one benefit **of higher education.** [The two phrases function as adjectives modifying the nouns *knowledge* and *benefit*, respectively.]

> Exercising your brain may safeguard **against atrophy.** [The phrase functions as an adverb modifying the verb *safeguard.*]

Table 45.5 Prepositions

aboard	back of	except for	near to	round
about	because of	excepting	notwithstanding	save
above	before	for	of	since
according to	behind	from	off	subsequent to
across	below	from among	on	through
across from	beneath	from between	on account of	throughout
after	beside	from under	on behalf of	till
against	besides	in	onto	to
along	between	in addition to	on top of	together with
alongside	beyond	in behalf of	opposite	toward
alongside of	but	in front of	out	under
along with	by	in place of	out of	underneath
amid	by means of	in regard to	outside	until
among	concerning	inside	outside of	unto
apart from	considering	inside of	over	up
around	despite	in spite of	over to	upon
as far as	down	instead of	owing to	up to
aside from	down from	into	past	with
at	during	like	prior to	within
away from	except	near	regarding	without

F.Y.I.

Prepositions often pair up with a verb and become part of an idiom, a slang expression, or a two-word verb.

CONJUNCTION

A **conjunction** connects individual words or groups of words.

> When we came back to Paris, it was clear **and** cold **and** lovely.
> —Ernest Hemingway

Coordinating Conjunctions

Coordinating conjunctions usually connect a word to a word, a phrase to a phrase, or a clause to a clause. The words, phrases, or clauses joined by a coordinating conjunction are equal in importance or are of the same type.

> Civilization is a race between education **and** catastrophe.   —H. G. Wells

Correlative Conjunctions

Correlative conjunctions are a type of coordinating conjunction used in pairs.

> There are two inadvisable ways to think: **either** believe everything **or** doubt everything.

Subordinating Conjunctions

Subordinating conjunctions connect two clauses that are not equally important. A subordinating conjunction connects a dependent clause to an independent clause.

> Experience is the worst teacher; it gives the test **before** it presents the lesson.
> [The clause *before it presents the lesson* is dependent. It connects to the independent clause *it gives the test*.]

Conjunctions

Coordinating: and, but, or, nor, for, so, yet

Correlative: either, or; neither, nor; not only, but (but also); both, and; whether, or

Subordinating: after, although, as, as if, as long as, because, before, even though, if, in order that, provided that, since, so that, than, that, though, unless, until, when, whenever, where, while

F.Y.I.

Relative pronouns and conjunctive adverbs can also connect clauses.

INTERJECTION

An **interjection** communicates strong emotion or surprise (*oh, ouch, hey,* and so on). Punctuation (often a comma or an exclamation point) is used to set off an interjection.

> **Hey! Wait! Well,** so much for catching the bus.

A CLOSER Look at the Parts of Speech

Noun A **noun** is a word that names something: a person, a place, a thing, or an idea.

> Toni Morrison [author] Lone Star [film]
> UC-Davis [university] Renaissance [era]
> A Congress of Wonders [book]

Pronoun A **pronoun** is a word used in place of a noun.

> I   my   that   themselves   which
> it   ours   they   everybody   you

Verb A **verb** is a word that expresses action, links words, or acts as an auxiliary verb to the main verb.

> are   break   drag   fly   run   sit   was
> bite   catch   eat   is   see   tear   were

Adjective An **adjective** describes or modifies a noun or pronoun. (The articles *a*, *an*, and *the* are adjectives.)

> **The carbonated** drink went down easy on **that hot, dry** day. [*The* and *carbonated* modify *drink*; *that*, *hot*, and *dry* modify *day*.]

Adverb An **adverb** describes or modifies a verb, an adjective, another adverb, or a whole sentence. An adverb generally answers questions such as *how, when, where, how often*, or *how much*.

> greatly   precisely   regularly   there   slowly   nearly
> here   today   partly   quickly   yesterday   loudly

Preposition A **preposition** is a word (or group of words) that shows the relationship between its object (a noun or pronoun that follows the preposition) and another word in the sentence. Prepositions introduce prepositional phrases.

> across   for   with   out   to   of

Conjunction A **conjunction** connects individual words or groups of words.

> and   because   but   for   or   since   so   yet

Interjection An **interjection** is a word that communicates strong emotion or surprise. Punctuation (often a comma or an exclamation point) is used to set off an interjection from the rest of the sentence.

> **Stop! No! What,** am I invisible?

In this chapter
Using Subjects and Predicates **715**
Using Phrases **718**
Using Clauses **720**
Using Sentence Variety **721**

CHAPTER
46

Constructing Sentences

A **sentence** is made up of one or more words that express a complete thought. Sentences are groups of words that make statements, ask questions, or express feelings.

> The Web delivers the universe in a box.

USING SUBJECTS AND PREDICATES

Sentences have two main parts: a **subject** and a **predicate.**

> Technology frustrates many people.

In the sentence above, *technology* is the subject—the sentence talks about technology. *Frustrates many people* is the complete predicate—it says something about the subject. (A predicate can also show action.)

The Subject

The **subject** names the person or thing either doing the action in a sentence or being talked about. The subject is most often a noun or a pronoun.

> **Technology** is an integral part of almost every business.
>
> **Manufacturers** need technology to compete in the world market.
>
> **They** could not go far without it.

A phrase or a clause may also function as a subject.

> **To survive without technology** is difficult. [infinitive phrase]
>
> **Downloading information from the Web** is easy. [gerund phrase]
>
> **That the information age would arrive** was inevitable. [noun clause]

To determine the subject of a sentence, ask yourself a question that begins with *who* or *what* and ends with the predicate. In most sentences, the subject comes before the verb; however, in questions and some exclamations, that order is reversed.

Simple Subject

A **simple subject** is the subject without the words that modify it.

> Thirty years ago, reasonably well-trained **mechanics** could fix any car on the road.

Complete Subject

A **complete subject** is the simple subject and the words that modify it.

> Thirty years ago, **reasonably well-trained mechanics** could fix any car on the road.

Compound Subject

A **compound subject** is composed of two or more simple subjects joined by a conjunction and sharing the same predicate(s).

> Today, **mechanics** and **technicians** would need to master a half million manual pages to fix every car on the road.
>
> **Dealerships** and their service **departments** must sometimes explain that to the customers.

Understood Subject

Sometimes a subject is **understood.** This means it is missing in the sentence, but a reader clearly understands what the subject is. An understood subject is most likely in an imperative sentence. (See page 721.)

> **[You]** Park on this side of the street. [The subject *you* is understood.]
>
> Put the CD player in the trunk.

Delayed Subject

In sentences that begin with *There is, It is, There was,* or *It was,* the subject usually follows the verb.

> **There were 70,000 fans in the stadium.** [The subject is *fans; were* is the verb. *There* is an expletive, an empty word.]
>
> It was a **problem** for stadium security. [*Problem* is the subject.]

The subject is also delayed in questions.

> Where was the **event?** [*Event* is the subject.]
>
> Was **Dave Matthews** playing? [*Dave Matthews* is the subject.]

In sentences that begin with *It is* or *It was* and describe the weather, distance, time, and some other conditions, the word *it* serves as the subject.

> **It** was raining.
>
> **It** is 90 miles from Chicago to Milwaukee.
>
> **It** is three o'clock.

The Predicate (Verb)

The **predicate,** which contains the verb, is the sentence part that either tells what the subject is doing or says something about the subject.

> Students **need technical skills as well as basic academic skills.**

Simple Predicate

A **simple predicate** is the verb without the words that describe or modify it.

> Today's workplace **requires** employees to have a range of skills.

Complete Predicate

A **complete predicate** is the verb and all the words that modify or explain it.

> Today's workplace **requires employees to have a range of skills.**

Compound Predicate

A **compound predicate** is composed of two or more verbs and all the words that modify or explain them.

> Engineers **analyze problems and calculate solutions.**

Direct Object

A **direct object** is the part of the predicate that receives the action of the verb. A direct object makes the meaning of the verb complete.

> Marcos visited several **campuses.** [The direct object *campuses* receives the action of the verb *visited* by answering the question "Marcos visited what?"]

A direct object may be compound.

> An admissions counselor explained the academic **programs** and the application **process.**

Indirect Object

An **indirect object** is the word(s) that tells *to whom/to what* or *for whom/for what* something is done. A sentence must have a direct object before it can have an indirect object.

> I showed our **children** my new school.

Use these questions to find an indirect object:

What is the verb?	*showed*
Showed what?	*school* (direct object)
Showed school to whom?	*children* (indirect object)

> I wrote **them** a note.

An indirect object may be compound.

> I gave the **instructor** and a few **classmates** my e-mail address.

USING PHRASES

A **phrase** is a group of related words that functions as a single part of speech. A phrase lacks a subject, a predicate, or both. There are three phrases in the following sentence.

> Examples of technology can be found in ancient civilizations.
>
> [**of technology:** prepositional phrase that functions as an adjective; no subject or predicate]
>
> [**can be found:** verb phrase; no subject]
>
> [**in ancient civilizations:** prepositional phrase that functions as an adverb; no subject or predicate]

Types of Phrases

There are several types of phrases: *verb, verbal, prepositional, appositive,* and *absolute.*

Verb Phrase

A **verb phrase** consists of a main verb, its helping verbs, and sometimes its modifiers.

> Students, worried about exams, **have camped at the library all week.**

Verbal Phrase

A **verbal phrase** is a phrase based on one of the three types of verbals: *gerund, infinitive,* or *participial.* See Table 46.1 to see how verbal phrases function.

A **gerund phrase** consists of a gerund and its modifiers. The whole phrase functions as a noun.

> **Becoming a marine biologist** is Rashanda's dream. [The gerund phrase is used as the subject of the sentence.]
>
> She has acquainted herself with the various methods for **collecting sea-life samples.** [The gerund phrase is the object of the preposition *for.*]

An **infinitive phrase** consists of the introductory word *to,* the fundamental form of a verb, and its modifiers. The whole phrase functions as a noun, an adjective, or an adverb.

> **To dream** is the first step in any endeavor. [The infinitive phrase functions as a noun used as the subject.]
>
> Remember **to make a plan to realize your dream.** [The infinitive phrase *to make a plan* functions as a noun used as a direct object; *to realize your dream* functions as an adjective modifying *plan.*]
>
> Finally, apply all of your talents and skills **to achieve your goals.** [The infinitive phrase functions as an adverb modifying *apply.*]

A **participial phrase** consists of a past or present participle (a verb form ending in *ing* or *ed*) and its modifiers. The phrase functions as an adjective.

> **Doing poorly in biology,** Theo signed up for a tutor. [The participial phrase modifies the noun *Theo.*]
>
> Some students **frustrated by difficult course work** don't seek help. [The participial phrase modifies the noun *students.*]

Table 46.1 Functions of Verbal Phrases

	Noun	Adjective	Adverb
Gerund	■		
Infinitive	■	■	■
Participial		■	

Prepositional Phrase

A **prepositional phrase** is a group of words beginning with a preposition and ending with a noun or a pronoun. Prepositional phrases are used mainly as adjectives and adverbs. See page 712 for a list of prepositions.

> Denying the existence **of exam week** hasn't worked **for anyone** yet. [The prepositional phrase *of exam week* is used as an adjective modifying the noun *existence; for anyone* is used as an adverb modifying the verb *has worked*.]
>
> Test days still dawn and GPAs still plummet **for the unprepared student.** [The prepositional phrase *for the unprepared student* is used as an adverb modifying the verbs *dawn* and *plummet*.]

A prepositional phrase may contain adjectives, but not adverbs. Do not mistake the following adverbs for nouns and incorrectly use them with a preposition: *here, there, everywhere, inside, outside, uptown, downtown.*

Appositive Phrase

An **appositive phrase,** which follows a noun or a pronoun and renames it, consists of a noun and its modifiers. An appositive adds new information about the noun or pronoun it follows.

> The Olympic-size pool, **a prized addition to the physical education building,** gets plenty of use. [The appositive phrase renames *pool*.]

Absolute Phrase

An **absolute phrase** consists of a noun and a participle (plus the participle's object, if there is one, and any modifiers). Because the noun acts like a subject and is followed by a verbal, an absolute phrase resembles a clause.

> **Their enthusiasm sometimes waning,** the students who cannot swim are required to take lessons. [The noun *enthusiasm* is modified by the present participle *waning;* the entire phrase modifies *students*.]

F.Y.I.

Phrases can add valuable information to sentences, but some phrases add nothing but "fat" to your writing.

USING CLAUSES

A **clause** is a group of related words that has both a subject and a predicate.

Independent/Dependent Clauses

An **independent clause** presents a complete thought and can stand alone as a sentence. A **dependent clause** (also called a subordinate clause) does not present a complete thought and cannot stand alone as a sentence.

> Although airplanes are twentieth-century inventions [dependent clause], people have always dreamed of flying [independent clause].

Types of Clauses

There are three basic types of dependent, or subordinate, clauses: *adverb, adjective,* and *noun.*

Adverb Clause

An **adverb clause** is used like an adverb to modify a verb, an adjective, or an adverb. All adverb clauses begin with subordinating conjunctions. (See page 713.)

> **Because Orville won a coin toss,** he got to fly the power-driven air machine first. [The adverb clause modifies the verb *got.*]

Adjective Clause

An **adjective clause** is used like an adjective to modify a noun or a pronoun. Adjective clauses begin with relative pronouns *(which, that, who).* (See page 701.)

> The men **who invented the first airplane** were brothers, Orville and Wilbur Wright. [The adjective clause modifies the noun *men. Who* is the subject of the adjective clause.]
>
> The first flight, **which took place December 17, 1903,** was made by Orville. [The adjective clause modifies the noun *flight. Which* is the subject of the adjective clause.]

Noun Clause

A **noun clause** is used in place of a noun. Noun clauses can appear as subjects, as direct or indirect objects, as predicate nominatives, or as objects of prepositions. They are introduced by subordinating words such as *what, that, when, why, how, whatever, who, whom, whoever,* and *whomever.*

> He wants to know **what made modern aviation possible.** [The noun clause functions as a direct object.]
>
> **Whoever invents an airplane with vertical takeoff ability** will be a hero. [The noun clause functions as the subject.]

F.Y.I.

If you can replace a whole clause with the pronoun *something* or *someone,* it is a noun clause.

Chapter 46 Constructing Sentences   721

USING SENTENCE VARIETY

A sentence can be classified according to the kind of statement it makes and according to the way it is constructed.

Kinds of Sentences

Sentences can make five basic kinds of statements: *declarative, interrogative, imperative, exclamatory,* or *conditional.*

Declarative Sentence

Declarative sentences make statements. They tell us something about a person, a place, a thing, or an idea.

> In 1955, Rosa Parks refused to follow segregation rules on a bus in Montgomery, Alabama.

Interrogative Sentence

Interrogative sentences ask questions.

> Do you think Ms. Parks knew she was making history?
>
> Would you have had the courage to do what she did?

Imperative Sentence

Imperative sentences give commands. They often contain an understood subject (*you*).

> Read Chapters 6 through 10 for tomorrow.

F.Y.I.

Imperative sentences with an understood subject are the only sentences in which it is acceptable to have no subjects stated.

Exclamatory Sentence

Exclamatory sentences communicate strong emotion or surprise.

> I simply can't keep up with these long reading assignments!
>
> Oh my gosh, you scared me!

Conditional Sentence

Conditional sentences express two circumstances. One of the circumstances depends on the other circumstance. The words *if, when,* or *unless* are often used in conditional statements.

> **If** you practice a few study-reading techniques, college reading loads will be manageable.
>
> **When** I manage my time, it seems I have more of it.
>
> Don't ask me to help you, **unless** you are willing to do the reading first.

Structure of Sentences

A sentence may be *simple, compound, complex,* or *compound-complex,* depending on the relationship between the independent and dependent clauses in it.

Simple Sentence

A **simple sentence** contains one independent clause. The independent clause may have compound subjects and predicates, and it may also contain phrases.

> My **back aches.** [single subject: *back;* single predicate: *aches*]
>
> My **teeth** and my **eyes hurt.** [compound subject: *teeth* and *eyes;* single predicate: *hurt*]
>
> My **memory** and my **logic come** and **go.** [compound subject: *memory* and *logic;* compound predicate: *come* and *go*]
>
> **I must be in need of a vacation.** [single subject: *I;* single predicate: *must be;* phrases: *in need, of a vacation*]

Compound Sentence

A **compound sentence** consists of two independent clauses. The clauses must be joined by a semicolon, by a comma and a coordinating conjunction *(and, but, or, nor, so, for, yet),* or by a semicolon and a conjunctive adverb *(besides, however, instead, meanwhile, then, therefore).*

> I had eight hours of sleep**, so** why am I so exhausted?
>
> I take good care of myself**;** I get enough sleep.
>
> I still feel fatigued**; therefore,** I must need more exercise.

Complex Sentence

A **complex sentence** contains one independent clause (in bold) and one or more dependent clauses (underlined).

> When I can, **I get eight hours of sleep.** [dependent clause; independent clause]
>
> When I get up on time, and if someone hasn't used up all the milk, **I eat breakfast.** [two dependent clauses; independent clause]

Compound-Complex Sentence

A **compound-complex sentence** contains two or more independent clauses (in bold type) and one or more dependent clauses (underlined).

> If I'm not in a hurry, **I take leisurely walks, and I try to spot some wildlife.** [dependent clause; two independent clauses]
>
> **I saw a hawk** when I was walking, **and other smaller birds were chasing it.** [independent clause; dependent clause; independent clause]

In this chapter

Period **723**	Apostrophe **731**	Hyphen **738**
Question Mark **724**	Colon **733**	Dash **740**
Exclamation Point **725**	Semicolon **734**	Brackets **741**
Parentheses **725**	Ellipsis **735**	Diagonal **741**
Comma **726**	Quotation Marks **736**	Italics (Underlining) **742**

CHAPTER

47

Using Punctuation

PERIOD

After Sentences

Use a **period** to end a sentence that makes a statement or gives a mild command.

> **STATEMENT:** The form of government in the United States is a republic.
>
> **MILD COMMAND:** Please read the instructions carefully.

It is *not* necessary to place a period after a statement that has parentheses around it and is part of another sentence.

> Think about joining a club **(the student affairs office has a list of organizations)** for fun and for leadership experience.

After Initials and Abbreviations

Use a period after an initial and some abbreviations.

| Mr. | Mrs. | B.C.E. | Ph.D. | Sen. Daniel K. Inouye |
| Dr. | M.A. | p.m. | U.S. | Dale Lake, Jr., D.D.S. |

When an abbreviation is the last word in a sentence, use only one period.

> Mikhail eyed each door until he found the name Rosa Lopez, **Ph.D.**

As Decimal Points

Use a period as a decimal point.

> The United States' public debt was nearly **$7.6** trillion by 2005.

QUESTION MARK

After Direct Questions

Use a **question mark** at the end of a direct question.

> What can I know**?** What ought I to do**?** What may I hope**?**   —Immanuel Kant
>
> Since when do you have to agree with people to defend them from injustice**?**
> —Lillian Hellman

Not After Indirect Questions

No question mark is used after an indirect question.

> After listening to Edgar sing, Mr. Noteworthy asked him if he had ever had formal voice training.

When a single-word question like *how*, *when*, or *why* is woven into the flow of a sentence, capitalization and special punctuation are not usually required.

> The questions we need to address at our next board meeting are not *why* or *whether*, but *how* and *when*.

After Quotations That Are Questions

When a question ends with a quotation that is also a question, use only one question mark, and place it within the quotation marks.

> Do you often ask yourself, "What should I be**?**"

To Show Uncertainty

Use a question mark within parentheses to show uncertainty about a word or phrase within a sentence.

> This July will be the 34th **(?)** anniversary of the first moon walk.

Do *not* use a question mark in this manner for formal writing.

For Questions in Parentheses or Dashes

A question within parentheses—or a question set off by dashes—is punctuated with a question mark unless the sentence ends with a question mark.

> You must consult your handbook **(what choice do you have?)** when you need to know a punctuation rule.
>
> Should I use your charge card **(you have one, don't you)**, or should I pay cash**?**
>
> Maybe somewhere in the pasts of these humbled people, there were cases of bad mothering or absent fathering or emotional neglect—**what family surviving the '50s was exempt?**—but I couldn't believe these human errors brought the physical changes in Frank.   —Mary Kay Blakely, *Wake Me When It's Over*

EXCLAMATION POINT

To Express Strong Feeling

Use an **exclamation point** to express strong feeling. It may be placed at the end of a sentence (or an elliptical expression that stands for a sentence). Use exclamation points sparingly.

> "That's not the point," said Wangero. "These are all pieces of dresses Grandma used to wear. She did all this stitching by hand. **Imagine!**"
> —Alice Walker, "Everyday Use"

> Su-su-something's crawling up the back of my neck**!**   —Mark Twain, *Roughing It*

> She was on tiptoe, stretching for an orange, when they heard, **"HEY YOU!"**
> —Beverley Naidoo, *Journey to Jo'burg*

PARENTHESES

To Enclose Explanatory or Supplementary Material

Use **parentheses** to enclose explanatory or supplementary material that interrupts the normal sentence structure.

> The RA **(resident assistant)** became my best friend.

To Set Off Numbers in a List

Use parentheses to set off numbers used with a series of words or phrases.

> Dr. Beck told us **(1)** plan ahead, **(2)** stay flexible, and **(3)** follow through.

For Parenthetical Sentences

When using a full "sentence" within another sentence, do not capitalize it or use a period inside the parentheses.

> Your friend doesn't have the assignment **(he was just thinking about calling you),** so you'll have to make a couple more calls.

When the parenthetical sentence comes after the main sentence, capitalize and punctuate it the same way you would any other complete sentence.

> But Mom doesn't say boo to Dad; she's always sweet to him. **(Actually she's sort of sweet to everybody.)**   —Norma Fox Mazer, *Up on Fong Mountain*

To Set Off References

Use parentheses to set off references to authors, titles, pages, and years.

> The statistics are alarming **(see page 9)** and demand action.

F.Y.I.

For unavoidable parentheses within parentheses (. . . [. . .] . . .), use brackets. Avoid overuse of parentheses by using commas instead.

COMMA

Between Independent Clauses

Use a **comma** between independent clauses that are joined by a coordinating conjunction (*and, but, or, nor, for, yet, so*).

> The most expensive film ever made was *Titanic,* **but** the largest makeup budget for any film was $1 million for *Planet of the Apes.*

Do not confuse a compound verb with a compound sentence.

> The $1 million makeup budget for *Planet of the Apes* shocked Hollywood **and** made producers uneasy. [compound verb]
>
> The $1 million makeup budget was 17 percent of the film's total production cost**, but** the film became a box-office hit and financial success. [compound sentence]

Between Items in a Series

Use commas to separate individual words, phrases, or clauses in a series. (A series contains at least three items.)

> Many college students must balance studying with **taking care of a family, working a job, getting exercise, and finding time to relax.**

Do *not* use commas when all the items in a series are connected with *or, nor,* or *and.*

> Hmm . . . should I study **or** do laundry **or** go out?

To Separate Adjectives

Use commas to separate adjectives that *equally* modify the same noun. Notice in the examples below that no comma separates the last adjective from the noun.

> You should exercise regularly and follow a **sensible, healthful** diet.
>
> A good diet is one that includes lots of **high-protein, low-fat** foods.

To Determine Equal Modifiers

To determine whether the adjectives in a sentence modify a noun *equally,* use these two tests.

1. Reverse the order of the adjectives. If the sentence is clear, the adjectives modify equally. (In the example below, *hot* and *crowded* can be reversed, and the sentence is still clear; *short* and *coffee* cannot.)

    Matt was tired of working in the **hot, crowded** lab and decided to take a **short coffee** break.

2. Insert *and* between the adjectives. If the sentence reads well, use a comma when *and* is omitted. (The word *and* can be inserted between *hot* and *crowded,* but *and* does not make sense between *short* and *coffee.*)

To Set Off Appositives

A specific kind of explanatory word or phrase called an **appositive** identifies or renames a preceding noun or pronoun.

> Albert Einstein, **the famous mathematician and physicist,** developed the theory of relativity.

Do *not* use commas with *restrictive appositives*. A restrictive appositive is essential to the basic meaning of the sentence.

> The famous mathematician and physicist **Albert Einstein** developed the theory of relativity.

To Set Off Clauses

Use a comma after most introductory clauses functioning as adverbs.

> **Although Charlemagne was a great patron of learning,** he never learned to write properly. [adverb clause]

Use a comma if the adverb clause following the main clause is not essential. Clauses beginning with *even though, although, while,* or another conjunction expressing a contrast are usually not needed to complete the meaning of a sentence.

> Charlemagne never learned to write properly, **even though he continued to practice.**

A comma is *not* used if the clause following the main clause is needed to complete the meaning of the sentence.

> Maybe Charlemagne didn't learn **because he had an empire to run.**

After Introductory Phrases

Use a comma after introductory phrases.

> **In spite of his practicing,** Charlemagne's handwriting remained poor.

A comma is usually omitted if the phrase follows the independent clause.

> Charlemagne's handwriting remained poor **in spite of his practicing.**

You may omit the comma after a short (four or fewer words) introductory phrase unless it is needed to ensure clarity.

> **At 6:00 a.m.** he would rise and practice his penmanship.

To Set Off Transitional Expressions

Use a comma to set off conjunctive adverbs and transitional phrases.

> Handwriting is not, **as a matter of fact,** easy to improve upon later in life; **however,** you can develop your handwriting if you are determined enough.

If a transitional expression blends smoothly with the rest of the sentence, the expression does not need to be set off.

> If you are in fact coming, I'll see you there.

To Set Off Items in Addresses and Dates

Use commas to set off items in an address and the year in a date.

> Send your letter to **1600 Pennsylvania Avenue, Washington, DC 20006, before January 1, 2006,** or send e-mail to president@whitehouse.gov.

No comma is placed between the state and ZIP code. Also, *no* comma separates the items if only the month and year are given: January 2006.

To Set Off Dialogue

Use commas to set off the words of the speaker from the rest of the sentence.

> **"Never be afraid to ask for help,"** advised Ms. Kane.
>
> **"With the evidence that we now have,"** Professor Thom said, **"many scientists believe there is life on Mars."**

To Separate Nouns of Direct Address

Use a comma to separate a noun of direct address from the rest of the sentence.

> **Jamie,** would you please stop whistling while I'm trying to work?

To Separate Interjections

Use a comma to separate a mild interjection from the rest of the sentence.

> **Okay,** so now what do I do?

F.Y.I.

Exclamation points are used after strong interjections: Wow! You're kidding!

To Set Off Interruptions

Use commas to set off a word, phrase, or clause that interrupts the movement of a sentence. Such expressions usually can be identified through the following tests: (1) They may be omitted without changing the meaning of a sentence; and (2) they may be placed nearly anywhere in the sentence without changing its meaning.

> For me, **well,** it was just a good job gone!
>
> —Langston Hughes, "A Good Job Gone"

> Lela, **as a general rule,** always comes to class ready for a pop quiz.

To Separate Numbers

Use commas to separate a series of numbers to distinguish hundreds, thousands, millions, and so on.

> Do you know how to write the amount **$2,025** on a check?
>
> 25,000   973,240   18,620,197

To Enclose Explanatory Words

Use commas to enclose an explanatory word or phrase.

> Time management, **according to many professionals,** is such an important skill that it should be taught in college.

To Separate Contrasted Elements

Use commas to separate contrasted elements within a sentence.

> We work to become, **not to acquire.**   —Eugene Delacroix
>
> Where all think alike, **no one thinks very much.** —Walter Lippmann

Before Tags

Use a comma before tags, which are short statements or questions at the ends of sentences.

> You studied for the test, **right?**

To Enclose Titles or Initials

Use commas to enclose a title or initials and given names that follow a surname.

> Until Martin, **Sr.,** was 15, he never had more than three months of schooling in any one year. —Ed Clayton, *Martin Luther King: The Peaceful Warrior*
>
> The genealogical files included the names Sanders, **L. H.,** and Sanders, **Lucy Hale.**

For Clarity or Emphasis

Use a comma for clarity or for emphasis. There will be times when none of the traditional rules call for a comma, but one will be needed to prevent confusion or to emphasize an important idea.

> What she does, does matter to us. [clarity]
>
> It may be those who do most, dream most. [emphasis] —Stephen Leacock

Avoid Overusing Commas

The commas (in red) below are used incorrectly. Do *not* use a comma between the subject and its verb or the verb and its object.

> Current periodicals on the subject of psychology, are available at nearly all bookstores.
>
> I think she should read, *Psychology Today.*

Do *not* use a comma before an indirect quotation.

> My roommate said, that she doesn't understand the notes I took.

A CLOSER Look
at Nonrestrictive and Restrictive Clauses and Phrases

Use Commas with Nonrestrictive Clauses and Phrases Use commas to enclose **nonrestrictive** (unnecessary) clauses and phrases. A nonrestrictive clause or phrase adds information that is not necessary to the basic meaning of the sentence. For example, if the clause or phrase (in **boldface**) were left out of the two examples below, the meaning of the sentences would remain clear. Therefore, commas are used to set them off.

> The locker rooms in Swain Hall, **which were painted and updated last summer,** give professors a place to shower. [clause]
>
> Work-study programs, **offered on many campuses,** give students the opportunity to earn tuition money. [phrase]

Don't Use Commas with Restrictive Clauses and Phrases Do *not* use commas to set off **restrictive** (necessary) clauses and phrases. A restrictive clause or phrase adds information that the reader needs to understand the sentence. For example, if the clause and phrase (in **boldface**) were dropped from the examples below, the meaning would be unclear.

> Only the professors **who run at noon** use the locker rooms in Swain Hall to shower. [clause]
>
> Tuition money **earned through work-study programs** is the only way some students can afford to go to college. [phrase]

Using *That* or *Which* Use *that* to introduce restrictive (necessary) clauses; use *which* to introduce nonrestrictive (unnecessary) clauses. When the two words are used in this way, the reader can quickly distinguish the necessary information from the unnecessary.

> Campus jobs **that are funded by the university** are awarded to students only. [restrictive]
>
> The cafeteria, **which is run by an independent contractor,** can hire nonstudents. [nonrestrictive]

Clauses beginning with *who* can be either restrictive or nonrestrictive.

> Students **who pay for their own education** are highly motivated. [restrictive]
>
> The admissions counselor, **who has studied student records,** said that many returning students earn high GPAs in spite of demanding family obligations. [nonrestrictive]

APOSTROPHE

In Contractions

Use an **apostrophe** to show that one or more letters have been left out of two words joined to form a contraction.

    don't → **o** is left out   she'd → **woul** is left out   it's → **i** is left out

An apostrophe is also used to show that one or more numerals or letters have been left out of numbers or words.

    class of '05 → **20** is left out   good mornin' → **g** is left out

To Form Plurals

Use an apostrophe and an *s* to form the plural of a letter, a number, a sign, or a word discussed as a word.

    A → **A's**   8 → **8's**   + → **+'s**

You use too many **and's** in your writing.

If two apostrophes are called for in the same word, omit the second one.

Follow closely the do's and **don'ts** [not **don't's**] on the checklist.

To Form Singular Possessives

The possessive form of singular nouns is usually made by adding an apostrophe and an *s*.

    Spock's ears   my **computer's** memory

When a singular noun of more than one syllable ends with an *s* or a *z* sound, the possessive may be formed by adding just an apostrophe—or an apostrophe and an *s*. When the singular noun is a one-syllable word, however, the possessive is usually formed by adding both an apostrophe and an *s*.

    Dallas' sports teams [or] **Dallas's** sports teams [two-syllable word]
    Kiss's last concert   my **boss's** generosity [one-syllable words]

To Form Plural Possessives

The possessive form of plural nouns ending in *s* is made by adding just an apostrophe.

    the **Joneses'** great-grandfather   **bosses'** offices

For plural nouns not ending in *s*, add an apostrophe and *s*.

    women's health issues   **children's** program

To Determine Ownership

You will punctuate possessives correctly if you remember that the word that comes immediately before the apostrophe is the owner.

    girl's guitar [*girl* is the owner]   **girls'** guitar [*girls* are the owners]
    boss's office [*boss* is the owner]   **bosses'** office [*bosses* are the owners]

To Show Shared Possession

When possession is shared by more than one noun, use the possessive form for the last noun in the series.

    Jason, Kamil, and **Elana's** sound system [All three own the same system.]
    Jason's, Kamil's, and Elana's sound systems [Each owns a separate system.]

In Compound Nouns

The possessive of a compound noun is formed by placing the possessive ending after the last word.

    his **mother-in-law's** name [singular]
    the **secretary of state's** career [singular]

    their **mothers-in-law's** names [plural]
    the **secretaries of state's** careers [plural]

With Indefinite Pronouns

The possessive form of an indefinite pronoun is made by adding an apostrophe and an *s* to the pronoun. (See page 701.)

    everybody's grades   **no one's** mistake   **one's** choice

In expressions using *else*, add the apostrophe and *s* after the last word.

    anyone else's   **somebody else's**

To Show Time or Amount

Use an apostrophe and an *s* with an adjective that is part of an expression indicating time or amount.

    yesterday's news   a **day's** wage   a **month's** pay

Punctuation Marks

´ (é)	Accent, acute	:	Colon	¶	Paragraph
` (è)	Accent, grave	,	Comma	( )	Parentheses
<>	Angle brackets	†	Dagger	.	Period
'	Apostrophe	—	Dash	?	Question mark
*	Asterisk	¨ (ä)	Dieresis	" "	Quotation marks
{ }	Braces	/	Diagonal/slash	§	Section
[ ]	Brackets	…	Ellipsis	;	Semicolon
^	Caret	!	Exclamation point	~ (ñ)	Tilde
ç	Cedilla	-	Hyphen	___	Underscore
ˆ (â)	Circumflex	.....	Leaders		

COLON

After Salutations
Use a **colon** after the salutation of a business letter.

> Dear Mr. Spielberg: Dear Professor Higgins: Dear Members:

Between Numbers Indicating Time or Ratios
Use a colon between the hours, minutes, and seconds of a number indicating time.

> 8:30 p.m.   9:45 a.m.   10:24:55

Use a colon between two numbers in a ratio.

> The ratio of computers to students is 1:20. [one to twenty]

For Emphasis
Use a colon to emphasize a word, a phrase, a clause, or a sentence that explains or adds impact to the main clause.

> **I have one goal for myself:** to become the first person in my family to graduate from college.

To Distinguish Parts of Publications
Use a colon between a title and a subtitle, volume and page, and chapter and verse.

> *Ron Brown: An Uncommon Life*   *Britannica* 4: 211   Psalm 23:1–6

To Introduce Quotations
Use a colon to introduce a quotation following a complete sentence.

> **John Locke is credited with this prescription for a good life:** "A sound mind in a sound body."
>
> **Lou Gottlieb, however, offered this version:** "A sound mind or a sound body—take your pick."

To Introduce a List
Use a colon to introduce a list following a complete sentence.

> **A college student needs a number of things to succeed:** basic skills, creativity, and determination.

Avoid Colon Errors
Do *not* use a colon between a verb and its object or complement.

> Dave likes: comfortable space and time to think. [incorrect]
>
> Dave likes two things: comfortable space and time to think. [correct]

SEMICOLON

To Join Two Independent Clauses

Use a **semicolon** to join two or more closely related independent clauses that are not connected with a coordinating conjunction. In other words, each of the clauses could stand alone as a separate sentence.

> I was thrown out of college for cheating on the metaphysics exam; I looked into the soul of the boy next to me.
> —Woody Allen

With Conjunctive Adverbs

Use a semicolon before a conjunctive adverb when the word connects two independent clauses in a compound sentence. A comma often follows the conjunctive adverb. Common conjunctive adverbs include *also, besides, however, instead, meanwhile, then,* and *therefore*.

> Many college freshmen are on their own for the first time; **however,** others are already independent and even have families.

With Transitional Phrases

Use a semicolon before a transitional phrase when the phrase connects two independent clauses in a compound sentence. A comma usually follows the transitional phrase.

> Pablo was born in the Andes; **as a result,** he loves mountains.

Transitional phrases

after all	at the same time	in addition	in the first place
as a matter of fact	even so	in conclusion	on the contrary
as a result	for example	in fact	on the other hand
at any rate	for instance	in other words	

To Separate Independent Clauses

Use a semicolon to separate independent clauses that contain internal commas, even when they are connected by a coordinating conjunction.

> Make sure your CD player, computer, bike, and other valuables are covered by a homeowner's insurance policy; and be sure to use the locks on your door, bike, and storage area.

To Separate Items in a Series That Contain Commas

Use a semicolon to separate items in a series that already contain commas.

> My favorite foods are pizza with pepperoni, onions, and olives; peanut butter and banana sandwiches; and liver with bacon, peppers, and onions.

ELLIPSIS

To Show Omitted Words

Use an **ellipsis** (three periods) to show that one or more words have been omitted in a quotation. When typing, leave one space before and after each period.

> **ORIGINAL:** We the people of the United States, in order to form a more perfect Union, establish justice, insure domestic tranquility, provide for the common defense, promote the general welfare, and secure the blessings of liberty to ourselves and our posterity, do ordain and establish this Constitution for the United States of America. —Preamble, U.S. Constitution
>
> **QUOTATION:** "We the people . . . in order to form a more perfect Union . . . establish this Constitution for the United States of America."

Omit internal punctuation (a comma, a semicolon, a colon, or a dash) on either side of the ellipsis marks unless it is needed for clarity.

To Use After Sentences

If words from a quotation are omitted at the end of a sentence, place the ellipsis after the period or other end punctuation.

> **QUOTATION:** "Five score years ago, a great American, in whose symbolic shadow we stand, signed the Emancipation Proclamation. . . . But one hundred years later, we must face the tragic fact that the Negro is still not free."
> —Martin Luther King, Jr., "I Have a Dream"

The first word of a sentence following a period and an ellipsis may be capitalized, even though it was not capitalized in the original.

> **QUOTATION:** "Five score years ago, a great American . . . signed the Emancipation Proclamation. . . . One hundred years later, . . . the Negro is still not free."

If the quoted material is a complete sentence (even if it was not in the original), use a period, then an ellipsis.

> **ORIGINAL:** I am tired; my heart is sick and sad. From where the sun now stands I will fight no more forever. —Chief Joseph of the Nez Percé
>
> **QUOTATION:** "I am tired. . . . I will fight no more forever."

To Show Pauses

Use an ellipsis to indicate a pause or to show unfinished thoughts.

> Listen . . . did you hear that?
>
> I can't figure out . . . this number doesn't . . . just how do I apply the equation in this case?

QUOTATION MARKS

To Punctuate Titles of Works Within Other Works

Use **quotation marks** to punctuate some titles. (Also, see page 742.)

"Two Friends" [short story]
"New Car Designs" [newspaper article]
"Desperado" [song]
"Multiculturalism and the Language Battle" [lecture title]
"The New Admissions Game" [magazine article]
"Reflections on Advertising" [chapter in a book]
"Force of Nature" [television episode from *Star Trek: The Next Generation*]
"Annabel Lee" [short poem]

For Special Words

Use quotation marks (1) to show that a word is being discussed as a word, (2) to indicate that a word is slang, or (3) to point out that a word is being used in a humorous or ironic way.

1. A commentary on the times is that the word **"honesty"** is now preceded by **"old-fashioned."** —Larry Wolters

2. I drank a Dixie and ate bar peanuts and asked the bartender where I could hear **"chanky-chank,"** as Cajuns call their music.
—William Least Heat-Moon, *Blue Highways*

3. To be popular, he works very hard at being **"cute."**

Placement of Periods or Commas

Always place periods and commas inside quotation marks.

"Dr. Slaughter wants you to have liquids, Will**,**" Mama said anxiously. "He said not to give you any solid food tonight**.**" —Olive Ann Burns, *Cold Sassy Tree*

Placement of Exclamation Points or Question Marks

Place an exclamation point or a question mark inside quotation marks when it punctuates both the main sentence and the quotation *or* just the quotation; place it outside when it punctuates the main sentence.

Do you often ask yourself, "What should I be**?**"

I almost croaked when he asked, "That won't be a problem, will it**?**"

Did he really say, "Finish this by tomorrow"**?**

Placement of Semicolons or Colons

Always place semicolons or colons outside quotation marks.

I just read "Computers and Creativity"**;** I now have some different ideas about the role of computers in the arts.

A CLOSER Look

at Marking Quoted Material

For Direct Quotations Use quotation marks before and after a direct quotation—a person's exact words.

> Sitting in my one-room apartment, I remember Mom saying, **"Don't go to the party with him."**

Do *not* use quotation marks for *indirect* quotations.

> I remember Mom saying **that I should not date him.** [These are not the speaker's exact words.]

For Quoted Passages Use quotation marks before and after a quoted passage. Any word that is not part of the original quotation must be placed inside brackets.

> ORIGINAL: First of all, it must accept responsibility for providing shelter for the homeless.

> QUOTATION: "First of all, it **[the federal government]** must accept responsibility for providing shelter for the homeless."

If you quote only part of the original passage, be sure to construct a sentence that is both accurate and grammatically correct.

> The report goes on to say that the federal government **"must accept responsibility for providing shelter for the homeless."**

For Long Quotations If more than one paragraph is quoted, quotation marks are placed before each paragraph and at the end of the last paragraph (Example A). Quotations that are five or more lines (MLA style) or forty words or more (APA style) are usually set off from the text by indenting ten spaces from the left margin (a style called "block form"). Do not use quotation marks before or after a block-form quotation (Example B), except in cases where quotation marks appear in the original passage (Example C).

Example A	Example B	Example C
"_____	_____	_____
_____.	_____	_____.
"_____	_____	_____
_____.	_____	"_____ ____
_____."	_____.	_____."

For Quoting Quotations Use single quotation marks to punctuate quoted material within a quotation.

> "I was lucky," said Jane. "The proctor announced, **'Put your pencils down,'** just as I was filling in the last answer."

HYPHEN

In Compound Words

Use a **hyphen** to make some compound words.

    great-great-grandfather [noun]   starry-eyed [adjective]
    mother-in-law [noun]   three-year-old [adjective]

Writers sometimes combine words in a new and unexpected way. Such combinations are usually hyphenated.

> And they pried pieces of **baked-too-fast** sunshine cake from the roofs of their mouths and looked once more into the boy's eyes.
>
> —Toni Morrison, *Song of Solomon*

F.Y.I.

Consult a dictionary to find how it lists a particular compound word. Some compound words *(living room)* do not use a hyphen and are written separately. Some are written solid *(bedroom)*. Some do not use a hyphen when the word is a noun *(ice cream)* but do use a hyphen when it is a verb or an adjective *(ice-cream sundae)*.

To Join Letters and Words

Use a hyphen to join a capital letter or a lowercase letter to a noun or a participle.

    T-shirt   U-turn   V-shaped   x-axis

To Join Words in Compound Numbers

Use a hyphen to join the words in compound numbers from *twenty-one* to *ninety-nine* when it is necessary to write them out. (See "Words Only," page 750.)

    Forty-two people found seats in the cramped classroom.

Between Numbers in Fractions

Use a hyphen between the numerator and denominator of a fraction, but not when one or both of these elements are already hyphenated.

    four-tenths   five-sixteenths   seven thirty-seconds (7/32)

In a Special Series

Use a hyphen when two or more words have a common element that is omitted in all but the last term.

    We have cedar posts in **four-**, **six-**, and **eight-**inch widths.

To Create New Words

Use a hyphen to form new words beginning with the prefixes *self, ex, all,* and *half*. Also use a hyphen to join any prefix to a proper noun, a proper adjective, or the official name of an office.

    post-Depression   mid-May   ex-mayor

To Prevent Confusion
Use a hyphen with prefixes or suffixes to avoid confusion or awkward spelling.

re-cover [not *recover*] the sofa   **shell-like** [not *shelllike*] shape

To Join Numbers
Use a hyphen to join numbers indicating a range, a score, or a vote.

Students study **30–40** hours a week. The final score was **84–82**.

To Divide Words
Use a hyphen to divide a word between syllables at the end of a line of print.

Guidelines for Word Division

1. Leave enough of the word at the end of the line to identify the word.
2. Never divide a one-syllable word: **rained, skills, through.**
3. Avoid dividing a word of five or fewer letters: **paper, study, July.**
4. Never divide a one-letter syllable from the rest of the word: **omit-ted**, not **o-mitted.**
5. Always divide a compound word between its basic units: **sister-in-law,** not **sis-ter-in-law.**
6. Never divide abbreviations or contractions: **shouldn't,** not **should-n't.**
7. When a vowel is a syllable by itself, divide the word after the vowel: **epi-sode,** not **ep-isode.**
8. Avoid dividing a numeral: **1,000,000;** not **1,000,-000.**
9. Avoid dividing the last word in a paragraph.
10. Never divide the last word in more than two lines in a row.
11. Check a dictionary.

To Form Adjectives
Use a hyphen to join two or more words that serve as a single-thought adjective before a noun.

In real life I am a large, **big-boned** woman with rough, **man-working** hands.

—Alice Walker, "Everyday Use"

Most single-thought adjectives are not hyphenated when they come after the noun.

In real life, I am large and **big boned.**

When the first of these words is an adverb ending in *ly,* do *not* use a hyphen. Also, do *not* use a hyphen when a number or a letter is the final element in a single-thought adjective.

fresh**ly** painted barn   grade **A** milk [letter is the final element]

DASH

To Set Off Nonessential Elements

Use a **dash** to set off nonessential elements—explanations, examples, or definitions—when you want to emphasize them.

> Near the semester's end—**and this is not always due to poor planning**—some students may find themselves in academic trouble.
>
> The term *caveat emptor*—**let the buyer beware**—is especially appropriate to Internet shopping.

F.Y.I.

A dash is indicated by two hyphens--with no spacing before or after--in typewriter-generated or handwritten material. Don't use a single hyphen when a dash (two hyphens) is required.

To Set Off an Introductory Series

Use a dash to set off an introductory series from the clause that explains the series.

> **Cereal, coffee, and a newspaper**—without these I can't get going in the morning.

To Show Missing Text

Use a dash to show that words or letters are missing.

> **Mr.**—won't let us marry.   —Alice Walker, *The Color Purple*

To Show Interrupted Speech

Use a dash (or ellipsis) to show interrupted or faltering speech in dialogue.

> Well, **I—ah—had** this terrible case of the flu, **and—then—ah—the** library closed because of that flash flood, **and—well—the** high humidity jammed my printer.
>
> —Excuse No. 101
>
> "You told me to tell her about the—"
> "Oh, just **stop**."
>
> —Joyce Carol Oates, "Why Don't You Come Live with Me It's Time"

For Emphasis

Use a dash in place of a colon to introduce or to emphasize a word, a series, a phrase, or a clause.

> **Jogging**—that's what he lives for.
>
> Life is like a grindstone—**whether it grinds you down or polishes you up depends on what you're made of.**
>
> This is how the world moves—**not like an arrow, but a boomerang.**
>
> —Ralph Ellison

BRACKETS

With Words That Clarify

Use **brackets** before and after words that are added to clarify what another person has said or written.

> "They'd **[the sweat bees]** get into your mouth, ears, eyes, nose. You'd feel them all over you."   —Marilyn Johnson and Sasha Nyary, "Roosevelts in the Amazon"

The brackets indicate that the words *the sweat bees* are not part of the original quotation but were added for clarification. (See page 737.)

Around Comments by Someone Other Than the Author

Place brackets around comments that have been added by someone other than the author or speaker.

> "In conclusion, *docendo discimus*. Let the school year begin!" **[Huh?]**

Around Editorial Corrections

Place brackets around an editorial correction.

> "Brooklyn alone has 8 percent of lead poisoning **[victims]** nationwide," said Marjorie Moore.   —Donna Actie, student writer

Around the Word *Sic*

Brackets should be placed around the word *sic* (Latin for "so" or "thus") in quoted material; the word indicates that an error appearing in the quoted material was made by the original speaker or writer.

> "There is a higher principal **[sic]** at stake here: Is the school administration aware of the situation?"

DIAGONAL

To Form Fractions or Show Choices

Use a **diagonal** (also called a *slash*) to form a fraction. Also place a diagonal between two words to indicate that either is acceptable.

> My **walking/running** shoe size is **5 1/2**; my dress shoes are **6 1/2**.

When Quoting Poetry

When quoting poetry, use a diagonal (with one space before and after) to show where each line ends in the actual poem.

> "A dryness is upon the house / My father loved and tended. / Beyond his firm and sculptured door / His light and lease have ended."
>   —Gwendolyn Brooks, "In Honor of David Anderson Brookes, My Father"

ITALICS (UNDERLINING)

In Handwritten and Printed Material

Italics is a printer's term for a style of type that is slightly slanted. In this sentence, the word *happiness* is printed in italics. In material that is handwritten or typed on a machine that cannot print in italics, underline each word or letter that should be in italics.

> In <u>The Road to Memphis,</u> racism is a contagious disease.
> [typed or handwritten]
>
> Mildred Taylor's *The Road to Memphis* exposes racism. [printed]

In Titles

Use italics to indicate the titles of magazines, newspapers, books, pamphlets, full-length plays, films, videos, radio and television programs, book-length poems, ballets, operas, lengthy musical compositions, cassettes, CDs, paintings and sculptures, legal cases, Web sites, and the names of ships and aircraft. (Also see page 736.)

> *Newsweek* [magazine]   *The Nutcracker* [ballet]
> *New York Times* [newspaper]   *Neverland* [film]
> *Sister Carrie* [book]   *The Thinker* [sculpture]
> *Othello* [play]   *Nightline* [television program]
> *Enola Gay* [airplane]   *GeoCities* [Web site]
> *The Joshua Tree* [CD]   *College Loans* [pamphlet]
> *ACLU v. the State of Ohio* [legal case]

When one title appears within another title, punctuate as follows:

> "The **Fresh Prince of Bel-Air** Rings True" is an article I read. [title of TV program in an article title]
>
> He wants to watch *Inside the "New York Times"* on PBS tonight. [title of newspaper in title of TV program]

For Key Terms

Italics are often used for a key term in a discussion or for a technical term, especially when it is accompanied by its definition. Italicize the term the first time it is used. Thereafter, put the term in Roman type.

> This flower has a **zygomorphic** (bilateral symmetry) structure.

For Foreign Words and Scientific Names

Use italics for foreign words that have not been adopted into the English language; italics are also used to denote scientific names.

> Say **arrivederci** to your fears and try new activities. [foreign word]
>
> The voyageurs discovered the shy **Castor canadensis,** or North American beaver. [scientific name]

In this chapter

Capitalization **743** Numbers **749** Acronyms and Initialisms **753**
Plurals **747** Abbreviations **751** Spelling **754**

CHAPTER 48

Checking Mechanics

CAPITALIZATION

Proper Nouns and Adjectives

Capitalize all proper nouns and all proper adjectives (adjectives derived from proper nouns). Table 48.1 provides a quick overview of capitalization rules. The following pages explain specific or special uses of capitalization.

Table 48.1 Capitalization at a Glance

Days of the week	Sunday, Monday, Tuesday
Months	June, July, August
Holidays, holy days	Thanksgiving, Easter, Hanukkah
Periods, events in history	Middle Ages, World War I
Special events	Tate Memorial Dedication Ceremony
Political parties	Republican Party, Socialist Party
Official documents	the Declaration of Independence
Trade names	Oscar Mayer hot dogs, Pontiac Firebird
Formal epithets	Alexander the Great
Official titles	Mayor John Spitzer, Senator Feinstein
Official state nicknames	the Badger State, the Aloha State
Geographical names	
Planets, heavenly bodies	Earth, Jupiter, the Milky Way
Continents	Australia, South America
Countries	Ireland, Grenada, Sri Lanka
States, provinces	Ohio, Utah, Nova Scotia
Cities, towns, villages	El Paso, Burlington, Wonewoc
Streets, roads, highways	Park Avenue, Route 66, Interstate 90
Sections of the United States and the world	the Southwest, the Far East
Landforms	the Rocky Mountains, the Kalahari Desert
Bodies of water	Nile and Ural Rivers, Lake Superior, Bee Creek
Public areas	Central Park, Yellowstone National Park

First Words

Capitalize the first word in every sentence and the first word in a full-sentence direct quotation.

> **Attending** the orientation for new students is a good idea.
>
> Max suggested, "**Let's** take the guided tour of the campus first."

Sentences in Parentheses

Capitalize the first word in a sentence that is enclosed in parentheses if that sentence is not contained within another complete sentence.

> The bookstore has the software. (**Now** all I need is the computer.)

Do *not* capitalize a sentence that is enclosed in parentheses and is located in the middle of another sentence.

> Your college will probably offer everything (**this** includes general access to a computer) that you'll need for a successful year.

Sentences Following Colons

Capitalize a complete sentence that follows a colon when that sentence is a formal statement, a quotation, or a sentence that you want to emphasize.

> Sydney Harris had this to say about computers: "**The** real danger is not that computers will begin to think like people, but that people will begin to think like computers."

Salutation and Complimentary Closing

In a letter, capitalize the first and all major words of the salutation. Capitalize only the first word of the complimentary closing.

> **Dear Personnel Director:**   **Sincerely** yours,

Sections of the Country

Words that indicate sections of the country are proper nouns and should be capitalized; words that simply indicate direction are not proper nouns.

> Many businesses move to the **South.** [section of the country]
>
> They move **south** to cut fuel costs and other expenses. [direction]

Languages, Ethnic Groups, Nationalities, and Religions

Capitalize languages, ethnic groups, nationalities, and religions.

> African American   Latino   Navajo   French   Islam

Nouns that refer to the Supreme Being and holy books are capitalized.

> God   Allah   Jehovah   the Koran   Exodus   the Bible

Titles

Capitalize the first word of a title, the last word, and every word in between except articles (*a, an, the*), short prepositions, and coordinating conjunctions. Follow this rule for titles of books, newspapers, magazines, poems, plays, songs, articles, films, works of art, and stories.

Going to Meet the Man *Chicago Tribune*
"Nothing Gold Can Stay" "Jobs in the Cyber Arena"
A Midsummer Night's Dream *The War of the Roses*

F.Y.I.

When citing titles in a bibliography, check the style manual you've been asked to follow. For example, in APA style, only the first word of a title is capitalized.

Organizations

Capitalize the name of an organization, or a team and its members.

American Indian Movement Republican Party
Tampa Bay Buccaneers Tucson Drama Club

Abbreviations

Capitalize abbreviations of titles and organizations. (Some other abbreviations are also capitalized. See pages 751–753.)

M.D. Ph.D. NAACP C.E. B.C.E. GPA

Letters

Capitalize letters used to indicate a form or shape.

U-turn **I**-beam **S**-curve **V**-shaped **T**-shirt

Words Used as Names

Capitalize words like *father, mother, uncle, senator,* and *professor* when they are parts of titles that include a personal name, or when they are substituted for proper nouns (especially in direct address).

Hello, **Senator** Feingold. [*Senator* is part of the name.]
Our **senator** is an environmentalist.

Who was your chemistry **professor** last quarter?
I had **Professor** Williams for Chemistry 101.

To test whether a word is being substituted for a proper noun, simply read the sentence with a proper noun in place of the word. If the proper noun fits in the sentence, the word being tested should be capitalized. Usually the word is not capitalized if it follows a possessive—*my, his, our, your,* and so on.

Did **Dad [Brad]** pack the stereo in the trailer? [*Brad* works in this sentence.]

Did your **dad [Brad]** pack the stereo in the trailer? [*Brad* does not work in this sentence; the word *dad* follows the possessive *your.*]

Titles of Courses

Words such as *technology, history,* and *science* are proper nouns when they are included in the titles of specific courses; they are common nouns when they name a field of study.

> Who teaches **Art History 202?** [title of a specific course]
>
> Professor Bunker loves teaching **history.** [a field of study]

The words *freshman, sophomore, junior,* and *senior* are not capitalized unless they are part of an official title.

> The **seniors** who maintained high GPAs were honored at the **Mount Mary Senior Honors Banquet.**

Internet, Web, and E-Mail

The words *Internet, Web,* and *World Wide Web* are always capitalized because they are considered proper nouns. When your writing includes a Web address (URL), capitalize any letters that the site's owner does (on printed materials or on the site itself). Not only is it respectful to reprint it exactly as it appears elsewhere, but, in fact, some Web addresses are case-sensitive and must be entered into a browser's address bar exactly as presented.

> When doing research on the **Internet,** be sure to record each site's **Web** address (**URL**) and each contact's **e-mail** address.

Some people include capital letters in their e-mail addresses to make certain features evident. Although e-mail addresses are not case-sensitive, repeat each letter in print just as its owner uses it.

Avoid Capitalization Errors

Do *not* capitalize any of the following:

- A prefix attached to a proper noun
- Seasons of the year
- Words used to indicate direction or position
- Common nouns and titles that appear near, but are not part of, a proper noun

Capitalize	Do not capitalize
American	un-American
January, February	winter, spring
The South is quite conservative.	Turn south at the stop sign.
Duluth City College	a Duluth college
Chancellor John Bohm	John Bohm, our chancellor
President Bush	the president of the United States
Earth (the planet)	earthmover
Internet	e-mail

PLURALS

Nouns Ending in a Consonant

Some nouns remain unchanged when used as plurals (*species, moose, halibut,* and so on), but the plurals of most nouns are formed by adding an *s* to the singular form.

   dorm—**dorms**   credit—**credits**   midterm—**midterms**

The plurals of nouns ending in *sh, ch, x, s,* and *z* are made by adding *es* to the singular form.

   lunch—**lunches**   wish—**wishes**   class—**classes**

Nouns Ending in *y*

The plurals of common nouns that end in *y*—preceded by a consonant—are formed by changing the *y* to *i* and adding *es*.

   dormitory—**dormitories**   sorority—**sororities**   duty—**duties**

The plurals of common nouns that end in *y* (preceded by a vowel) are formed by adding only an *s*.

   attorney—**attorneys**   monkey—**monkeys**   toy—**toys**

The plurals of all proper nouns ending in *y* (whether preceded by a consonant or a vowel) are formed by adding an *s*.

   the three **Kathys**   the five **Faheys**

Nouns Ending in *o*

The plurals of words ending in *o* (preceded by a vowel) are formed by adding an *s*.

   radio—**radios**   cameo—**cameos**   studio—**studios**

The plurals of most nouns ending in *o* (preceded by a consonant) are formed by adding *es*.

   echo—**echoes**   hero—**heroes**   tomato—**tomatoes**

Musical terms always form plurals by adding an *s*; check a dictionary for other words of this type.

   alto—**altos**   banjo—**banjos**   solo—**solos**   piano—**pianos**

Nouns Ending in *f* or *fe*

The plurals of nouns that end in *f* or *fe* are formed in one of two ways: If the final *f* sound is still heard in the plural form of the word, simply add *s*; if the final sound is a *v* sound, change the *f* to *ve* and add an *s*.

   Plural ends with *f* sound: roof—**roofs**   chief—**chiefs**

   Plural ends with *v* sound: wife—**wives**   loaf—**loaves**

The plurals of some nouns that end in *f* or *fe* can be formed by either adding *s* or changing the *f* to *ve* and adding an *s*.

   Plural ends with either sound: hoof—**hoofs, hooves**

Irregular Spelling

Many foreign words (as well as some of English origin) form a plural by taking on an irregular spelling; others are now acceptable with the commonly used *s* or *es* ending. Take time to check a dictionary.

    child—**children**   alumnus—**alumni**   syllabus—**syllabi, syllabuses**

    goose—**geese**   datum—**data**   radius—**radii, radiuses**

Words Discussed as Words

The plurals of symbols, letters, figures, and words discussed as words are formed by adding an apostrophe and an *s*.

    Many colleges have now added **A/B's** and **B/C's** as standard grades.

You can choose to omit the apostrophe when the omission does not cause confusion.

    YMCA's or **YMCAs**   **CD's** or **CDs**

Nouns Ending in *ful*

The plurals of nouns that end with *ful* are formed by adding an *s* at the end of the word.

    three **teaspoonfuls**   two **tankfuls**   four **bagfuls**

Compound Nouns

The plurals of compound nouns are usually formed by adding an *s* or an *es* to the important word in the compound.

    brothers-in-law   **maids** of honor   **secretaries** of state

Collective Nouns

Collective nouns do not change in form when they are used as plurals.

    class [a unit—singular form]

    class [individual members—plural form]

Because the spelling of the collective noun does not change, it is often the pronoun used in place of the collective noun that indicates whether the noun is singular or plural. Use a singular pronoun (*its*) to show that the collective noun is singular. Use a plural pronoun (*their*) to show that the collective noun is plural.

    The class needs to change **its** motto. [The writer is thinking of the group as a unit.]

    The class brainstormed with **their** professor. [The writer is thinking of the group as individuals.]

F.Y.I.

To determine whether a plural requires the article *the*, you must first determine whether it is definite or indefinite. Definite plurals use *the*, whereas indefinite plurals do not require any article.

NUMBERS

Numerals or Words

Numbers from one to one hundred are usually written as words; numbers 101 and greater are usually written as numerals. Hyphenate numbers written as two words if less than one hundred.

> two   seven   ten   twenty-five   106   1,079

The same rule applies to the use of ordinal numbers.

> second   tenth   twenty-fifth   ninety-eighth   106th   333rd

If numbers greater than 101 are used infrequently in a piece of writing, you may spell out those that can be written in one or two words.

> two hundred   fifty thousand   six billion

You may use a combination of numerals and words for very large numbers.

> 1.5 million   3 billion to 3.2 billion   6 trillion

Numbers being compared or contrasted should be kept in the same style.

> **8** to **11** years old *or* **eight** to **eleven** years old

Particular decades may be spelled out or written as numerals.

> the **'80s** and **'90s** *or* the **eighties** and **nineties**

Numerals Only

Use numerals for the following forms: decimals, percentages, pages, chapters (and other parts of a book), addresses, dates, telephone numbers, identification numbers, and statistics.

> **26.2**   **8** percent   Chapter **7**
> pages **287–289**   Highway **36**   **(212) 555-1234**
> **398-55-0000**   a vote of **23** to **4**   May **8, 1999**

Abbreviations and symbols are often used in charts, graphs, footnotes, and so forth, but typically are not used in texts.

> He is **five feet one inch** tall and **ten years old**.
>
> She walked **three and one-half miles** to work through **twelve inches** of snow.

However, abbreviations and symbols may be used in scientific, mathematical, statistical, and technical texts.

> Between **20%** and **23%** of the cultures yielded positive results.
>
> Your **245B** model requires **220V**.

Always use numerals with abbreviations and symbols.

> **5'4"**   **8%**   **10** in.   **3** tbsp.   **6** lb. **8** oz.   **90**°F

Use numerals after the name of local branches of labor unions.

> The Office and Professional Employees International Union, Local **8**

Hyphenated Numbers

Hyphens are used to form compound modifiers indicating measurement. They are also used for inclusive numbers and written-out fractions.

    a **three-mile** trip   the **2001–2005** presidential term

    a **2,500-mile** road trip   **one-sixth** of the pie

    a **thirteen-foot** clearance   **three-eighths** of the book

Time and Money

If time is expressed with an abbreviation, use numerals. If time is expressed in words, spell out the number.

    4:00 a.m. *or* **four** o'clock [not 4 o'clock]

    the **5:15** p.m. train

    a **seven o'clock** wake-up call

If money is expressed with a symbol, use numerals. If the currency is expressed in words, spell out the number.

    $20 *or* **twenty** dollars [not 20 dollars]

Abbreviations of time and of money may be used in text.

    The concert begins at **7:00** p.m., and tickets cost **$30**.

Words Only

Use words to express numbers that begin a sentence.

    Fourteen students "forgot" their assignments.

    Three hundred contest entries were received.

Change the sentence structure if this rule creates a clumsy construction.

    CLUMSY: **Six hundred thirty-nine** students are new to the campus this fall.

    BETTER: This fall, **639** students are new to the campus.

Use words for numbers that precede a compound modifier that includes a numeral. (If the compound modifier uses a spelled-out number, use numerals in front of it.)

    She sold **twenty 35-millimeter** cameras in one day.

    The chef prepared **24 eight-ounce** filets.

Use words for the names of numbered streets of one hundred or less.

    Ninth Avenue   **123 Forty-fourth** Street

Use words for the names of buildings if that name is also its address.

    One Thousand State Street   Two Fifty Park Avenue

Use words for references to particular centuries.

    the twenty-first century   the fourth century B.C.E.

ABBREVIATIONS

An **abbreviation** is the shortened form of a word or a phrase. These abbreviations are always acceptable in both formal and informal writing:

Mr. Mrs. Ms. Dr. Jr. a.m. (A.M.) p.m. (P.M.)

In formal writing, do not abbreviate the names of states, countries, months, days, units of measure, or courses of study. Do not abbreviate the words *Street, Road, Avenue, Company,* and similar words when they are part of a proper name. Also, do not use signs or symbols (%, &, #, @) in place of words. (The dollar sign, however, is appropriate when numerals are used to express an amount of money.)

When abbreviations are called for (in charts, lists, bibliographies, notes, and indexes, for example), standard abbreviations are preferred. Reserve the postal abbreviations for ZIP code addresses. (See Table 48.2.)

Table 48.2 Standard and Postal Abbreviations

States/Territories

	Standard	Postal
Alabama	Ala.	AL
Alaska	Alaska	AK
Arizona	Ariz.	AZ
Arkansas	Ark.	AR
California	Cal.	CA
Colorado	Colo.	CO
Connecticut	Conn.	CT
Delaware	Del.	DE
District of Columbia	D.C.	DC
Florida	Fla.	FL
Georgia	Ga.	GA
Guam	Guam	GU
Hawaii	Hawaii	HI
Idaho	Idaho	ID
Illinois	Ill.	IL
Indiana	Ind.	IN
Iowa	Ia.	IA
Kansas	Kans.	KS
Kentucky	Ky.	KY
Louisiana	La.	LA
Maine	Me.	ME
Maryland	Md.	MD
Massachusetts	Mass.	MA
Michigan	Mich.	MI
Minnesota	Minn.	MN
Mississippi	Miss.	MS
Missouri	Mo.	MO
Montana	Mont.	MT
Nebraska	Neb.	NE
Nevada	Nev.	NV
New Hampshire	N.H.	NH
New Jersey	N.J.	NJ
New Mexico	N.Mex.	NM
New York	N.Y.	NY
North Carolina	N.C.	NC
North Dakota	N.Dak.	ND
Ohio	Ohio	OH
Oklahoma	Okla.	OK
Oregon	Ore.	OR
Pennsylvania	Pa.	PA
Puerto Rico	P.R.	PR
Rhode Island	R.I.	RI
South Carolina	S.C.	SC
South Dakota	S.Dak.	SD
Tennessee	Tenn.	TN
Texas	Tex.	TX
Utah	Utah	UT
Vermont	Vt.	VT
Virginia	Va.	VA
Virgin Islands	V.I.	VI
Washington	Wash.	WA
West Virginia	W.Va.	WV
Wisconsin	Wis.	WI
Wyoming	Wyo.	WY

Canadian Provinces

	Standard	Postal
Alberta	Alta.	AB
British Columbia	B.C.	BC
Labrador	Lab.	NL
Manitoba	Man.	MB
New Brunswick	N.B.	NB
Newfoundland	N.F.	NL
Northwest Territories	N.W.T.	NT
Nova Scotia	N.S.	NS
Nunavut		NU
Ontario	Ont.	ON
Prince Edward Island	P.E.I.	PE
Quebec	Que.	QC
Saskatchewan	Sask.	SK
Yukon Territory	Y.T.	YT

Address Abbreviations

	Standard	Postal
Apartment	Apt.	APT
Avenue	Ave.	AVE
Boulevard	Blvd.	BLVD
Circle	Cir.	CIR
Court	Ct.	CT
Drive	Dr.	DR
East	E.	E
Expressway	Expy.	EXPY
Freeway	Frwy.	FWY
Heights	Hts.	HTS
Highway	Hwy.	HWY
Junction	Junc.	JCT
Lake	L.	LK
Lane	Ln.	LN
Meadows	Mdws.	MDWS
North	N.	N
Palms	Palms	PLMS
Park	Pk.	PK
Parkway	Pky.	PKY
Place	Pl.	PL
Plaza	Plaza	PLZ
Post Office Box	P.O. Box	PO BOX
Ridge	Rdg.	RDG
River	R.	RV
Road	Rd.	RD
Room	Rm.	RM
Rural	R.	R
Rural Route	R.R.	RR
Shore	Sh.	SH
South	S.	S
Square	Sq.	SQ
Station	Sta.	STA
Street	St.	ST
Suite	Ste.	STE
Terrace	Ter.	TER
Turnpike	Tpke.	TPKE
Union	Un.	UN
Village	Vil.	VLG

Common Abbreviations

abr. abridged, abridgment
AC, ac alternating current, air-conditioning
ack. acknowledgment
AM amplitude modulation
A.M., a.m. before noon (Latin *ante meridiem*)
AP advanced placement
ASAP as soon as possible
avg., av. average
B.A. bachelor of arts degree
BBB Better Business Bureau
B.C.E. before common era
bibliog. bibliography
biog. biographer, biographical, biography
B.S. bachelor of science degree
C 1. Celsius 2. centigrade 3. coulomb
c. 1. *circa* (about) 2. cup(s)
cc 1. cubic centimeter 2. carbon copy 3. community college
CDT, C.D.T. central daylight time
C.E. common era
CEEB College Entrance Examination Board
chap. chapter(s)
cm centimeter(s)
c/o care of
COD, c.o.d. 1. cash on delivery 2. collect on delivery
co-op cooperative
CST, C.S.T. central standard time
cu 1. cubic 2. cumulative
D.A. district attorney
d.b.a., d/b/a doing business as
DC, dc direct current
dec. deceased
dept. department
disc. discount
DST, D.S.T. daylight saving time
dup. duplicate

ed. edition, editor
e.g. for example (Latin *exempli gratia*)
EST, E.S.T. eastern standard time
etc. and so forth (Latin *et cetera*)
F Fahrenheit, French, Friday
FM frequency modulation
F.O.B., f.o.b. free on board
FYI for your information
g 1. gravity 2. gram(s)
gal. gallon(s)
gds. goods
gloss. glossary
GNP gross national product
GPA grade point average
hdqrs. headquarters
HIV human immunodeficiency virus
hp horsepower
Hz hertz
ibid. in the same place (Latin *ibidem*)
id. the same (Latin *idem*)
i.e. that is (Latin *id est*)
illus. illustration
inc. incorporated
IQ, I.Q. intelligence quotient
IRS Internal Revenue Service
ISBN International Standard Book Number
JP, J.P. justice of the peace
K 1. kelvin (temperature unit) 2. Kelvin (temperature scale)
kc kilocycle(s)
kg kilogram(s)
km kilometer(s)
kn knot(s)
kw kilowatt(s)
l lake
L liter(s)
lat. latitude
l.c. lowercase
lit. literary; literature
log logarithm, logic
long. longitude

Ltd., ltd. limited
m meter(s)
M.A. master of arts degree
man. manual
Mc, mc megacycle
MC master of ceremonies
M.D. doctor of medicine (Latin *medicinae doctor*)
mdse. merchandise
mfg. manufacture, manufacturing
mg milligram(s)
mi. 1. mile(s) 2. mill(s) (monetary unit)
misc. miscellaneous
mL milliliter(s)
mm millimeter(s)
mpg, m.p.g. miles per gallon
mph, m.p.h. miles per hour
MS 1. manuscript 2. multiple sclerosis
Ms. title of courtesy for a woman
M.S. master of science degree
MST, M.S.T. mountain standard time
NE northeast
neg. negative
N.S.F., n.s.f. not sufficient funds
NW northwest
oz, oz. ounce(s)
PA public-address system
pct. percent
pd. paid
PDT, P.D.T. Pacific daylight time
PFC, Pfc. private first class
pg., p. page
Ph.D. doctor of philosophy
P.M., p.m. after noon (Latin *post meridiem*)
POW, P.O.W. prisoner of war
pp. pages
ppd. 1. postpaid 2. prepaid
PR, P.R. public relations

PSAT Preliminary Scholastic Aptitude Test
psi, p.s.i. pounds per square inch
PST, P.S.T. Pacific standard time
PTA, P.T.A. Parent-Teacher Association
R.A. residence assistant
RF radio frequency
R.P.M., rpm revolutions per minute
R.S.V.P., r.s.v.p. please reply (French *répondez s'il vous plaît*)
SAT Scholastic Aptitude Test
SE southeast
SOS 1. international distress signal 2. any call for help
Sr. 1. senior (after surname) 2. sister (religious)
SRO, S.R.O. standing room only
std. standard
SW southwest
syn. synonymous, synonym
tbs., tbsp. tablespoon(s)
TM trademark
UHF, uhf ultrahigh frequency
v 1. physics: velocity 2. volume
V electricity: volt
VA Veterans Administration
VHF, vhf very high frequency
VIP informal: very important person
vol. 1. volume 2. volunteer
vs. versus, verse
W 1. electricity: watt(s) 2. physics: (also **w**) work 3. west
whse., whs. warehouse
whsle. wholesale
wkly. weekly
w/o without
wt. weight
www World Wide Web

ACRONYMS AND INITIALISMS

Acronym

An **acronym** is a word formed from the first (or first few) letters of words in a set phrase. Even though acronyms are abbreviations, they require no periods.

radar	radio detecting and ranging
CARE	Cooperative for Assistance and Relief Everywhere
NASA	National Aeronautics and Space Administration
VISTA	Volunteers in Service to America
FICA	Federal Insurance Contributions Act

Initialism

An **initialism** is similar to an acronym except that the initials used to form this abbreviation are pronounced individually.

CIA	Central Intelligence Agency
FBI	Federal Bureau of Investigation
FHA	Federal Housing Administration

Common Acronyms and Initialisms

AIDS	acquired immunodeficiency syndrome	OSHA	Occupational Safety and Health Administration
APR	annual percentage rate	PAC	political action committee
CAD	computer-aided design	PIN	personal identification number
CAM	computer-aided manufacturing	POP	point of purchase
CETA	Comprehensive Employment and Training Act	PSA	public service announcement
		REA	Rural Electrification Administration
FAA	Federal Aviation Administration		
FCC	Federal Communications Commission	RICO	Racketeer Influenced and Corrupt Organizations (Act)
FDA	Food and Drug Administration	ROTC	Reserve Officers' Training Corps
FDIC	Federal Deposit Insurance Corporation	SADD	Students Against Destructive Decisions
FEMA	Federal Emergency Management Agency	SASE	self-addressed stamped envelope
		SPOT	satellite positioning and tracking
FHA	Federal Housing Administration	SSA	Social Security Administration
FTC	Federal Trade Commission	SUV	sport-utility vehicle
IRS	Internal Revenue Service	SWAT	Special Weapons and Tactics
MADD	Mothers Against Drunk Driving	TDD	telecommunications device for the deaf
NAFTA	North American Free Trade Agreement	TMJ	temporomandibular joint
NATO	North Atlantic Treaty Organization	TVA	Tennessee Valley Authority
		VA	Veterans Administration
OEO	Office of Economic Opportunity	WHO	World Health Organization
ORV	off-road vehicle		

SPELLING

Write *i* Before *e*

Write *i* before *e* except after *c*, or when sounded like *a* as in *neighbor* and *weigh*.

    believe   relief   receive   eight

This sentence contains eight exceptions:

    Neither sheik dared **leisurely seize either weird species** of **financiers.**

Words with Consonant Endings

When a one-syllable word (*bat*) ends in a consonant (*t*) preceded by one vowel (*a*), double the final consonant before adding a suffix that begins with a vowel (*batting*).

    sum—**summary**   god—**goddess**

When a multisyllable word (*control*) ends in a consonant (*l*) preceded by one vowel (*o*), the accent is on the last syllable (*con trol´*), and the suffix begins with a vowel (*ing*)—the same rule holds true: Double the final consonant (*controlling*).

    prefer—**preferred**   begin—**beginning**   gallop—**galloping**
    forget—**forgettable**   admit—**admittance**   hammer—**hammered**

Words with a Final Silent *e*

If a word ends with a silent *e*, drop the *e* before adding a suffix that begins with a vowel. Do *not* drop the *e* when the suffix begins with a consonant.

    state—**stating**—**statement**   like—**liking**—**likeness**
    use—**using**—**useful**   nine—**ninety**—**nineteen**

Exceptions are **judgment, truly, argument,** and **ninth.**

Words Ending in *y*

When *y* is the last letter in a word and the *y* is preceded by a consonant, change the *y* to *i* before adding any suffix except those beginning with *i*.

    fry—**fries, frying**   hurry—**hurried, hurrying**
    lady—**ladies**   ply—**pliable**
    happy—**happiness**   beauty—**beautiful**

When forming the plural of a word that ends with a *y* that is preceded by a vowel, add *s*.

    toy—**toys**   play—**plays**   monkey—**monkeys**

> **F.Y.I.**
> Never trust your spelling to even the best spell checker. Carefully proofread and use a dictionary for words you know your spell checker does not cover.

CHAPTER 49

Using the Right Word

a, an Use *a* as the article before words that begin with consonant sounds and before words that begin with the long vowel sound *u* (yü). Use *an* before words that begin with other vowel sounds.

> **An** older student showed Kris **an** easier way to get to class.
>
> **A** uniform is required attire for **a** cafeteria worker.

a lot, alot, allot *Alot* is not a word; *a lot* (two words) is a vague descriptive phrase that should be used sparingly, especially in formal writing. *Allot* means to give someone a share.

> Prof Dubi **allots** each of us five spelling errors per semester and he thinks that's **a lot.**

accept, except The verb *accept* means "to receive or believe"; the preposition *except* means "other than."

> The instructor **accepted** the student's story about being late, but she wondered why no one **except** him had forgotten about the change to daylight savings time.

adapt, adopt, adept *Adapt* means "to adjust or change to fit"; *adopt* means "to choose and treat as your own" (a child, an idea). *Adept* is an adjective meaning "proficient or well trained."

> After much thought and deliberation, we agreed to **adopt** the black Lab from the shelter. Now we have to agree on how to **adapt** our lifestyle to fit our new roommate.

adverse, averse *Adverse* means "hostile, unfavorable, or harmful." *Averse* means "to have a definite feeling of distaste—disinclined."

> Groans and other **adverse** reactions were noted as the new students, **averse** to strenuous exercise, were ushered past the X-5000 pump-and-crunch machine.

advice, advise *Advice* is a noun meaning "information or recommendation"; *advise* is a verb meaning "to recommend."

> Successful people will often give you sound **advice**, so I **advise** you to listen.

affect, effect *Affect* means "to influence"; the noun *effect* means "the result"; and the verb *effect* means "to bring about."

> The employment growth in a field will **affect** your chances of getting a job. The **effect** [noun] may be a new career choice. However, career changes can **effect** [verb] challenges.

aid, aide As a verb, *aid* means "to help"; as a noun, *aid* means "the help given." An *aide* is a person who acts as an assistant.

all of *Of* is seldom needed after *all*.

>**All** the reports had an error in them.
>
>**All** the speakers spoke English.
>
>**All of** us voted to reschedule the meeting. [Here *of* is needed for the sentence to make sense.]

all right, alright *Alright* is the incorrect form of *all right*. (*Note:* The following are spelled correctly: *always, altogether, already, almost*.)

allude, elude *Allude* means "to indirectly refer to or hint at something"; *elude* means "to escape attention or understanding altogether."

>Ravi often **alluded** to wanting a supper invitation by mentioning the "awful good" smells from the kitchen. These hints never **eluded** Ma's good heart.

allusion, illusion *Allusion* is an indirect reference to something or someone, especially in literature; *illusion* is a false picture or idea.

>Did you recognize the **allusion** to David in the reading assignment? Until I read that part, I was under the **illusion** that the young boy would run away from the bully.

already, all ready *Already* is an adverb meaning "before this time" or "by this time." *All ready* is an adjective form meaning "fully prepared." (*Note:* Use *all ready* if you can substitute *ready* alone in the sentence.)

>By the time I was a junior in high school, I had **already** taken my SATs. That way, I was **all ready** to apply early to college.

altogether, all together *Altogether* means "entirely." *All together* means "in a group" or "all at once." (*Note:* Use *all together* if you can substitute *together* alone in the sentence.)

>**All together** there are 35,000 job titles to choose from. That's **altogether** too many to even think about.

among, between *Among* is typically used when emphasizing distribution throughout a body or a group of three or more; *between* is used when emphasizing distribution to individuals.

>There was discontent **among** the relatives after learning that their aunt had divided her entire fortune **between** a canary and a favorite waitress at the local cafe.

amoral, immoral *Amoral* means "neither moral (right) nor immoral (wrong)"; *immoral* means "wrong, or in conflict with traditional values."

>Carnivores are **amoral** in their hunt; poachers are **immoral** in theirs.

amount, number *Amount* is used for bulk measurement. *Number* is used to count separate units. (See also "fewer.")

>The **number** of new instructors hired next year will depend on the **amount** of revenue raised by the new sales tax.

and etc. Don't use *and* before *etc.*

>Did you remember your textbook, notebook, handout, **etc.?**

annual, biannual, semiannual, biennial, perennial An *annual* event happens once every year. A *biannual* event happens twice a year (*semiannual* is the same as *biannual*). A *biennial* event happens every two years. A *perennial* event happens throughout the year, every year.

anxious, eager Both words mean "looking forward to," but *anxious* also connotes fear or concern.

> The professor is **eager** to move into the new building, but she's a little **anxious** that students won't be able to find her new office.

anymore, any more *Anymore* means "any longer"; *any more* means "any additional."

> We won't use that textbook **anymore**; please call if you have **any more** questions.

any one (of), anyone *Any one* means "any one of a number of people, places, or things"; *anyone* is a pronoun meaning "any person."

> Choose **any one** of the proposed weekend schedules. **Anyone** wishing to work on Saturday instead of Sunday may do so.

appraise, apprise *Appraise* means "to determine value." *Apprise* means "to inform."

> Because of the tax assessor's recent **appraisal** of our home, we were **apprised** of an increase in our property tax.

as Don't use *as* in place of *whether* or *if*.

> **INCORRECT:** I don't know **as** I'll accept the offer.
>
> **CORRECT:** I don't know **whether** I'll accept the offer.

Don't use *as* when it is unclear whether it means *because* or *when*.

> **UNCLEAR:** We rowed toward shore **as** it started raining.
>
> **CORRECT:** We rowed toward shore **because** it started raining.

assure, ensure, insure (See "insure.")

bad, badly *Bad* is an adjective, used both before nouns and after linking verbs. *Badly* is an adverb.

> Christina felt **bad** about serving us **bad** food.
>
> Larisa played **badly** today.

beside, besides *Beside* means "by the side of." *Besides* means "in addition to."

> **Besides** the two suitcases you've already loaded into the trunk, remember the smaller one **beside** the van.

between, among (See "among.")

bring, take *Bring* suggests the action is directed toward the speaker; *take* suggests the action is directed away from the speaker.

> If you're not going to **bring** the video to class, **take** it back to the resource center.

can, may In formal contexts, *can* is used to mean "being able to do"; *may* is used to mean "having permission to do."

> **May** I borrow your bicycle to get to the library? Then I **can** start working on our group project.

capital, capitol The noun *capital* refers to a city or to money. The adjective *capital* means "major or important" or "seat of government." *Capitol* refers to a building.

> The **capitol** is in the **capital** city for a **capital** reason. The city government contributed **capital** for the building expense.

cent, sent, scent *Cent* is a coin; *sent* is the past tense of the verb "send"; *scent* is an odor or a smell.

> For thirty-seven **cents**, I **sent** my friend a love poem in a perfumed envelope.
>
> She adored the **scent** but hated the poem.

chord, cord *Chord* may mean "an emotion or a feeling," but it also may mean "the combination of three or more tones sounded at the same time," as with a guitar *chord*. A *cord* is a string or a rope.

> The guitar player strummed the opening **chord,** which struck a responsive **chord** with the audience.

chose, choose *Chose* (chōz) is the past tense of the verb *choose* (chüz).

> For generations, people **chose** their careers based on their parents' careers; now people **choose** their careers based on the job market.

climactic, climatic *Climactic* refers to the climax, or high point, of an event; *climatic* refers to the climate, or weather conditions.

> Because we are using the open-air amphitheater, **climatic** conditions in these foothills will just about guarantee the wind gusts we need for the **climactic** third act.

coarse, course *Coarse* means "of inferior quality, rough, or crude"; *course* means "a direction or a path taken." *Course* also means "a class or a series of studies."

> A basic writing **course** is required of all students.
>
> Due to years of woodworking, the instructor's hands are rather **coarse.**

compare with, compare to Things in the same category are *compared with* each other; things in different categories are *compared to* each other.

> **Compare** Christopher Marlowe's plays **with** William Shakespeare's plays.
>
> My brother **compared** reading *The Tempest* **to** visiting another country.

complement, compliment *Complement* means "to complete or go well with." *Compliment* means "to offer an expression of admiration or praise."

> We wanted to **compliment** Zach on his decorating efforts; the bright yellow walls **complement** the purple carpet.

comprehensible, comprehensive *Comprehensible* means "capable of being understood"; *comprehensive* means "covering a broad range, or inclusive."

> The theory is **comprehensible** only to those who have a **comprehensive** knowledge of physics.

comprise, compose *Comprise* means "to contain or consist of"; *compose* means "to create or form by bringing parts together."

> Fruitcake **comprises** a variety of nuts, candied fruit, and spice.
>
> Fruitcake is **composed** of [not *comprised of*] a variety of flavorable ingredients.

conscience, conscious A *conscience* gives one the capacity to know right from wrong. *Conscious* means "awake or alert, not sleeping or comatose."

> Your **conscience** will guide you, but you have to be **conscious** to hear what it's "saying."

continual, continuous *Continual* often implies that something is happening often, recurring; *continuous* usually implies that something keeps happening, uninterrupted.

> The **continuous** loud music during the night gave the building manager not only a headache, but also **continual** phone calls.

counsel, council, consul When used as a noun, *counsel* means "advice"; when used as a verb, *counsel* means "to advise." *Council* refers to a group that advises. A *consul* is a government official appointed to reside in a foreign country.

> The city **council** was asked to **counsel** our student **council** on running an efficient meeting. Their **counsel** was very helpful.

decent, descent, dissent *Decent* means "good." *Descent* is the process of going or stepping downward. *Dissent* means "disagreement."

> The food was **decent**.
>
> The elevator's fast **descent** clogged my ears.
>
> Their **dissent** over the decisions was obvious in their sullen expressions.

desert, dessert *Desert* is barren wilderness. *Dessert* is food served at the end of a meal. The verb *desert* means "to abandon."

different from, different than Use *different from* in formal writing; use either form in informal or colloquial settings.

> Rafael's interpretation was **different from** Andrea's.

discreet, discrete *Discreet* means "showing good judgment, unobtrusive, modest"; *discrete* means "distinct, separate."

> The essay question had three **discrete** parts.
>
> Her roommate had apparently never heard of quiet, **discreet** conversation.

disinterested, uninterested Both words mean "not interested." However, *disinterested* is also used to mean "unbiased or impartial."

> A person chosen as an arbitrator must be a **disinterested** party.
>
> Professor Eldridge was **uninterested** in our complaints about the assignment.

effect, affect (See "affect.")

elicit, illicit *Elicit* is a verb meaning "to bring out." *Illicit* is an adjective meaning "unlawful."

> It took two quick hand signals from the lookout at the corner to **elicit** the **illicit** exchange of cash for drugs.

eminent, imminent *Eminent* means "prominent, conspicuous, or famous"; *imminent* means "ready or threatening to happen."

> With the island's government about to collapse, assassination attempts on several **eminent** officials seemed **imminent**.

ensure, insure (See "insure.")

except, accept (See "accept.")

explicit, implicit *Explicit* means "expressed directly or clearly defined"; *implicit* means "implied or unstated."

> The professor **explicitly** asked that the experiment be wrapped up on Monday, **implicitly** demanding that her lab assistants work on the weekend.

farther, further *Farther* refers to a physical distance; *further* refers to additional time, quantity, or degree.

> **Further** research showed that walking **farther** rather than faster would improve his health.

fewer, less *Fewer* refers to the number of separate units; *less* refers to bulk quantity.

> Because of spell checkers, students can produce papers containing **fewer** errors in **less** time.

figuratively, literally *Figuratively* means "in a metaphorical or analogous way—describing something by comparing it to something else"; *literally* means "actually."

> The lab was **literally** filled with sulfurous gases—**figuratively** speaking, dragon's breath.

first, firstly Both words are adverbs meaning "before another in time" or "in the first place." However, do not use *firstly*, which is stiff and unnatural sounding.

> **INCORRECT: Firstly** I want to see the manager.
>
> **CORRECT: First** I want to see the manager.

Note: When enumerating, use the forms *first, second, third, next, last*—without the *ly*.

fiscal, physical *Fiscal* means "related to financial matters"; *physical* means "related to material things."

> The school's **fiscal** work is handled by its accounting staff.
>
> The **physical** work is handled by its maintenance staff.

for, fore, four *For* is a conjunction meaning "because," or a preposition used to indicate the object or recipient of something; *fore* means "earlier" or "the front"; *four* is the word for the number 4.

> The crew brought treats **for** the barge's **four** dogs, who always enjoy the breeze at the **fore** of the vessel.

former, latter When two things are being discussed, *former* refers to the first thing, and *latter* to the second.

> Our choices are going to a movie or eating at the Pizza Palace: The **former** is too expensive, and the **latter** too fattening.

good, well *Good* is an adjective; *well* is nearly always an adverb. (When used to indicate state of health, *well* is an adjective.)

> A **good** job offers opportunities for advancement, especially for those who do their jobs **well**.

heal, heel *Heal* means "to mend or restore to health"; a *heel* is the back part of a human foot.

healthful, healthy *Healthful* means "causing or improving health"; *healthy* means "possessing health."

> **Healthful** foods and regular exercise build **healthy** bodies.

I, me *I* is a subject pronoun; *me* is used as an object of a preposition, a direct object, or an indirect object. (A good way to know if *I* or *me* should be used in a compound subject is to eliminate the other subject; the sentence should make sense with the pronoun—*I* or *me*—alone.)

> **INCORRECT:** My roommate and **me** went to the library last night.
>
> **CORRECT:** My roommate and **I** went to the library last night. [Eliminate *My roommate and*; the sentence still makes sense.]
>
> **INCORRECT:** Rasheed gave the concert tickets to Erick and **I**.
>
> **CORRECT:** Rasheed gave the concert tickets to Erick and **me**. [Eliminate *Erick and*; the sentence still makes sense.]

illusion, allusion (See "allusion.")

immigrate (to), emigrate (from) *Immigrate* means "to come into a new country or environment." *Emigrate* means "to go out of one country to live in another."

> **Immigrating** to a new country is a challenging experience.
>
> People **emigrating** from their homelands face unknown challenges.

imminent, eminent (See "eminent.")

imply, infer *Imply* means "to suggest without saying outright"; *infer* means "to draw a conclusion from facts." (A writer or a speaker *implies*; a reader or a listener *infers*.)

> Dr. Rufus **implied** I should study more; I **inferred** he meant my grades had to improve or I'd be repeating the class.

ingenious, ingenuous *Ingenious* means "intelligent, discerning, clever"; *ingenuous* means "unassuming, natural, showing childlike innocence and candidness."

> Gretchen devised an **ingenious** plan to work and receive college credit for it.
>
> Ramón displays an **ingenuous** quality that attracts others.

insure, ensure, assure *Insure* means "to secure from financial harm or loss," *ensure* means "to make certain of something," and *assure* means "to put someone's mind at rest."

> Plenty of studying generally **ensures** academic success.
>
> Nicole **assured** her father that she had **insured** her new car.

interstate, intrastate *Interstate* means "existing between two or more states"; *intrastate* means "existing within a state."

irregardless, regardless *Irregardless* is the substandard synonym for *regardless*.

> **INCORRECT: Irregardless** of his circumstance, José is cheerful.
>
> **CORRECT: Regardless** of his circumstance, José is cheerful.

it's, its *It's* is the contraction of "it is." *Its* is the possessive form of "it."

> **It's** not hard to see why my husband feeds that alley cat; **its** pitiful limp and mournful mewing would melt any heart.

later, latter *Later* means "after a period of time." *Latter* refers to the second of two things mentioned.

> The **latter** of the two restaurants you mentioned sounds good.
>
> Let's meet there **later**.

lay, lie *Lay* means "to place." *Lay* is a transitive verb. (See page 704.) Its principal parts are *lay*, *laid*, and *laid*.

> If you **lay** another book on my table, I won't have room for anything else.
>
> Yesterday, you **laid** two books on the table.
>
> Over the last few days, you must have **laid** at least 20 books there.

Lie means "to recline." *Lie* is an intransitive verb. (See page 704.) Its principal parts are *lie*, *lay*, and *lain*.

> The cat **lies** down anywhere it pleases.
>
> It **lay** down yesterday on my tax forms.
>
> It has **lain** down many times on the kitchen table.

learn, teach *Learn* means "to acquire information"; *teach* means "to give information."

> Sometimes it's easier to **teach** someone else a lesson than it is to **learn** one yourself.

leave, let *Leave* means "to allow something to remain behind." *Let* means "to permit."

> Please **let** me help you carry that chair; otherwise, **leave** it for the movers to pick up.

lend, borrow *Lend* means "to give for temporary use"; *borrow* means "to receive for temporary use."

> I asked Haddad to **lend** me $15 for a CD, but he said I'd have to find someone else to **borrow** from.

less, fewer (See "fewer.")

liable, libel *Liable* is an adjective meaning "responsible according to the law" or "exposed to an adverse action"; the noun *libel* is a written defamatory statement about someone, and the verb *libel* means "to publish or make such a statement."

> Supermarket tabloids, **liable** for ruining many a reputation, make a practice of **libeling** the rich and the famous.

liable, likely *Liable* means "responsible according to the law" or "exposed to an adverse action"; *likely* means "in all probability."

> Rain seems **likely** today, but if we cancel the game, we are still **liable** for paying the referees.

like, as *Like* should not be used in place of *as*. *Like* is a preposition, which is followed by a noun, a pronoun, or a noun phrase. *As* is a subordinating conjunction, which introduces a clause. Avoid using *like* as a subordinating conjunction. Use *as* instead.

> **INCORRECT:** You don't know her **like** I do.
>
> **CORRECT:** You don't know her **as** I do.
>
> **CORRECT: Like** the others in my study group, I do my work **as** any serious student would—CORRECT: carefully and thoroughly.

literally, figuratively (See "figuratively.")

loose, lose, loss The adjective *loose* (lüs) means "free, untied, unrestricted"; the verb *lose* (lüz) means "to misplace or fail to find or control"; the noun *loss* (lòs) means "something that is misplaced and cannot be found."

> Her sadness at the **loss** of her longtime companion caused her to **lose** weight, and her clothes felt uncomfortably **loose.**

may, can (See "can.")

maybe, may be Use *maybe* as an adverb; use *may be* as a verb phrase.

> She **may be** the computer technician we've been looking for. **Maybe** she will upgrade the software and memory.

miner, minor A *miner* digs in the ground for ore. A *minor* is a person who is not legally an adult. The adjective *minor* means "of no great importance."

> The use of **minors** as coal **miners** is no **minor** problem.

number, amount (See "amount.")

oral, verbal *Oral* means "uttered with the mouth"; *verbal* means "relating to or consisting of words and the comprehension of words."

> The actor's **oral** abilities were outstanding, and her pronunciation and intonation impeccable; however, I doubted the playwright's **verbal** skills after trying to decipher the play's meaning.

OK, okay This expression, spelled either way, is appropriate in informal writing; however, avoid using it in papers, reports, or formal correspondence of any kind.

> Your proposal is satisfactory [not *okay*] on most levels.

passed, past *Passed* is a verb. *Past* can be used as a noun, an adjective, or a preposition.

> That little pickup truck **passed** my 'Vette! [verb]
>
> My stepchildren hold on dearly to the **past.** [noun]
>
> I'm sorry, but my **past** life is not your business. [adjective]
>
> The officer drove **past** us, not noticing our flat tire. [preposition]

peace, piece *Peace* means "tranquility or freedom from war." A *piece* is a part or fragment.

> Someone once observed that **peace** is not a condition, but a process—a process of building goodwill one **piece** at a time.

people, person Use *people* to refer to human populations, races, or groups; use *person* to refer to an individual or the physical body.

> What the American **people** need is a good insect repellent.
>
> The forest ranger recommends that we check our **persons** for wood ticks when we leave the woods.

percent, percentage *Percent* means "per hundred"; for example, 60 percent of 100 jelly beans would be 60 jelly beans. *Percentage* refers to a portion of the whole. Generally, use the word *percent* when it is preceded by a number. Use *percentage* when no number is used.

> Each person's **percentage** of the reward amounted to $125—25 **percent** of the $500 offered by Crime Stoppers.

personal, personnel *Personal* means "private." *Personnel* are people working at a particular job.

> Although choosing a major is a **personal** decision, it can be helpful to consult with guidance **personnel.**

perspective, prospective *Perspective* is a person's point of view or the capacity to view things realistically; *prospective* is an adjective meaning "expected in or related to the future."

> From my immigrant neighbor's **perspective,** any job is a good job.
>
> **Prospective** students wandered the campus on visitors' day.

pore, pour, poor The noun *pore* is an opening in the skin; the verb *pore* means "to gaze intently." *Pour* means "to move with a continuous flow." *Poor* means "needy or pitiable."

> **Pour** hot water into a bowl, put your face over it, and let the steam open your **pores.** Your **poor** skin will thank you.

precede, proceed To *precede* means "to go or come before"; *proceed* means "to move on after having stopped" or "go ahead."

> Our biology instructor often **preceded** his lecture with these words: "Okay, sponges, **proceed** to soak up more fascinating facts!"

principal, principle As an adjective, *principal* means "primary." As a noun, it can mean "a school administrator" or "a sum of money." A *principle* is an idea or a doctrine.

> His **principal** gripe is lack of freedom. [adjective]
>
> My son's **principal** expressed his concerns to the teachers. [noun]
>
> After 20 years, the amount of interest was higher than the **principal.** [noun]
>
> The **principle** of *caveat emptor* guides most consumer groups.

quiet, quit, quite *Quiet* is the opposite of noisy. *Quit* means "to stop or give up." *Quite* means "completely" or "to a considerable extent."

> The meeting remained **quite quiet** when the boss told us he'd **quit.**

quote, quotation *Quote* is a verb; *quotation* is a noun.

> The **quotation** I used was from Woody Allen. You may **quote** me on that.

real, very, really Do not use the adjective *real* in place of the adverbs *very* or *really*.

> My friend's cake is usually **very** [not *real*] fresh, but this cake is **really** stale.

right, write, wright, rite *Right* means "correct or proper"; it also refers to that which a person has a legal claim to, as in *copyright*. *Write* means "to inscribe or record." A *wright* is a person who makes or builds something. *Rite* is a ritual or ceremonial act.

> Did you **write** that it is the **right** of the **shipwright** to perform the **rite** of christening—breaking a bottle of champagne on the bow of the ship?

scene, seen *Scene* refers to the setting or location where something happens; it also may mean "sight or spectacle." *Seen* is the past participle of the verb "see."

> An exhibitionist likes to be **seen** making a **scene.**

set, sit *Set* means "to place." *Sit* means "to put the body in a seated position." *Set* is a transitive verb; *sit* is an intransitive verb.

> How can you just **sit** there and watch as I **set** the table?

sight, cite, site *Sight* means "the act of seeing" or "something that is seen." *Cite* means "to quote" or "to summon to court." *Site* means "a place or location" or "to place on a site."

After **sighting** the faulty wiring, the inspector **cited** the building contractor for breaking two city codes at a downtown work **site**.

some, sum *Some* refers to an unknown thing, an unspecified number, or a part of something. *Sum* is a certain amount of money or the result of adding numbers together.

Some of the students answered too quickly and came up with the wrong **sum**.

stationary, stationery *Stationary* means "not movable"; *stationery* refers to the paper and envelopes used to write letters.

Odina uses **stationery** that she can feed through her portable printer. Then she drops the mail into a **stationary** mail receptacle at the mall.

take, bring (See "bring.")

teach, learn (See "learn.")

than, then *Than* is used in a comparison; *then* tells when.

Study more **than** you think you need to. **Then** you will probably be satisfied with your grades.

their, there, they're *Their* is a possessive personal pronoun. *There* is a pronoun used as a function word to introduce a clause or an adverb used to point out location. *They're* is the contraction for "they are."

Look over **there**.

There is a comfortable place for students to study for **their** exams, so **they're** more likely to do a good job.

threw, through *Threw* is the past tense of "throw." *Through* means "from one side of something to the other."

In a fit of frustration, Sachiko **threw** his cell phone right **through** the window.

to, too, two *To* is a preposition that can mean "in the direction of." *To* is also used to form an infinitive. *Too* means "also" or "very." *Two* is the number 2.

Two causes of eye problems among students are lights that fail **to** illuminate properly and computer screens with **too** much glare.

vain, vane, vein *Vain* means "valueless or fruitless"; it may also mean "holding a high regard for one's self." *Vane* is a flat piece of material set up to show which way the wind blows. *Vein* refers to a blood vessel or a mineral deposit.

The weather **vane** indicates the direction of the wind; the blood **vein** determines the direction of flowing blood; and the **vain** mind moves in no particular direction, content to think only about itself.

vary, very *Vary* means "to change"; *very* means "to a high degree."

To ensure the **very** best employee relations, the workloads should not **vary** greatly from worker to worker.

verbal, oral (See "oral.")

waist, waste *Waist* is the part of the body just above the hips. The verb *waste* means "to squander" or "to wear away, decay"; the noun *waste* refers to material that is unused or useless.

> His **waist** is small because he **wastes** no opportunity to exercise.

wait, weight *Wait* means "to stay somewhere expecting something." *Weight* refers to a degree or unit of heaviness.

> The **weight** of sadness eventually lessens; one must simply **wait** for the pain to dissipate.

ware, wear, where *Ware* refers to a product that is sold; *wear* means "to have on or to carry on one's body"; *where* asks the question "In what place?" or "In what situation?"

> The designer boasted, "**Where** can one **wear** my **wares**? Anywhere."

weather, whether *Weather* refers to the condition of the atmosphere. *Whether* refers to a possibility.

> **Weather** conditions affect nearly all of us, **whether** we are farmers, pilots, or plumbers.

well, good (See "good.")

which, that (See page 730.)

who, which, that *Who* refers to people. *Which* refers to nonliving objects or to animals. (*Which* should never refer to people.) *That* may refer to animals, people, or nonliving objects.(See page 730.)

who, whom *Who* is used as the subject of a verb; *whom* is used as the object of a preposition or as a direct object.

> Captain Mather, to **whom** the survivors owe their lives, is the man **who** is being honored today.

who's, whose *Who's* is the contraction for "who is." *Whose* is a possessive pronoun.

> **Whose** car are we using, and **who's** going to pay for the gas?

your, you're *Your* is a possessive pronoun. *You're* is the contraction for "you are."

> If **you're** like most Americans, you will have held eight jobs by **your** fortieth birthday.

In this chapter

The Parts of Speech **767**

Understanding Sentence Basics **779**

Sentence Problems **780**

Numbers, Word Parts, and Idioms **782**

CHAPTER

50

Addressing ESL Issues

English may be your second, third, or fifth language. As a multilingual learner, you bring to your writing the culture and knowledge of the languages you use. This broader perspective enables you to draw on many experiences and greater knowledge as you write and speak. Whether you are an international student or someone who has lived here a long time and is now learning more about English, this chapter provides you with important information about writing in English.

THE PARTS OF SPEECH

Noun

Count Nouns

Count nouns refer to things that can be counted. They can have *a, an, the,* or *one* in front of them. One or more adjectives can come between the *a, an, the,* or *one* and the singular count noun.

> an apple   one orange   a plump, purple plum

Count nouns can be singular, as in the examples above, or plural, as in the examples below.

> plums, apples, oranges

When count nouns are plural, they can have the word *the,* a number, or a demonstrative adjective in front of them. (See page 776.)

> I used **the** plums to make a pie.
>
> He placed **five** apples on my desk.
>
> **These** oranges are so juicy!

The *number* of a noun refers to whether it names a single thing (book), in which case its number is *singular,* or whether it names more than one thing (books), in which case the number of the noun is *plural.*

Noncount Nouns

Noncount nouns refer to things that cannot be counted. Do not use *a*, *an*, or *one* in front of them. These nouns have no plural form, so they always take a singular verb. Some nouns that end in *s* are not plural; they are noncount nouns.

fruit, furniture, rain, thunder, advice, mathematics, news

Abstract nouns name ideas or conditions rather than people, places, or objects. Many abstract nouns are noncount nouns.

The students had **fun** at the party. Good **health** is a wonderful gift.

Collective nouns name a whole category or group and are often noncount nouns.

homework, furniture, money, faculty, committee, flock

The parts or components of a group or category named by a noncount noun are often count nouns. For example, *report* and *assignment* are count nouns that are parts of the collective, noncount noun *homework*.

Two-Way Nouns

Some nouns can be used as either count or noncount nouns, depending on what they refer to.

I would like a **glass** of water. [count noun]

Glass is used to make windows. [noncount noun]

Articles and Other Noun Markers

Specific Articles

Use articles and other noun markers or modifiers to give more information about nouns. The **specific** (or **definite**) **article** *the* is used to refer to a specific noun.

I found **the** book I misplaced yesterday.

Indefinite Articles

Use the **indefinite article** *a* or *an* to refer to a nonspecific noun. Use *an* before singular nouns beginning with the vowels *a, e, i, o,* and *u*. Use *a* before nouns beginning with all other letters of the alphabet, the consonants. Exceptions do occur: *a unit; a university*.

I always take **an** apple to work.

It is good to have **a** book with you when you travel.

Indefinite pronouns can also mark nonspecific nouns—*all, any, each, either, every, few, many, more, most, neither, several, some* (for singular and plural count nouns); *all, any, more, most, much, some* (for noncount nouns).

Every student is encouraged to register early.

Most classes fill quickly.

Determining Whether to Use Articles

Listed below are a number of guidelines to help you determine whether to use an article and which one to use.

Use *a* or *an* with singular count nouns that do not refer to one specific item.

> **A zebra** has black and white stripes. **An apple** is good for you.

Do not use *a* or *an* with plural count nouns.

> **Zebras** have black and white stripes. **Apples** are good for you.

Do not use *a* or *an* with noncount nouns.

> **Homework** needs to be done promptly.

Use *the* with singular count nouns that refer to one specific item.

> **The apple** you gave me was delicious.

Use *the* with plural count nouns.

> **The zebras** at Brookfield Zoo were healthy.

Use *the* with noncount nouns.

> **The money** from my uncle is a gift.

Do not use *the* with most singular proper nouns.

> **Mother Theresa** loved the poor and downcast.

F.Y.I.

There are many exceptions: the Sahara Desert, the University of Minnesota, the Fourth of July.

Use *the* with plural nouns.

> **the Joneses** [both Mr. and Mrs. Jones], **the Rocky Mountains, the United States**

Possessive Adjectives

Possessive nouns and pronouns are used to mark nouns.

> **Possessive Nouns:** *Tanya's, father's, store's*
> The car is **Tanya's,** not her **father's.**
> **Possessive Pronouns:** *my, your, his, her, its, our*
> **My** hat is purple.

Demonstrative Adjectives

Demonstrative pronouns can mark nouns.

> **Demonstrative Pronouns:** *this, that, these, those* (for singular and plural count nouns); *this, that* (for noncount nouns)
> **Those** chairs are lovely. Where did you buy **that** furniture?

Quantifiers

Expressions of quantity and measure are often used with nouns. Below are some of these expressions and guidelines for using them.

The following expressions of quantity can be used with count nouns: *each, every, both, a couple of, a few, several, many, a number of.*

> We enjoyed **both** concerts we attended. **A couple of** songs performed were familiar to us.

Use a number to indicate a specific quantity of a continuum.

> I saw **fifteen** cardinals in the park.

To indicate a specific quantity of a noncount noun, use *a* + quantity (such as *bag, bottle, bowl, carton, glass,* or *piece*) + *of* + noun.

> I bought **a carton of milk, a head of lettuce, a piece of cheese,** and **a bag of flour** at the grocery store.

The following expressions can be used with noncount nouns: *a little, much, a great deal of.*

> We had **much** wind and **a little** rain as the storm passed through yesterday.

The following expressions of quantity can be used with both count and noncount nouns: *no/not any, some, a lot of, lots of, plenty of, most, all, this, that.*

> I would like **some** apples [*count noun*] and **some** rice [*noncount noun*], please.

Verb

As the central part of the predicate, a verb conveys much of a sentence's meaning. Using verb tenses and forms correctly ensures that your readers will understand your sentences as you intend them to. For a more thorough review of verbs, see pages 704–709.

Progressive (Continuous) Tenses

Progressive or continuous tense verbs express action that is in progress.

To form the **present continuous** tense, use the helping verb *am, is,* or *are* with the *ing* form of the main verb.

> He **is washing** the car right now.
>
> Kent and Chen **are studying** for a test.

To form the **past continuous** tense, use the helping verb *was* or *were* with the *ing* form of the main verb.

> Yesterday he **was working** in the garden all day.
>
> Julia and Juan **were watching** a movie.

To form the **future continuous** tense, use *will* or a phrase that indicates the future, the helping verb *be,* and the *ing* form of the main verb.

> Next week he **will be painting** the house.
>
> He **plans to be painting** the house soon.

Note that some verbs are generally not used in the continuous tenses, such as the following groups of frequently used verbs:

- Verbs that express thoughts, attitudes, and desires: *know, understand, want, prefer*
- Verbs that describe appearances: *seem, resemble*
- Verbs that indicate possession: *belong, have, own, possess*
- Verbs that signify inclusion: *contain, hold*

    Kala **knows** how to ride a motorcycle.

    NOT: Kala **is knowing** how to ride a motorcycle.

Verb Complements

Verb complements are words used to complete the meaning of transitive verbs. A verb complement can be a direct object (sometimes with an indirect object), an object complement, or a subject complement in the case of a linking verb.

Verb complements include verb forms called verbals. There are three kinds of verbals: infinitives, gerunds, and participles. There are no grammar rules describing which verbs accompany which complements, so take note of the following information.

Infinitives as Complements Infinitives can follow many verbs, including these: *agree, appear, attempt, consent, decide, demand, deserve, endeavor, fail, hesitate, hope, intend, need, offer, plan, prepare, promise, refuse, seem, tend, volunteer, wish.*

    He **promised to bring** some samples.

The following verbs are among those that can be followed by a noun or pronoun plus the infinitive: *ask, beg, choose, expect, intend, need, prepare, promise, want.*

    I **expect you to be** there on time.

Except in the passive voice, the following verbs must have a noun or pronoun before the infinitive: *advise, allow, appoint, authorize, cause, challenge, command, convince, encourage, forbid, force, hire, instruct, invite, order, permit, remind, require, select, teach, tell, tempt, trust.*

    I will **authorize Emily to use** my credit card.

Unmarked infinitives (no *to*) can follow these verbs: *make, have, let, help.*

    These glasses **help me see** the board.

Gerunds as Complements Gerunds can follow these verbs: *admit, avoid, consider, deny, discuss, dislike, enjoy, finish, imagine, miss, postpone, quit, recall, recommend, regret.*

    I **recommended hiring** Ian for the job.

Infinitives or Gerunds as Complements Either gerunds or infinitives can follow these verbs: *begin, continue, hate, like, love, prefer, remember, start, stop, try.*

> I **hate having** cold feet. I **hate to have** cold feet.

Sometimes the meaning of a sentence will change depending on whether you use a gerund or an infinitive.

> I **stopped to smoke.** [I *stopped* weeding the garden to *smoke* a cigarette.]
>
> I **stopped smoking.** [I no longer smoke.]

Common Modal Auxiliary Verbs

Modal auxiliary verbs are a kind of auxiliary verb. (See page 704.) They help the main verb express meaning. Modals are sometimes grouped with other helping or auxiliary verbs.

Modal verbs must be followed by the base form of a verb without *to* (not by a gerund or an infinitive). Also, modal verbs do not change form; they are always used as they appear in Table 50.1.

Table 50.1 Modal Auxiliary Verbs

Modal	Expresses	Sample Sentence
can	ability	I **can** program a VCR.
could	ability	I **could** baby-sit Tuesday.
	possibility	He **could** be sick.
might	possibility	I **might** be early.
may, might	possibility	I **may** sleep late Saturday.
	request	**May** I be excused?
must	strong need	I **must** study more.
have to	strong need	I **have to** (have got to) exercise.
ought to	feeling of duty	I **ought to** (should) help Dad.
should	advisability	She **should** retire.
	expectation	I **should** have caught that train.
shall	intent	**Shall** I stay longer?
will	intent	I **will** visit my grandma soon.
would	intent	I **would** live to regret my offer.
	repeated action	He **would** walk in the meadow.
would + you	polite request	**Would you** help me?
could + you	polite request	**Could you** type this letter?
will + you	polite request	**Will you** give me a ride?
can + you	polite request	**Can you** make supper tonight?

Common Two-Word Verbs

Table 50.2 lists some common verbs in which two words—a verb and a preposition—work together to express a specific action. A noun or pronoun is often inserted between the parts of the two-word verb when it is used in a sentence: break *it* down, call *it* off.

Table 50.2 Two-Word Verbs

break down	to take apart or fall apart
call off	cancel
call up	make a phone call
clear out	leave a place quickly
cross out	draw a line through
do over	repeat
figure out	find a solution
fill in/out	complete a form or an application
fill up	fill a container or tank
*find out	discover
*get in	enter a vehicle or building
*get out of	leave a car, a house, or a situation
*get over	recover from a sickness or a problem
give back	return something
give in/up	surrender or quit
hand in	give homework to a teacher
hand out	give someone something
hang up	put down a phone receiver
leave out	omit or don't use
let in/out	allow someone or something to enter or go out
look up	find information
mix up	confuse
pay back	return money or a favor
pick out	choose
point out	call attention to
put away	return something to its proper place
put down	place something on a table, the floor, or other location
put off	delay doing something
shut off	turn off a machine or light
*take part	participate
talk over	discuss
think over	consider carefully
try on	put on clothing to see if it fits
turn down	lower the volume
turn up	raise the volume
write down	write on a piece of paper

*These two-word verbs should not have a noun or pronoun inserted between their parts.

Spelling Guidelines for Verb Forms

The same spelling rules that apply when adding a suffix to other words apply to verbs as well. Most verbs need a suffix to indicate tense or form. The third-person singular form of a verb, for example, usually ends in *s,* but it can also end in *es.* Formation of *ing* and *ed* forms of verbs and verbals needs careful attention, too. Consult the rules below to determine which spelling is correct for each verb. (For general spelling guidelines, see page 754.)

> **F.Y.I.**
> There may be exceptions to these rules when forming the past tense of irregular verbs because the verbs are formed by changing the word itself, not merely by adding *d* or *ed.* (See Table 45.3, which lists irregular verbs, on page 709.)

Past Tense: Adding *ed*

Add *ed*

- When a verb ends with two consonants:

   touch—**touched**, ask—**asked**, pass—**passed**

- When a verb ends with a consonant preceded by two vowels:

   heal—**healed**, gain—**gained**

- When a verb ends in *y* preceded by a vowel:

   annoy—**annoyed**, flay—**flayed**

- When a multisyllable verb's last syllable is not stressed (even when the last syllable ends with a consonant preceded by a vowel):

   budget—**budgeted**, enter—**entered**, interpret—**interpreted**

Change *y* to *i* and add *ed* when a verb ends in a consonant followed by *y*:

   liquefy—**liquefied**, worry—**worried**

Double the final consonant and add *ed*

- When a verb has one syllable and ends with a consonant preceded by a vowel:

   wrap—**wrapped**, drop—**dropped**

- When a multisyllable verb's last syllable (ending in a consonant preceded by a vowel) is stressed:

   admit—**admitted**, confer—**conferred**, abut—**abutted**

Past Tense: Adding *d*

Add *d*

- When a verb ends with *e:*

   chime—**chimed**, tape—**taped**

- When a verb ends with *ie:*

   tie—**tied**, die—**died**, lie—**lied**

Present Tense: Adding *s* or *es*

Add *es*

- When a verb ends in *ch, sh, s, x,* or *z:*
  watch—**watches**, fix—**fixes**
- To *do* and *go:*
  do—**does**, go—**goes**

Change *y* to *i* and add *es* when the verb ends in a consonant followed by *y:*
  liquefy—**liquefies**, quantify—**quantifies**

Add *s* to most other verbs, including those already ending in *e* and those that end in a vowel followed by *y:*
  write—**writes**, buy—**buys**

Present Tense: Adding *ing*

Drop the *e* and add *ing* when the verb ends in *e:*
  drive—**driving**, rise—**rising**

Double the final consonant and add *ing*

- When a verb has one syllable and ends with a consonant preceded by a vowel:
  wrap—**wrapping**, sit—**sitting**
- When a multisyllable verb's last syllable (ending in a consonant preceded by a vowel) is stressed:
  forget—**forgetting**, begin—**beginning**, abut—**abutting**

Change *ie* to *y* and add *ing* when a verb ends with *ie:*
  tie—**tying**, die—**dying**, lie—**lying**

Add *ing*

- When a verb ends with two consonants:
  touch—**touching**, ask—**asking**, pass—**passing**
- When a verb ends with a consonant preceded by two vowels:
  heal—**healing**, gain—**gaining**
- When a verb ends in *y:*
  buy—**buying**, study—**studying**, cry—**crying**
- When a multisyllable verb's last syllable is not stressed (even when the last syllable ends with a consonant preceded by a vowel):
  budget—**budgeting**, enter—**entering**, interpret—**interpreting**

F.Y.I.

Never trust your spelling to even the best computer spell checker. Carefully proofread. Use a dictionary for questionable words your spell checker may miss.

Adjective

Placing Adjectives

You probably know that an adjective often comes before the noun it modifies. When several adjectives are used in a row to modify a single noun, it is important to arrange the adjectives in the well-established sequence used in English writing and speaking. The following list shows the usual order of adjectives when you use more than one.

First, place

1. articles   a, an, the
   demonstrative adjectives   that, those
   possessives   my, her, Misha's

Then, place words that

2. indicate time   first, next, final
3. tell how many   one, few, some
4. evaluate   beautiful, dignified, graceful
5. tell what size   big, small, short, tall
6. tell what shape   round, square
7. describe a condition   messy, clean, dark
8. tell what age   old, young, new, antique
9. tell what color   blue, red, yellow
10. tell what nationality   English, Chinese, Mexican
11. tell what religion   Islam, Jewish, Protestant
12. tell what material   satin, velvet, wooden

Finally, place nouns

13. used as adjectives   computer [monitor], spice [rack]

   my second try [1 + 2 + noun]

   gorgeous young white swans [4 + 8 + 9 + noun]

Present and Past Participles as Adjectives

Both the **present participle,** which always ends in *ing,* and the **past participle** can be used as adjectives. Exercise care in choosing whether to use the present or the past participle. A participle can come either before a noun or after a linking verb.

A **present participle** used as an adjective should describe a person or thing that is causing a feeling or situation.

   His **annoying** comments made me angry.

A **past participle** should describe a person or thing that experiences a feeling or situation.

   He was **annoyed** because he had to wait so long.

Within each of the following pairs, the present (*ing* form) and past participles (*ed* form) have different meanings.

   annoying/annoyed   depressing/depressed   fascinating/fascinated
   boring/bored   exciting/excited   surprising/surprised

Nouns as Adjectives

Nouns sometimes function as adjectives by modifying another noun. When a noun is used as an adjective, it is always singular.

> Many European cities have **rose** gardens.
>
> Marta recently joined a **book** club.

Try to avoid using more than two nouns as adjectives for another noun. These "noun compounds" can get confusing. Prepositional phrases may get the meaning across better than long noun strings.

> Omar is a **crew** member in the **restaurant** kitchen during **second** shift.
>
> **NOT:** Omar is a **second-shift restaurant kitchen crew** member.

Adverb

Placing Adverbs

Consider the following guidelines for placing adverbs correctly. See page 711 for more information about adverbs.

Place **adverbs that tell how often** (*frequently, seldom, never, always, sometimes*) after a helping verb and before the main verb. In a sentence without a helping verb, adverbs that tell *how often* are placed before an action verb but after a "be" verb.

> The salesclerk will **usually** help me.

Place **adverbs that tell when** (*yesterday, now, at five o'clock*) at the end of a sentence.

> Auntie El came home **yesterday.**

Adverbs that tell where (*upside-down, around, downstairs*) usually follow the verb they modify. Many prepositional phrases (*at the beach, under the stairs, below the water*) are used as adverbs that tell *where*.

> We waited **on the porch.**

Adverbs that tell how (*quickly, slowly, loudly*) can be placed either at the beginning, in the middle, or at the end of a sentence—but not between a verb and its direct object.

> **Softly** he called my name. He **softly** called my name. He called my name **softly.**

Place **adverbs that modify adjectives** directly before the adjective.

> That is a **most** unusual dress.

Adverbs that modify clauses are most often placed in front of the clause, but they can also go inside or at the end of the clause.

> **Fortunately,** we were not involved in the accident.
>
> We were not involved, **fortunately,** in the accident.
>
> We were not involved in the accident, **fortunately.**

Adverbs that are used with verbs that have objects must *not* be placed between the verb and its object.

> Luis **usually** catches the most fish. **Usually,** Luis catches the most fish.
>
> **NOT:** Luis catches **usually** the most fish.

Preposition

A **preposition** combines with a noun to form a prepositional phrase, which usually acts as an adverb or adjective. See page 712 for more information, including Table 45.5, which lists common prepositions.

Using *in, on, at,* and *by*

In, on, at, and *by* are four common prepositions that refer to time and place. Here are some examples of how these prepositions are used in each case.

To show time

on a specific day or date: *on* June 7, *on* Wednesday

in part of a day: *in* the afternoon

in a year or month: *in* 2003, *in* April

in a period of time: completed *in* an hour

by a specific time or date: *by* noon, *by* the fifth of May

at a specific time of day or night: *at* 3:30 this afternoon

To show place

at a meeting place or location: *at* school, *at* the park

at the edge of something: standing *at* the bar

at the corner of something: turning *at* the intersection

at a target: throwing a dart *at* the target

on a surface: left *on* the floor

on an electronic medium: *on* the Internet, *on* television

in an enclosed space: *in* the box, *in* the room

in a geographic location: *in* New York City, *in* Germany

in a print medium: *in* a journal

by a landmark: *by* the fountain

Do not insert a preposition between a transitive verb and its direct object. Intransitive verbs, however, are often followed by a prepositional phrase (a phrase that begins with a preposition).

I **cooked** hot dogs on the grill. [transitive verb]

I **ate** in the park. [intransitive verb]

Phrasal Prepositions

Some prepositional phrases begin with more than one preposition. These **phrasal prepositions** are commonly used in both written and spoken communication. A list of common phrasal prepositions follows:

according to	because of	in case of	on the side of
across from	by way of	in spite of	up to
along with	except for	instead of	with respect to

UNDERSTANDING SENTENCE BASICS

Simple sentences in the English language follow the five basic patterns shown below. (See Chapter 46, "Constructing Sentences," for more information.)

Subject + Verb

S — V
Naomie winked.

Some verbs like *winked* are intransitive. Intransitive verbs *do not* need a direct object to express a complete thought. (See page 704.)

Subject + Verb + Direct Object

S — V — DO
Harris grinds his teeth.

Some verbs like *grinds* are transitive. Transitive verbs *do* need a direct object to express a complete thought. (See page 704.)

Subject + Verb + Indirect Object + Direct Object

S — V — IO — DO
Elena offered her friend an anchovy.

The direct object names who or what receives the action; the indirect object names to whom or for whom the action was done.

Subject + Verb + Direct Object + Object Complement

S — V — DO — OC
The chancellor named Ravi the outstanding student of 2002.

The object complement renames or describes the direct object.

Subject + Linking Verb + Predicate Noun (or Predicate Adjective)

S — LV — PN   S — LV — PA
Paula is a computer programmer.   Paula is very intelligent.

A linking verb connects the subject to the predicate noun or predicate adjective. The predicate noun renames the subject; the predicate adjective describes the subject.

Inverted Order

In the sentence patterns above, the subject comes before the verb. In a few types of sentences, such as those below, the subject comes *after* the verb.

LV — S — PN
Is Larisa a poet? [a question]

LV — S
There was a meeting. [a sentence beginning with *there*]

SENTENCE PROBLEMS

This section looks at potential trouble spots and sentence problems. For more information about English sentences, their parts, and how to construct them see Chapter 46, "Constructing Sentences." This page and the next cover the types of problems and errors found in English writing.

Double Negatives

When making a sentence negative, use *not* or another negative adverb (*never, rarely, hardly, seldom,* and so on), but not both. Using both results in a double negative.

Subject-Verb Agreement

Be sure the subject and verb in every clause agree in person and number.

> The **student was** rewarded for her hard work.
>
> The **students were** rewarded for their hard work.
>
> The **instructor,** as well as the students, **is** expected to attend the orientation.
>
> The **students,** as well as the instructor, **are** expected to attend the orientation.

Omitted Words

Do not omit subjects or the expletives *there* or *it*. In all English clauses and sentences (except imperatives, where the subject *you* is understood), there must be a subject.

> Your mother was very quiet; **she** seemed to be upset.
> **NOT:** Your mother was very quiet; seemed to be upset.
>
> **There** is not much time left.
> **NOT:** Not much time left.
>
> **It** is well known that fruits and grains are good for you.
> **NOT:** Well known that fruits and grains are good for you.

Repeated Words

Do not repeat the subject of a clause or sentence.

> The doctor prescribed an antibiotic.
> **NOT:** The doctor, **she** prescribed an antibiotic.

Do not repeat an object in an adjective clause.

> I forgot the flowers that I intended to give to my hosts.
> **NOT:** I forgot the flowers that I intended to give **them** to my hosts.

Sometimes, the beginning relative pronoun is omitted but understood.

> I forgot the flowers I intended to give to my hosts. [The relative pronoun *that* is omitted.]

Conditional Sentences

Conditional sentences express a situation requiring that a condition be met to be true. Selecting the correct verb tense for use in the two clauses of a conditional sentence can be problematic. Below are three types of conditional sentences.

Factual Conditionals

The conditional clause begins with *if, when, whenever,* or a similar expression. Furthermore, the verbs in the conditional clause and the main clause should be in the same tense.

> **Whenever** we **had** time, we **took** a break and **went** for a swim.

Predictive Conditionals

These sentences express future conditions and possible results. The conditional clause begins with *if* or *unless* and has a present tense verb. The main clause uses a modal (*will, can, should, may, might*) plus the base form of the verb.

> **Unless** we **find** a better deal, we **will buy** this sound system.

Hypothetical Past Conditionals

These sentences describe a situation that is unlikely to happen or that is contrary to fact. To describe situations in the past, the verb in the conditional clause is in the past perfect tense, and the verb in the main clause is formed from *would have, could have,* or *might have* plus the past participle.

> **If** we **had started out** earlier, we **would have arrived** on time.

If the hypothetical situation is a present or future one, the verb in the conditional clause is in the past tense, and the verb in the main clause is formed from *would, could,* or *might* plus the base form of the verb.

> **If** we **bought** groceries once a week, we **would** not **have** to go to the store so often.

Quoted and Reported Speech

Quoted speech is the use of exact words from another source in your own writing; you must enclose these words in quotation marks. It is also possible to report nearly exact words without quotation marks. This is called **reported speech,** or indirect quotation. (See pages 736–737 for a review of the use of quotation marks.)

> **DIRECT QUOTATION:** Felicia said, "Don't worry about tomorrow."
>
> **INDIRECT QUOTATION:** Felicia said that you don't have to worry about tomorrow.

In the case of a question, when a direct quotation is changed to an indirect quotation, the question mark is not needed.

> **DIRECT QUOTATION:** Ahmad asked, "Which of you will give me a hand?"
>
> **INDIRECT QUOTATION:** Ahmad asked which of us would give him a hand.

WRITE Connection

In academic writing, the use of another source's spoken or written words in one's own writing without proper acknowledgment is called *plagiarism.* Plagiarism is severely penalized in academic situations.

NUMBERS, WORD PARTS, AND IDIOMS

Numbers

As a multilingual/ESL learner, you may be accustomed to a way of writing numbers that is different from the way it is done in North America. Become familiar with the North American conventions for writing numbers. Pages 749–750 show you how numbers are written and punctuated in both word and numeral form.

Using Punctuation with Numerals

The **period** is used to express percentages (5.5%, 75.9%) and the **comma** is used to organize large numbers into units (7,000; 23,100; 231,990,000). Commas are not used, however, in writing the year (2002).

Cardinal Numbers

Cardinal numbers are used when counting a number of parts or objects. Cardinal numbers can be used as nouns (she counted to *ten*), pronouns (I invited many guests, but only *three* came), or adjectives (there are *ten* boys here).

Write out in words the numbers one through one hundred. Numbers 101 and greater are often written as numerals.

Ordinal Numbers

Ordinal numbers show place or succession in a series: the fourth row, the twenty-first century, the tenth time, and so on. Ordinal numbers are used to talk about the parts into which a whole can be divided, such as a fourth or a tenth, and as the denominator in fractions, such as one-fourth or three-fifths. Written fractions can also be used as nouns (I gave him *four-fifths*) or as adjectives (a *four-fifths* majority).

Table 50.3 shows names and symbols of the first twenty-five ordinal numbers. Consult a college dictionary for a complete list of cardinal and ordinal numbers.

Table 50.3 Ordinal Numbers

First	1st	Tenth	10th	Nineteenth	19th
Second	2nd	Eleventh	11th	Twentieth	20th
Third	3rd	Twelfth	12th	Twenty-first	21st
Fourth	4th	Thirteenth	13th	Twenty-second	22nd
Fifth	5th	Fourteenth	14th	Twenty-third	23rd
Sixth	6th	Fifteenth	15th	Twenty-fourth	24th
Seventh	7th	Sixteenth	16th	Twenty-fifth	25th
Eighth	8th	Seventeenth	17th		
Ninth	9th	Eighteenth	18th		

Prefixes, Suffixes, and Roots

Tables 50.4 and 50.5 list many common word parts and their meanings. Learning them can help you determine the meaning of unfamiliar words as you come across them in your reading. For instance, if you know that *hemi* means half, you can conclude that *hemisphere* means "half of a sphere."

Table 50.4 Prefixes and Suffixes

Prefixes	Meaning	Suffixes	Meaning
a, an	not, without	able, ible	able, can do
anti, ant	against	age	act of, state of
co, con, com	together, with	al	relating to
di	two, twice	ate	cause, make
dis, dif	apart, away	en	made of
ex, e, ec, ef	out	ence, ency	action, quality
hemi, semi	half	esis, osis	action, process
il, ir, in, im	not	ice	condition, quality
inter	between	ile	relating to
intra	within	sion, tion	act of, state of
multi	many	ish	resembling
non	not	ment	act of, state of
ob, of, op, oc	toward, against	ology	study, theory
per	throughout	ous	full of, having
post	after	some	like, tending to
super, supr	above, more	tude	state of
trans, tra	across, beyond	ward	in the direction of
tri	three		
uni	one		

Table 50.5 Word Roots

Roots	Meaning	Roots	Meaning
acu	sharp	ject	throw
am, amor	love, liking	log, ology	word, study, speech
anthrop	man	man	hand
aster, astr	star	micro	small
auto	self	mit, miss	send
biblio	book	nom	law, order
bio	life	onym	name
capit, capt	head	path, pathy	feeling, suffering
chron	time	rupt	break
cit	to call, start	scrib, script	write
cred	believe	spec, spect, spic	look
dem	people	tele	far
dict	say, speak	tempo	time
erg	work	tox	poison
fid, feder	faith, trust	vac	empty
fract, frag	break	ver, veri	true
graph, gram	write, written	zo	animal

Idioms

Idioms are phrases that are used in a special way. An idiom can't be understood just by knowing the meaning of each word in the phrase. It must be learned as a whole. For example, the idiom *to bury the hatchet* means "to settle an argument," even though the individual words in the phrase mean something much different. These pages list some of the common idioms in American English.

a bad apple	One troublemaker on a team may be called **a bad apple.** [*a bad influence*]
an axe to grind	Mom has **an axe to grind** with the owners of the dog that dug up her flower garden. [*a problem to settle*]
as the crow flies	She lives only two miles from here **as the crow flies.** [*in a straight line*]
beat around the bush	Dad said, "Where were you? Don't **beat around the bush.**" [*avoid getting to the point*]
benefit of the doubt	Ms. Hy gave Henri the **benefit of the doubt** when he explained why he fell asleep in class. [*another chance*]
beyond the shadow of a doubt	Salvatore won the 50-yard dash **beyond the shadow of a doubt.** [*for certain*]
blew my top	When my money got stolen, **I blew my top.** [*showed great anger*]
bone to pick	Nick had a **bone to pick** with Adrian when he learned they both liked the same girl. [*problem to settle*]
break the ice	Shanta was the first to **break the ice** in the room full of new students. [*start a conversation*]
burn the midnight oil	Carmen had to **burn the midnight oil** the day before the big test. [*work late into the night*]
chomping at the bit	Dwayne was **chomping at the bit** when it was his turn to bat. [*eager, excited*]
cold shoulder	Alicia always gives me the **cold shoulder** after our disagreements. [*ignores me*]
cry wolf	If you **cry wolf** too often, no one will come when you really need help. [*say you are in trouble when you aren't*]
drop in the bucket	My donation was a **drop in the bucket.** [*a small amount compared to what's needed*]
face the music	José had to **face the music** when he got caught cheating on the test. [*deal with the punishment*]
flew off the handle	Tramayne **flew off the handle** when he saw his little brother playing with matches. [*became very angry*]
floating on air	Teresa was **floating on air** when she read the letter. [*feeling very happy*]

Idiom	Example
food for thought	The coach gave us some **food for thought** when she said that winning isn't everything. [*something to think about*]
get down to business	In five minutes you need to **get down to business** on this assignment. [*start working*]
get the upper hand	The other team will **get the upper hand** if we don't play better in the second half. [*probably win*]
hit the ceiling	Rosa **hit the ceiling** when she saw her sister painting the television. [*was very angry*]
hit the hay	Patrice **hit the hay** early because she was tired. [*went to bed*]
in a nutshell	**In a nutshell,** Coach Roby told us to play our best. [*to summarize*]
in the nick of time	Zong grabbed his little brother's hand **in the nick of time** before he touched the hot pan. [*just in time*]
in the same boat	My friend and I are **in the same boat** when it comes to doing Saturday chores. [*have the same problem*]
iron out	Jamil and his brother were told to **iron out** their differences about cleaning their room. [*solve, work out*]
it stands to reason	**It stands to reason** that if you keep lifting weights, you will get stronger. [*it makes sense*]
knuckle down	Grandpa told me to **knuckle down** at school if I want to be a doctor. [*work hard*]
learn the ropes	Being new in school, I knew it would take some time to **learn the ropes.** [*get to know how things are done*]
let's face it	"**Let's face it!**" said Mr. Sills. "You're a better long distance runner than a sprinter." [*let's admit it*]
let the cat out of the bag	Tia **let the cat out of the bag** and got her sister in trouble. [*told a secret*]
lose face	If I strike out again, I will **lose face.** [*be embarrassed*]
nose to the grindstone	If I keep my **nose to the grindstone,** I will finish my homework in one hour. [*working hard*]
on cloud nine	Walking home from the party, I was **on cloud nine.** [*feeling very happy*]
on pins and needles	I was **on pins and needles** as I waited to see the doctor. [*feeling nervous*]
over and above	**Over and above** the assigned reading, I read two library books. [*in addition to*]
put his foot in his mouth	Chivas **put his foot in his mouth** when he called his teacher by the wrong name. [*said something embarrassing*]

put your best foot forward	Grandpa said that whenever you do something, you should **put your best foot forward**. [*do the best that you can do*]
rock the boat	The coach said, "Don't **rock the boat** if you want to stay on this team." [*cause trouble*]
rude awakening	I had a **rude awakening** when I saw the letter *F* at the top of my Spanish quiz. [*sudden, unpleasant surprise*]
save face	Grant tried to **save face** when he said he was sorry for making fun of me in class. [*fix an embarrassing situation*]
see eye to eye	My sister and I finally **see eye to eye** about who gets to use the phone first after school. [*are in agreement*]
sight unseen	Grandma bought the television **sight unseen.** [*without seeing it first*]
take a dim view	My brother will **take a dim view** if I don't help him at the store. [*disapprove*]
take it with a grain of salt	If my sister tells you she has no homework, **take it with a grain of salt.** [*don't believe everything you're told*]
take the bull by the horns	This team needs to **take the bull by the horns** to win the game. [*take control*]
through thick and thin	Max and I will be friends **through thick and thin.** [*in good times and in bad times*]
time flies	When you're having fun, **time flies.** [*time passes quickly*]
time to kill	We had **time to kill** before the ballpark gates would open. [*extra time*]
to go overboard	The teacher told us not **to go overboard** with fancy lettering on our posters. [*to do too much*]
under the weather	I was feeling **under the weather,** so I didn't go to school. [*sick*]
word of mouth	We found out who the new teacher was by **word of mouth.** [*talking to other people*]

F.Y.I.

Like idioms, **collocations** are groups of words that often appear together. They may help you identify different senses of a word; for example, "old" means slightly different things in these collocations: *old* man, *old* friends. You will find sentence construction easier if you check for collocations.

Index

A
A, an, 755, 768–770
Abbreviations, 287, 723, 751–752
 capitalization of, 745
 Postal Service, 390, 751
Absolute phrases, 719
Abstract nouns, 698
Abstracts
 APA format for, 158
 research using, 141
Academic writing, 9
Accept, except, 755
Acceptance messages, 412
Accessibility
 for technical writing, 111
 testing for, 68
Accusatory words, 256
Acronyms, 287, 753
Action verbs, 704
Active voice, 114, 266, 282, 707
Adapt, adopt, adept, 755
Addresses
 commas in, 728
 in letters, 380
Adjective clauses, 720
Adjectives, 710
 –adverb errors, 277
 commas between, 285, 726
 definition of, 274
 ESL issues with, 776–777
 hyphens with, 739
Adverb clauses, 720
Adverbs, 711
 –adjective errors, 277
 conjunctive, 734
 definition of, 274
 ESL issues with, 777
Adverse, averse, 755
Advertisements, job, 630–631
Advice, advise, 755
Advice messages, 401
Affect, effect, 755
Ageism, word choice to avoid, 82
Aid, aide, 756
AIDA (attention, interest, desire, and action) formula, 453

All of, 756
All right, alright, 756
Allude, elude, 756
Allusion, illusion, 756
Allusions, 684
Almanacs, 140
A lot, alot, allot, 755
Already, all ready, 756
Altogether, all together, 756
American, 79
American Psychological Association. *See* APA (American Psychological Association) documentation system
Among, between, 756
Amoral, immoral, 756
Amount, number, 756
Analogies, 191, 684
Analysis, 191
And etc., 756
Anecdotes, 191, 684
Announcement messages, 402
 of negative changes, 441
Annual, biannual, semiannual, biennial, perennial, 757
Antecedents, pronoun agreement with, 280, 700
Antithesis, 684
Anxious, eager, 757
Anymore, any more, 757
Any one, anyone, 757
APA (American Psychological Association) documentation system, 158–166
 in-text citations in, 159–160
 paper format in, 158
Apologies, 419
Apostrophes, 285, 731–732
Appeals to ignorance, 176
Appendixes, 158
Application letters, for jobs, 328–329
Application process, for jobs, 305–352
 correspondence in, 327–340
 interviews in, 341–352
 job-search process and, 307–318
 résumés in, 319–326

Appositive phrases, 719
 commas with, 285
Appositives, 727
Appraise, apprise, 757
Argumentation. *See* Persuasion
Articles (parts of speech), 768–770
Artifacts, research with, 136
As, 757
As, like, 762
Assistance, requests for, 456
Assumptions, cultural diversity and, 72
Assure, ensure, insure, 761
At, by, in, on, 778
Atlases, 141
Attacks against persons, 177
Attitude. *See also* Tone
 revising for, 31
 "you," 240–241
Attribution, 153
Audience/readers
 answering questions of, 190
 design for, 289–304
 determining who will be, 14
 document design and, 295–296
 ethics and, 174–177
 intercultural, 74–77
 organizing for, 216
 profiling, 18
 teamwork and, 54
 for technical writing, 110–111
 testing documents with, 68–69
Author-date citation system, 158, 159–160
Authority, 54, 659
Authors
 APA in-text citation of, 159–160
 in APA references lists, 161–166
Automated functions, in word-processing, 48
Auxiliary verbs, 704

B
Bad, badly, 757
Bad-news messages, 359, 361
 BEBE formula for, 427
 checklist for, 450

Bad-news messages *(continued)*
  claims or complaints, 446–449
  denying requests, 428–435
  problem explanations, 440–443
  rejecting suggestions, proposals, or bids, 436–439
  resignations, 444–445
  tact in, 426–427
Bandwagon mentality, 175–176
Bare assertions, 175
Bar graphs, 95–97
BEBE (buffer, explanation, bad news, exit) formula, 427
Believability, 194
Benchmarks for writing, 181–186
Beside, besides, 757
Between, among, 756
Bibliographies
  APA format for, 158, 161–166
  audiovisual sources in, 164–166
  books in, 161–162
  electronic sources in, 164–166
  periodicals in, 163
  reference works, 141
  research using, 141
  working, 124
Bids
  forms for, 530
  major proposals, 532
  rejecting, 436, 439
Binding methods, 293, 294
Biographical resources, 141
Body, of documents
  drafting, 26
  of graphics, 87
  in letters, 380
  paragraphs in, 226, 227
  in presentations, 681
Books
  APA citation style for, 159–160
  in APA references lists, 161–162
  reference entries for, 161–162
Boolean searches, 123
Borrow, lend, 762
Brackets, 741
  with quotations, 155
Brainstorming, 22, 657
Bring, take, 757
Brochures, 552, 554–559
Building-block organization, 111
Business English, 252–253
Business writing. *See* Writing, workplace
By, in, on, at, 778

C
Call numbers, 138
Call reports, 504–507
Can, may, 757
Capital, capitol, 758
Capitalization, 286
Captions, for graphics, 87
Cardinal numbers, 782
Career planning, 312–314. *See also* Job-search process
  checklist for, 318
  requesting raises or promotions, 470–471

Case
  of nouns, 699
  of pronouns, 703
Cause and effect
  analyzing, 210
  diagrams, 21
  faulty conclusions and, 176
  organization in, 222
Cent, sent, scent, 758
Centralized routing, 66
Chain of command, 6
Channels, communication, 7
Charts, 99–101. *See also* Graphics
  in prewriting, 20–21
  software for, 88–89
Chord, cord, 758
Chose, choose, 758
Chronological order
  in process descriptions, 200–201
  in résumés, 321–323
Circular reasoning, 175
Citations, APA format for, 158, 159–160
Cite, site, sight, 765
Claims
  acceptance messages for, 416–417
  guidelines for making, 446–449
  negative responses to, 433
  writing, 446–449
Claims of truth, value, or policy, 188–189
Clarity
  editing for, 33
  revising for, 29
  transition words for, 231
  in word choice, 243–258
Classification, 200–201, 206
Classification diagrams, 21
Clauses, 720
  commas with, 727
  definition of, 274
  nonrestrictive, 730
  subordinate, 265
  wordiness and, 244–245
Clichés, 251
Climactic, climatic, 758
Clip art, 89
Closings
  drafting, 27
  in letters, 380
  in memos, 366
Clothing, 598
Clustering, 22
Coarse, course, 758
Coherence
  paragraph, 230
Collection letters, 466–469
Colons, 733
  capitalization with, 744
  with quotation marks, 735
Color
  document design and, 296
  of paper, 295
  in Web pages, 576
Commas, 726–730
  with numbers, 782
  problems with, 284–285
  with quotation marks, 735
Comma splices, 281

Commenting tools, 48
Common nouns, 698
Communication, 4–8, 645–676
  basics of, 647–654
  business organization and, 6–7
  challenges and strategies in, 8
  checklist for, 12, 654
  context in, 73
  group, 655–664
  intercultural, 71–84
  in meetings, 665–676
  process of, 4–5
  resolving conflict and, 653
  understanding conflict and, 652
Company culture, 6, 603
Company profiles, 620–623
Compare and contrast charts, 20
Compare with, compare to, 758
Comparisons
  contrasts and, 209
  developing, 193
  organization in, 223
  testing documents with, 68
  unclear, 277
  weak or misleading, 177
  words for, 231
Competencies, employee, 605
Complaints
  acceptances messages for, 416–417
  guidelines for making, 446–449
  negative responses to, 433
  writing, 446–449
Complement, compliment, 758
Completeness, 229
  sentence, 281
Compound predicates, 262, 717
Comprehensible, comprehensive, 758
Comprehension, testing for, 68
Comprise, compose, 758
Computers. *See also* Web
  databases, 51
  graphics programs, 88–89
  in library research, 137–144
  presentations using, 691
  writing with, 45–52
Concept proposals, 524
Conciseness, 246–247. *See also* Wordiness
Conclusions
  drawing faulty, 176
  paragraphs in, 226
  in presentations, 682
  from research, 151
  transition words for, 231
Concrete nouns, 698
Conditional sentences, 781
Confidentiality
  abusing, 173
  policy statement on, 617
Confident tone, 239
Confirmation messages, 403
Conflict, 652
  in groups, 663
  resolving, 653
  teamwork and, 54
Congratulations messages, 420
Conjunctions, 274, 713
  coordinating, 713

Index

Connotations, 255, 256
Conscience, conscious, 759
Consensus decision making, 659
Content
 accessibility of technical, 111
 revising for, 29
Context
 cultural diversity and, 73
 prewriting and, 18
 in research, 136
Continual, continuous, 759
Continuous tenses, 770–771
Contractions, 731
Contrasts, 209. *See also* Comparisons
 commas with, 729
Control, of group-writing projects, 64
Control studies, 69
Conversational voice, 184, 233–242
Cooperation, requests for, 408, 455
Coordinating conjunctions, 713
Coordination, combining sentences with, 262
Copy-and-paste, plagiarism and, 157
Copy, correct (trait 6 of good writing), 184, 273–288
Correctness, 273–288
 error checklist for, 288
 proofreading for, 35, 60–61
 unclear wording and, 275–277
 word choice and, 250
Correlative conjunctions, 263, 713
Correspondence, 353–478
 application letters, 328–329
 application-related, 327–340
 bad-news, 359, 425–450
 basics for, 355–362
 catalog of, 361
 checklist for, 362
 e-mail, 369–373, 376
 faxes, 374–376
 form messages, 473–478
 forms of address in, 391–395
 good-news, 358, 397–424
 interview follow-up letters, 348–349, 352
 job-acceptance letters, 334–335
 job-rejection letters, 336–337
 letters, 377–396
 medium selection for, 357
 memos, 363–368
 with multilingual readers, 77
 neutral, 358, 397–424
 persuasive, 360, 451–472
 recommendation-request letters, 330–331
 sensitive, 256
 successful, 356
 thank-you letters, 338–339
Counsel, council, consul, 759
Cover messages, 404
Cover sheets, 374–375
Creative thinking, 192–193
Credibility, 452, 532
Credit applications
 approval letters for, 413
 negative responses to, 434–435
 persuasive, 459
Crisis management messages, 443

Criticism
 constructive vs. destructive, 239
 giving and taking, 651
 teamwork and, 54
Cultural barriers, 8
Cultural diversity, 7
 checklist for, 84
 language selection and, 74
 resources on, 82
 showing respect for, 78–82
 technical writing and, 113
 writing for, 71–84

D

Dashes, 740
Databases, 51
 in APA references lists, 164
 library catalogs, 137–139
 in periodical research, 142–144
Deadwood, 244
Decent, descent, dissent, 759
Decimal points, 723
Decision making, 194
 in groups, 659
Deductive logic, 208
Deep Web tools, 149
Definition diagrams, 21
Definitions, 202–203
 italics for, 742
 negative, 685
Delegating work, 601
Demonstrations, 191
Demonstrative pronouns, 702
Dependent/independent clauses, 720, 726, 734
 semicolons with, 734
Descriptions
 organization in, 222
 of people, places, or things, 198–199
 of processes, 200–201
Desert, dessert, 759
Design. *See also* Document design
 for graphics, 86
Desktop publishing software, 49
 graphics in, 89
Diagonals, 741
Different from, different than, 759
Directness, 398–399
Direct objects, 699, 717
Disabilities or impairments, word choice and, 78
Disagreement, in groups, 663. *See also* Conflict
Discrete, discrete, 759
Discrimination, dealing with, 607
Disinterested, uninterested, 759
Dissertations, in APA references lists, 166
Diversity. *See* Cultural diversity
Documentation of sources, 152–155
 APA system for, 158–166
 in bibliographies, 124
 ethics and, 172
 plagiarism and, 156–157
 when to provide, 152
Document design, 289–304
 for brochures, 554–559
 checklist for, 304
 effective, 184

 format development in, 294–296
 ineffective, 182
 integrating graphics into text and, 90–91
 for letters, 379
 for newsletters, 561
 page layout in, 297–301
 planning, 293
 principles in, 292
 problems in, 301
 reader-friendly (trait 7 of good writing), 184, 289–304
 refining, 36
 for report forms, 539–544
 for résumés, 321–325
 sample of, 40
 typography in, 302–303
 weak vs. strong, 290–291
Donation requests, 458
 negative responses to, 432
Double negatives, 286, 780
Drafting (step 2 in the writing process), 17, 24–27
 closings, 27
 goals of, 17
 in group projects, 65
 openings, 25
 peer review and, 56–59
 sample of, 38
 teamwork in, 65
 techniques in, 24
Drafting programs, 89
Dress, appropriate, 598

E

Eating and drinking etiquette, 597
Editing (step 4 in the writing process), 17, 32–36
 goals of, 17
 in group projects, 66
 peer, 60–62
 proofreading and, 35
 sample of, 39
 for sentence smoothness, 34
 teamwork in, 66
 techniques for, 32
 for word choice, 33
 in word-processing software, 47–48
Effect, affect, 755
80–20 rule, 600
Either/or thinking, 175
Elbow, Peter, 57
Electronic documents. *See also* E-mail; Faxes
 in APA references lists, 164–165
 report forms, 543
 résumés, 325
Elicit, illicit, 759
Ellipses, 735
 with quotations, 155
E-mail, 369–373. *See also* Correspondence
 advantages and disadvantages of, 372
 in APA references lists, 165
 capitalization in, 746
 checklist for, 376
 etiquette for, 373
 format for, 371

E-mail *(continued)*
  form messages, 473–478
  guidelines for, 370
  multilingual readers and, 77
  résumé design for, 325
  subject lines in, 221
  when to use, 357
Emigrate, immigrate, 761
Eminent, imminent, 759
Emoticons, 373
Emotional appeals, 684–685
  abuse of, 176
Emphasis, 269, 729, 733, 740
  transition words for, 231
Employees
  delegating work to, 601
  developing, 604–606
  discrimination and, 607
  evaluation of, 636–639
  human resources writing and, 625–644
  recommendations for, 640–643
Enclosure notes, 380
Encyclopedias, 140
Energy, revising for, 31. *See also* Tone; Voice
English as a Second Language. *See* ESL issues
Ensure, insure, 761
Envelopes, 388–390
ESL issues, 767–786. *See also* Cultural diversity
  with adjectives, 776–777
  with adverbs, 777
  with articles, 768–770
  with idioms, 784–786
  intercultural audiences and, 74–77
  with numbers, 782
  with parts of speech, 767–778
  with prepositions, 778
  with sentences, 779–781
  with verbs, 770–775
  with word parts, 783
Essays, application, 332–333
Ethics, 11, 169–178
  checklist for, 178
  e-mail and, 373
  guidelines for, 170
  information, 171–173
  of persuasion, 174–177
  sensitive documents and, 256
Ethnocentrism, 72
Etiquette
  appropriate dress and, 598
  of eating and drinking, 597
  in e-mail, 373
  of introductions, 594–595
  management and, 589, 594–598
Euphemisms, 251
Evaluation
  tables, 21
  as thinking pattern, 211
Evidence, misusing, 177
Examples, 191
Except, accept, 755
Exclamation points, 725
  with quotation marks, 735
Executive summaries, 514–515

Expense reimbursement, policy statement on, 616
Expert testimony, 191
Explanations of problems, 440–443. *See also* Instructions
Expletives, 267
Explicit, implicit, 760
External messages, 7

F
Facts
  errors in, 277
  ideas vs., 188
  provable vs. not provable, 190
Fact sheets, 620–623
False causes, 176
Farther, further, 760
Faxes, 369, 374–376
Fewer, less, 760, 762
Figuratively, literally, 760
Figures, 92
First, firstly, 760
Fiscal, physical, 760
Five W's charts, 20
Flowcharts
  designing, 100
  in prewriting, 20
  for research overview, 118
Flowery phrases, 251
Flyers, 552–553
Focus
  finding, 23
  in paragraphs, 228
  for prewriting, 23
  revising for, 29
Focus groups, 69
Fonts, 302–303
For, fore, four, 760
Forecasts, information flow and, 217
Foreign words, 742
Format, 36. *See also* Document design
  APA style, 158
  developing, 294–296
  for e-mail, 371
  multilingual readers and, 77
  planning, 293
  in word-processing software, 47
Former, latter, 760
Form messages, 473–478
  checklist for, 478
  guide, 477
  guidelines for, 474
  menu, 476
  negative responses to inquiries, 431
  sales, 462
  standard, 475
Forms
  report, 539–544
  on Web pages, 577
Fractions, 738
Freewriting, 22
Full-block letter format, 384, 386
Funding-request denials, 429
Fund-raising messages, 458
Further, farther, 760

G
Gantt charts, 101
Gender
  of nouns, 698
  of pronouns, 703
Gender-neutral language, 80–81
Gerund phrases, 718–719
Gerunds, 708
  as verb complements, 771–772
Gobbledygook, 177
Good, well, 760
Good-news or neutral messages, 358, 361
  checklist for, 424
  claim acceptances, 416–417
  directness in, 398–399
  goodwill, 418–423
  informative, 400–405
  order placements, 414–415
  positive responses, 410–413
  routine inquiries/requests, 406–409
  SEA formula for, 399
Goodwill messages, 418–423
Government publications
  in APA references lists, 162, 165
  on diversity, 82
Grammar, 697–714
  basic errors in, 286
  correctness in, 273–288
  error checklist, 288
  word-processing tools for, 32, 48
Grant proposals, 533–535
Graphics
  charts, 99–101
  checklist for, 106
  choosing, 92
  computer editing of, 49
  design guidelines for, 86
  graphs, 94–98
  instructions containing, 572–573
  integrating into text, 90–91
  in newsletters, 561–563
  page layout and, 298
  in partitioning, 204
  parts of, 87
  in prewriting, 20–21
  software for, 88–89
  tables, 93
  visuals, 102–105
  in Web pages, 576
Graphs, 94–98
Grouping principles, 206
Group projects, 55
  checklist for, 66
  working on, 63–67
Groups
  building, 656
  checklist for, 664
  communicating in, 655–664
  conflict in, 663
  decision making in, 659
  listening in, 660
  responding in, 661
  roles in, 662
  working in, 657–658
Groupware, 50
Guides, research with, 141

Index

H
Half-truths, 177
Handbooks, 141
Hasty conclusions, 176
Headers and footers, 295
Headings
  APA style for, 158
  in document design, 295
  in letters, 380
  in memos, 366
  in newsletters, 561
  in organization, 221
  tombstones, 301
  in Web pages, 576
Heal, heel, 760
Healthful, healthy, 761
Helping verbs, 704
Highlighting, 300
Honesty, 174, 426
Human resources writing, 625–644
  checklist for, 644
  employee evaluations, 636–639
  employee recommendations, 640–643
  job advertisements, 630–631
  job application follow-up letters, 632–635
  job descriptions, 626–629
  job offers, 635
Humor, 75, 238
Hyperlinks, 146, 577
Hyphens, 738–739
  in numbers, 750

I
I, me, 761
Icons, 296
Ideas
  checklist for, 214
  creative thinking and, 192–193
  qualifying, 189
  revising for, 29
  stating clearly, 188–189
  strong (trait 1 of good writing) 184, 187–214
  subordination and, 264–265
  supporting, 190–191
  vague, 182
Idioms, 784–786
Illustrations, 191
Image editing, 49
Immigrate, emigrate, 761
Imminent, eminent, 759
Implicit, explicit, 760
Imply, infer, 761
In, on, at, by, 778
Incident reports, 486–489
Indefinite pronouns, 701, 732
Indexes, 296
  research using, 141
Indirect objects, 699, 717
Inductive logic, 208
Infer, imply, 761
Infinitive phrases, 718–719
  as verb complements, 771, 772
Infinitives, 708
Information. *See also* Research
  classifying, 206
  competence with, 605
  controlling the flow of, 217
  ethics and, 171–173
  gathering for presentations, 679
  gathering in prewriting, 19–22
  ranking by importance, 207
  requests for, 407
  transition words for, 231
Ingenious, ingenuous, 761
In-house public-relations writing, 547
  newsletters, 560–563
Initialisms, 287, 373, 753
Initials, periods with, 723
Innovation, 192–193
Inquiries
  negative responses to, 430–431
  routine, 406–409
Instructions, 565–574
  checklist for, 574
  with drawings, 573
  giving and taking, 650
  guidelines for, 568–569
  with lists of materials, 570
  with photographs, 572
  for procedures, 571
  safety-information guidelines and, 567
  tips for writing, 567
  types of, 566
Insure, ensure, assure, 761
Intensive pronouns, 700
Intercultural communication, 71–84. *See also* Cultural diversity
  checklist for, 84
Interjections, 713–714
  definition of, 274
Internal messages, 7
Internet
  business Web sites, 145
  capitalization of, 746
  diversity-related resources on, 82
  job-search resources on, 315, 317
  locating information on, 146–150
  research on, 145–150
  resources on, 51
  search engines, 149–150
  writing for, 575–586
Interpersonal skills, 605
Interrogative pronouns, 702
Interruptions, 660
  commas in, 728
Interstate, intrastate, 761
Interviews
  in APA references lists, 166
  common questions in, 346–347
  conducting, 350–351
  designing questions for, 351
  follow-up to, 343, 348–349, 352
  inappropriate or illegal questions in, 344–345
  job-search, 341–352
  preparing for, 342–343, 350
  as research, 132–133
In-text citations, APA format for, 158, 159–160
Intransitive verbs, 704
Introductions, in documents
  to major reports, 516
  paragraphs in, 225
  in presentations, 680
Introductions, of people, 594–595
Investigative reports, 490–495
Invitations, 409
  persuasive, 457
Irony, 685
Irregardless, regardless, 761
Italics/underlining, 742
Its, it's, 761

J
Jargon, 112, 250
Job-acceptance letters, 334–335
Job advertisements, 630–631
Job descriptions, 626–629
Job-offer letters, 635
Job-rejection letters, 336–337
Job-search process, 307–318. *See also* Application process; Human resources writing
  career planning and, 312–314, 318
  job market assessment in, 309–311
  overview of, 308
  researching organizations in, 316
  resignation letters and, 444–445
  resources in, 315, 317
Journals. *See* Periodicals
Justification proposals, 528–529

K
Karmaloop, 583
Kennedy, John F., 684–685
KeyTronic Corporation, 582
Keyword searches, 123
Known/new principle, 217

L
Language selection, 74
  ethics and, 177
Later, latter, 762
Latter, former, 760
Lay, lie, 762
Learn, teach, 762
Learning a Living: SCANS Report, 604–606, 655
Leave, let, 762
Legal issues
  in interview questions, 344–345
  in language selection, 77
Lend, borrow, 762
Less, fewer, 760, 762
Let, leave, 762
Letters, 377–396. *See also* Correspondence
  appearance of, 379
  basic, 380–381
  checklist for, 396
  collection, 466–469
  credit approval, 413
  envelopes and, 388–390
  expanded, 382–383
  folding, 388
  form, 431, 462, 473–478
  formats for, 384–387
  forms of address in, 391–395
  guidelines for, 378
  request or proposal acceptance, 412
  of resignation, 444–445

Letters *(continued)*
  sales, 460–465
  when to use, 357
Liable, libel, 762
Liable, likely, 762
Library research, 137–144, 146
Lie, lay, 762
Like, as, 762
Line drawings, 104–105
Line formatting, 300
Line graphs, 94
Linking verbs, 705
Listening, 595, 647, 649
  active, 660
  to criticism, 651
  in groups, 660
  to instructions, 650
Lists, 224
  colons with, 733
  parentheses in, 725
  for presentations, 683
  in Web pages, 576
Literally, figuratively, 760
Location, transition words that show, 231
Logic, 194
  deductive and inductive, 208
  fallacies in, 175–177
  order-of-importance, 207
  in organization, 215–232
Loose, lose, loss, 763
Lose-lose situations, 652

M

Magazines. *See* Periodicals
Majority decision making, 659
Management, 587–644
  appropriate dress checklist for, 598
  checklist for, 608
  delegating and, 601
  discrimination and, 607
  effective, 599–608
  employee development and, 604–606
  etiquette and, 495–598
  of group-writing projects, 63–67
  human resources writing and, 625–644
  problem solving and, 602
  time, 589–593
  workplace climate and, 603
  writing by, 609–624
  of writing tasks, 600
Management writing, 609–624
  checklist for, 624
  company profiles, 620–623
  mission statements, 610–611
  policy statements, 614–617
  position statements, 612–613
  procedures, 618–619
Manners, 596
Maps, 102, 141
Margins
  APA style for, 158
  document design and, 299
  in letters, 381
May, can, 757
Maybe, may be, 763
Me, I, 761

Meal etiquette, 597
Mechanics, 743–754. *See also* Copy; Grammar; Punctuation; Sentences; Word choice
  abbreviations, 751–752
  acronyms and initialisms, 753
  capitalization, 743–746
  correcting problems with, 286–287
  definition of, 274
  in effective writing, 184
  error checklist, 288
  numbers, 749–750
  plurals, 747–748
  spelling, 754
  in technical writing, 113
Meetings, 665–676
  formal, 666, 667
  formal vs. informal, 666
  making motions in, 669
  minutes of, 672–675, 676
  officers in, 670–671
  order of business in, 668
  parliamentary procedure in, 667, 675
Memos, 363–368
  checklist for, 368
  expanded, 366–367
  guidelines for, 364
  resignation, 445
  sample of basic, 365
  when to use, 357
Menus, Web page, 577
Merge/Compare functions, 48
Metasearch tools, 149–150
Middle paragraphs, drafting, 26. *See also* Body, of documents
Milestone charts, 101
Miner, minor, 763
Minority decision making, 659
Minutes of meetings, 672–675, 676
Mission statements, 578, 610–611
Modal auxiliary verbs, 772
Modifiers
  consistency with, 263
  dangling, 276
  equal, 726
  misplaced, 276
  squinting, 276
  strong, 249
  in technical writing, 115
  wordiness and, 244
Mood, of verbs, 707
Motion pictures, in APA references lists, 166
Motions, making in meetings, 669, 675
Motivational appeals, 684–689
Multilingual readers, 75–77

N

Names
  capitalization of, 286
  positive word choice and, 255
  sensitivity in handling, 240, 255
Negative change announcements, 441
Negative writing/words, 33, 238–239, 255–257
  gender and, 80–81
  sentence structure and, 267
  technical writing and, 113
Neither/nor, 267

NetLibrary, 139
Neutral messages. *See* Good-news or neutral messages
Newsletters, 560–563
Newspapers. *See* Periodicals
News releases, 548–551
Nonrestrictive clauses and phrases, 730
Notes, with graphics, 87
Note-taking, 124, 126–127
Notices, 402
Noun clauses, 720
Nouns, 274, 697–699
  as adjectives, 777
  choosing specific, 248
  count/noncount, 767–768
  definition of, 274
  ESL issues with, 767–768
  plurals of, 747–748
Number, amount, 756
Numbers and numerals, 287, 749–750
  cardinal vs. ordinal, 782
  commas in, 728
  ESL issues with, 782
  fractions, 738
  hyphens with, 738, 739
  of nouns, 698
  of pronouns, 702
  as support for claims, 191
  of verbs, 705

O

OAQS (observe, appreciate, question, suggest) scheme, 57
Obfuscation, 177
Observation
  in research, 130–131
  strategies for, 198
  as support in writing, 191
Observe, Appreciate, Question, Suggest (OAQS), 57
Officers, in meetings, 670–671
OK, okay, 763
Oklahoma Trucking Association (OTA), 584
Omissions, 735, 740
  ESL issues with, 780
On, at, by, in, 778
Openings, drafting, 25. *See also* Introductions, in documents
Operational improvement proposals, 524–529
Opposition, treating fairly, 174
Options, evaluating, 211
Oral, verbal, 763
Ordering correspondence, 414–415
Order of business, 668
Order-of-importance logic, 207
Ordinal numbers, 782
Oregon Labor Market Information System (OLMIS), 585
Organization
  checklist for, 232
  of correspondence, 356
  direct vs. indirect, 219
  effective, 184
  ineffective, 182
  lists and, 224
  logical (trait 2 of good writing), 184, 215–232

outlines in, 220
paragraphs in, 225–231
patterns for, 222–223
for presentations, 680–682
for prewriting, 23
for proposals, 523
for report forms, 541
for reports, 511
of research findings, 151
revising for, 30
strategies for, 216–217, 218–224
for technical writing, 111
three-part structure, 218
titles and headings in, 221
Organizational charts, 99
Organizational structure, communication and, 6–7
Orphans, 301
Outlines, 23, 220
for presentations, 679, 683, 686–687
Oversimplification, 175

P
Page layout, 36, 297–301. *See also* Document design
newsletter, 561
planning, 293
Page numbers, 296
APA style for, 158
Paper choice, 295
for letters, 379
Paradigm shifts, 192
Paragraphs, 225–231
coherent, 230
complete, 229
focus and unity in, 228
in letters, 381
middle, 26
opening, 25
structure of, 227
testing supporting, 227
transition/linking, 226, 231
types of, 225–226
Parallelism
faulty, 282
in presentations, 685
in sentences, 263
in technical writing, 114
Paraphrasing, 129
Parentheses, 725
questions in, 723
Parliamentary procedure, 667, 675
Participation, requests for, 457
Participial phrases, 718–719
Participles, 708
Partitioning, 204–205
Parts of speech, 274, 713
adjectives, 710
adverbs, 711
articles, 768–770
conjunctions, 274, 713
ESL issues with, 767–778
interjections, 713–714
nouns, 274, 697–699
prepositions, 274, 712
pronouns, 274, 275, 280, 283, 700–703
verbals, 708

verbs, 267, 278–279, 282, 567, 704–709, 770–775
Passed, past, 763
Passive voice, 114, 266, 282, 707
Past participles, 776
Peace, piece, 763
Peer editing, 60–62
checklist for, 69
Peer review, 56–59
checklist for, 69
People, person, 763
Percent, percentage, 763
Periodicals
in APA references lists, 163, 164
electronic, 164
reference entries for, 163
research with, 142–144
Periodic reports, 496–499
Periods, 723
with numbers, 782
with quotation marks, 735
Person
of pronouns, 702
shifts in, 283
of verbs, 705
Person, people, 763
Personal, personnel, 764
Personality, 595
Personal pronouns, 700, 702–703
sensitivity in handling, 240
Personal qualities, 604
Perspective, 54
Perspective, prospective, 764
Persuasion
AIDA formula for, 453
checklist for, 472
in collection letters, 466–469
in correspondence, 360, 361
ethics in, 174–177
message writing for, 451–472
organization in, 223
in requesting raises or promotions, 470–471
in sales messages, 460–465
in special requests and promotional messages, 454–459
strategies in, 452–453
Photographs, 103
software for editing, 89
Phrasal prepositions, 778
Phrases
commas with, 727
definition of, 274
nonrestrictive, 730
prepositional, 712
types of, 718–719
wordiness and, 244–245
Physical, fiscal, 760
Piece, peace, 763
Pie graphs, 98
Plagiarism, 156–157, 172
Planners, prewriting, 19
Planning
document design, 293
for group writing, 63–64
group-writing projects, 63–64
for presentations, 679
research, 119–122
Plurals, 731, 747–748

Poetry, 741
Pointing, 57
Policy procedure instructions, 619
Policy statements, 614–617
Poor results explanations, 442
Popular sentiment, appeals to, 176
Pore, pour, poor, 764
Position statements, 612–613
Positive adjustment letters, 417
Positive response messages, 410–413
Positive writing/words, 33, 238–239, 255–257
gender and, 80–81
sentence structure and, 267
technical writing and, 113
word choice and, 255–257
Possessives, 731–732
apostrophes and, 285
Postscripts, 380
Precede, proceed, 764
Predicate adjectives, 710
Predicates, 274, 717
compound, 262, 717
definition of, 274
mismatched subjects and, 283
Predictions, 191
Prefixes, 783
Prejudices, 72
Prepositional phrases, 719
Prepositions, 712, 778
definition of, 274
double, 286
Presentations, 677–694
beginning, 678
checklist for, 694
computer, 691
group, 658
in manuscript form, 683, 688–689
motivational appeals in, 684–689
organization for, 680–682
in outline form, 683, 686–687
planning, 679
practicing, 692
Q & A sessions in, 682
software for, 50
stage fright and, 693
typography for, 303
visuals in, 690
writing, 683
Present participles, 776
Prewriting (step 1 in the writing process), 17, 18–23
goals of, 17
in group projects, 65
information gathering for, 19–22
organization for, 23
planners for, 19
sample of, 37
teamwork in, 65
techniques in, 18
Primary sources, 122
Principal, principle, 764
Problem/solution diagrams, 20
Problem solving, 194, 602
group, 658
Procedures
guidelines for, 618–619
instructions in, 571

Processes
  describing, 200–201
  instructions for, 566
  organization for, 222
  partitioning, 204–205
Progressive tenses, 770–771
Progress reports, 500–503
Promotional messages, 454–459
Promotions (personal)
  employee recommendations for, 640–643
  requesting, 470–471
Pronouns, 700–703
  antecedent agreement with, 280
  definition of, 274
  gender-neutral, 81
  person shifts in, 283
  unclear references to, 275
Proofreading, 35
  symbols for, 60–61
Proper adjectives, 710
Proper nouns, 697, 743
Proposals, 481–484, 521–538
  acceptance messages for, 412
  checklist for, 484, 538
  grant, 533–535
  guidelines for, 522–523
  justification, 528–529
  operational improvement, 524–529
  organizing, 523
  rejecting, 436, 438
  research, 533, 535–537
  sales or client, 530–532
  troubleshooting, 525–527
  types of, 483
  writing successful, 482–483
Prospective, perspective, 764
Protocol testing, 69
Public-relations writing, 547–564
  brochures, 552, 554–559
  checklist for, 564
  flyers, 552–553
  in-house, 547
  newsletters, 560–563
  news releases, 548–551
  public, 547
Punctuation, 274, 284–285, 722–742
  apostrophes, 285, 731–732
  brackets, 155, 741
  colons, 733
  commas, 726–730
  dashes, 740
  definition of, 274
  diagonals, 741
  ellipses, 155, 735
  exclamation points, 725
  hyphens, 738–739
  italics/underlining, 742
  with numbers, 782
  parentheses, 725
  periods, 723
  question marks, 724
  quotation marks, 154, 736–737
  semicolons, 734
Purchase orders, 415

Q
Qualitative testing, 68–69
Quality, judgments of, 211
Quality control, policy statement on, 615
Quantifiers, 770
Quantitative testing, 69
Question-and-answer sessions, 682
Question marks, 724
  with quotation marks, 735
Questions, rhetorical, 685
Quiet, quit, quite, 764
Quotation marks, 154, 736–737
Quotations. *See also* Summarizing
  colons with, 733
  direct vs. indirect, 781
  ESL issues with, 781
  indicating beginning and ending of, 153–154
  indicating changes to, 155
  integrating, 154
  omissions from, 735
  plagiarism and, 157
  of poetry, 741
  in presentations, 685
  question marks with, 724
  quotation marks with, 154
  from research, 129
Quote, quotation, 764
Quoted speech, 781

R
Race and ethnicity, word choice for, 79. *See also* Cultural diversity
Radio broadcasts, in APA references lists, 165–166
Raises, requesting, 470–471
Ranking information, 207
Real, very, really, 764
Reasonableness, 194
Recommendations
  for employees, 640–643
  letters requesting, 330–331
Recordings, in APA references lists, 166
Red herrings, 176
Redundancy, 245
References, personal
  recommendation-request letters and, 330–331
  updates to, 338–339
References lists, APA style for, 158, 161–166. *See also* Bibliographies
Reference works, 140–141
Refining. *See* Editing
Reflexive pronouns, 700
Regardless, irregardless, 761
Relative pronouns, 701
Reliability, 194
Repetition
  ESL issues with, 780
  positive, 269
  in technical writing, 114
Reported speech, 781
Report forms, 539–544
  checklist for, 544
  electronic, 543
  guidelines for, 540
  organizing, 541
  paper, 542
Reports, 479–520
  in APA references lists, 162
  back matter in, 511, 519
  call, 504–507
  checklist for, 484, 508, 520
  executive summaries in, 514–515
  findings in major, 511, 517
  forms for, 539–544
  front matter in, 511–513
  guidelines for, 510–519
  incident, 486–489
  investigative, 490–495
  key points in, 511, 518
  major, 509–520
  organizing major, 511
  periodic, 496–499
  progress, 500–503
  short, 485–508
  trip, 504–507
  types of, 483
  writing successful, 482–483
Requests
  denying, 428–435
  persuasive messages for, 454–459
  routine, 406–409
    acceptance messages for, 412
    acceptances of, 412
    positive replies to, 411
Requests for proposals (RFPs), 532
Research, 117–168
  documentation in, 152–155, 158–166
  evaluating sources in, 125
  flowchart for, 118
  information sites for, 121
  intercultural, 73
  Internet, 145–150
  job-search, 315–316
  library, 137–144
  managing, 123–129
  organizing findings in, 151
  plagiarism and, 156–157
  planning, 119–122
  primary, 130–136
  primary and secondary sources in, 122
  proposals, 533, 535–537
  resources in, 120
  using and integrating sources in, 152–155
Resignation letters, 444–445
Resource skills, 605
Respect
  for diversity, 78–82
  forms of address and, 391–395
  multilingual readers and, 77–82
  tone and, 237
Responsibility, judgments of, 211
Results, explanations of poor, 442
Résumés, 319–326
  chronological, 321–323
  electronic, 325
  functional, 324
  guidelines for, 320–325
Review tools, in software, 48
Revision (step 3 in the writing process), 17, 28–31
  goals of, 17
  in group projects, 66
  for ideas, 29
  for organization, 30

teamwork in, 66
techniques in, 28
  for voice, 31
Rhetorical questions, 685
Right, write, wright, rite, 764
Robert's Rules of Order, 667
Round-robin routing, 66
Routing documents, 66
Run-on sentences, 274

S
Safety-information guidelines, 567
Sales messages, 460–465
  proposals, 530–532
SCANS Report, 604–606, 655
Scene, seen, 764
Scientific method, 194
Scientific names, italics for, 742
Search engines, 149–150
SEA (situation, explanation, action)
  formula, 399
Secondary sources, 122
Semiblock letter format, 384–385
Semicolons, 734
  with quotation marks, 735
Sentences, 269–272
  capitalization in, 744
  checklist for, 272
  choppy, 34, 261, 262–265
  combining, 262–265
  complex, 722
  compound, 262, 722
  compound-complex, 722
  concluding, 227
  conditional, 721, 781
  constructing, 715–722
  correcting errors in, 278–283
  declarative, 721
  editing for smoothness, 34
  effective, 184
  elements of, 274
  energizing, 266–269
  ESL issues with, 779–781
  exclamatory, 721
  fragments, 281
  imperative, 721
  ineffective, 182
  interrogative, 721
  problems in, 261
  rambling, 34, 270
  run-on, 281
  shifts in, 282–283
  simple, 722
  smooth (trait 5 of good writing),
    184, 259–272
  structure of, 722
  subjects in, 274, 283, 715–716
  supporting, 227
  technical writing and, 114–115
  tired, 34
  variety in, 268–269, 721–722
Set, sit, 764
Sexist language, 80–81
Sheets, in word-processing software, 48
Sic, 741
Sight, cite, site, 765
Simple predicates, 717
Sit, set, 764
Site maps, 579

Slang, 252
Slashes, 741
Slippery slope fallacy, 176
Software
  in APA references lists, 165
  checklist for, 52
  desktop publishing, 49
  graphics, 88–89
  groupware, 50
  image editing, 49
  learning new, 46
  presentation, 50
  word-processing, 47–48
Some, sum, 765
Speaking. *See also* Presentations
  advantages/disadvantages of, 14
  communication basics and, 647–654
  effectively, 648
  giving criticism, 651
  giving instructions, 650
  introducing speakers and, 595
  writing vs., 14–15
Spelling
  verb forms, 774–775
  word-processing tools for, 32, 48
Spreadsheets, in time management,
    592–593
Standard functions, in word-
    processing, 47
Stationary, stationery, 765
Statistical resources, 140
Statistics, as support for claims, 191
Stereotypes, 72
Straw man fallacy, 175
Subjects, sentence, 274, 715–717
  compound, 262
  definition of, 274
  double, 286
  mismatched predicates with, 283
  predicates in, 717
  verb agreement with, 278–279, 780
Subject trees, 147–148
Subordinate clauses, 265
Subordinating conjunctions, 713
Subordination, combining sentences
    through, 264–265
Suffixes, 783
Suggestions
  proposals of, 524
  rejecting, 436, 437
Summarizing, 57
  information flow and, 217
  in reports, 511, 514–515
  in research, 128–129
  transition words for, 231
Support for ideas, 190–191
  paragraphs in, 227
  for presentations, 679
Supportive climate, 603
Surveys, as research, 134–135
Sympathy messages, 421
Synergy, 192
Synthesis, 212–213
Systems skills, 605

T
Tables, 92, 93
  in Web pages, 576
Tables of contents, 296, 513

Tact, 239, 426–427
Take, bring, 757
Teach, learn, 762
Teamwork, 53–70
  checklist for, 66
  foundations of, 54
  peer review in, 56–59
  project management and, 63–64
  types of, 55
Technical writing, 107–116
  checklist for, 116
  effective vs. ineffective, 109
  sentence style in, 114–115
  strategies for, 110–111
  word choice in, 112–113
Technology skills, 605
Television broadcasts, in APA
  references lists, 165–166
Telling, 57
Templates, in word-processing
  software, 48
Tenses, verb, 706, 770–771, 774–775
Testimonials, unreliable, 177
Testimony, expert, 191
Testing documents with readers,
    68–69
Than, then, 765
Thank-you letters, 422
  in job searches, 338–339
That, which, 730
That, who, which, 766
The, a, an, 768–770
Their, there, they're, 765
Thesaurus function, in word process-
    ing, 48
Thinking
  cause and effect analysis and, 210
  combining patterns in, 196–197
  compare/contrast pattern in, 209
  definitions and, 202–203
  descriptions and, 198–201
  evaluating pattern options in, 211
  information classification pattern in,
    206
  logical, 175–177, 194, 208
  partitioning pattern in, 204–205
  patterns in, 195–213
  ranking information and, 207
  skills in, 604
  synthesis of material and, 212–213
Three-part structure, 218
Threw, through, 765
Time, transition words that show, 231
Time management, 589–593
  etiquette and, 596
Title pages, APA format for, 158
Titles, of works
  capitalization of, 286, 745
  of graphics, 87
  italics for, 742
  in organization, 221
  quotation marks for, 735–736
  reports, 512
Titles, personal
  capitalization of, 743, 745
  commas with, 729
  gender-neutral, 80–81
  in letters, 391–395
To, too, two, 765

Tone, 236–237. *See also* Voice
  confident vs. doubtful, 239
  conversational voice (trait 3 of good writing), 184, 233–242
  revising for, 31
Track Changes, 48
Transitions
  coherence and, 230–231
  commas with, 727
  information flow and, 217
  paragraphs for, 226
  semicolons with, 734
  words for, 231
Transitive verbs, 704
Translations, 74
Transparency, revising for, 31
Trip reports, 504–507
Troubleshooting proposals, 525–527
Typefaces, 302
Typography, 36, 302–303
  for headings, 221
  justification in, 381
  planning, 293

U

Unity, 228
Unpublished papers, in APA references lists, 166
Update messages, 405
URLs, 146. *See also* Internet
U.S. Labor Department, 604–606, 655
U.S. Postal Service
  abbreviations list, 390, 751
  envelope guidelines, 389
Usability testing, 68
Usage, definition of, 274. *See also* Mechanics

V

Vagueness, 277
Vain, vane, vein, 765
Values, cultural diversity and, 72
Vary, very, 765
Venn diagrams, 20
Verbal, oral, 763
Verbal phrases, 718–719, 719
Verbals, 708
Verb complements, 771–772
Verb phrases, 718
Verbs, 704–709
  active, 266
  attributive, 153
  command, 567
  ESL issues with, 770–775
  in instructions, 567
  irregular, 709
  sentence energy and, 267
  strong, 267
  subject agreement with, 278–279, 780
  tense shifts, 282
  tenses of, 706, 770–771, 774–775
  two-word, 773

vivid, 249
voice shifts, 282
Very, really, real, 764
Very, vary, 765
Visuals, 92, 102–105
  in presentations, 690
Vocabulary resources, 140
Voice, 233–242
  active vs. passive, 114, 266, 282, 707
  awkward, 182
  checklist for, 242
  conversational (trait 3 of good writing), 184, 233–242
  effective, 184
  positive vs. negative, 238–239
  revising for, 31
  shifts in, 282
  strong, 235
  in technical writing, 114
  tone of, 236–237
  of verbs, 707
  weak, 234
  "you attitude," 240–241

W

Waist, waste, 766
Wait, weight, 766
Ware, wear, where, 766
Weather, whether, 766
Web sites/pages, 575–586. *See also* Internet
  in APA references lists, 164–165
  in business research, 145
  checklist for, 586
  development strategies for, 578–581
  elements of, 576
  functions of, 577
  government agency, 585
  library, 146
  manufacturing company, 582
  material for, 580
  menus in, 577
  nonprofit agency, 584
  retail business, 583
  sample, 582–585
  site maps, 579
  testing, 581
Welcome messages, 423
Well, good, 760
Which, that, 730
Which, that, who, 766
White papers, 612–613
White space, 299, 379
Who, which, that, 766
Who, whom, 766
Who's, whose, 766
Widows, 301
Win-lose situations, 652
Win-win situations, 652
Word choice (trait 4 of good writing), 33, 243–258
  Business English, 252–253

checklist for, 258
common errors in, 755–766
conciseness in, 246–247
correctness in, 250
disabilities or impairments and, 78
documents requiring sensitive, 256
effective, 184
ethnicity/race and, 79
exactness in, 248–249
fresh, 254
gender-neutral language and, 80–81
ineffective, 182
jargon and, 112, 250
multilingual readers and, 75
negative vs. positive, 80–81, 255–257
for persuasion, 452
removing unnecessary words in, 244–247
showing respect for diversity through, 78–82
slanted, 177
for technical writing, 112–113
for transitions, 231
Word division, 739
Wordiness, 33, 113, 244–247
  conciseness vs., 246–247
Word parts, 783
Word-processing programs, 47–48
  graphics in, 88
Word roots, 783
Workplace climate, 603
Workplace etiquette, 596
Write, wright, rite, right, 764
Writer's block, 42
Writing, workplace
  academic vs., 9
  advantages/disadvantages of, 14
  career plans, 312–314
  checklist for, 43
  code of ethics for, 11
  for diversity, 71–84
  ethics in, 169–178
  example of, 37–41
  management, 609–624
  managing, 600
  prewriting, 18–23
  principles in, 10
  process for, 13–44
  research for, 117–168
  résumés, 319–326
  speaking vs., 14–15
  teamwork in, 53–70
  technical, 107–116
  technology and, 45–52
  traits of effective, 181, 184–185
  traits of ineffective, 182–183
Writing Without Teachers (Elbow), 57

Y

Yearbooks, 140
"You attitude," 240–241
Your, you're, 766